MW01106115

Advances in Experimental Medicine and Biology

More information about this series at http://www.springer.com/series/5584

Vasilis Vasiliou • Samir Zakhari
Helmut K. Seitz • Jan B. Hoek
Editors

Biological Basis of Alcohol-Induced Cancer

Editors
Vasilis Vasiliou
Department of Environmental Health Sciences
Yale School of Public Health
New Haven, CT, USA

Helmut K. Seitz
Centre of Alcohol Research
University of Heidelberg
Germany and Salem Medical Centre
Heidelberg, Germany

Samir Zakhari
Distilled Spirits Council of the United States
Washington, DC, USA

Jan B. Hoek
Department of Pathology, Anatomy, & Cell
Thomas Jefferson University
Philadelphia, PA, USA

ISSN 0065-2598 ISSN 2214-8019 (electronic)
ISBN 978-3-319-09613-1 ISBN 978-3-319-09614-8 (eBook)
DOI 10.1007/978-3-319-09614-8
Springer Cham Heidelberg New York Dordrecht London

Library of Congress Control Number: 2014956111

Printed on acid-free paper

Springer is part of Springer Science+Business Media (www.springer.com)

Preface

Chronic heavy alcohol consumption is a major health issue worldwide and may lead to addiction and damage of almost every organ of the body. Alcohol accounts for approximately 2 million deaths per year (more than 3 % of all deaths). One of the most severe diseases caused by chronic alcohol consumption is cancer. Alcohol causes cancer of the upper alimentary (oral cavity, pharynx, esophagus) and respiratory tract (larynx), the liver, the colorectum, and the female breast.

Various workshops of the NIH/NIAAA in 1978, 2004, and 2010 have focused on alcohol as a cancer-causing agent. In addition, in 2007 the International Agency for Research on Cancer (IARC) in Lyon, France has invited an international group of experts to evaluate the role of alcohol in cancer development. This group came to the conclusion on the basis of epidemiologic data that alcohol is carcinogenic in humans, based on carcinogenicity of acetaldehyde, the first metabolite of ethanol oxidation.

As epidemiology identified alcohol as a cancer risk factor in countless publications, more and more publications on molecular mechanisms of alcohol in carcinogenesis appeared since then. Therefore in 2010 it was time to start a new series of symposia on alcohol and cancer initiated by the European Society for Biomedical Research on Alcoholism (ESBRA). In September 2010 more than 100 specialists in the field met at the German Cancer Research Centre (DKFZ) in Heidelberg, Germany for the first International Congress on ALCOHOL AND CANCER. As a result of this congress various new approaches to the old topic originated due to multiple international co-operations. Thus, it was only a question of time when this success story could be continued.

In May 2013 the second international congress on ALCOHOL AND CANCER took place in Breckenridge, Colorado, USA. This book represents a summary of the presentations given during this congress. In general, a significant progress in the understanding of the mechanisms by which alcohol effects carcinogenesis was noted. We are convinced that the reader of this book benefits from its content, and we hope due to the increasing interest in this field of various researchers from different areas in cancer research that this may further lead to elucidate the mechanisms by which alcohol acts as a carcinogen possibly resulting in therapeutic approaches. Finally, we hope to continue these Symposia in the future on a biennial basis.

New Haven, CT — Vasilis Vasiliou
Washington, DC — Samir Zakhari
Heidelberg, Germany — Helmut K. Seitz
Philadelphia, PA — Jan B. Hoek

Contents

Contributors

Rajesh Agarwal Department of Pharmaceutical Sciences, University of Colorado Denver, Aurora, CO, USA

University of Colorado Cancer Center, University of Colorado Denver, Aurora, CO, USA

J. Arcaroli Division of Medical Oncology, University of Colorado School of Medicine, Aurora, CO, USA

Silvia Balbo Masonic Cancer Center, University of Minnesota, Minneapolis, MN, USA

Philip J. Brooks Division of Metabolism and Health Effects, Laboratory of Neurogenetics, National Institute on Alcohol Abuse and Alcoholism, Bethesda, MD, USA

Terence N. Bukong Department of Medicine, University of Massachusetts Medical School, Worcester, MA, USA

Carol A. Casey Research Service (151), Veterans Affairs Nebraska-Western Iowa Health Care System, Omaha, NE, USA

The Department of Internal Medicine, University of Nebraska Medical Center, Omaha, NE, USA

Arthur I. Cederbaum Department of Pharmacology and Systems Therapeutics, Mount Sinai School of Medicine, New York, NY, USA

Ying Chen Department of Pharmaceutical Sciences, School of Pharmacy, University of Colorado Denver Anschutz Medical Campus, Aurora, CO, USA

Wendie S. Cohick Department of Animal Sciences, Rutgers, The State University of New Jersey, New Brunswick, NJ, USA

Catina Crismale-Gann Department of Animal Sciences, Rutgers, The State University of New Jersey, New Brunswick, NJ, USA

Hongbin Dong Department of Pharmaceutical Sciences, School of Pharmacy, University of Colorado Denver Anschutz Medical Campus, Aurora, CO, USA

C.J. Peter Eriksson, Ph.D. Department of Public Health, Hjelt Institute, University of Helsinki, Helsinki, Finland

Department of Alcohol, Drugs and Addiction, National Institute for Health and Welfare, University of Helsinki, Helsinki, Finland

Douglas E. Feldman, Ph.D. Southern California Research Center for ALPD and Cirrhosis, Keck School of Medicine of the University of Southern California, Los Angeles, CA, USA

Department of Pathology, Keck School of Medicine of the University of Southern California, Los Angeles, CA, USA

Maura Gasparetto Department of Medicine, University of Colorado, Aurora, CO, USA

Stephen R. Gasparovich Biloxi, MS, USA

Frank J. Gonzalez Laboratory of Metabolism, Center for Cancer Research, National Cancer Institute, Bethesda, MD, USA

Claire Heit Department of Pharmaceutical Sciences, School of Pharmacy, University of Colorado Denver Anschutz Medical Campus, Aurora, CO, USA

L. Hennings Department of Pathology, University of Arkansas for Medical Sciences, Little Rock, AR, USA

Jan B. Hoek Department of Pathology, Anatomy and Cell Biology, Thomas Jefferson University, Philadelphia, PA, USA

John R. Humphries, B.S. Department of Biology, UNC Charlotte, Charlotte, NC, USA

Ling Jin Department of Oral Biology, College of Dentistry, University of Illinois at Chicago, Chicago, IL, USA

Craig Jordan Department of Medicine, University of Colorado, Aurora, CO, USA

Tiffany A. Katz Hillman Cancer Center, University of Pittsburg Cancer Institute, Pittsburgh, PA, USA

Kathryn E. Lazure Research Service (151), Veterans Affairs Nebraska-Western Iowa Health Care System, Omaha, NE, USA

The Department of Internal Medicine, University of Nebraska Medical Center, Omaha, NE, USA

Yongke Lu Department of Pharmacology and Systems Therapeutics, Mount Sinai School of Medicine, New York, NY, USA

Keigo Machida, Ph.D. Southern California Research Center for ALPD and Cirrhosis, Keck School of Medicine of the University of Southern California, Los Angeles, CA, USA

Department of Molecular Microbiology and Immunology, University of Southern California, Los Angeles, CA, USA

Soumen K. Manna Laboratory of Metabolism, Center for Cancer Research, National Cancer Institute, Bethesda, MD, USA

Iain H. McKillop, Ph.D. Department of Surgery, Carolinas Medical Center, Charlotte, NC, USA

Department of Biology, UNC Charlotte, Charlotte, NC, USA

Department of Surgery, Cannon Research Center, Charlotte, NC, USA

Benita McVicker, Ph.D. Research Service (151), Veterans Affairs Nebraska-Western Iowa Health Care System, Omaha, NE, USA

Gary G. Meadows Department of Pharmaceutical Sciences, College of Pharmacy, Washington State University, Spokane, WA, USA

K.E. Mercer Arkansas Children's Nutrition Center, University of Arkansas for Medical Sciences, Little Rock, AR, USA

Department of Pediatrics, University of Arkansas for Medical Sciences, Little Rock, AR, USA

W. Messersmith Division of Medical Oncology, University of Colorado School of Medicine, Aurora, CO, USA

Takeshi Mizukami National Hospital Organization Kurihama Medical and Addiction Center, Kanagawa, Japan

Sebastian Mueller Department of Internal Medicine, Salem Medical Center and Center for Alcohol Research, University of Heidelberg, Heidelberg, Germany

Gary J. Murray, Ph.D. Division of Metabolism and Health Effects, National Institute on Alcohol Abuse and Alcoholism, National Institutes of Health, Bethesda, MD, USA

David J. Niemeyer, M.D. Department of Surgery, Carolinas Medical Center, Charlotte, NC, USA

Sylvia I. Pavlova Department of Oral Biology, College of Dentistry, University of Illinois at Chicago, Chicago, IL, USA

Dennis R. Petersen Department of Pharmaceutical Sciences, University of Colorado Denver Anchutz Medical Campus, Aurora, CO, USA

Daniel A. Pollyea Department of Medicine, University of Colorado, Aurora, CO, USA

Vanessa Rausch Department of Internal Medicine, Salem Medical Center and Center for Alcohol Research, University of Heidelberg, Heidelberg, Germany

M.J.J. Ronis Arkansas Children's Nutrition Center, University of Arkansas for Medical Sciences, Little Rock, AR, USA

Department of Pediatrics, University of Arkansas for Medical Sciences, Little Rock, AR, USA

Banishree Saha Department of Medicine, University of Massachusetts Medical School, Worcester, MA, USA

Dipak K. Sarkar Department of Animal Sciences, Rutgers, The State University of New Jersey, New Brunswick, NJ, USA

Joel Schwartz Department of Oral Medicine and Diagnostic Sciences, College of Dentistry, University of Illinois at Chicago, Chicago, IL, USA

Robert A. Sclafani University of Colorado Cancer Center, University of Colorado Denver, Aurora, CO, USA

Department of Biochemistry and Molecular Genetics, University of Colorado Denver, Aurora, CO, USA

Helmut K. Seitz, M.D., A.G.A.F. Centre of Alcohol Research, University of Heidelberg, Germany and Salem Medical Centre, Heidelberg, Germany

Yatrik M. Shah Department of Molecular and Integrative Physiology, University of Michigan, Ann Arbor, MI, USA

Colin T. Shearn, Ph.D. Department of Pharmaceutical Sciences, University of Colorado Denver Anchutz Medical Campus, Aurora, CO, USA

Sangeeta Shrotriya Department of Pharmaceutical Sciences, University of Colorado Denver, Aurora, CO, USA

David Sindram, M.D., Ph.D. Department of Surgery, Carolinas Medical Center, Charlotte, NC, USA

S. Singh Department of Pharmaceutical Sciences, University of Colorado Anschutz Medical Campus, Aurora, CO, USA

Clay Smith, M.D. Department of Medicine, University of Colorado, Aurora, CO, USA

Division of Hematology, University of Colorado, Aurora, CO, USA

Felix Stickel, M.D. Department of Clinical Research, University of Bern, Bern, Germany

Hepatology Unit, Clinic Beau-Site, Bern, Germany

Hillary Stires Department of Animal Sciences, Rutgers, The State University of New Jersey, New Brunswick, NJ, USA

Gyongyi Szabo, M.D., Ph.D. Department of Medicine, University of Massachusetts Medical School, Worcester, MA, USA

Lin Tao Department of Oral Biology, College of Dentistry, University of Illinois at Chicago, Chicago, IL, USA

Peter Thomas Department of Surgery and Biomedical Sciences, Creighton University School of Medicine, Omaha, NE, USA

David C. Thompson Department of Clinical Pharmacy, University of Colorado School of Medicine, Aurora, CO, USA

Department of Clinical Pharmacy, School of Pharmacy, University of Colorado Anschutz Medical Campus, Aurora, CO, USA

Kyle J. Thompson, Ph.D. Department of Surgery, Carolinas Medical Center, Charlotte, NC, USA

Matthew D. Thompson Laboratory of Metabolism, Center for Cancer Research, National Cancer Institute, Bethesda, MD, USA

Hidekazu Tsukamoto, D.V.M., Ph.D. Southern California Research Center for ALPD and Cirrhosis, Keck School of Medicine of the University of Southern California, Los Angeles, CA, USA

Department of Pathology, Keck School of Medicine of the University of Southern California, Los Angeles, CA, USA

Department of Veterans Affairs Greater Los Angeles Healthcare System, Los Angeles, CA, USA

Dean J. Tuma Research Service (151), Veterans Affairs Nebraska-Western Iowa Health Care System, Omaha, NE, USA

The Department of Internal Medicine, University of Nebraska Medical Center, Omaha, NE, USA

V. Vasiliou Department of Environmental Health Sciences, Yale School of Public Health, New Haven, CT, USA

Xiaodong Wang Department of Pharmacology and Systems Therapeutics, Mount Sinai School of Medicine, New York, NY, USA

Defeng Wu Department of Pharmacology and Systems Therapeutics, Mount Sinai School of Medicine, New York, NY, USA

Akira Yokoyama, M.D. National Hospital Organization Kurihama Medical and Addiction Center, Kanagawa, Japan

Tetsuji Yokoyama Department of Health Promotion, National Institute of Public Health, Saitama, Japan

Samir Zakhari Formerly Division of Metabolism and Health Effects, NIAAA, NIH, Bethesda, MD, USA

SVP, Science, DISCUS, Washington, DC, USA

Faya Zhang Department of Pharmaceutical Sciences, College of Pharmacy, Washington State University, Spokane, WA, USA

Hui Zhang Department of Pharmaceutical Sciences, College of Pharmacy, Washington State University, Spokane, WA, USA

Zhaohui Zhu Department of Pharmaceutical Sciences, College of Pharmacy, Washington State University, Spokane, WA, USA

Chapter 1
Introduction

Gary J. Murray

Alcohol use disorders are a serious health problem having not only a profound primary effect on individuals, their life expectancy, and general health but also significant secondary effects on families and friends and a tertiary effect on society through the increased cost of health care and losses in productivity. The consumption of alcoholic beverages has long been an integral part of celebrations throughout the world. Some have even suggested a historic and evolutionary connection between humans and alcohol metabolism linking survival of our species with the ability to consume rotting fruit and eliminate alcohol [1]. In the early history of mankind, when much of the water may have been toxic or potentially filled with pathogens, consumption of alcohol may have been safer than consumption of water. The use and abuse of alcohol, whether to celebrate, grieve, or just cope with life's ups and downs, has led to a culture that condones and encourages its use and yet in recent years has become increasingly aware of the dangers of overindulgence.

To some, the consequences of consuming too much alcohol may be mild discomfort associated with the "morning after," perhaps headache, malaise, some gastric distress, and inability to concentrate. To others especially for those who consume excessive quantities of alcohol either chronically or as binge drinkers, the manifestation of symptoms may be much more severe and potentially include serious tissue damage and life-threatening organ failure. Although moderate alcohol consumption has been associated with health benefits—e.g., decrease in the risk of coronary artery disease and ischemic stroke—the World Health Organization has identified chronic alcohol consumption as one of the top ten risk factors in terms of the years of life lost to premature mortality and years lived with disability [2]. The most onerous among these adverse health effects may be the development of cancer.

Gary J. Murray, Ph.D. (✉)
Division of Metabolism and Health Effects, National Institute on Alcohol Abuse and Alcoholism, National Institutes of Health, Bethesda, MD, USA
e-mail: gary.murray@nih.gov

V. Vasiliou et al. (eds.), *Biological Basis of Alcohol-Induced Cancer*, Advances in Experimental Medicine and Biology 815,
DOI 10.1007/978-3-319-09614-8_1

Despite a suspected association between excessive use of alcohol and death due to cancer reported in an epidemiological study as early as 1903 [3], it took until 1988 for the research community through the International Agency for Research on Cancer (IARC) to agree on the potential risk [4]. Alcohol was subsequently designated as a type 1 carcinogen in 2010 [5]. Although the association between smoking and cancer risk has been publicly accepted for many years, a similar recognition of the risks of alcohol consumption has not been part of the public perception of the risks of drinking.

Most epidemiological studies report a J-shaped risk profile for the development of cancers of any kind such that the risk increases dramatically at higher levels of alcohol consumption. This also means that below a threshold level, usually set at the level considered as "moderate drinking," the risk of developing cancer appears to be less than or equal to the risk for nondrinkers. The NIAAA and many international bodies have agreed that moderate alcohol consumption may be defined as up to one drink per day for women and up to two drinks per day for men, whereas heavy alcohol drinking is defined as having more than three drinks on any day or more than seven drinks per week for women and more than four drinks on any day or more than 14 drinks per week for men. The type of alcohol consumed shows no correlation with the development of cancer; however, the pattern of drinking appears to have a strong influence, with binge alcohol consumption considered to be far more likely to produce alcohol-related pathology. This may, over time, develop into one of the many forms of cancer.

Clear patterns have emerged between alcohol consumption and the development of head and neck cancers, particularly cancers of the oral cavity, pharynx and larynx, and esophagus [6]. Often, these are associated most strongly with individuals with combined risk factors such as coincident alcohol and tobacco use [7]. Excessive alcohol use is an independent risk factor for and a primary cause of hepatocellular carcinoma. A number of contributing factors have been identified including chronic infection and infections with hepatitis B and hepatitis C viruses [8]. Numerous epidemiological studies have found an association between alcohol consumption and the risk of breast cancer in women. For example, a slightly higher risk of breast cancer at low-to-moderate levels of alcohol consumption was identified in the Million Women Study in the United Kingdom [9], a study that included more than 28,000 women with breast cancer. However, these results have been called into question because they report lifetime breast cancer risk related to alcohol consumption reported at a single point in time [10] and thus may overestimate the risk of cancer among light-moderate drinkers due to underreporting of intake [11]. As with most cancers, tracking exposure as a precipitating event in the development of breast cancer is not generally possible since an individual's disease is the sum of a lifetime of contributory exposures. For instance, the pattern of drinking and age of onset of drinking behavior must be considered to determine if drinking at a particularly susceptible age has a critical influence on the results [12]. While alcohol exposures during childhood and adolescence are likely to affect a woman's long-term risk of breast cancer, these have received far less research attention than exposures that occur later in life. Study participants are often stratified into high- and low-risk drinking groups solely on the basis of recent alcohol use. This can potentially lead to erroneous conclusions since the origination of the cancer, the "precipitating event," may be temporally far removed from the survey date and the effects may be long-lived. Indeed, damage caused by exposure to alcohol or any other carcinogen may initially be seen only at a subcellular or molecular level but continue to exert an influence on risk for many years after cessation of drinking. This has been clearly demonstrated for alcohol-associated head and neck cancers. Pooled analyses of multiple case-controlled studies of cancer of the oral cavity and pharynx and for esophageal cancer have shown that the alcohol-associated cancer risk did not begin to decrease until at least 10 years after stopping alcohol drinking and did not approach that of never drinkers for at least 15 years after cessation of alcohol drinking [13]. More attention needs to be paid to the patterns of alcohol consumption throughout an individual's life in order to properly assign risk.

Not all cancers are identical in initiation, promotion, and progression, and many hypotheses have been proposed by the scientific community to explain the mechanism(s) by which alcohol consumption may be linked to the development of cancer. Epidemiologic evidence emphasizes the significant risk for development of various forms of cancer associated with the consumption of alcohol. It has also been shown that the risk increases due to synergy when alcohol acts in concert with other major risk factors including viral hepatitis B and C, smoking, and obesity. One of the best examples for this is the observation that especially high risks of cancer exist in individuals who both drink and smoke. Synergy between these two apparently independent risk factors has led to the hypothesis that alcohol induces enzyme systems capable of generating higher levels of carcinogens from tobacco smoke [13] or that alterations in the microbiome due to smoking and poor oral hygiene are responsible for increased acetaldehyde generated in the UADT [14].

Multiple mechanisms exist by which alcohol may contribute to the initiation or progression of carcinogenesis. The development of cancer may be promoted by any or all of the following mechanisms: the metabolism of alcohol to acetaldehyde and more toxic downstream metabolites; changes in the oxidation state (to result in the generation of reactive oxygen species, ROS); alterations in folate metabolism and disturbances in the methionine cycle leading to epigenetic aberrations and changes in gene expression; and alcohol-induced disruption of gut barrier with dysregulation of the immune system and chronic inflammation. One area of consensus in the scientific community seems to be that alcohol rarely acts alone in the development of cancer but requires a confluence of host genetic and environmental factors to result in the development or proliferation of cancer.

The IARC has classified both alcohol and the acetaldehyde metabolite as type 1A carcinogens [5]. This conclusion is based on the weight of the genetic and epidemiological evidence combined with the readily demonstrated cytotoxic properties and its ability to form DNA-acetaldehyde adducts and to generate numerous additional mutagenic species at concentrations attainable in vivo [14]. These may cause cancer by damaging DNA and arresting the repair mechanisms operative within the cell as will be discussed in greater detail in Chap. 5 in this volume (Balbo and Brooks). The best evidence for a role for acetaldehyde in a development of cancer comes from epidemiological data from subjects with an inactive form of aldehyde dehydrogenase (ALDH2). Individuals who are heterozygous for this mutation (*ALDH2*1/2*2*) have a greatly increased risk of developing various forms of head and neck cancers as a result of increased exposure to acetaldehyde [15].

Metabolism of ethanol at each site of exposure occurs by stepwise conversion first to acetaldehyde and then to acetate. Following ingestion, the highest direct exposure to both alcohol and acetaldehyde is most probably at the port of entry, the oropharyngeal tract. After drinking, alcohol may be present at high millimolar concentrations in the UADT and in the GI tract. In spite of a relatively short time of exposure, direct exposure to alcohol at high levels in the oral and gastric mucosa resulting in the local generation of acetaldehyde is the simplest and most likely explanation for the elevation in cancer risk to the upper aerodigestive tract. A small fraction of the ingested alcohol is metabolized at the point of ingestion, but much of the alcohol is delivered to the systemic circulation where it is rapidly distributed throughout the aqueous compartment in the body. Absorption into the systemic circulation results in lower but still significant exposures to both alcohol and acetaldehyde. Because of the high concentration of alcohol in many tissues, it may act nonspecifically on a variety of targets in the brain and peripheral organs. In individuals with optimally functioning metabolic and clearance systems and who do not drink excessively, the concentrations of ethanol and its metabolites are lower, minimizing the probability of serious consequences.

The cellular metabolism of alcohol and acetaldehyde and the generation of toxic by-products are dependent on the distribution of these substrates and the cell- and tissue-specific distribution of the enzymes responsible, the human alcohol and aldehyde dehydrogenases and the various forms of the cytochrome P450 enzymes (CYP2E1), and the individual kinetic parameters of enzyme isoforms. In this context, it is important to note that both acute and chronic alcohol consumption results in oxidative stress (through the generation of ROS and depletion of reduced glutathione) and the induction of CYP2E1 expression, all of which can lead to tissue damage and cancer. Acetaldehyde is produced by the metabolism of ethanol by the combination of these endogenous enzymes and the enzymes present in the microbiome and the bacteria found in the UADT, gut, and digestive tract. The latter vary from person to person and within an individual according to both location along the digestive tract and diet. The contribution of the microbiome to the process of converting ethanol into acetaldehyde and this capacity to generate carcinogens in situ represent an understudied area.

The importance of ethanol's influence on epigenetics and the relationship to the etiology of alcohol-induced cancer need to be emphasized [16]. Alcohol's influence on one-carbon metabolism, the folate and methionine cycles, induces epigenetic alterations that may include DNA methylation, histone modification, and RNA-mediated gene silencing. As a general mechanism, methylation in the promoter region results in gene silencing and thus may reduce the risk of cancer; however, alcohol may induce both global hypomethylation and focal hypermethylation. Contributing complexity to the control of gene expression, hypomethylation of certain oncogenes may lead to increases in gene expression, dedifferentiation, and proliferation [17, 18], whereas specific (hypermethylation of tumor suppressor genes to decrease their expression may also increase the risk for cancer [19].

Chronic inflammation and cytokine signaling have been associated with the development of many serious adverse consequences of alcohol abuse and identified in epidemiological studies as risk factor for developing cancer [20]. Multiple mechanisms exist by which cytokines may promote immune and nonimmune cells to survive and proliferate, to induce angiogenesis and tissue remodeling, or to migrate, thus contributing to invasiveness. Elucidation of alcohol's effects on signaling pathways involved in cell proliferation, survival, cell death, angiogenesis, and motility, the cascade of events leading to the chronic inflammatory state, the development of fibrosis and cirrhosis, and the specific triggers that result in the progression from fatty liver, through the various stages of alcoholic liver disease leading to hepatocellular carcinoma, will provide insights into the relative importance of each in alcohol-induced tumorigenesis.

At the molecular level, each cancer type may be characterized by genetic alterations involving the mutation of proto-oncogenes into oncogenes and/or silencing of tumor suppressor genes. Failures in the control of growth and survival in normal cells by genetic and epigenetic alterations transform cells such that they display the hallmarks frequently shared by most, if not all, human cancers. Alcohol may influence the initiation, progression, and metastases of cancer by any of the mechanisms mentioned here or explored more fully in the remaining chapters. Rarely is it sufficient to have a defect in a single gene, nor is it essential for a large number of genetic loci to be altered, but there may be some support for the need for more than one system failure to develop cancer.

Elucidation of the cellular and molecular mechanism(s) by which alcohol exerts its carcinogenic effect relative to cancer initiation, progression, and metastases will undoubtedly generate useful therapeutic strategies; however, the global battle against cancer will not be won with treatment alone. These, coupled with effective prevention measures, are urgently needed to prevent a cancer crisis. It is clear that alcohol can cause cancer but that retrospective, self-reported data may introduce uncertainty in the level of alcohol consumption associated with risk. Epidemiological studies have led and continue to lead the way and have provided a very useful road map for focusing the types of cancer for which alcohol may be suspected and even indicted; however, the level of culpability and guilt remains to be determined by the quality of the evidence and a more rigorous scientific jury. At present, our understanding of the role of alcohol in cancer induction or promotion is a work in progress as investigators worldwide continue to expand our knowledge base and integrate epidemiological observations with studies on mechanism and clinical care. Each advance in this area holds the promise of the development of new therapeutic strategies to prevent or treat one or more of the alcohol-related cancers.

References

1. Dudley R (2000) Evolutionary origins of human alcoholism in primate frugivory. Q Rev Biol 75(1):3–15
2. Whiteford HA et al (2013) Global burden of disease attributable to mental and substance use disorders: findings from the Global Burden of Disease Study 2010. Lancet 382(9904):1575–1586

3. Newsholme A (1903) The possible association of the consumption of alcohol with excessive mortality from cancer. Br Med J 2(2241):1529–1531
4. IARC (1988) Alcohol drinking. Biological data relevant to the evaluation of carcinogenic risk to humans. IARC Monogr Eval Carcinog Risks Hum 44:101–152
5. IARC Working Group on the Evaluation of Carcinogenic Risks to Humans, World Health Organization, International Agency for Research on Cancer (2010) Alcohol consumption and ethyl carbamate, vol ix, IARC monographs on the evaluation of carcinogenic risks to humans. 2010, Lyon, France . International Agency for Research on Cancer, Lyon, 1424p; Distributed by WHO Press
6. Baan R et al (2007) Carcinogenicity of alcoholic beverages. Lancet Oncol 8(4):292–293
7. Hashibe M et al (2009) Interaction between tobacco and alcohol use and the risk of head and neck cancer: pooled analysis in the International Head and Neck Cancer Epidemiology Consortium. Cancer Epidemiol Biomarkers Prev 18(2):541–550
8. Grewal P, Viswanathen VA (2012) Liver cancer and alcohol. Clin Liver Dis 16(4):839–850
9. Allen NE et al (2009) Moderate alcohol intake and cancer incidence in women. J Natl Cancer Inst 101(5):296–305
10. Brooks PJ, Zakhari S (2013) Moderate alcohol consumption and breast cancer in women: from epidemiology to mechanisms and interventions. Alcohol Clin Exp Res 37(1):23–30
11. Klatsky AL et al (2014) Moderate alcohol intake and cancer: the role of underreporting. Cancer Causes Control 25(6):693–699
12. Liu Y et al (2012) Intakes of alcohol and folate during adolescence and risk of proliferative benign breast disease. Pediatrics 129(5):e1192–e1198
13. Rehm J, Patra J, Popova S (2007) Alcohol drinking cessation and its effect on esophageal and head and neck cancers: a pooled analysis. Int J Cancer 121(5):1132–1137
14. Marietta C et al (2009) Acetaldehyde stimulates FANCD2 monoubiquitination, H2AX phosphorylation, and BRCA1 phosphorylation in human cells in vitro: implications for alcohol-related carcinogenesis. Mutat Res 664(1–2):77–83
15. Yokoyama A, Omori T (2005) Genetic polymorphisms of alcohol and aldehyde dehydrogenases and risk for esophageal and head and neck cancers. Alcohol 35(3):175–185
16. Shukla SD et al (2008) Emerging role of epigenetics in the actions of alcohol. Alcohol Clin Exp Res 32(9):1525–1534
17. Wainfan E et al (1989) Rapid appearance of hypomethylated DNA in livers of rats fed cancer-promoting, methyl-deficient diets. Cancer Res 49(15):4094–4097
18. Shen L et al (1998) Correlation between DNA methylation and pathological changes in human hepatocellular carcinoma. Hepatogastroenterology 45(23):1753–1759
19. Davis CD, Uthus EO (2004) DNA methylation, cancer susceptibility, and nutrient interactions. Exp Biol Med (Maywood) 229(10):988–995
20. McClain CJ et al (1998) Tumor necrosis factor and alcoholic liver disease. Alcohol Clin Exp Res 22(5 Suppl):248S–252S

Chapter 2
Alcohol and Breast Cancer: Reconciling Epidemiological and Molecular Data

Samir Zakhari and Jan B. Hoek

Abstract Breast cancer is the most diagnosed cancer in women worldwide. Epidemiological studies have suggested a possible causative role of alcohol consumption as a risk factor for breast cancer. However, such conclusions should be interpreted with considerable caution for several reasons. While epidemiological studies can help identify the roots of health problems and disease incidence in a community, they are by necessity associative and cannot determine cause and effect relationships. In addition, all these studies rely on self-reporting to determine the amount and type of alcoholic beverage consumed, which introduces recall bias. This is documented in a recent study which stated that the apparent increased risk of cancer among light-moderate drinkers may be "substantially due to underreporting of intake." Another meta-analysis about alcohol and breast cancer declared "the modest size of the association and variation in results across studies leave the causal role of alcohol in question." Furthermore, breast cancer develops over decades; thus, correlations between alcohol consumption and breast cancer cannot be determined in epidemiological studies with windows of alcohol exposure that captures current or recent alcohol intake, after clinical diagnosis.

Numerous risk factors are involved in breast carcinogenesis; some are genetic and beyond the control of a woman; others are influenced by lifestyle factors. Breast cancer is a heterogeneous and polygenic disease which is further influenced by epigenetic mechanisms that affect the transciptomes, proteomes and metabolomes, and ultimately breast cancer evolution. Environmental factors add another layer of complexity by their interactions with the susceptibility genes for breast cancer and metabolic diseases. The current state-of-knowledge about alcohol and breast cancer association is ambiguous and confusing to both a woman and her physician.

S. Zakhari (✉)
Former Director, Division of Metabolism and Health Effects, NIAAA, NIH,
Bethesda, MD 20852, USA

SVP, Science, DISCUS, Washington, DC 20005, USA
e-mail: szakhari@discus.org

J.B. Hoek
Department of Pathology, Anatomy and Cell Biology, Thomas Jefferson University,
Philadelphia, PA 19107, USA

V. Vasiliou et al. (eds.), *Biological Basis of Alcohol-Induced Cancer*,
Advances in Experimental Medicine and Biology 815,
DOI 10.1007/978-3-319-09614-8_2

Confronting the huge global breast cancer issue should be addressed by sound science.

It is advised that women with or without a high risk for breast cancer should avoid overconsumption of alcohol and should consult with their physician about risk factors involved in breast cancer. Since studies associating moderate alcohol consumption and breast cancer are contradictory, a woman and her physician should weigh the risks and benefits of moderate alcohol consumption.

Keywords Breast cancer • Epidemiology • Alcohol • Acetaldehyde • Reactive oxygen species • Estrogen • Folate • Metabolism • Epigenetics • Alcohol dehydrogenase • Aldehyde dehydrogenase • BRCA1 • BRCA2

2.1 Introduction

Cancer is the leading cause of death in developed countries; worldwide, it is estimated that cancer could result in 12 million deaths in 2030 [1]. The most common cancers worldwide are lung, breast, colorectal, stomach, and prostate. In women, the leading causes of cancer death are lung, breast, and colorectal cancers. In an annual report by the National Cancer Institute [2], overall cancer death rates continued to decrease in the USA in the period between 1975 and 2010; most declines were observed in female breast, prostate, lung, and colorectal cancers. The sharp decrease in breast cancer between 2002 and 2003 was attributed most likely to the reductions in the use of postmenopausal hormone-replacement therapy (HRT) [3]. Notwithstanding the significant decline in breast cancer mortality rates in the industrialized nations since 1990, breast cancer represents the most common female malignancy worldwide and is one of the primary causes of death among women globally [4].

There are a multitude of underlying etiological risk factors for breast cancer, enumerated below, including the use of HRT. However, before discussing risk factors, it is imperative to understand how breast cancer develops.

2.2 The Biology of Breast Cancer

Breast development starts by the rapid division of stem cells at puberty and continues through woman's first full-term pregnancy. After birth, the hormonal milieu (estrogen, progesterone, growth hormone, prolactin) and cell fate-determining signaling pathways transform a high percentage of mother's breast cells into mature, differentiated milk-producing cells. Breast cell division is controlled by signals, such as estrogen, that allow cells to enter the cell cycle and promote cell division. Many proto-oncogenes code for the signals that control the cell cycle. Certain mutations in proto-oncogenes can result in oncogenes that code for protein signals that cause overexpression of growth factors or their receptors, resulting in uncontrolled cell division and growth. For instance, erbB2, a member of the epidermal growth factor (EGF) receptor family, also known as HER-2 (for Human Epidermal Growth Factor Receptor 2) or HER2/neu is a receptor tyrosine kinase protein that promotes cell proliferation. HER2 itself does not bind growth factors, but it can heterodimerize with other members of the EGF receptor family and channel EGF and growth factor signals into more effective growth-promoting pathways. Overexpression of HER2 can thereby enhance the growth and proliferation of cancer cells; HER-2-positive breast cancers are more aggressive than other types of breast cancer. Other oncogenes that influence breast cancer include many other members of the tyrosine kinase family as well as cell cycle regulatory proteins, such as c-myc, cyclin D-1, and the cyclin regulator CDK-1. Opposing the oncogenes are tumor suppressor genes, such as p53 which recognizes cells with mutated DNA and causes apoptosis to these cells. Mutations in the p53 gene result in the continuous reproduction of cells with damaged DNA, enhancing cancer development [5].

Breast cancer is a heterogeneous disease that encompasses more than 20 different subtypes. Numerous molecular, cellular, and pathological processes are involved in the transformation of healthy tissue to preinvasive lesions, such as ductal carcinoma in situ (DCIS), to invasive breast cancer. More than 70 % of DCIS lesions express estrogen receptors, and about 50 % of the lesions overexpress the *HER2/neu* proto-oncogene [6], In addition, the p53 tumor suppressor gene is mutated in roughly 25 % of lesions [7]. Based on molecular characteristics and clinical outcome, subtypes of breast cancer are defined by gene expression profiles including evaluation of estrogen receptor (ER), progesterone receptor (PR), and HER2 receptor, all of which affect the tumor growth rate and its metastatic potential, reflected in the disease grading [8, 9].

In addition to ER and PR, studies have revealed the presence and potential importance of several nuclear receptors in breast cancer, including receptors for steroid hormones (androgen, corticosteroids), vitamins A and D, fatty acids, and food-derived xenobiotic lipids [10]. Among other major signaling pathways involved in mammary carcinogenesis is increased Wnt signaling [11]. The Wnt signaling pathway controls the stability and activity of β-catenin, a transcription factor that drives the expression of a large number of proliferation promoting signals, as well as signaling pathways that control the activity of mTOR, a critical junction in the cell-growth control. Wnt signaling is an important factor during mammary development and is involved in stem cell fate determination. Wnt also determines the differentiation of cancer stem cells, and its unregulated activation can promote tumorigenesis. In particular, there is evidence that Wnt activation is involved in triple-negative breast tumors, i.e., breast tumors that are not characterized by overexpression of HER2, ER, or PR. The role of aberrant Wnt signaling in breast carcinogenesis is further highlighted by the finding that knockdown of the tumor-suppressor gene PTEN (*phosphatase and tensin homolog deleted on chromosome 10*) resulted in the activation of the Wnt/β-catenin pathway in human breast cells [12]. In addition to Wnt signaling, notch signaling regulates mammary stem and progenitor cell activity in breast tissue and commits stem cells to the luminal cell lineage [13].

2.3 Known Risk Factors for Breast Cancer

To fully understand the findings and ramifications of epidemiological studies on alcohol and breast cancer, it is essential to consider the range of known risk factors involved in breast cancer development. Many of the primary risk factors for breast cancer cannot be readily modified. These include the strong risk factors aging, genetics (inherited changes in certain genes and family history of breast cancer), risk caused by prenatal history (e.g., daughters born to mothers who used diethylstilbestrol (DES) during pregnancy have increased risk), and reproductive parameters which determine the cumulative lifetime estrogen exposure (early menarche, before age 12; delayed menopause, after age 55; delayed child bearing; first full-term pregnancy after age 30; miscarriage; abortion). Modifiable lifestyle risk factors include dietary habits (consumption of polyunsaturated fats and excessive alcohol), smoking, exposure to radiation or synthetic estrogens, viral infection, physical inactivity, use of HRT, obesity, diabetes, breast implants, and even changes in circadian rhythm homeostasis, such as night-shift work. Needless to say, other risk factors may exist that are not yet fully understood or even known.

2.4 Alcohol as a Risk Factor for Breast Cancer

Chronic heavy alcohol consumption (drinking too much too often) and binge drinking (too much too fast) are risky drinking behaviors that could promote various pathological conditions, including cancer. More recently, some epidemiological studies have suggested that even moderate alcohol consumption can increase the risk of breast cancer by a small extent [14]. By contrast, others reported a decrease in breast cancer risk due to moderate drinking [15]. Equally, the molecular basis of alcohol use as a risk factor remains disputed. In view of the contradictory results of the epidemiological as well as the molecular studies on alcohol and breast cancer, the landscape of available information will be discussed under these two categorizations: (a) *epidemiological studies*, which cover case–control or cohort studies, conducted in various countries and with a wide range of cohort size, and (b) studies addressing the *molecular basis* that might contribute to the influence of alcohol on breast cancer risk. We will then consider to what extent information on molecular and cellular actions of alcohol can account for the epidemiological findings on alcohol as a risk factor for breast cancer.

2.4.1 Epidemiological Studies

A large number of prospective studies and some case–control studies on alcohol use as a risk factor for breast cancer have been reported over the past decades. Although there is consensus that heavy alcohol use can be a significant risk factor, the findings are more controversial with regard to moderate alcohol use. These studies use a wide range of different sample sizes and methodologies, various definitions of a "drink," and diverse criteria of moderate or heavy drinking and consider different times of drinking in a woman's life. A constant feature is that essentially all studies obtain alcohol use data by self-reporting, the reliability of which is often problematic, particularly for longer time intervals. All these factors can explain at least part of the divergent findings and confusion. Studies discussed below are not intended to present a comprehensive review of epidemiological studies, rather a sampling of various studies, in different countries, with diverse methodologies and sample size, varied dietary intake, and different results. For clarity of the discussion that will follow, these studies are enumerated below, and combined comments on them are discussed in the concluding remarks section.

1. The Nurses' Health Study initiated in the USA in 1980 administered a dietary questionnaire (including the use of beer, wine, and spirits) to 89,538 nurses between the age of 34 and 59, with no history of cancer. During the ensuing 4 years, 601 cases of breast cancer were diagnosed. (In this study, a drink was defined by rather nonstandard criteria with inaccurate estimates of their alcohol content). The study reported relative risk (RR) of 1.3 for women consuming one-third to one drink/day (compared to nondrinkers RR = 1.0), which went up to 1.6 for those consuming more than one drink/day, although, ironically, RR was not further increased in those consuming 1.8 or more drinks/day, and there was no increase in risk for those who drank less than 1/3 of a drink/day [16]. The study noted the potential impact of various other risk factors, such as body weight, cigarette smoking, and being nulliparous, but stated that these were not themselves associated with breast cancer risk in these studies. However, several other studies have reported that these are risk factors for breast cancer (see below). Moreover, the combined risk of alcohol use with these other factors could not be resolved.
2. An update of the Nurses' Health Study (1) was published in 2011 [17]. Cumulative average alcohol intake in 1994, the midpoint of the follow-up period, was used to assess RR for breast cancer. Compared with women who never consumed alcohol, those who consumed 5–9.9 g per day (equivalent to 3–6 drinks per week) had a modest increase in risk (RR = 1.15); little difference was found between risk and various alcoholic drinks (RR per 10 g/day was 1.12 for wine and 1.09 for beer or liquor). Women who on average consumed at least 30 g of alcohol/day (slightly over two drinks per day) had a greater risk of breast cancer (RR = 1.51). Alcohol consumption seemed to be more strongly associated with the risk of ER+ status, PR+ status, or both for women who drank 10 or more g/day; this interaction did not reach statistical significance though. The authors of this study highlighted the importance of considering lifetime exposure when evaluating the effect of alcohol. However, determining lifetime alcohol use by average use/day may miss important patterns of alcohol use that may have an influence on the outcome, as was recognized in some other studies.

3. In 1994, Longnecker [18] reported on a meta-analysis of 29 case–control and 9 follow-up studies from the USA, Australia, Italy, France, Greece, the Netherlands, Canada, England, Sweden/Norway, Denmark, New Zealand, and Argentina. Daily consumption of a drink was associated with an 11 % increase in breast cancer risk compared to nondrinkers. However, the author reported that the "slope of the dose response curve was quite modest" and "the modest size of the association and variation in results across studies leave the causal role of alcohol in question." Needless to say, these studies were conducted in different countries with wide variations in their dietary habits, environmental factors, smoking, and genetic background.
4. In a case–control study of 890 cases of Black and White women, 20–74 years old, in the USA, the odds ratio (OR) to develop breast cancer for women who have a recent consumption of 1 or 2 drinks/day, compared to nondrinkers, was 1.4; intriguingly, consumption of two or more drinks/day resulted in OR of 1.0 (i.e., no increase in risk). In addition, average lifetime consumption of 91 g/week (about 6.5 drinks) resulted in a "nonsignificant increased" risk (OR = 1.5) in women reporting binge drinking [19]. Also, ORs did not differ by race, age, menopausal status, use of HRT, or body mass index (BMI). Obviously, these correlations are intrinsically questionable. It is hard to see how recent consumption can be a causal factor in breast cancer that probably has started at least 20 years earlier. Also, it is difficult to correlate average lifetime consumption in a meaningful manner with the molecular events leading to breast cancer.
5. A small case–control study in France involving 437 women between the age of 25 and 85, reported a decrease in risk for breast cancer for women consuming less than 1.5 drinks/day; OR = 0.58, after adjustment for BMI, parity, breastfeeding, physical activity, history of breast cancer, diet, and duration of ovulation [20]. Three patterns of alcohol consumption were identified (abstinent, sporadic, and frequent drinkers). Sporadic drinkers comprised women who drank four times per week or less, while frequent drinkers were defined as those who consumed alcohol five times a week or more. Alcohol consumption was recorded as units (one unit = 10 g of ethanol in 4.2 oz of wine, 11 oz of beer, or 1 oz of spirits). No association was found between the pattern of total alcohol consumption and breast cancer risk. The study noted that drinking pattern could change during the period under consideration. For example, "a woman who claimed not to be drinking at the time of interview could, in fact, have been at some previous point alcoholic or could have had sporadic alcoholism that motivated the cessation of drinking. In such cases, the longest typical phase of consumption during that individual's history was used for the study."
6. The risk of breast cancer due to total caloric intake, coffee and alcohol consumption was studied in 280 breast cancer French Canadian women who were noncarriers of six specific mutations in *BRCA1/2* genes found more frequently in families of French Canadian descent. They were compared with 280 matching women without breast cancers who were not carriers of these mutations [21]. Data were obtained by using a food-frequency questionnaire (FFQ) that

"covered the period 2 years prior to the diagnosis for cases and a corresponding period for the controls." In addition, "alcohol-related beverages consumed were summed to obtain the total amount drunk per week." Average alcohol consumption was 9.8 and 6.3 g/day for cases and controls, respectively. The study concluded that "more than two bottles of beer per week" increased breast cancer risk by 34 %, whereas >10 oz of wine or >6 oz of spirit per week increased cancer risk by 16 % and 9 %, respectively. The study acknowledged "recall bias" as a limitation.

7. A case–control study involving 1,728 women 20–49 years of age, in Los Angeles County, California, administered a questionnaire about early, lifetime, or recent alcohol consumption [22]. The study reported that alcohol intake "during the recent 5 year period before the breast cancer diagnosis was associated with increased breast cancer risk" and that "intake of two or more alcoholic drinks per day during this 5 year period was associated with an 82 % increase in breast cancer risk relative to never drinkers." Ironically, there was no risk increase for "lifetime alcohol intake."
8. On the other hand, a population-based study (1,508 cases) collected information on alcohol intake throughout life. Consumption of 15–30 g/day (approximately 1–2 drinks) throughout life was associated with a modest 33 % increase in risk particularly among women with low BMI (<25) and those diagnosed with estrogen receptor-positive tumors; but heavier consumption (>30 g per day) was not. Risk did not vary with alcohol type or by patterns of use (recent use, intake prior to age 20 years) [23].
9. Another population-based case–control study about lifetime alcohol consumption did not find an increase in breast cancer risk among women younger than 50 years of age; however, among those over 50 years of age, ever drinking conferred a relative risk of 1.8. Information about alcohol intake was obtained using a questionnaire from women 40–75 years old who participated in a screening program in central Sweden [24].
10. Breast density is a risk factor for breast cancer. The impact of alcohol consumption on mammographic density was assessed for 1,207 cases from three populations (Japan, Hawaii, California) [25]; alcohol intake was estimated from "self-administered questionnaire" and recorded as "ever vs. never," and for Hawaii and Japan only, the "ever drinkers were divided into ≤1 and >1 drink/day." Results showed that alcohol consumption did not significantly modify the effect of mammographic density on breast cancer risk "in this pooled analysis." The study stated that "whereas the dichotomous model did not indicate an association between alcohol drinking and breast cancer, the relative risk was elevated for women consuming >1 drink/day without reaching statistical significance." The study invoked "recall bias" and stated that "as in all epidemiological studies, alcohol intake may have been underreported" and "this analysis had limited ability to model the exact relations between alcohol intake, mammographic density, and breast cancer risk and the findings need to be interpreted with caution."

In short, these various studies highlight the intrinsic problem of assessing long-term or even lifetime drinking patterns through a recall questionnaire approach.

Another question is whether specific subtypes of breast cancer show enhanced risk related to past or current alcohol use and whether alcohol use synergizes with other breast cancer risk factors. In recent years, a number of large cohort studies were conducted that provided the opportunity to assess alcohol use history and other breast cancer risk factors for some of the major cancer subtypes.

11. The Million Women Study [26] conducted in the United Kingdom and published in 2009 calculated the RR for 21 site-specific cancers due to beverage alcohol consumption, including breast cancer, based on a questionnaire asking about the average alcohol consumption per week. Of the 1,280,296 women recruited, data from 708,265 women from a follow-up survey three years later were used. The study reported that women who drank alcohol were "likely to be younger, leaner, more affluent, and to do strenuous exercise more frequently" and more likely to "have ever used oral contraceptives and to be currently using hormone replacement therapy" than nondrinkers. Also, among drinkers, "the proportion of current smokers increased with increasing alcohol intake." The RR of breast cancer was 1.08, 1.13, and 1.29 for women who drank 3–6, 7–14, and 15 or more drinks/week, respectively. The study estimated a 12 % increase in breast cancer risk per 10 g increment of alcohol intake.
12. The Women's Health Study in the USA conducted a 10-years follow-up on 38,454 women 45 years or older who were free of cancer and cardiovascular disease at baseline and provided detailed dietary information, including alcohol consumption [27]. High alcohol consumption (30 g/day—over two drinks) was associated with a modest increase in breast cancer risk (RR = 1.32) that was limited to ER+ and PR+ tumors. The RR for an increment of 10 g/day of alcohol were 1.11 for ER+/PR+ tumors, 1.00 for ER+/PR− tumors, and 0.99 for ER−/PR− tumors. The association seemed strongest among those taking HRT currently, albeit statistically not significant. In addition, the RR of breast cancer for a 10 g/day increment was similar for different beverages (1.15 for beer, 1.13 for white wine, and 1.08 for red wine or liquor).
13. In a population-based Swedish Mammography Cohort study, self-reported data on alcohol consumption were collected in 1987 and 1997 from 51,847 postmenopausal women [28]. After adjusting for age; family history of breast cancer; BMI; parity; age at menarche, first birth, and menopause; diet; and HRT use, alcohol consumption was associated with an increased risk for the development of ER+ tumors, irrespective of PR status, especially in women using HRT. Consumption of 10 g or more of alcohol/day increased RR to 1.35 for ER+/PR+, 2.36 for ER+/PR−, 0.62 for ER−/PR+, and 0.80 for ER−/PR− tumors versus nondrinkers. However, the link between alcohol use and ER+ or PR+ status was not consistent across different studies. The study by Terry et al. [23] mentioned earlier reported that alcohol consumption increases the risk in ER+/PR+ breast cancer but not in ER−/PR− [23]; among postmenopausal

women no statistically significant differences were observed in the risk factor profiles for ER+ PR+ and ER−PR− breast cancer [29]. An additional case–control study showed that alcohol increases risk in ER+/PR+ tumors, but not for ER+/PR− and ER−/PR− tumors [30]. Alcohol use appears to be more strongly associated with risk of lobular carcinomas and hormone receptor-positive tumors than it is with other types of breast cancer [31]. To add to the confusion, another study reported an increased risk for ER+/PR+, ER−/PR−, and ER−/PR+ tumors, but not for ER+/PR− in women 20–44 years of age [32]. Two additional case–control studies reported positive associations of alcohol consumption with risk of either ER+ or ER− tumors [33, 34] and one with ER+ tumors only [35]. Finally, two studies in which alcohol consumption was categorized into only "ever vs. never" reported no association irrespective of joint ER and PR status [36, 37].

14. In contrast to the Swedish study, the Iowa Women's Health Study found that alcohol intake was most strongly associated with ER−/PR− tumors in following 37,105 cancer-free women 55–69 years of age, who filled out a questionnaire by mail, and were followed up for 7 years [38]. Alcohol consumption over the past year was self-reported and was averaged as g/week. The study reported that there was a 55 % increased risk for ER−/PR− tumors in "women who had ever drunk alcohol"; however, alcohol consumption was not quantified.
15. The interactions between alcohol consumption and HRT was studied in 40,680 postmenopausal California teachers using a questionnaire for alcohol consumption during the past year and HRT use for the past 5 years [39]. Subjects are grouped into three categories: nondrinkers, those consuming <20 g/day of alcohol, and those who consume ≥20 g/day. Increased breast cancer risk associated with alcohol consumption was observed among postmenopausal women who were current users of HRT (RR = 1.60 for those consuming <20 g/day and RR = 2.11 for consumers of ≥20 g/day). Alcohol did not increase risk among women who had stopped using HRT within 3 years. Results were similar for ER+ and ER+/PR+, and no increase in risk was observed in ER− tumors.
16. A study on 989 cases of breast cancer in women aged 23–74 years in three Italian areas investigated the role of alcohol according to ER and PR status by collecting information on lifetime alcohol consumption using FFQ [40]. The weekly number of drinks was calculated, taking into account that one drink corresponds to approximately 125 mL of wine, 330 mL of beer, and 30 mL of hard liquor, each containing about 15 g of ethanol (30 mL of 80 proof liquor contains only 9.6 g of ethanol). The study reported that consumption of ≥13.8 g/day increased the risk of ER+ tumors (OR = 2.16), ER− (OR = 1.36), ER+/PR+ (OR = 2.34), ER−/PR− (OR = 1.25) and concluded that alcohol is more strongly associated with ER+ and ER+/PR+ than ER− breast tumors.
17. The National Institutes of Health-AARP Diet and Health Study obtained information from 184,418 postmenopausal women aged 50–71 years, about their alcohol use and diet, through a mailed questionnaire at baseline [41], Breast cancer cases and ER and PR status were identified through linkage to state cancer registries. The authors reported that "Moderate consumption of alcohol

was associated with breast cancer, especially hormone receptor-positive tumors." However, a closer analysis of the data indicated that the RR did not reach significance for both light (0.4–0.7 self-reported drinks/day) and moderate (0.7–1.4 drinks/day) alcohol use (RR of 1.13 and 1.07, respectively, for ER+/PR+ cancers, with 95 % confidence intervals ranging from 0.89 to 1.38) and even self-reported drinking at higher levels (1.4–2.5 drinks/day) with an RR of 1.34 and CI 1.06–1.69 was based on only 89 cases. Other cancer subtypes (ER+/PR− or ER−/PR−) did not show significant increases and were based on even lower incidence. Therefore, the conclusion of the authors that moderate drinking was associated with breast cancer is not supported by data.

18. The Women's Health Initiative-Observational Study enrolled 87,724 postmenopausal women aged 50–79 years, without a history of breast cancer between 1993 and 1998, who self-reported their alcohol use histories [42]. In a follow-up through 2005, a total of 2,944 patients with invasive breast cancer were diagnosed. The study reported that the RR in women who consumed seven or more alcoholic beverages/week was 1.82 for hormone receptor-positive invasive lobular carcinoma and a statistically nonsignificant 1.14 for hormone receptor-positive invasive ductal carcinoma. Women who reported drinking one or more alcoholic beverage/day were more likely to be nulliparous, with low BMI, currently use HRT, and smoke. Alcohol use was assessed only at baseline, and the authors stated that "Extensive measurement errors or changes in alcohol use could affect the study conclusions."
19. In 1966, Doll and colleagues reported on breast cancer incidence in five continents where the USA was reported to be 4–7 times higher than in Asian populations [43]. Almost half of the East Asian population is deficient in the mitochondrial enzyme that metabolizes acetaldehyde, the first metabolite of alcohol and a suspect in breast carcinogenesis. Although the drinking history of breast cancer patients was not assessed in this study, this observation suggests that acetaldehyde metabolism may not be a dominant determinant in breast cancer risk. To test any association between acetaldehyde and breast cancer, the effect of alcohol consumption on breast cancer incidence rates was studied in 597 Chinese, Japanese, and Filipino women living in San Francisco–Oakland, Los Angeles, and Oahu, Hawaii [44]. Breast cancer risk was not significantly associated with alcohol drinking (OR = 0.9) in Asian American women. Furthermore, a prospective study performed in Japan using data from 35,844 women who completed a self-administered questionnaire found that consuming <15 g/day did not significantly increase the risk for breast cancer. However, risk was significantly increased in women who consumed ≥15 g/day [45]. To add to the confusion, the Miyagi Cohort Study in Japan involving 19,227 women found that consuming ≥15 g/day of alcohol "had no significant relation to breast cancer risk" [46].
20. To test whether alcohol-induced facial flushing (i.e., women who have the defective ALDH2*2 gene thereby cannot effectively metabolize acetaldehyde further to acetate) modifies the risk for breast cancer, a prospective study was undertaken by Japan Public Health Center on 50,757 pre- and postmenopausal

women aged 40–69 years, using self-reported questionnaire [47]. After 13.8 years of follow-up, 572 cases of invasive breast cancer were diagnosed. The study reported that, compared to never drinkers, regular alcohol drinkers (>150 g ethanol/week—about 2 drinks/day, which is higher than the definition of moderate drinking) have 78 % increased risk for breast cancer in premenopausal women and 21 % increase in postmenopausal women. Consumption of 10 g/day of alcohol was associated with 6 % increase in risk for overall breast cancer (compared to 12 % in the Million Women Study discussed above). This effect was not modified by alcohol-induced facial flushing, by folate intake, by smoking, by BMI, nor by exogenous estrogen use by postmenopausal women. There was no statistically significant association between alcohol intake and ER+ tumors. A previous study also showed no association between polymorphism of ALDH enzyme and risk of breast cancer [48]. Furthermore, a review of epidemiological evidence in Japanese populations using three cohort studies and eight case–control studies by Nagata and colleagues reported that "epidemiologic evidence on the association between alcohol drinking and breast cancer remains insufficient in terms of both the number and methodological quality of studies among the Japanese population" [49].

21. A case–control study conducted in China involved 1,009 cancer cases, in which alcohol consumption data were obtained in a face-to-face interview within three months after diagnosis. Tumors' ER/PR status was obtained from pathology reports. The study reported that low-moderate alcohol consumption was inversely associated with breast cancer risk: adjusted odds ratio (OR) for women who consumed <5 g/day was 0.4 and 0.62 in post- and premenopausal women, respectively, compared to nondrinkers [15]. OR was low across hormone receptor status groups even for those consuming <15 g/day for postmenopausal women (OR = 0.36–0.56) and premenopausal women (OR = 0.57–0.64). Consuming >15 g/day increased OR in postmenopausal women regardless of the hormone receptor status. Apart from the wide range of participants' age (20–87 years)—which influences the relative contribution of various risk factors for breast cancer-quantification of alcohol consumption was haphazard. For instance, the study stated that "Standard drinking vessels used by Zhejiang residents were displayed during the interview to increase the accuracy of measurement" without stating the volume or alcohol content. Furthermore, alcohol consumption was based on a "reference" recall period one year "before diagnosis." Consumption of ≥15 g/day appeared to increase breast cancer in postmenopausal women with ER+/PR– or ER–/PR+.
22. Similarly, a study on 712 breast cancer cases, aged 30–74 years from the New Mexico Tumor Registry, collected data on recent and past alcohol intake via in-person interview. Compared to nondrinkers, low recent alcohol intake (<148 g/week, ~10.5 drinks) was associated with reduced risk of breast cancer for non-Hispanic Whites (OR = 0.49) independent of hormone receptor status for both pre- (OR = 0.29) and postmenopausal women (OR = 0.56). Past alcohol intake did not demonstrate association with breast cancer, and trends were nonsignificant [50].

Several studies explored the relationship between alcohol use and folate deficiency as related risk factors in the development of breast cancer.

23. A study conducted on 1,000 Mexican women [51] with breast cancer using "in-person interviews" determined "recent alcohol intake" and whether the patient is an "ever drinker" or "never drinker." It concluded that "any alcohol intake increases risk of breast cancer," and "insufficient intake of folate may further elevate risk for developing breast cancer." "Ever drinking was associated with a twofold increase in the odds of breast cancer" reported the study. However, the definition of ever drinking was a "yes" or "no" without quantification, and the authors declared that "recall bias is a concern."
24. Another study, the Women's Health Initiative-Observational Study, gathered baseline questionnaires which addressed alcohol and folate intake from 88,530 postmenopausal women 50–79 years [52] and found no evidence for folate attenuating alcohol's effect on breast cancer risk. Similarly, the American Cancer Society Cancer Prevention Study II Nutrition Cohort [53] examined the relationship between alcohol, dietary intake of folate and methionine, and breast cancer risk in 66,561 postmenopausal women. Women who consumed 15 or more grams of ethanol/day had increased risk of breast cancer (RR = 1.26) compared with nonusers. However, no association between risk of breast cancer and dietary folate, total folate, or methionine intake was found, and there was no evidence of an interaction between dietary folate or total folate and alcohol.
25. Possible interaction between alcohol and folate was investigated in 24,697 postmenopausal women in the "Diet, Cancer and Health" follow-up study which included 388 cases of breast cancer and 388 randomly selected controls to estimate the breast cancer incidence rate ratio (IRR) in conditional logistic regression analysis [54]. Alcohol intake was associated with risk of breast cancer mainly among women with folate intake below 300 μg (IRR = 1.19 per 10 g average daily alcohol intake); no association between alcohol and breast cancer risk was found among women with a folate intake higher than 350 μg (e.g., folate intake >400 μg; IRR = 1.01). The authors concluded that adequate folate intake may attenuate the risk of breast cancer associated with high alcohol intake.
26. A case–control study in pre- and postmenopausal Japanese women including 1,754 breast cancer patients aged 20–79 years found that self-reported alcohol consumption was associated with the risk of breast cancer [55]. Consuming ≥23 g/day of alcohol increased the risk by 39 % compared to nondrinkers. However, no significant positive association was observed among premenopausal women. High folate intake was associated with a lower risk of developing breast cancer in pre- but not postmenopausal women. In addition, high folate intake reduced the risk of breast cancer in women consuming ≥23 g/day of alcohol only in post- but not premenopausal women. Determining the risk based on the tumor receptor status was misleading and confusing. For example, in premenopausal women with ER+/PR+/HER2+ tumors, the odds ratio (OR) of developing breast cancer for those drinking 1 to ≤5 g alcohol/day,

5 to ≤23 g/day, and ≥23 g/day were 0.84, 1.61, and 0.84, respectively. For ER−/PR−/HER2+ OR was 0.7, 1.92, and 0.52, respectively. Examination of data revealed that the ORs for ER−/PR−/HER2+ tumors were based on 4, 7, and 1 patients, respectively. Similarly, for ER−/PR−/HER2− tumors, ORs were 0.47, 2.47, and 1.39 based on 2, 5, and 1 patients, respectively.

27. The relation between alcohol intake and the risk of breast cancer was investigated in 274,688 women participating in the European Prospective Investigation into Cancer and Nutrition study (EPIC). Alcohol information was obtained by self-reports. The IRR per 10 g/day of continuous higher recent alcohol intake was 1.03. No association was seen between lifetime alcohol intake and risk of breast cancer. No difference in risk was shown between users and nonusers of HRT, and there was no significant interaction between alcohol intake and BMI, HRT, or dietary folate [56].

In summarizing the main outcomes of the epidemiological studies, despite the indications suggested by many of these studies that there is some relationship between alcohol use history and the risk for developing breast cancer, the nature of that relationship remains poorly characterized. Major open questions are what aspects of a woman's drinking history influence breast cancer risk, whether different subtypes of breast cancer account for the increased risk, and how an individual's physiological response to alcohol and its metabolites could interact with other breast cancer risk factors to promote disease onset or progression. A better understanding of the molecular basis by which alcohol use is thought to enhance cancer risk is needed. The following section will explore the information available from molecular and cellular studies that have addressed these questions.

2.4.2 *Molecular Studies*

Although epidemiological studies about alcohol and breast cancer resulted in controversial results, identified no causal association, and at low to moderate levels of drinking correlations were tenuous at best, experimental studies suggested possible mechanisms that could be invoked, including estrogen metabolism and response, acetaldehyde-induced cell mutation, oxidative stress, and epigenetic modifications involving one-carbon metabolism pathways. These mechanisms are elegantly reviewed by Seitz and colleagues [57] and by Dumitrescu and Shields [58].

2.4.2.1 Estrogen Metabolism

Estrogen plays an important role in breast cell division and hence carcinogenesis. It has been postulated that prolonged exposure of mammary tissue to estrogen and progesterone, due to early menarche and/or delayed menopause, may contribute to higher breast cancer risk. In postmenopausal women, estrogen levels are maintained

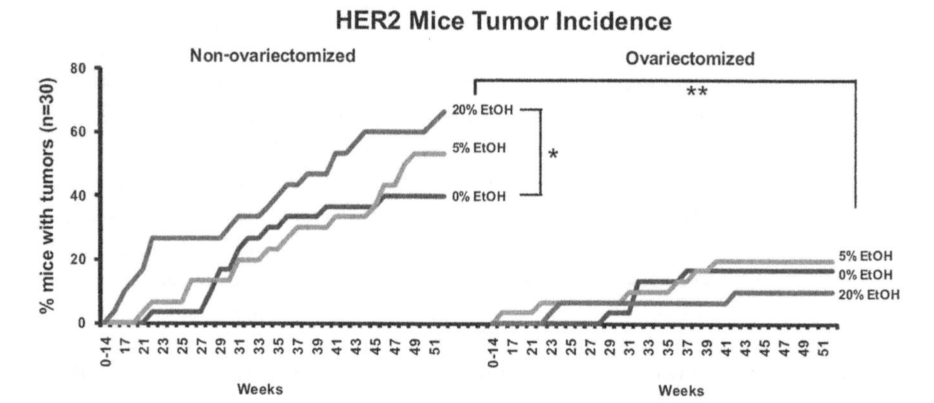

Fig. 2.1 Consumption of 20 % alcohol enhanced tumorigenesis in mice. Ovariectomy abolished this effect despite alcohol-induced increase in estrogen levels (From Wong et al. [61])

mostly by the activity of the aromatase enzyme which catalyzes the last step in estrogen biosynthesis from androgens (i.e., androstenedione to estrone and testosterone to estradiol) [59].

In one experimental study on alcohol alone and breast cancer, 20 female ICR mice were given 10–15 % ethanol solution as the *sole* drinking fluid for 25 months, with ad libitum solid diet [60]. Approximately, 45 % of mice developed either papillary or medullary adenocarcinoma of glandular epithelial origin. However, the relevance of this model for human alcohol consumption is questionable. Taking into account the average life span of ICR mouse, which is 2–2.5 years, these animals were given alcohol solution as the only drink available for about as much as 85 % of their life, at a rate equivalent to nonstop binge drinking, a situation that is neither physiological nor normal for humans.

Another study used transgenic mice that overexpress the HER2 protein (encoded by the proto-oncogene HER2/neu) in the mammary epithelium, resulting in the development of estrogen receptor alpha (ERα)-negative mammary tumors, similar to those of patients with HER2+ breast cancer [61]. Non-ovariectomized (NOVX) and ovariectomized (OVX) mice were exposed to 0, 5, and 20 % ethanol in the drinking water at 9 weeks of age till the endpoint (week 52), when serum was collected to determine estrogen levels. Tumor incidence in the 5 and 20 % alcohol-consuming NOVX mice was 53.33 and 66.67 %, respectively, compared to 40 % in the control mice; however, tumor incidence reached statistical significance only in mice consuming 20 % alcohol. Increase in tumor incidence was associated with increased systemic estrogen levels, increased expression of aromatase, and increased expression of ER-α in the tumors of 20 % alcohol-consuming mice. Additionally, ovariectomy blocked the effects of 20 % alcohol on tumor development (Fig. 2.1) despite the increase in estrogen levels due to alcohol. The authors concluded that "alcohol promotes mammary tumor development only in the presence of normal systemic estrogen levels, which the OVX animals lack," and "alcohol consumption promotes HER2 breast cancer development via the estrogen signaling pathway." While 20 % alcohol consumption increased estrogen levels in OVX mice, the estrogen levels were still significantly lower than those of NOVX control mice. Also, OVX mice failed to develop tumors in numbers comparable to NOVX mice, which led the authors to state "estrogen may be important for the tumor model in general and that failure to see tumor promotion with alcohol is a secondary effect."

The results of this study highlight the importance of assessing the HER2 status in addition to that of ER and PR. To translate these results to humans, women who take estrogen-containing HRT could have an increase in breast cancer risk due to the combined effects of HRT and alcohol. However, the epidemiological study (#26 above) that took into account the HER2 status found that in premenopausal women with ER+/PR+/HER2+ tumors, the risk of developing breast cancer for those drinking 5 to ≤23 g/day was increased by 61 % and for ER−/PR−/HER2+ by 92 % for women drinking the same amount. It is apparent that the results of this epidemiological study do not dovetail in a straightforward manner with the mouse study, suggesting that the relationships between these variables are more complex.

The use of HRT that contains estrogen adds to the complexity of the interaction between various risk factors. For example, the Women's Health Initiative (WHI) reported that women who received ≥5 years of continuous treatment with estrogen and progestin have increased risk of breast cancer [62]. Similar results were reported by studies #12, #13, and #15 above, but not by #4. In the same WHI study, postmenopausal women with prior hysterectomy who received estrogen alone showed a statistically significant decrease in breast cancer risk [63]. In addition, women in the French observational E3N study who received estrogen alone or estrogens combined with micronized progesterone showed no increase in breast cancer risk; however, those who received estrogens and androgenic progestins, or who were on HRT for long time, were at increased risk [64]. In a Finnish study [65], postmenopausal women using estradiol (E2)-progestogen therapy showed no increase in breast cancer incidence within the first 3 years of use.

Since supraphysiological estrogen doses caused mammary adenocarcinomas in rats [66], and alcohol consumption increased plasma estrogen levels (not to a supraphysiological level) in human female volunteers [67], it was postulated that alcohol use should be more strongly associated with ER+ than ER− tumors. However, epidemiological studies that assessed the risk of alcohol consumption based on tumor status were contradictory. For example, while some epidemiological studies showed a modest increase in ER+ tumors with the consumption of 15–30 g/day, there was no association with consumption of >30 g/day (see study #8 above). The link between alcohol use and ER+ or PR+ status was not consistent across different studies (see discussion under #13). For instance, studies showed statistically nonsignificant associations with either or both ER+, PR+ for women consuming ≥10 g/day (study #2). Consumption of 30 g/day was associated with a modest increase in risk of ER+/PR+ tumors, but there was no increase in risk for ER+/PR− tumors (study #12). Other studies showed that 10 g/day of alcohol increased the risk in either ER+ or ER− (study #13), or mostly in ER−/PR− (study #14). Study #17 showed that consumption of 10–20 g/day was associated with 7 and

28 % increase in risk for ER+/PR+ and ER−/PR− tumors, respectively. Finally, meta-analysis of 4 prospective and 16 case–control studies [68] showed that an increase in alcohol consumption of 10 g per day was associated with increased risks for ER+/PR+ (11 %) and ER+/PR− (15 %). The authors concluded that the observed positive associations with alcohol for ER+/PR+ and ER+/PR− tumors cannot be explained by estrogen-dependent pathway only.

In addition, estrogen status is influenced by numerous exogenous factors. For instance, persistent exposure of mammary gland stem and progenitor cells to different environmental factors such as xenoestrogens (bisphenol A, phthalates, ethinyl estradiol, phytoestrogens) alters their epigenetic reprogramming during epithelial differentiation [69]. This is mediated, in part, through ERα nuclear receptors which activate or silence the transcription of target genes [70]. Interactions between ERα and various enzymes involved in histone modifications (histone acetyltransferases, histone deacetylases, histone methyltransferases, histone demethylases), co-activators, and co-repressors have introduced another layer of complexity in the epigenetic regulation of breast carcinogenesis [71]. Furthermore, women who were exposed in utero to diethylstilbestrol (DES), a synthetic estrogen, are at greater risk of developing breast cancer in their 40s (1.8–2.5-fold increased risk) and in their 50s (threefold increased risk) [72]. These environmental and epigenetic factors involving estrogen need to be taken into consideration in epidemiological studies.

Estrogen levels are intertwined with obesity to influence breast cancer risk. While some studies showed no effect of BMI on risk for breast cancer (e.g., study #20 above), dysregulation of sex hormones, hyperinsulinemia, and inflammatory cytokines in obese women are factors that could influence the risk for breast cancer. Obesity is significantly associated with low plasma levels of sex-hormone-binding globulin (SHBG), which increases the bioavailability of estrogens and androgens [73]. Thus, many established risk factors for breast cancer may function through an endocrine mechanism.

2.4.2.2 Alcohol Metabolism

The liver is the major organ for metabolizing ethanol mainly by oxidative pathway which involves cytosolic alcohol dehydrogenase (ADH), of which multiple isoenzymes exist—e.g., in humans, class I ADH is composed of three genes (ADH1A, ADH1B, and ADH1C)—to produce acetaldehyde, a highly reactive molecule. ADH acts on a wide range of substrates including retinol. The cytochrome P450 isozymes, mainly CYP2E1, predominantly present in the endoplasmic reticulum, also contribute to ethanol oxidation to acetaldehyde in the liver, particularly at higher alcohol concentrations. CYP2E1-dependent ethanol oxidation may occur in other tissues where ADH activity is low. CYP2E1 also produces highly reactive oxygen species (ROS), including hydroxyethyl, superoxide anion, and hydroxyl radicals. Acetaldehyde, produced by ethanol oxidation through any of these mechanisms, is rapidly metabolized mainly by mitochondrial aldehyde dehydrogenase (ALDH2) to form acetate and NADH and to a much lesser extent by ALDH1 in the cytosol.

Mitochondrial NADH is oxidized by the mitochondrial electron transport chain. Chronic alcohol consumption renders mitochondrial oxidative phosphorylation inefficient by interfering with the main respiratory complexes (Complex I, III, IV, and V) of the electron transport system encoded on mitochondrial DNA (mtDNA), resulting in the formation of the superoxide anion. In breast cancer, like in other cancers, mitochondrial function is severely impaired [74]. One early event in breast carcinogenesis can be mutations in mtDNA that destabilize the oxidative phosphorylation system (OXPHOS) which can result in a shift in energy metabolism toward enhanced aerobic glycolysis. Alcohol metabolism could influence breast carcinogenesis by generating acetaldehyde and ROS and by interfering with retinol metabolism.

Acetaldehyde is suspected in playing a role in breast carcinogenesis. Since blood acetaldehyde levels are very low or undetectable after alcohol consumption in humans [75], the human breast would not be exposed to significant levels of exogenous acetaldehyde. Thus, ADH activity and in situ generation of acetaldehyde in the human breast tissue after ethanol consumption is of potential significance. In mammary tissue of rats, cytosolic ADH and ALDH1 activities were 5.8 and 8.3 % of that in the liver of the same animals, respectively [76]. Similarly, mitochondrial ALDH2 activity in breast tissue was 7.1 % of that in the liver. In humans, studies on normal and neoplastic breast tissue showed that class I, but not class IV, ADH is expressed in human mammary epithelium, which can support ADH-mediated oxidation of ethanol; however, the expression of class I ADH is dramatically reduced or abrogated in invasive breast cancers [77]. The authors opined that this "virtual abrogation of expression of class I ADH in invasive breast cancer suggests that the enzyme has some 'tumor suppressor' function in the mammary epithelium." However, whether the reduction in class I ADH activity was causally related to tumor formation or was merely a bystander effect was not considered. To investigate acetaldehyde formation by the cytosolic pathway and the microsomal fraction in the mammary tissue, Sprague–Dawley female rats were injected intraperitoneally with 0.8 mL ethanol/kg/day for four consecutive days [78]. Mammary microsomal metabolism of alcohol to acetaldehyde by CYP2E1 was not induced after ethanol (or acetone) treatment, despite reports that CYP2E1 is expressed in normal and cancerous breast tissue [79]. In contrast, the cytosolic fraction of alcohol treated animals showed higher concentrations of acetaldehyde.

In humans, the interaction between alcohol consumption and ADH2 polymorphism with respect to breast cancer risk was reported in 278 German women with invasive breast cancer [80]. The authors stated that breast cancer risk associated with alcohol consumption may vary according to ADH2 polymorphism, probably due to differences in alcohol metabolism.

Variations in ADH and ALDH activities were reported to influence the risk of breast cancer. For example, while the *ADH1B* genotype [81] was not associated with breast cancer risk in a German population, a role for the *ADH1C* genotype has been suggested. This genotype, which is expressed mainly in the liver but also in breast tissue, has three polymorphic genes: *ADH1C1*1* and *ADH1C1*2* genotypes which result in enzymes with fast and intermediate turnover rates and which increase the risk of breast cancer in Chinese drinkers, compared to *ADH1C2*2* which results in an enzyme with a slow rate of metabolism [82]. Similar results were obtained in a Long Island Breast Cancer Study which genotyped 1,047 breast cancer cases. Consumption of 15–30 g/day was associated with OR of 2.0, 1.5, and 1.3 in *ADH1C1*1*, *ADH1C1*2*, and *ADH1C2*2* genotypes, respectively [83]. Ironically, another study in Caucasian postmenopausal women found an association between risk of breast cancer and the slow metabolizing variant, which led the authors to conclude that "ethanol rather than acetaldehyde is related to breast cancer risk" [84]. However, two studies found no association between breast cancer risk and functional allelic variants of the *ADH1B* and *1C* genes [85] and *ADH1B* and *ALDH2*. The authors concluded that "our findings do not support the hypothesis that acetaldehyde is the main contributor to the carcinogenesis of alcohol-induced breast cancer" [86].

Acetaldehyde and NADH produced by alcohol metabolism can be substrates for xanthine oxidoreductase (XOR), which is inducible by alcohol and produces ROS, especially superoxide anion [87]. To add to the complexity, XOR also metabolizes (activates) nitrofurans and nitroimidazoles, chemicals that are used in veterinary medicine and by beekeepers in honey-producing hives. Therefore, residues of these compounds could exist in animal-derived foods and honey and might be involved in the associated mammary carcinogenic effects [88].

To explain acetaldehyde's role in carcinogenesis, scientists proposed a model in which acetaldehyde reacts with DNA to generate DNA lesions that form interstrand cross-links (ICLs). Cells are protected against replication blocking DNA lesions and ICLs through the Fanconi anemia-BRCA (FANC-BRCA) DNA-damage response network. Mutations in two major susceptibility genes, *BRCA1 and BRCA2*, which are involved in the maintenance of genomic integrity and DNA repair, were identified as major risk factors for breast cancer [89, 90]. The role of high levels of acetaldehyde in activating the FANC-BRCA network was discussed elsewhere [91, 92]. Furthermore, polymorphisms in the DNA repair gene *XRCC1* was associated with increased breast cancer risk in African-American women [93].

The role of acetaldehyde in breast carcinogenesis has not been evident in epidemiological studies. For example, study #19 above did not find a significant association between breast cancer risk and alcohol consumption in Asian American women, almost half of whom are deficient in ALDH2, the mitochondrial enzyme that metabolizes acetaldehyde. Furthermore, study #20 reported that the increase in risk for breast cancer in Japanese population was not modified by alcohol-induced facial flushing, which means the risk was not modified in women who have the defective ALDH2*2.

To examine alcohol effects on oxidative stress in the mammary tissue, female Sprague–Dawley rats were fed alcohol for 28 days [78]. An increase in hydroperoxide, but not the lipid peroxidation product malondialdehyde (MDA) concentration, and a significant reduction in glutathione in mammary tissue were observed. A study by Li and colleagues comparing breast cancer patients with cancer-free women reported that the levels of hydroxyl radical-DNA adducts and MDA-DNA adducts were ninefold higher in patient's normal breast tissue adjacent to tumor tissue than in breast tissue from cancer-free controls [94]. These reports highlight the potential that oxidative stress may lead to DNA damage in cancer patients that is not evident in healthy women.

Class I ADH has the potential to catalyze the oxidation of retinol (vitamin A) to retinal [95], the first step in the biosynthesis of retinoic acid (RA), the principal mediator for maintaining epithelia in a differentiated state. Chronic and excessive alcohol intake interferes with retinoid metabolism and results in reduced RA. Alcohol acts as a competitive inhibitor of oxidation of vitamin A to RA (which involves ADH and ALDH) and induces CYP2E1, which can enhance catabolism of vitamin A and RA. The biological activity of RA is primarily mediated by nuclear retinoid receptors which are involved in the antitumor activity of retinoids. Studies indicate cross-talks between classic retinoids and various intracellular pathways controlling the growth, survival, and invasive/metastatic behavior of breast cancer cells [96]. Impaired RA homeostasis interferes with signaling (e.g., downregulates retinoid target gene expression) and with "crosstalk" with the mitogen-activated protein kinase signaling pathway (MAPK), including Jun N-terminal kinase and p38 kinase [97]. These observations could have implications for breast cancer prevention. However, better understanding of the alcohol–retinoid interaction and the molecular mechanisms involved is needed before it would be justified to pursue retinoids in the prevention of breast cancer. Nonetheless, retinoids could be potential components of innovative and rational therapeutic combinations for breast cancer. Yet, it is important to evaluate the responsiveness of ER+ tumors to retinoids and whether HER2 expression always plays a negative role in modulating retinoid sensitivity of HER2+/ER+ mammary tumors, as suggested by some studies [98].

2.4.2.3 Folate Metabolism/Epigenetic Factors

Mutations in oncogenes and tumor suppressor genes result in specific gene expression profiles that are involved in the regulation of cellular homeostasis, including cell proliferation and DNA repair and survival. However, differentiation of mammary stem cells to primitive progenitor cells is under epigenetic control. Epigenetic mechanisms, which result in changes in gene expression patterns without altering DNA sequence, partake in mammary glands developmental phases from in utero to menopause, as well as in breast carcinogenesis. One of these epigenetic mechanisms is DNA methylation [99].

DNA methylation involves the transfer of a methyl group from S-adenosylmethionine (SAM)—by DNA methyltransferases (DNMTs)—onto the 5′-position of the cytosine residue found in cytosine guanosine dinucleotide pairs (CpG). SAM is generated from methionine. After the methyl transfer reaction, SAM forms S-adenosylhomocysteine, which is then broken down to homocysteine. The latter can be remethylated to form methionine, by transferring a methyl group either from N5-methyltetrahydrofolate (THF) by methionine synthase or from betaine by betaine-homocysteine methyl transferase. Hypermethylation of CpG groups renders affected loci inaccessible to transcription factors resulting in transcriptional

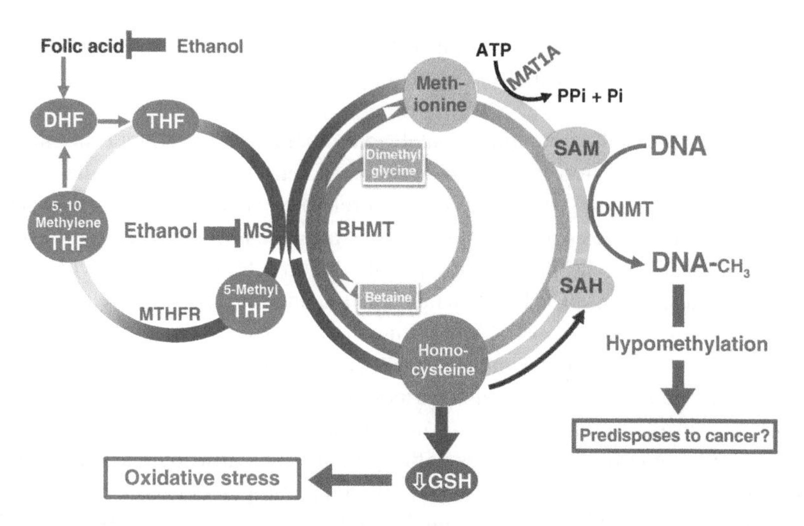

Fig. 2.2 Alcohol's effects on homocysteine/methionine metabolism and DNA methylation. *MTHFR* methylene tetrahydrofolate reductase, *MAT* methionine adenosyltransferase, *HCC* hepatocellular carcinoma, *BHMT* betaine homocysteine methyltransferases, *GSH* glutathione, *ATP* adenosine triphosphate, *Pi* inorganic phosphate

silencing. Importantly, CpG methylation in promoter regions of tumor-suppressor genes (e.g., *BRCA1*) leads to the inactivation of these cancer-preventing proteins. Similarly, hypermethylation of numerous genes, whose biological function include hormone regulation, DNA repair, cell cycle regulation, tissue remodeling, apoptosis, cell adhesion and invasion, cell growth inhibition, and angiogenesis, has been identified in breast tumors [100]. Furthermore, DNA hypermethylation results in aberrant regulation of the Wnt pathway in breast cancer [101]; BRCA1 expression is suppressed by a combination of gene mutation and DNA hypermethylation [102]. One epidemiological study of the interactions between alcohol consumption and breast cancer risk in *BRCA1* and *BRCA2* mutation carriers reported no significant interaction with *BRCA1* mutations but a greater risk of alcohol-associated breast cancer in women with *BRCA2* mutations [103]. In fact, the same investigators reported an inverse association between breast cancer and current alcohol consumption in women with a *BRCA1* mutation [104]. Another study reported no association between alcohol intake and breast cancer risk for women with *BRCA1* or *BRCA2* mutations and suggested a possible reduction in risk in *BRCA2* mutation carriers with "modest" alcohol intake [105].

DNA hypomethylation can also contribute to breast carcinogenesis [106]. For example, promoter hypomethylation could reactivate the expression of certain proto-oncogenes (such as *synuclein γ*) which are associated with tumor metastasis [107].

Chronic heavy alcohol consumption leads to substantial DNA hypomethylation as a result of significant reduction in tissue SAM (Fig. 2.2). Additionally, alcohol perturbs the folate cycle that is involved in the methionine metabolic pathway, which is integral to supplying the methyl groups necessary for DNA methylation. Folate is an important nutrient required for DNA synthesis, and at least 30 different enzymes are involved in the complex folate cycle including methylenetetrahydrofolate reductase (MTHFR), methionine synthase (MTR), and methionine synthase reductase (MTRR). Defects or polymorphic variations in the folate metabolic pathway may influence cancer susceptibility. However, a study on 1,063 women with breast cancer found no association between MTHFR genotype and risk for breast cancer, and there was no evidence of an interaction of genotype and alcohol consumption in premenopausal women. However, in postmenopausal women, there was an increase in breast cancer risk in those who were homozygote *TT* for *MTHFR C677T* and drank >1.9 drinks/day [108].

Chronic heavy alcohol consumption can cause relative folate deficiency due to the negative effects of alcohol on folate metabolism, including malabsorption, increased excretion, or enzymatic suppression. Based on the above interactions between folate and alcohol, it would be expected that high folate intake should ameliorate the association between alcohol intake and risk for breast cancer that is caused by this mechanism. Examination of epidemiological studies revealed inconsistencies in the findings. For example, studies discussed under #25 and #26 showed that folate intake attenuated the alcohol-associated risk for breast cancer, so did another epidemiological study in Anglo-Australian [109] women aged 40–69 years; whereas studies discussed under #20 and #23 showed no association. Other studies reported that high intake of folic acid increased the risk of breast cancer in postmenopausal women enrolled in the Prostate, Lung, Colorectal, and Ovarian Cancer Screening Trial [110] cohort in the USA. Additionally, a case–control study in 570 Thai women concluded that genetic polymorphisms in folate and alcohol metabolic pathways may contribute to the etiology of breast cancer among Thai women [111]. In conclusion, the impact of folate supplementation on the risk for alcohol-induced breast cancer probably is affected by a wide range of other factors that are not well understood.

2.5 Concluding Remarks

Breast cancer is the most diagnosed cancer in women worldwide; it is one of the primary causes of death among women globally. Women, particularly if they have known genetic susceptibilities, should consult with their physician about risk factors involved in breast cancer. Overconsumption of alcohol is a risk factor for many diseases and one that women with or without a high risk for breast cancer are well advised to avoid,

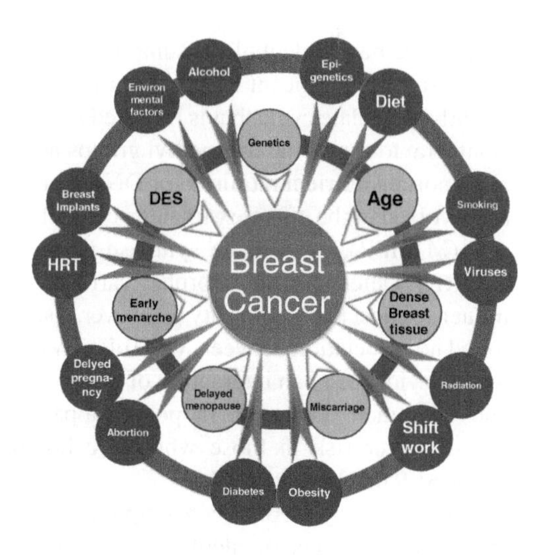

Fig. 2.3 Multiple risk factors associated with breast cancer

The question whether a woman should not drink at all in order to reduce the risk of breast cancer or may drink moderately without undue risk is not settled. The reasons are threefold: (1) there are at least 20 recognized risk factors that can affect the onset and outcome of breast cancer (Fig. 2.3) and the overall risk depends on the interactions of all these factors, including those that have not yet been identified; (2) epidemiological studies resulted in contradictory associations between the amount of alcohol consumed and risk for breast cancer; and (3) discrepancies exist between epidemiological and molecular studies.

2.5.1 Multiple Risk Factors

Numerous risk factors are involved in breast carcinogenesis; only a few of them will be discussed below to illustrate the complex interactions between these risk factors. Some of the risk factors associated with breast cancer are genetic and beyond the control of a woman. Mutations in *BRCA1 and BRCA2*, were identified as major risk factors for breast cancer [89, 90]. However, the incomplete penetrance of these mutations suggests that other factors, environmental and hormonal, may modify that risk, for example, a study in which monozygotic (MZ) twins who carried identical *BRCA1* gene mutation resulted in discordant phenotypes; one suffered from breast cancer twice in 27 years while her MZ twin remained healthy [112].

The majority of breast cancers are not hereditary. Most late-onset breast cancer occurs in the absence of a first-degree family history of breast cancer. Genome-wide association studies have identified genetic susceptibility variants of medium-penetrance [113–116] (which confer risk of 2–3-fold per allele) and modest penetrance (which increase the risk 1.1–1.3-fold per allele). However, these variants could explain only

20–25 % of familial breast cancer risk [117]. These findings led to the hypothesis that susceptibility to breast cancer is polygenic, i.e., conferred by a large number of loci, each with limited contribution to breast cancer risk [118].

Age is another important nonmodifiable risk factor. Invasive breast cancer or its precursor lesion DCIS occurs at an exponential rate until about age 50 (menopause) followed by a slower rate of increase [119], supporting the notion that breast cancer biology is age-dependent. Early-onset breast cancer, therefore, could largely represent inherited mutations (*BRCA1, BRCA2, TP53, ATM, or PTEN*) or early life transforming events that affect the immature mammary cells [120]. In contrast, late-onset breast cancer could be due to an early mutagenic initiating event, which is then subjected to later life exposure to endogenous or exogenous promoting agent(s) and further compounded by age-related impairments in macromolecular repair, immune surveillance, or xenobiotic detoxification. This could explain the increased risk from HRT.

Risk factors are greatly influenced by epigenetic mechanisms that change gene expression patterns for cell differentiation, proliferation, and apoptosis. These epigenetic mechanisms include DNA methylation, histone modifications, and the effects of noncoding RNAs such as microRNAs (miRNA) [99]. Epigenetic changes that influence the transciptomes, proteomes, and metabolomes and ultimately breast cancer evolution, are brought about by numerous endogenous (e.g., hormonal, microbiota, aging, inflammation, inherited diseases such as type 1 diabetes) and exogenous (diet, smoking, infection, alcohol, obesity, radiation, shift work, circadian rhythm disturbances) etiological factors that result in the vast heterogeneity of breast cancers. In fact, a study on the clinically relevant subtypes, luminal A and basal-like breast cancer revealed that distinct molecular mechanisms might have been preprogrammed at an early stage of the disease and that these breast tumor subtypes represent biologically distinct disease entities that may require different therapeutic strategies [121].

The interactions between genes and the environment are crucial. Environmental factors add another layer of complexity by their interactions with the susceptibility genes for breast cancer and for metabolic diseases. For example, in a twin study in Finland [122], the probability that a co-twin would develop breast cancer (given that one twin already had breast cancer) was 10 % for monozygotic and 8 % for dizygotic twins, suggesting that combined environmental effects are dominant in the development of breast cancer. In addition, in a subset of the Million Women Study (study #11 above), the strongest suggestion of a gene–environment interaction was between the high-risk common variant CASP8-rs1045485 and alcohol consumption; the per-allele relative risk of breast cancer for CASP8-rs1045485 was not increased by consuming <1 drink/day, but was increased by 23 % (nonsignificant) in women who reported consuming one or more alcoholic drinks/day [123]. Another study found no interaction between the breast cancer susceptibility locus CASP8-rs17468277 and consuming <20 g/day of alcohol; however risk was increased by 45 % in those who drank ≥20 g/day [124]. Therefore, epidemiological studies should focus on gene–environment interactions rather than singling out individual risk factors as if they operate in isolation.

2.5.2 Epidemiological Studies

While epidemiological studies, when conducted properly, can be effective tools to identify the root of health problems and disease outbreaks in a community, epidemiological studies that deal with alcohol consumption and its consequences should be interpreted with considerable caution. Such studies can only highlight associations and cannot determine cause-and-effect relationships. Considering the wide range of variations in the size of studies, measurement errors in input and outcome variables, and individual variations in genetic background and life style factors, many of which were not taken into account, it is not surprising that the results have been contradictory. The epidemiological studies on alcohol and breast cancer discussed above point to the following important points as possible sources of the variability in outcomes:

1. All studies use self-report to determine the amount and type of alcoholic beverage consumed. Most of the studies acknowledged "recall bias" that makes the alcohol consumption variable notoriously inaccurate, and drinking pattern could change during the period under consideration. In fact, a recent study on moderate alcohol intake and cancer stated that the apparent increased risk of cancer among light-moderate drinkers may be "substantially due to underreporting of intake" [125]. There is now strong evidence that recall over prolonged periods tends to grossly underestimate actual drinking frequency. This was highlighted dramatically in a recent study by Stockwell et al. [126] who compared Quantity-Frequency (QF) recall with beverage-specific "yesterday" (BSY) consumption reports and with alcoholic beverage sales reports in a Canadian population to demonstrate that underreporting of alcohol consumption was considerable (2–3-fold) and varied by age and consumption level. Particularly relevant is the finding that moderate drinkers underreport their drinking much more than more frequent drinkers. In many of the epidemiological studies summarized above, an underreporting by a factor of 2–3 would increase the alcohol consumption associated with increased risk of breast cancer into the recognized risky drinking category. Other studies used as input of alcohol consumption "ever" vs. "never" without even quantification, and others use inaccurate information to characterize drink size (e.g., study #1 characterizes a drink of a liquor as one ounce containing 15.1 g of alcohol). The definition of a standard drink according to the National Institute on Alcohol Abuse and Alcoholism [127] and the Dietary Guidelines for Americans is: 12 oz of beer (5 % alcohol), 5 oz of wine (12 % alcohol), and 1.5 oz of distilled spirit (40 % alcohol). In addition, the Dietary Guidelines for Americans define moderate drinking as consuming no more than one drink/day for a woman, and no more than 2 drinks/day for a man. Although some studies used FFQ and claim it as valid, lack of consumption ascertainment can result in confusing results. For example, in study #4 above, consumption of 1 or 2 drinks/day increased breast cancer risk by 40 %, whereas consumption of two or more drinks/day resulted in no increase in risk. A 42 % decrease in breast cancer risk was associated with drinking <1.5 drink/day (study #5),

whereas drinking 3–6 drinks/week was associated with a 15 % increase in risk (study #2). In fact, a meta-analysis reported by Longnecker (#3) stated "the modest size of the association and variation in results across studies leave the causal role of alcohol in question."

2. Patterns of consumption are rarely analyzed. The majority of studies use average weekly consumption. Drinking seven drinks on a Saturday night and nothing the rest of the week is not, health-wise, the same as having one drink every day of the week.
3. Breast cancer develops over a relatively long period of time, often more than 20 years [128], and thus, correlations between alcohol consumption and breast cancer cannot be determined in epidemiological studies with windows of alcohol exposure that captures current or recent alcohol intake, after clinical diagnosis; for example, studies # 4, 6, 14, and 21 assessed recent alcohol consumption, consumption 2 years prior to diagnosis, past-year drinking, or consumption 3 months after diagnosis, respectively.
4. Despite the fact that age per se is a risk factor for breast cancer and during aging other environmental factors may compound or modify the risk, most epidemiological studies stratify data based on pre- or postmenopausal status alone. Thus, including women (ages 20–74, 25–85, 40–75, 23–74, and 20–87 in studies # 4, 5, 7,16, and 21, respectively)—who have various degrees of exposure to environmental and lifestyle factors that influence the genesis of breast cancer—without stratification could be confounding. Similarly, studies that took into account other risk factors such as smoking, BMI, use of HRT, etc., resulted in contradictory findings (see the above cited studies).
5. Results of the association between breast cancer risk and alcohol consumption based on ER, PR, and HER2 status are very contradictory and vary between different studies. Therefore, correlations with HRT use cannot be deciphered, based on currently available data.
6. Rarely does an epidemiological study differentiate between specific subtypes of breast cancer (there are about 20 different types) and risk of alcohol consumption. These different subtypes, as described above, are heterogeneous, have distinct molecular mechanisms, are associated with unique risk factors, and might have been preprogrammed at an early stage of the disease.
7. In a recent article, Ogino and colleagues [129] advocated the use of "Molecular Pathological Epidemiology" (MPE) to understand the interplay between etiological factors, cellular molecular characteristics, and disease evolution in multifactorial diseases such as cancer. While conventional molecular epidemiology generally considers a disease as a single entity, MPE integrates analyses of population studies together with the macroenvironment and molecular and microenvironment. This approach will allow scientists to investigate the relationships between potential etiological agent and disease subtypes based on molecular signatures. This concept is similar to systems biology approach and, according to the authors, "enables us to link potential etiological factors to specific molecular pathology, and gain novel pathogenic insights on causality." This is a very important concept that needs to be applied to alcohol and breast cancer.

2.5.3 Discrepancies Between Epidemiological and Molecular Studies

A major challenge in understanding the epidemiological findings is to elucidate the biological basis underlying the association. Several molecular mechanisms have been postulated, including formation of acetaldehyde and ROS, epigenetic effect through the folate cycle, and estrogen formation. However, there seems to be discord between molecular and epidemiological studies.

(a) Acetaldehyde formation

Molecular studies have demonstrated the presence of alcohol-metabolizing enzymes ADH, ALDH1, and ALDH2 in breast tissue. Since acetaldehyde can form adducts with DNA and result in DNA lesions, it would seem logical that genes that encode for fast metabolizing ADH or defective ALDH2*2 enzymes that result in acetaldehyde accumulation would increase the risk for breast cancer. While some studies such as the Long Island Breast Cancer Study showed an increased risk in fast metabolizers; other epidemiological studies in Asian American women and in Japan (# 19, 20) showed that breast cancer risk was not significantly associated with alcohol drinking and that facial flushing associated with the defective ALDH2*2 did not modify risk for breast cancer.

(b) Folate metabolism

Alcohol's effects on folate absorption and metabolism are well documented. Consequently, it is expected that levels of folate consumption can have a protective effect on breast cancer among women who consume alcohol. While some studies concluded that folate has protective effect on breast cancer in alcohol-consuming women, the Women's Health Initiative-Observational Study found no evidence for folate attenuating alcohol's effect on breast cancer risk. Similarly, the American Cancer Society Cancer Prevention Study II Nutrition Cohort reported no association between risk of breast cancer and dietary folate, total folate, or methionine intake, and there was no evidence of an interaction between dietary folate or total folate and alcohol. For more discussion, the reader is referred to Sect. 2.4.2.3 under molecular mechanisms.

(c) Estrogen metabolism

See discussion above about estrogen metabolism.

In summary, despite all the effort, there is no solid evidence associating moderate alcohol consumption with an increased incidence of breast cancer. A woman and her physician should weigh the risks and benefits of moderate alcohol consumption, which could be a part of a healthy life style. This is especially important in light of the fact that moderate alcohol consumption has been associated with potential health benefits, including decreased risk of coronary artery disease and overall mortality, protection against congestive heart failure, decreased risk of ischemic stroke, and protection against type 2 diabetes and rheumatoid arthritis.

Confronting the huge global breast cancer issue should be based on sound science. The lack of unambiguous reliable information about alcohol and breast cancer has opened the door to various unsubstantiated opinions on the subject and to errant hypotheses about causes and prevention. That caused a great deal of confusion among women about the subject. Thus, confirming association between breast cancer and alcohol consumption, if any, should be a high priority. Systematic studies should be based on a large cohort; ascertain alcohol consumption by reliable and validated methods, preferably over prolonged periods, take into account the interaction with the multitude of other risk factors; and incorporate detailed biological information, including breast cancer subtypes. Public health policies must be rooted in impeccable science.

References

1. http://www.cancer.gov/researchandfunding/priorities/global-research-activities
2. Edwards BK, Noone A-M, Mariotto AB, Simard AP, Boscoe FP, Henley SJ, Jemal A, Cho H, Anderson RN, Kohler BA, Eheman CR, Ward EM (2014) Annual report to the nation on the status of cancer, 1975–2010, featuring prevalence of comorbidity and impact on survival among persons with lung, colorectal, breast, or prostate cancer. Cancer 120(9):1290–1314
3. Ravdin PM, Cronin KA, Howlader N et al (2007) The decrease in breast cancer incidence in 2003 in the United States. N Engl J Med 356:1670–1674
4. Benson JR, Jatol I (2012) The global breast cancer burden. Future Oncol 8:697–702
5. Dairkee SH, Smith HS (1996) The genetic analysis of breast cancer progression. J Mammary Gland Biol Neoplasia 1:139–149
6. Allred DC, Clark GM, Molina R et al (1992) Overexpression of HER-2/neu and its relationship with other prognostic factors change during the progression of in situ to invasive breast cancer. Hum Pathol 23:974–979
7. Rudas M, Neumayer R, Grant MFX, Mittelbok M, Jakesz R, Reiner A (1997) p53 protein expression, cell proliferation and steroid hormone receptors in ductal and lobular in situ carcinomas of the breast. Eur J Cancer 33:39–44
8. Sotiriou C, Puszati L (2009) Gene expression signatures in breast cancer. N Engl J Med 360:790–800
9. Cancer Genome Atlas Network (2012) Comprehensive molecular portraits of human breast tumours. Nature 490:61–70
10. Conzen SD (2008) Nuclear receptors and breast cancer. Mol Endocrinol 22:2215–2228
11. Visvader JE (2009) Keeping abreast of the mammary epithelial hierarchy and breast tumorigenesis. Genes Dev 23:2563–2577
12. Korkaya H et al (2009) Regulation of mammary stem/progenitor cells by PTEN/Akt/b-catenin signaling. PLoS Biol 7:e1000121
13. Bouras T et al (2008) Notch signaling regulates mammary stem cell function and luminal cell-fate commitment. Cell Stem Cell 3:429–441
14. Boyle P, Boffetta P (2009) Alcohol consumption and breast cancer risk. http://breast-cancer-research.com/supplements/11/S3/S3
15. Zhang M, Holman CDJ (2011) Low-to-moderate alcohol intake and breast cancer risk in Chinese women. Br J Cancer 105:1089–1095
16. Willett WC, Stampfer MJ, Colditz GA, Rosner BA, Hennekens CH, Speizer FE (1987) Moderate alcohol consumption and the risk of breast cancer. N Engl J Med 316:1174–1180
17. Chen WY, Rosner B, Hankinson SE, Colditz GA, Willett WC (2011) Moderate alcohol consumption during adult life, drinking patterns, and breast cancer risk. JAMA 306:1884–1890
18. Longnecker MP (1994) Alcoholic beverage consumption in relation to risk of breast cancer: meta-analysis and review. Cancer Causes Control 5:73–82
19. Kinney AY, Millikan RC, Lin YH, Moorman PG, Newman B (2000) Alcohol consumption and breast cancer among black and white women in North Carolina (United States). Cancer Causes Control 11:345–357

20. Bessaoud F, Daurès JP (2008) Patterns of alcohol (especially wine) consumption and breast cancer risk: a case–control study among a population in Southern France. Ann Epidemiol 18:467–475
21. Bissonauth V, Shatenstein B, Fafard E, Maugard C et al (2009) Risk of breast cancer among French-Canadian women, noncarriers of more frequent BRCA1/2 mutations and consumption of total energy, coffee, and alcohol. Breast J 15(1):S63–S71
22. Berstad P, Ma H, Bernstein L, Ursin G (2008) Alcohol intake and breast cancer risk among young women. Breast Cancer Res Treat 108:113–120
23. Terry MB, Zhang FF, Kabat G, Britton JA et al (2006) Lifetime alcohol intake and breast cancer risk. Ann Epidemiol 16:230–240
24. Holmberg L, Baron JA, Byers T, Wolk A, Ohlander EM, Zack M, Adami HO (1995) Alcohol intake and breast cancer risk: effect of exposure from 15 years of age. Cancer Epidemiol Biomarkers Prev 4:843–847
25. Conroy SM, Koga K, Woolcott CG, Dahl T et al (2012) Higher alcohol intake may modify the association between mammographic density and breast cancer: an analysis of three case–control studies. Cancer Epidemiol 36:458–460
26. Allen NE, Beral V, Casabonne D, Kan SW, Reeves GK, Brown A, Green J, Million Women Study Collaborators (2009) Moderate alcohol intake and cancer incidence in women. J Natl Cancer Inst 101:296–305
27. Zhang SM, Lee IM, Manson JE, Cook NR, Willett WC, Buring JE (2007) Alcohol consumption and breast cancer risk in the Women's Health Study. Am J Epidemiol 165:667–676
28. Suzuki R, Ye W, Rylander-Rudqvist T, Saji S, Colditz GA, Wolk A (2005) Alcohol and postmenopausal breast cancer risk defined by estrogen and progesterone receptor status: a prospective cohort study. J Natl Cancer Inst 97:1601–1608
29. Cotterchio M, Kreiger N, Theis B et al (2003) Hormonal factors and the risk of breast cancer according to estrogen- and progesterone receptor subgroup. Cancer Epidemiol Biomarkers Prev 12:1053–1060
30. Enger SM, Ross RK, Paganini-Hill A et al (1999) Alcohol consumption and breast cancer oestrogen and progesterone receptor status. Br J Cancer 79:1308–1314
31. Li CI, Malone KE, Porter PL et al (2003) The relationship between alcohol use and risk of breast cancer by histology and hormone receptor status among women 65–79 years of age. Cancer Epidemiol Biomarkers Prev 12:1061–1066
32. Britton JA, Gammon MD, Schoenberg JB et al (2002) Risk of breast cancer classified by joint estrogen receptor and progesterone receptor status among women 20–44 years of age. Am J Epidemiol 156:507–516
33. McTiernan A, Thomas DB, Johnson LK et al (1986) Risk factors for estrogen receptor-rich and estrogen receptor-poor breast cancer. J Natl Cancer Inst 77:849–854
34. Cooper JA, Rohan TE, Cant EL et al (1989) Risk factors for breast cancer by estrogen receptor status: a population-based case–control study. Br J Cancer 59:119–125
35. Nasca PC, Liu S, Baptiste MS et al (1994) Alcohol consumption and breast cancer: estrogen receptor status and histology. Am J Epidemiol 140:980–988
36. Yoo KY, Tajima K, Miura S et al (1997) Breast cancer risk factors according to combined estrogen and progesterone receptor status: a case–control analysis. Am J Epidemiol 146: 307–314
37. Huang WY, Newman B, Millikan RC et al (2000) Hormone-related factors and risk of breast cancer in relation to estrogen receptor and progesterone receptor status. Am J Epidemiol 151:703–714
38. Potter JD, Cerhan JR, Sellers TA, McGovern PG, Drinkard C, Kushi LR, Folsom AR (1995) Progesterone and estrogen receptors and mammary neoplasia in the Iowa Women's Health Study: how many kinds of breast cancer are there? Cancer Epidemiol Biomarkers Prev 4:319–326
39. Horn-Ross PL, Canchola AJ, Bernstein L et al (2012) Alcohol consumption and breast cancer risk among postmenopausal women following the cessation of hormone therapy use: the California Teachers Study. Cancer Epidemiol Biomarkers Prev 21:2006–2013

40. Deandrea S, Talamini R, Foschi R et al (2008) Alcohol and breast cancer risk defined by estrogen and progesterone receptor status: a case–control study. Cancer Epidemiol Biomarkers Prev 17:2025–2028
41. Lew JQ, Freedman ND, Leitzmann MF, Brinton LA, Hoover RN, Hollenbeck AR, Schatzkin A, Park Y (2009) Alcohol and risk of breast cancer by histologic type and hormone receptor status in postmenopausal women: the NIH-AARP Diet and Health Study. Am J Epidemiol 170:308–317
42. Li CI, Chlebowski RT, Freiberg M, Johnson KC, Kuller L, Lane D, Lessin L, O'Sullivan MJ, Wactawski-Wende J, Yasmeen S, Prentice R (2010) Alcohol consumption and risk of post-menopausal breast cancer by subtype: the women's health initiative observational study. J Natl Cancer Inst 102:1422–1431
43. Doll R, Payne P, Waterhouse J (eds) (1966) Cancer incidence in five continents, vol 1 UICC Tech Rep. Springer-Verlag, Berlin
44. Brown LM, Gridley G, Wu AH et al (2010) Low level alcohol intake, cigarette smoking and risk of breast cancer in Asian-American women. Breast Cancer Res Treat 120:203–210
45. Lin Y, Kikuchi S, Tamakoshi K, Wakai K, Kondo T et al (2005) Prospective study of alcohol consumption and breast cancer risk in Japanese women. Int J Cancer 116:779–783
46. Kawai M, Minami Y, Kakizaki M, Kakugawa Y et al (2011) Alcohol consumption and breast cancer risk in Japanese women: the Miyagi Cohort study. Breast Cancer Res Treat 128: 817–825
47. Suzuki R, Iwasaki M, Inoue M, Sasazuki S, Sawada N, Yamaji T, Shimazu T, Tsugane S, Japan Public Health Center-Based Prospective Study Group (2010) Alcohol consumption-associated breast cancer incidence and potential effect modifiers: the Japan Public Health Center-based Prospective Study. Int J Cancer 127:685–695
48. Choi JY, Abel J, Neuhaus T, Ko Y et al (2003) Role of alcohol and genetic polymorphisms of CYP2E1 and ALDH2 in breast cancer development. Pharmacogenetics 13:67–72
49. Nagata C, Mizoue T, Tanaka K, Tsuji I, Wakai K, Inoue M, Tsugane S, Research Group for the Development and Evaluation of Cancer Prevention Strategies in Japan (2007) Alcohol drinking and breast cancer risk: an evaluation based on a systematic review of epidemiologic evidence among the Japanese population. Jpn J Clin Oncol 37:568–574
50. Baumgartner KB, Annegers JF, McPherson RS, Frankowski RF, Gilliland FD, Samet JM (2002) Is alcohol intake associated with breast cancer in Hispanic women? The New Mexico Women's Health Study. Ethn Dis 12:460–469
51. Beasley JM, Coronado GD, Livaudais J et al (2010) Alcohol and risk of breast cancer in Mexican women. Cancer Causes Control 21:863–870
52. Duffy CM, Assaf A, Cyr M et al (2008) Alcohol and folate intake and breast cancer risk in the WHI observational study. Breast Cancer Res Treat 116:551–562
53. Feigelson HS, Jonas CR, Robertson AS et al (2003) Alcohol, folate, methionine, and risk of incident breast cancer in the American Cancer Society Cancer Prevention Study II Nutrition Cohort. Cancer Epidemiol Biomarkers Prev 12:161–164
54. Tjønneland A, Christensen J, Olsen A, Stripp C et al (2006) Folate intake, alcohol and risk of breast cancer among postmenopausal women in Denmark. Eur J Clin Nutr 60:280–286
55. Islam T, Ito H, Sueta A, Hosono S, Hirose K, Watanabe M et al (2013) Alcohol and dietary folate intake and the risk of breast cancer: a case–control study in Japan. Eur J Cancer Prev 22:358–366
56. Tjønneland A, Christensen J, Olsen A, Stripp C, Thomsen BL et al (2007) Alcohol intake and breast cancer risk: the European Prospective Investigation into Cancer and Nutrition (EPIC). Cancer Causes Control 18:361–373
57. Seitz HK, Pelucchi C, Bagnardi V, LaVecchia C (2012) Epidemiology and pathophysiology of alcohol and breast cancer: update 2012. Alcohol Alcohol 47:204–212
58. Dumitrescu RG, Shields PG (2005) The etiology of alcohol-induced breast cancer. Alcohol 35:213–225
59. Auchus ML, Auchus RJ (2012) Human steroid biosynthesis for the oncologist. J Investig Med 60:495–503

60. Watabiki T, Okii Y, Tokiyasu T, Yoshimura S, Yoshida M et al (2000) Long-term ethanol consumption in ICR mice causes mammary tumor in females and liver fibrosis in males. Alcohol Clin Exp Res 24(4):117S–122S
61. Wong AW, Dunlap SM, Holcomb VB, Nunez NP (2012) Alcohol promotes mammary tumor development via the estrogen pathway in estrogen receptor alpha-negative HER2/neu mice. Alcohol Clin Exp Res 36:577–587
62. Anderson GL et al (2006) Prior hormone therapy and breast cancer risk in the Women's Health Initiative randomized trial of estrogen plus progestin. Maturitas 55:103–115
63. Chlebowski RT, Adderson GL (2012) Changing concepts: menopausal hormone therapy and breast cancer. J Natl Cancer Inst 104:517–527
64. Fournier A, Mesrine S, Boutron-Ruault MC, Clavel-Chapelon F (2009) Estrogen–progestagen menopausal hormone therapy and breast cancer: does delay from menopause onset to treatment initiation influence risks? J Clin Oncol 27:5138–5143
65. Lyytinen H, Pukkala E, Ylikorkala O (2009) Breast cancer risk in postmenopausal women using estradiol-progestogen therapy. Obstet Gynecol 113:65–73
66. Russo IH, Russo J (1996) Mammary gland neoplasia in long-term rodent studies. Environ Health Perspect 104:938–967
67. Seitz HK, Maurer B (2007) The relationship between alcohol metabolism, estrogen levels and breast cancer risk. Alcohol Res Health 30:42–43
68. Suzuki R, Orsini N, Mignone L, Saji S, Wolk A (2008) Alcohol intake and risk of breast cancer defined by estrogen and progesterone receptor status—a meta-analysis of epidemiological studies. Int J Cancer 122:1832–1841
69. Cheng AS, Culhane AC, Chan MW et al (2008) Epithelial progeny of estrogen-exposed breast progenitor cells display a cancer-like methylome. Cancer Res 68:1786–1796
70. Hsu PY, Deatherage DE, Rodriguez BA et al (2009) Xenoestrogen-induced epigenetic repression of microRNA-9-3 in breast epithelial cells. Cancer Res 69:5936–5945
71. Hervouet E, Cartron P-F, Jouvenot M, Delage-Mourroux R (2013) Epigenetic regulation of estrogen signaling in breast cancer. Epigenetics 8:237–245
72. Palmer JR et al (2006) Prenatal diethylstilbestrol exposure and risk of breast cancer. Cancer Epidemiol Biomarkers Prev 15:1509, http://cebp.aacrjournals.org/content/15/8/1509.full.pdf+html
73. Key TJ, Appleby PN, Reeves GK, Roddam A et al (2003) Body mass index, serum sex hormones, and breast cancer risk in postmenopausal women. J Natl Cancer Inst 95: 1218–1226
74. Mullen AR, Wheaton WW, Jin ES, Chen PH, Sullivan LB, Cheng T, Yang Y, Linehan WM, Chandel NS, DeBerardinis RJ (2012) Reductive carboxylation supports growth in tumour cells with defective mitochondria. Nature 481:385–388
75. Eriksson CJ (2007) Measurement of acetaldehyde: what levels occur naturally and in response to alcohol? Novartis Found Symp 285:247–255; discussion 256–60
76. Castro GD, Delgado de Layño AMA, Fanelli SL et al (2008) Acetaldehyde accumulation in rat mammary tissue after an acute treatment with alcohol. J Appl Toxicol 28:315–321
77. Triano EA, Slusher LB, Atkins TA, Beneski JT et al (2003) Class I alcohol dehydrogenase is highly expressed in normal human mammary epithelium but not in invasive breast cancer: implications for breast carcinogenesis. Cancer Res 63:3092–3100
78. Fanelli SL, Maciel MF, Díaz Gómez MI et al (2011) Further studies on the potential contribution of acetaldehyde accumulation and oxidative stress in rat mammary tissue in the alcohol drinking promotion of breast cancer. J Appl Toxicol 31:11–19
79. Kapucuoglu N, Coban T, Raunio H, Pelkonen O et al (2003) Immunohistochemical demonstration of the expression of CYP2E1 in human breast tumour and nontumour tissues. Cancer Lett 196:153–159
80. Sturmer T, Wang-Gohrke S, Arndt V, Boeing H et al (2002) Interaction between alcohol dehydrogenase II gene, alcohol consumption, and risk for breast cancer. Br J Cancer 87: 519–523

81. Lilla C, Koehler T, Kropp S et al (2005) Alcohol dehydrogenase 1B (ADH1B) genotype, alcohol consumption and breast cancer risk by age 50 years in a German case–control study. Br J Cancer 92:2039–2041
82. Mao Q, Gao L, Wang H, et al (2012) The alcohol dehydrogenase 1C(rs698) genotype and breast cancer: a meta analysis. Asia Pacific J Public Health. doi:10.1177/1010539512446962
83. Terry MB, Gammon MD, Zhang FF et al (2006) ADH3 genotype, alcohol intake and breast cancer risk. Carcinogenesis 27:840–847
84. Larsen SB, Vogel U, Christensen J et al (2010) Interaction between ADH1C Arg272Gln and alcohol intake in relation to breast cancer risk suggests that ethanol is the causal factor in alcohol related breast cancer. Cancer Lett 295:191–197
85. Visvanathan K, Crum RM, Strickland PT et al (2007) Alcohol dehydrogenase polymorphisms, low-to-moderate alcohol consumption, and risk of breast cancer. Alcohol Clin Exp Res 31:467–476
86. Kawase T, Matsuo K, Hiraki A et al (2009) Interactions of the effects of alcohol drinking and polymorphisms in alcohol-metabolizing enzymes on the risk of female breast cancer in Japan. J Epidemiol 19:244–250
87. Wright RM, McManaman JL, Repine JE (1999) Alcohol-induced breast cancer: a proposed mechanism. Free Radic Biol Med 26:348–354
88. Bartel LC, Montalto de Mecca M, Castro JA (2009) Nitroreductive metabolic activation of some carcinogenic nitro heterocyclic food contaminants in rat mammary tissue cellular fractions. Food Chem Toxicol 47:140–144
89. Miki Y et al (1994) A strong candidate for the breast and ovarian cancer susceptibility gene BRCA1. Science 266:66–71
90. Wooster R et al (1995) Identification of the breast cancer susceptibility gene BRCA2. Nature 378:789–792
91. Marietta C, Thompson LH, Lamerdin JE, Brooks PJ (2009) Acetaldehyde stimulates FANCD2 monoubiquitination, H2AX phosphorylation, and BRCA1 phosphorylation in human cells in vitro: implications for alcohol-related carcinogenesis. Mutat Res 664: 77–83
92. Abraham J, Balbo S, Crabb D, Brooks PJ (2011) Alcohol metabolism in human cells causes DNA damage and activates the Fanconi anemia-breast cancer susceptibility (FA-BRCA) DNA damage response network. Alcohol Clin Exp Res 35:2113–2120
93. Duell EJ, Millikan RC, Pittman GS, Winkel S et al (2001) Polymorphisms in the DNA repair gene XRCC1 and breast cancer. Cancer Epidemiol Biomarkers Prev 10:217–222
94. Li D, Zhang W, Sahin AA, Hittelman WN (1999) DNA adducts in normal tissue adjacent to breast cancer: a review. Cancer Detect Prev 23:454–462
95. Duester G (2000) Families of retinoid dehydrogenases regulating vitamin A function: production of visual pigment and retinoic acid. Eur J Biochem 267:4315–4324
96. Garattini E, Bolis M, Garattini SK, Fratelli M et al (2014) Retinoids and breast cancer: from basic studies to the clinic and back again. Cancer Treat Rev 40:739–749
97. Wang XD (2005) Alcohol, vitamin A, and cancer. Alcohol 35:251–258
98. Siwak DR, Mendoza-Gamboa E, Tari AM (2003) HER2/neu uses Akt to suppress retinoic acid response element binding activity in MDA-MB-453 breast cancer cells. Int J Oncol 23:1739–1745
99. Jones PA, Baylin SB (2007) The epigenomics of cancer. Cell 128:683–692
100. Lo P-K, Sukumar S (2008) Epigenomics and breast cancer. Pharmacogenomics 9:1879–1902
101. Klarmann G, Decker A, Farrar WL (2008) Epigenetic gene silencing in the Wnt pathway in breast cancer. Epigenetics 3:59–63
102. Nowsheen S, Aziz K, Tran PT et al (2012) Epigenetic inactivation of DNA repair in breast cancer. Cancer Lett 342(2):213–222, http://dx.doi.org/10.1016/j.canlet.2012.05.015
103. Dennis J, Krewski D, Côté FS, Fafard E, Little J, Ghadirian P (2011) Breast cancer risk in relation to alcohol consumption and BRCA gene mutations—a case-only study of gene–environment interaction. Breast J 17:477–484

104. Dennis J, Ghadirian P, Little J, Lubiniski J et al (2010) Alcohol consumption and the risk of breast cancer among *BRCA1* and *BRCA2* mutation carriers. Breast 19:479–483
105. McGuire V, John EM, Felberg A et al (2006) No increased risk of breast cancer associated with alcohol consumption among carriers of BRCA1 and BRCA2 mutations ages <50 years. Cancer Epidemiol Biomarkers Prev 15:1565–1567
106. Locke WJ, Clark SJ (2012) Epigenome remodelling in breast cancer: insights from an early in vitro model of carcinogenesis. Breast Cancer Res 14:215–229
107. Fan M, Yan PS, Hartman-Frey C et al (2006) Diverse gene expression and DNA methylation profiles correlate with differential adaptation of breast cancer cells to the antiestrogen tamoxifen and fulvestrant. Cancer Res 66:11954–11966
108. Platek ME, Shields PG, Marian C, McCann SE et al (2009) Alcohol consumption and genetic variation in methylenetetrahydrofolate reductase and 5-methylenetetrahydrofolate-homocysteine methyltransferases in relation to breast cancer risk. Cancer Epidemiol Biomarkers Prev 18:2453–2459
109. Baglietto L, English DR, Gertig DM et al (2005) Does dietary folate intake modify effect of alcohol consumption on breast cancer risk? Prospective cohort study. BMJ 331(7520):807. doi:10.1136/bmj.38551.446470.06
110. Stolzenberg-Solomon RZ, Chang S-C, Leitzmann MF et al (2006) Folate intake, alcohol use, and postmenopausal breast cancer risk in the Prostate, Lung, Colorectal, and Ovarian Cancer Screening Trial. Am J Clin Nutr 83:895–904
111. Sangrajrang S, Sato Y, Sakamoto H, Ohnami S et al (2010) Genetic polymorphisms in folate and alcohol metabolism and breast cancer risk: a case–control study in Thai women. Breast Cancer Res Treat 123:885–893
112. Lasa A, Ramon y Cajal T, Llort G, Suela J, Cigudosa JC, Cornet M, Alonso C, Barnadas A, Baiget M (2010) Copy number variations are not modifiers of phenotypic expression in a pair of identical twins carrying a BRCA1 mutation. Breast Cancer Res Treat 123:901–905
113. Erkko H et al (2007) A recurrent mutation in PALB2 in Finnish cancer families. Nature 446:316–319
114. The CHEK2 Breast Cancer Case–Control Consortium (2004) CHECK2*1100delC and susceptibility to breast cancer: a collaborative analysis involving 10,860 breast cancer cases and 9,065 controls from ten studies. Am J Hum Genet 74:1175–1182
115. Renwick A et al (2006) ATM mutations that cause ataxia-telangiectasia are breast cancer susceptibility alleles. Nat Genet 38:837–875
116. Seal S et al (2006) Truncating mutations in the Fanconi anemia gene BRIP1 are low-penetrance breast cancer susceptibility alleles. Nat Genet 38:1239–1241
117. Mavaddat N, Antoniou AC, Easton DF, Garcia-Closas M (2010) Genetic susceptibility to breast cancer. Mol Oncol 4:174–191
118. Pharoah PDP et al (2002) Polygenic susceptibility to breast cancer and implications for prevention. Nat Genet 31:33–36
119. Quong J, Eppenberger-Castori S, Moore D 3rd et al (2002) Age-dependent changes in breast cancer hormone receptors and oxidant stress markers. Breast Cancer Res Treat 76:221–236
120. Anderson WF, Pfeiffer RM, Dores GM, Sherman ME (2006) Comparison of age-distribution patterns of different histopathologic types of breast cancer. Cancer Epidemiol Biomarkers Prev 15:1899–1905
121. Sørlie T, Wang Y, Xiao C, Johnsen H et al (2006) Distinct molecular mechanisms underlying clinically relevant subtypes of breast cancer: gene expression analyses across three different platforms. BMC Genomics 7:127
122. Verkasalo PK, Kaprio J, Koskenvuo M, Pukkala E (1999) Genetic predisposition, environment and cancer incidence: a nationwide twin study in Finland, 1976–1995. Int J Cancer 83:743–749
123. Travis RC, Reeves GK, Green J, Bull D et al (2010) Gene–environment interactions in 7610 women with breast cancer: prospective evidence from the Million Women Study. Lancet 375:2143–2151

124. Nickels S, Truong T, Hein R, Stevens K et al (2013) Evidence of gene–environment interactions between common breast cancer susceptibility loci and established environmental risk factors. PLoS Genet 9(3):e1003284
125. Klatsky AL, Udaltsova N, Li Y, Baer D et al (2014) Moderate alcohol intake and cancer: the role of underreporting. Cancer Causes Control. doi:10.1007/s10552-014-0372-8
126. Stockwell T, Zhao J, Macdonald S (2014) Who under-reports their alcohol consumption in telephone surveys and by how much? An application of the 'yesterday method' in a national Canadian substance use survey. Addiction 109(10):1657–1666. doi:10.1111/add.12609
127. http://niaaa.nih.gov/alcohol-health/overview-alcohol-consumption/standard-drink
128. Brooks PJ, Zakhari S (2013) Moderate alcohol consumption and breast cancer in women: from epidemiology to mechanisms and interventions. Alcohol Clin Exp Res 37:23 30
129. Ogino S, Lochhead P, Chan AT, Nishihara R et al (2013) Molecular pathological epidemiology of epigenetics: emerging integrative science to analyze environment, host and disease. Mod Pathol 26(4):465–484

Chapter 3
Genetic–Epidemiological Evidence for the Role of Acetaldehyde in Cancers Related to Alcohol Drinking

C.J. Peter Eriksson

Abstract Alcohol drinking increases the risk for a number of cancers. Currently, the highest risk (Group 1) concerns oral cavity, pharynx, larynx, esophagus, liver, colorectum, and female breast, as assessed by the International Agency for Research on Cancer (IARC). Alcohol and other beverage constituents, their metabolic effects, and alcohol-related unhealthy lifestyles have been suggested as etiological factors. The aim of the present survey is to evaluate the carcinogenic role of acetaldehyde in alcohol-related cancers, with special emphasis on the genetic–epidemiological evidence. Acetaldehyde, as a constituent of alcoholic beverages, and microbial and endogenous alcohol oxidation well explain why alcohol-related cancers primarily occur in the digestive tracts and other tissues with active alcohol and acetaldehyde metabolism. Genetic–epidemiological research has brought compelling evidence for the causality of acetaldehyde in alcohol-related cancers. Thus, IARC recently categorized alcohol-drinking-related acetaldehyde to Group 1 for head and neck and esophageal cancers. This is probably just the tip of the iceberg, since more recent epidemiological studies have also shown significant positive associations between the aldehyde dehydrogenase *ALDH2 (rs671)*2* allele (encoding inactive enzyme causing high acetaldehyde elevations) and gastric, colorectal, lung, and hepatocellular cancers. However, a number of the current studies lack the appropriate matching or stratification of alcohol drinking in the case-control comparisons, which has led to erroneous interpretations of the data. Future studies should consider these aspects more thoroughly. The polymorphism phenotypes (flushing and nausea) may provide valuable tools for future successful health education in the prevention of alcohol-drinking-related cancers.

C.J.P. Eriksson, Ph.D. (✉)
Department of Public Health, Hjelt Institute, University of Helsinki, Mannerheimintie 172, 00014 Helsinki, Finland

Department of Alcohol, Drugs and Addiction, National Institute for Health and Welfare, P.O. Box 30, 00271 Helsinki, Finland
e-mail: peter.cj.eriksson@helsinki.fi

V. Vasiliou et al. (eds.), *Biological Basis of Alcohol-Induced Cancer*, Advances in Experimental Medicine and Biology 815,
DOI 10.1007/978-3-319-09614-8_3

Keywords Cancer • Acetaldehyde • Alcohol • Aldehyde dehydrogenase • Genetic epidemiology, digestive tracts • Pancreas • Lung • Liver • Breast

3.1 Introduction

An association between alcohol drinking and risk of cancer has been reported already more than 100 years ago [1]. Throughout the years, alcohol drinking has been linked to a number of cancers in different organs, such as oral cavity, pharynx, larynx, esophagus, stomach, colon, rectum, liver, pancreas, lung, breast, kidney, urinary bladder, prostate, endometrium, ovaries, testes, and brain. Extensive evaluations and reviews of the alcohol-related cancers have been published in the Monographs of the International Agency for Research on Cancer (IARC) [2, 3]. Currently, the most convincing evidence (Group 1) for alcohol-drinking-related cancers is targeted to oral cavity, pharynx, larynx, esophagus, colorectum, liver, and female breast [4]. Alcohol and other beverage constituents, their metabolic effects, and alcohol-related unhealthy lifestyles have been suggested as etiological carcinogenic factors. More specifically, a number of hypotheses, with more or less relevance, have been proposed, including oxidative stress [5], toxicokinetics (alcohol inhibits the breakdown of carcinogens) [6], vitamin A (e.g., alcohol-mediated depletion and/or excess of retinol) [7], folate deficiency by unhealthy lifestyle and by alcohol-mediated inhibition of folate uptake [8], insulin-like growth factors [9], immunodeficiency and immunosuppression [10], alcohol and sex hormone elevations [11], and alcoholic cirrhosis [12]. The fundamental problem in determining the etiology of alcohol-drinking-related cancers has been the lack of carcinogenicity of ethanol per se.

The aim of the present survey is to review the progress in collecting direct and indirect evidence for the role of acetaldehyde in the carcinogenicity of alcohol drinking, with special emphasis on the genetic–epidemiological evidence.

3.2 General Evidence for the Carcinogenicity of Acetaldehyde

The general evidence and indications for acetaldehyde being directly or indirectly coupled to the etiology of cancer are briefly outlined in the following.

In contrast to ethanol, acetaldehyde is a cytotoxic, genotoxic, mutagenic, and clastogenic compound already at low concentrations (below 1 mM) [13]. Acetaldehyde has been shown to form sister-chromatid exchanges in human and mammalian cells in vitro [14] and to elevate micronuclei formation (general tumor biomarker) in human lymphocytes [15]. These aforementioned effects by acetaldehyde have been proposed to originate from a variety of DNA–acetaldehyde adducts [16]. Recently, these acetaldehyde adducts (potential future biomarkers for carcinogenicity) have been determined also during moderate acute alcohol intake in humans [17].

Sufficient evidence for a direct carcinogenic effect of acetaldehyde has been reported in experimental animal studies [18].

The locations of alcohol-drinking-derived cancers are located in regions with active endogenous (liver, breast, and pancreas) and microbial (upper aerodigestive and gastrointestinal tracts) alcohol oxidation (i.e., acetaldehyde formation).

Most, if not all, locations, which are in direct contact with the first passage of the alcohol consumed (i.e., primarily the digestive tracts), are susceptible to alcohol-drinking-related cancers. These associations indicate an additional carcinogenic role of external acetaldehyde as a constituent in alcoholic beverages. For example, indirect evidence for the carcinogenicity of beverage-containing acetaldehyde has been observed in the northwest regions of France, where the use of apple brandy Calvados (with high acetaldehyde content) is positively associated with increased risk for esophageal cancer [19, 20].

It is well known that high risk of upper aerodigestive tract, especially esophageal, cancers is positively correlated to alcohol-drinking-induced "flushing" and nausea in East-Asian populations [21]. These phenotypes are typical effects of elevated acetaldehyde levels during alcohol intoxication [22].

3.3 Genetic–Epidemiological Evidence for the Carcinogenicity of Acetaldehyde

The balance between acetaldehyde formation (i.e., alcohol oxidation) and elimination rates determines human tissue acetaldehyde levels during alcohol intoxication. The key enzymes for human alcohol oxidation are considered to be the Class 1 alcohol dehydrogenases consisting of three subunits encoded by the genes *ADH1A*, *ADH1B*, and *ADH1C*. In addition alcohol is also oxidized by the inducible cytochrome P450 enzymes (especially by the form encoded by the gene *CYP2E1*) and to a minor extent by catalase. Acetaldehyde oxidation is primarily catalyzed with the aldehyde dehydrogenase encoded by the gene *ALDH2* and to some extent with the enzyme encoded by the gene *ALDH1A1*.

The difficulties in assessing the genetic–epidemiological evidence and the role of alcohol- and acetaldehyde-metabolizing enzyme polymorphisms (primarily concerning the *ALDH2 rs671**2 allele) in the acetaldehyde-mediated carcinogenesis in different tissues are reviewed in the following.

3.3.1 Basic Difficulties in Assessing the Genetic–Epidemiological Evidence

In case-control genotype comparisons, it is important to match the control population with the case group for all the factors which may affect the genotype distributions. Thus, concerning alcohol-related cancers, most of which, if not all, are positively correlated with the overall alcohol consumption, matching and/or stratification for the dose and time factors is crucial. Here the general rule is that the higher the matched alcohol consumption is in cases and controls, the higher is the probability to get valid data about possible genotype frequency differences. Consequently, with lower alcohol consumption, less impact is obtained from the alcohol-related cancers in comparison to other reasons for cancer development. Thus, the more non- to low-alcohol drinkers are included in the control population, the more it hampers the stratifications and adjustments for alcohol consumption and the possibility to get valid significant genotype differences.

An underestimated problem with East-Asian populations is the fact that the *ALDH2 (rs671)*2* allele on one hand is the probable source for carcinogenicity (and thus the *2 allele should be enriched in the case population) but on the other hand this allele prevents alcohol drinking [22] and consequently the development of cancer. Thus, the same allele is also enriched in the non- to low-drinking control population, which artificially diminishes the chances of getting appropriate significance. The normal *ALDH2 (rs671)*2* allelic variation is reported to be 0–40 % in Chinese and 0–30 % in the rest of Asian populations [23]. Thus, these limits should be used as the cutoff for the validity of normal Asian control populations. It should be noted that if significant differences have been reached with control populations exceeding these limits, the real significance would probably have been even higher.

In resemblance with the *ALDH2 (rs671)*2* allele, but not to the same extent, is the case with the *ADH1 (rs1229984)*2* allele (encoding overactive alcohol dehydrogenase), which also is reported to attenuate alcohol drinking [24].

Other problematic factors to consider are the global distributional differences of the *ALDH* and *ADH* polymorphisms [23, 25] and environmental effects on the relation between genotype and alcohol drinking. For example, during the years there has been a trend for an increase in alcoholism frequency in Japanese heterozygotes (*ALDH2 (rs671)*1/*2*) [26]. Also adjustment or matching for smoking is often neglected. On the other hand, due to the strong association between drinking and smoking, adjustment may mask the real truth. Active endogenous alcohol and acetaldehyde metabolism may reduce the odds ratios for certain cancers, especially cancer in the colorectum, which already without alcohol drinking is in an acetaldehyde-containing environment due to microbial metabolism.

3.3.2 Upper Aerodigestive Tract Cancer

The most compelling evidence for the carcinogenic role of elevated acetaldehyde related to alcohol drinking is the genetic–epidemiological results, linking deficient aldehyde dehydrogenase activity [encoded by the *ALDH2 (rs671)*2* allele] to the etiology of upper aerodigestive tract (UADT) cancers. At least 57 studies display significant positive associations between the *2 allele and UADT cancers, all, except one, in East-Asian populations [27–83]. The exception was a study of Black South Africans [62], in which the *ALDH2*2* allele was claimed to be *rs671*. However, this allele was not observed in a recent study of Black South Africans [84].

In three studies no significant differences were obtained for the role of the *ALDH2 (rs671)*2* allele. Factors like lack of control reference [85], too high frequency (45 %) of nondrinkers plus lack of adjustment for smoking [86], and huge age differences between groups plus insufficient adjustment for overall alcohol consumption [87] may explain these results. In two studies the association was negative, most likely due to too high frequency (>50 %) of *2 allele individuals plus history of esophageal cancer in the control group [88] and too high frequency (>50 %) of never drinkers [89].

In addition to the Asian rs671 data, four studies on non-Asian populations also display positive associations between more recently discovered new *ALDH2*2* single nucleotide polymorphisms (SNPs) and UADT cancers [84, 90–92]. All of the new SNPs are in close linkage disequilibrium (LD) with each other [90, 91, 93], and responsible minor alleles are from SNPs rs886205 [84, 90], rs440 and rs441 [84, 90], rs4767364, rs4648328, and rs737289 [91], and rs2238151 [92]. One Asian study also confirms the positive association between rs886205 and UADT in a Chinese population, but in this polymorphism, it is the major allele which was the carcinogenic agent [94]. The close genomic proximity to the Asian rs671 SNP and the same phenotype outcome, i.e., alcohol-drinking-related carcinogenicity, indicate that these new SNPs may also encode deficient ALDH activity (but to a lesser extent compared with the rs671), which would cause systemic acetaldehyde elevations. However, this important aspect remains to be settled by future studies.

Also the *ADH1B (rs1229984)*1/*1* genotype encoding less active alcohol dehydrogenase has convincingly been associated with UADT cancers in a number of studies, including both Asian and non-Asian populations [95]. This has been explained by studies showing that less active alcohol dehydrogenase causes slower alcohol oxidation, which prolongs the appearance of affecting acetaldehyde [96].

As the consequence of the overwhelming genetic–epidemiological data in support for the carcinogenic role of acetaldehyde, IARC decided to categorize alcohol-drinking-related acetaldehyde to Group 1 regarding esophageal and head and neck cancers [4].

3.3.3 Gastric Cancer

Currently, at least nine studies have been published on the association between gastric cancer and the *ALDH (rs671)*2* allele. These studies include East-Asian populations, four Japanese [32, 38, 58, 97], three Korean [98–100], and two Chinese [101, 102]. However, the overall significance ($P=0.000$) is calculated only from five studies [58, 97, 98, 100, 102] (Table 3.1), because most likely the same participants have been, at least partly, used in other smaller studies by the same authors [32, 58, 99]. One Chinese study [101] is omitted from the table because the precancerous group was compared with a reference group consisting of chronic atrophic gastritis patients, the condition which also has been shown to be positively associated with the *ALDH (rs671)*2* allele [96]. The overall significance of the five studies shown in Table 3.1 most likely represents an underestimation, because the two studies, which only displayed tendencies,

Table 3.1 Case–control studies on the association between *ALDH2* (rs671) and gastric cancer

	ALDH2 genotypes (*n*)					
	**1/*1*		**1/*2* + **2/*2*			
Authors	Case	Control	Case	Control	OR (95 % CI)	*P*
Nan et al. (2005)[a] [98]	286	462	135	168	1.30 (0.99–1.70)	0.061
Yokoyama et al. (2007)[b] [58]	29	242	16	39	3.42 (1.70–6.88)	0.001
Cao et al. (2010)[c] [102]	50	51	15	6	2.53 (0.86–7.49)	0.092
Shin et al. (2011)[d] [100]	53	44	15	3	4.26 (1.10–16.47)	0.036
Matsuo et al. (2013)[e] [97]	87	145	44	32	2.29 (1.35–3.88)	0.002
Combined significance						0.000

Part of the calculations has not been displayed in the publications and is marked here as "calculated from the original data"

[a]Korean men and women, not adjusted for alcohol drinking, *P* calculated from the original data

[b]Japanese alcoholic men, OR (95 % CI) calculated from the original data, *P* adjusted for alcohol drinking and smoking

[c]Chinese male drinkers, ≥2.5 kg*year, not adjusted for alcohol drinking, *P* calculated from the original data

[d]Korean male and female heavy drinkers, ≥144 g ethanol per week, OR and *P* adjusted for smoking

[e]Japanese male and female heavy drinkers, ≥115 g ethanol per week, OR and *P* calculated from the original data

were not well stratified and adjusted for the alcohol consumption and also included too many nondrinkers in their control populations.

In congruence with the European studies on the positive association between the *ALDH rs886205* C allele and UADT cancers [84, 90], the same allele and another minor allele of the rs16941667 SNP have also been associated to gastric cancer in European populations [103, 104]. These findings strengthen the evidence for a general carcinogenic effect of acetaldehyde also in non-Asian populations.

The overall significance (Table 3.1) demonstrates that there is sufficient evidence that acetaldehyde is a functional carcinogen in the stomach after chronic alcohol drinking, at least in East-Asian individuals carrying the *ALDH2 (rs671)*2* allele.

3.3.4 Colorectal Cancer

Currently, at least 11 studies have been published on the association between colorectal cancer and the *ALDH (rs671)*2* allele (Table 3.2) [32, 105–114]. Increased overall risk by the *2 allele was observed in a Japanese [32] and a Chinese [105] study. Increased risks were also shown in Japanese studies specifically for rectum [106] and colon [107]. Similarly, increased risks in Japanese populations were indicated by a gene–gene interaction with the *ALDH2 (rs671)*2* allele combined with the *ADH1B*2* [111] or *CD36*C* [113] alleles. On the other hand, the *ALDH*2* allele was also associated with significantly lower risk in another Chinese study [114]. The four remaining studies displayed nonsignificant results [108–110, 112].

Table 3.2 Case–control studies on the association between the *ALDH2* (rs671)*2 allele and colorectal cancer

Authors	*1/*2+*2/*2 vs. *1/*1 OR (95 % CI)	*P*	Proportion of abstainers or low (<15 g/day) drinkers in the whole population (%)
Chiang et al. (2012)[a] [105]	1.76 (1.15–1.17)	0.010	No information
Yokoyama et al. (1998)[b] [32]	3.35 (1.51–7.45)	0.017	0
Matsuo et al. (2002)[c] [106]	1.80 (0.54–5.96)	0.387[f]	0
Murata et al. (1999)[d] [107]	1.20 (0.77–1.87)	0.437[g]	39
Yin et al. (2007)[e] [108]	1.07 (0.64–1.79)	0.448	0
Otani et al. (2005)[d] [109]	1.16 (0.73–1.86)	0.549	32
Miyasaka et al. (2010)[d] [110]	0.80 (0.41–1.48)	0.528	54
Matsuo et al. (2006)[d] [111]	0.99 (0.74–1.31)	0.987[h]	40
Yang et al. (2009)[a] [112]	0.92 (0.72–1.17)	0.494	78
Kuriki et al. (2005)[d] [113]	0.80 (0.52–1.24)	0.376[i]	52
Gao et al. (2008)[a] [114]	0.54 (0.30–0.98)	0.050	48

Part of the calculations has not been displayed in the publications and is marked here as "calculated from the original data"

[a]Chinese men and women, OR and *P* calculated from the original data

[b]Japanese alcoholic men, OR adjusted for alcohol drinking and smoking, *P* calculated from the original data

[c]Japanese male and female drinkers, ≥39 g ethanol once a week or more frequently, OR and *P* calculated from the original data

[d]Japanese men and women, OR and *P* calculated from the original data

[e]Japanese male and female drinkers, ≥40 g ethanol per day, OR and *P* calculated from the original data

[f]Increased risk, $P=0.10$ (specifically for rectum)

[g]Increased risk, $P=0.04$ (specifically for colon)

[h]Increased risk ($P<0.05$) by gene–gene interaction with *ALDH2* (rs671)*2 and *ADH1B*1*

[i]Increased risk ($P<0.05$) by gene–gene interaction with *CD36*C* allele combinations

In addition to the Asian studies, two European investigations have assessed the role of other *ALDH2* alleles, namely, the rs441 [115] and the *rs886205* and *rs440* alleles [116], but without obtaining any significant results.

The seemingly discrepant data may partly be explained by the choice of control populations. As seen in Table 3.2, the higher the proportion of nondrinkers to very low drinkers is, the less indication of increased risk of the *ALDH (rs671)*2* allele is achieved. As explained in Sect. 3.3.1, there are problems in including nondrinking control populations, because of the enrichment of the *2 allele. In addition the problem of active microbial endogenous alcohol metabolism in the gut may mask some of the effects by the alcohol-drinking-related acetaldehyde. Altogether, it seems that acetaldehyde here is a player, but in order to get sufficient evidence, future studies should better stratify and match the alcohol drinking of the control populations with the drinking of cancer patients.

3.3.5 Pancreatic Cancer

Little information (only three Japanese studies, two of which include about the same population) is available on the role of *ALDH2 (rs671)*2* allele on pancreatic cancer [110, 117, 118]. After calculation of the original data, only nonsignificant results with higher frequency of the *1/*2+*2/*2 vs. *1/*1 genotypes emerge, with ORs (95 % CI) 1.50 (0.95–2.38), $P=0.088$ [118], and 1.15 (0.85–1.55), $P=0.401$ [110]. The study displaying the trend [118] included only alcohol drinkers, while the study with $P=0.401$ contained 54 % nondrinkers [110], which is in congruence regarding the studies on gastric and colorectal cancers (high proportion of nondrinkers reduces the significance). At this point there is no valid genetic–epidemiological data assessing a significant role of acetaldehyde in the etiology of pancreatic cancer.

3.3.6 Liver Cancer

Currently, at least 12 studies, all on East-Asian populations (ten Japanese and two Chinese), have been published on the association between hepatocellular cancer (HCC) and the *ALDH (rs671)*2* allele [32, 119–129], nine of which are displayed in Table 3.3 [119–127]. The three studies not presented in Table 3.3 are inconclusive

Table 3.3 Case–control studies on the association between the *ALDH2* (rs671)*2 allele and hepatocellular cancer (HCC)

Authors	*1/*2+*2/*2* vs. **1/*1* OR (95 % CI)	*P*	Proportion of non-drinkers in the whole population (%)
Kato et al. (2003)[a] [119]	5.36 (2.10–14.02)	0.000	No information
Tomoda et al. (2012)[b] [120]	1.73 (1.19–2.51)	0.005	No information
Sakamoto et al. (2006)[c] [121]	2.13 (1.20–3.78)	0.014	0
Munaka et al. (2003)[d] [122]	9.77 (1.63–58.60)	<0.05	38
Ding et al. (2008)[e] [123]	1.97 (0.79–4.89)	0.185	0
Takeshita et al. (2000)[f] [124]	1.46 (0.57–3.74)	0.486	0
Koide et al. (2000)[b] [125]	0.79 (0.43–1.45)	0.536	41
Yu et al. (2002)[g] [126]	0.74 (0.46–1.20)	0.269	59
Shibata et al. (1998)[h] [127]	0.44 (0.21–0.92)	0.044	86

Part of the calculations has not been displayed in the publications and is marked here as "calculated from the original data"

[a]Japanese men and women, **1/*1+*1/*2* vs. **2/*2* genotypes, OR no adjustments, *P* calculated from original data

[b]Japanese men and women, OR and *P* calculated from original data

[c]Japanese male and female drinkers, drinking more than or once per week, OR and *P* calculated from original data

[d]Japanese men and women, OR and *P* adjusted for drinking, smoking, hepatitis B and C virus, age and gender

[e]Chinese male and female drinkers, >300 g ethanol per day, OR and *P* calculated from original data

[f]Japanese male drinkers, ≥24 g per day, OR and *P* calculated from original data

[g]Chinese men and women, OR and *P* calculated from original data

[h]Japanese men, OR and *P* calculated from original data

because of low number of cases, less than 25 per study and very few individuals carrying the *ALDH (rs671)*2* allele (four individuals altogether). At first hand looking at Table 3.3, the data of the nine studies seem inconclusive with four studies showing a significant positive association between the *2 allele and risk for HCC [119–122], nonsignificance in four studies [123–130], and one significant negative association [129]. However, a closer view of the data demonstrates again the same phenomenon as for the gastric and colorectal cancers (see previous Sects. 3.3.3 and 3.3.4), i.e., the nonsignificance and negative correlation are most likely the consequence of a too high proportion of nondrinkers to low drinkers in the control group. The problem is the enrichment of the *2 allele individuals in the control population. Thus, it is hardly a coincidence that the three studies displaying negative associations are the studies in which the frequency of *2 allele individuals is highest compared with the *1/*1 genotype frequency of the control population, 62 %, 57 %, and 49 % in the studies [125–127], respectively. This means that most likely the role of acetaldehyde in the alcohol-related HCC has been underestimated.

3.3.7 *Lung Cancer*

Currently, the association between the *ALDH2 (rs671)*2* allele and lung cancer has been reported in two Japanese [32, 130] and two Korean [131, 132] studies (Table 3.4). Altogether, the combined data indicate a significant role of acetaldehyde in the etiology of lung cancer. The odds ratios and probabilities may be, at least to some extent, underestimated. For example, in the largest study of those displayed in Table 3.4, the frequency of the *2 allele individuals was considerably higher (67 %) compared to the *1/*1 carriers (33 %) in the control population [132], indicating an overrepresentation of the *2 allele in the reference group. New studies with better matched control groups will be needed for an even more valid assessment of the carcinogenic role of acetaldehyde in lung cancer in East Asia.

Table 3.4 Case–control studies on the association between *ALDH2* (rs671) and lung cancer

	ALDH2 genotypes (*n*)					
	**1/*1*		**1/*2 + *2/*2*			
Authors	Case	Control	Case	Control	OR (95 % CI)	*P*
Yokoyama et al. (1998)[a] [32]	5	443	2	44	8.20 (1.27–53.15)	0.132
Minegishi et al. (2007)[b] [130]	163	54	68	18	1.25 (0.68–2.29)	0.550
Eom et al. (2009)[c] [131]	112	103	20	13	1.69 (0.73–3.91)	0.201
Park et al. (2010)[d] [132]	322	688	396	728	1.16 (0.97–1.39)	0.108
Combined significance						<0.010

Part of the calculations has not been displayed in the publications and is marked here without adjustments as "calculated from the original data"

[a]Japanese alcoholic men, OR adjusted for alcohol drinking and smoking, *P* calculated from the original data

[b]Japanese male and female heavy drinkers, average >221 g ethanol per week, OR and *P* calculated from the original data

[c]Korean male and female drinkers, ≥108 g per week, OR and *P* adjusted for smoking

[d]Korean men and women, OR and *P* calculated from the original data

In addition to the role of *ALDH2 (rs671)*2* allele in East-Asian lung cancer, one Norwegian study displays a higher risk for lung cancer by the minor allele of the *ALDH2 rs4646777* SNP [133]. The role of this SNP as well as of other possibly important European SNP candidates needs to be evaluated in future studies.

3.3.8 Breast Cancers

All studies (one Korean [134], one Japanese [135], one Thai [136], and one Spanish [137]) on the role of ALDH2 in breast cancer have been nonconclusive. The Asian studies concerned the *ALDH2 (rs671)*2* allele and the European study the minor alleles of *ALDH2 rs737280*, *rs2238151*, *rs11066028*, and *rs11066034*.

3.4 Summary and Conclusions

The general evidence for the carcinogenicity of acetaldehyde related to alcohol drinking is based on the cytotoxic, genotoxic, mutagenic, and clastogenic properties of acetaldehyde: acetaldehyde may form sister-chromatid exchanges and elevate micronuclei formation (general tumor biomarker) with a variety of DNA–acetaldehyde adducts; acetaldehyde adducts have been determined after alcohol intake in humans; acetaldehyde carcinogenicity has been shown in experimental animal studies; proximity between the locations of microbial activity, acetaldehyde exposure, and alcohol-drinking-derived cancers; and indication of carcinogenic effects by external acetaldehyde from the alcoholic beverage. This acetaldehyde should be recognized and considered for future assessments on limits for safe acetaldehyde concentrations in the alcoholic beverage. Currently, there is a need for developing new secure acetaldehyde regulations and directives for the alcohol industry.

The genetic–epidemiological evidence for the carcinogenicity of acetaldehyde is based on the positive association between the *ALDH2 (rs671)*2* allele and cancer in the upper aerodigestive tract, stomach, colorectum, liver, and the lungs. Regarding the pancreas and breast, the data is still nonconclusive. The results of a number of investigations are hampered by choosing non- to low-alcohol drinkers as the control and/or reference group in comparing the *rs671* genotype distribution in cases and controls. The thorough matching or stratification of alcohol drinking has also commonly been neglected. The result has been an erroneous interpretation of the data. Considering these difficulties clearly indicates that acetaldehyde, in addition to the already officially accepted carcinogenic effect in the upper aerodigestive tract, is most likely a responsible factor in the etiology of gastric, colorectal, hepatocellular, and lung cancers. Since the phenotypes (flushing and nausea) of the *rs671* polymorphism are clearly expressed, they provide a valuable tool for future successful health education in the prevention of alcohol-drinking-related cancers.

Acknowledgment This work has been supported by the Magnus Ehrnrooth Foundation and the Liv och Hälsa medical association.

References

1. Newsholme A (1903) The possible association of the consumption of alcohol with excessive mortality from cancer. Br Med J 2241:1529–1531
2. IARC (2010) Alcohol consumption and ethyl carbamate. IARC Monogr Eval Carcinog Risks Hum 96:1–1424
3. IARC (2012) personal habits and indoor combustions. A review of human carcinogens. IARC Monogr Eval Carcinog Risks Hum 100E:1–575
4. Secretan B, Straif K, Baan R et al (2009) A review of human carcinogens—Part E: tobacco, areca nut, alcohol, coal smoke, and salted fish. Lancet Oncol 10(11):1033–1034
5. Sanchez-Alvarez R, Martinez-Outschoorn UE, Lin Z et al (2013) Ethanol exposure induces the cancer-associated fibroblast phenotype and lethal tumor metabolism: implications for breast cancer prevention. Cell Cycle 12(2):289–301. doi:10.4161/cc.23109
6. Lieber CS, Baraona E, Leo MA et al (1987) International Commission for Protection against Environmental Mutagens and Carcinogens. ICPEMC Working Paper No. 15/2. Metabolism and metabolic effects of ethanol, including interaction with drugs, carcinogens and nutrition. Mutat Res 186(3):201–233
7. Leo MA, Lieber CS (1999) Alcohol, vitamin A, and beta-carotene: adverse interactions, including hepatotoxicity and carcinogenicity. Am J Clin Nutr 69(6):1071–1085
8. Hamid A, Wani NA, Kaur J (2009) New perspectives on folate transport in relation to alcoholism-induced folate malabsorption—association with epigenome stability and cancer development. FEBS J 276(8):2175–2191. doi:10.1111/j.1742-4658.2009.06959
9. Yu H, Rohan T (2000) Role of the insulin-like growth factor family in cancer development and progression. J Natl Cancer Inst 92(18):1472–1489
10. Watson RR, Nixon P, Seitz HK et al (1994) Alcohol and cancer. Alcohol Alcohol Suppl 2:453–455
11. Singletary KW, Gapstur SM (2001) Alcohol and breast cancer: review of epidemiologic and experimental evidence and potential mechanisms. JAMA 286(17):2143–2151
12. Stickel F, Schuppan D, Hahn EG et al (2002) Cocarcinogenic effects of alcohol in hepatocarcinogenesis. Gut 51(1):132–139
13. Obe G, Ristow H (1979) Mutagenic, cancerogenic and teratogenic effects of alcohol. Mutat Res 65(4):229–259
14. Obe G, Anderson D, International Commission for Protection against Environmental Mutagens and Carcinogens (1987) ICPEMC Working Paper No. 15/1. Genetic effects of ethanol. Mutat Res 186(3):177–200
15. Ishikawa H, Yamamoto H, Tian Y et al (2003) Effects of ALDH2 gene polymorphisms and alcohol-drinking behavior on micronuclei frequency in non-smokers. Mutat Res 541(1–2): 71–80
16. Brooks PJ, Theruvathu JA (2005) DNA adducts from acetaldehyde: implications for alcohol-related carcinogenesis. Alcohol 35(3):187–93
17. Balbo S, Meng L, Bliss RL et al (2012) Kinetics of DNA adduct formation in the oral cavity after drinking alcohol. Cancer Epidemiol Biomarkers Prev 21(4):601–8. doi:10.1158/1055-9965
18. Soffritti M, Belpoggi F, Lambertin L et al (2002) Results of long-term experimental studies on the carcinogenicity of formaldehyde and acetaldehyde in rats. Ann N Y Acad Sci 982:87–105
19. Launoy G, Milan C, Day NE et al (1997) Oesophageal cancer in France: potential importance of hot alcoholic drinks. Int J Cancer 71(6):917–923
20. Linderborg K, Joly JP, Visapää JP et al (2008) Potential mechanism for Calvados-related oesophageal cancer. Food Chem Toxicol 2:476–479

21. Brooks PJ, Enoch MA, Goldman D et al (2009) The alcohol flushing response: an unrecognized risk factor for esophageal cancer from alcohol consumption. PLoS Med 6(3):e50. doi:10.1371/journal.pmed.1000050
22. Eriksson CJP (2001) The role of acetaldehyde in the actions of alcohol (update 2000). Alcohol Clin Exp Res 5 Suppl ISBRA):15S–32S
23. Li H, Borinskaya S, Yoshimura K et al (2009) Refined geographic distribution of the oriental ALDH2*504Lys (nee 487Lys) variant. Ann Hum Genet 73(Pt 3):335–345. doi:10.1111/j.1469-1809.2009.00517
24. Carr LG, Foroud T, Stewart T et al (2002) Influence of ADH1B polymorphism on alcohol use and its subjective effects in a Jewish population. Am J Med Genet 112(2):138–143
25. Li H, Mukherjee N, Soundararajan U et al (2007) Geographically separate increases in the frequency of the derived ADH1B*47His allele in eastern and western Asia. Am J Hum Genet 81(4):842–846
26. Higuchi S (1994) Polymorphisms of ethanol metabolizing enzyme genes and alcoholism. Alcohol Alcohol Suppl 2:29–34
27. Yokoyama A, Muramatsu T, Ohmori T et al (1996) Esophageal cancer and aldehyde dehydrogenase-2 genotypes in Japanese males. Cancer Epidemiol Biomarkers Prev 5(2):99–102
28. Yokoyama A, Muramatsu T, Ohmori T et al (1996) Multiple primary esophageal and concurrent upper aerodigestive tract cancer and the aldehyde dehydrogenase-2 genotype of Japanese alcoholics. Cancer 77(10):1986–1990
29. Yokoyama A, Ohmori T, Muramatsu T et al (1996) Cancer screening of upper aerodigestive tract in Japanese alcoholics with reference to drinking and smoking habits and aldehyde dehydrogenase-2 genotype. Int J Cancer 68(3):313–316
30. Hori H, Kawano T, Endo M et al (1997) Genetic polymorphisms of tobacco- and alcohol-related metabolizing enzymes and human esophageal squamous cell carcinoma susceptibility. J Clin Gastroenterol 25(4):568–575
31. Yokoyama A, Ohmori T, Muramatsu T et al (1998) Short-term follow-up after endoscopic mucosectomy of early esophageal cancer and aldehyde dehydrogenase-2 genotype in Japanese alcoholics. Cancer Epidemiol Biomarkers Prev 7(6):473–476
32. Yokoyama A, Muramatsu T, Ohmori T et al (1998) Alcohol-related cancers and aldehyde dehydrogenase-2 in Japanese alcoholics. Carcinogenesis 19(8):1383–1387
33. Tanabe H, Ohhira M, Ohtsubo T et al (1999) Genetic polymorphism of aldehyde dehydrogenase 2 in patients with upper aerodigestive tract cancer. Alcohol Clin Exp Res 23(4 Suppl):17S–20S
34. Yokoyama A, Muramatsu T, Omori T et al (1999) Alcohol and aldehyde dehydrogenase gene polymorphisms influence susceptibility to esophageal cancer in Japanese alcoholics. Alcohol Clin Exp Res 23(11):1705–1710
35. Nomura T, Noma H, Shibahara T et al (2000) Aldehyde dehydrogenase 2 and glutathione S-transferase M 1 polymorphisms in relation to the risk for oral cancer in Japanese drinkers. Oral Oncol 36(1):42–46
36. Muto M, Hitomi Y, Ohtsu A et al (2000) Association of aldehyde dehydrogenase 2 gene polymorphism with multiple oesophageal dysplasia in head and neck cancer patients. Gut 47(2):256–261
37. Chao YC, Wang LS, Hsieh TY et al (2000) Chinese alcoholic patients with esophageal cancer are genetically different from alcoholics with acute pancreatitis and liver cirrhosis. Am J Gastroenterol 95(10):2958–2964
38. Yokoyama A, Muramatsu T, Omori T et al (2001) Alcohol and aldehyde dehydrogenase gene polymorphisms and oropharyngolaryngeal, esophageal and stomach cancers in Japanese alcoholics. Carcinogenesis 22(3):433–439
39. Matsuo K, Hamajima N, Shinoda M et al (2001) Gene-environment interaction between an aldehyde dehydrogenase-2 (ALDH2) polymorphism and alcohol consumption for the risk of esophageal cancer. Carcinogenesis 22(6):913–916
40. Itoga S, Nomura F, Makino Y et al (2002) Tandem repeat polymorphism of the CYP2E1 gene: an association study with esophageal cancer and lung cancer. Alcohol Clin Exp Res 26(8 Suppl):15S–19S

41. Yokoyama A, Watanabe H, Fukuda H et al (2002) Multiple cancers associated with esophageal and oropharyngolaryngeal squamous cell carcinoma and the aldehyde dehydrogenase-2 genotype in male Japanese drinkers. Cancer Epidemiol Biomarkers Prev 9:895–900
42. Watanabe S, Sasahara K, Kinekawa F et al (2002) Aldehyde dehydrogenase-2 genotypes and HLA haplotypes in Japanese patients with esophageal cancer. Oncol Rep 9(5):1063–1068
43. Muto M, Nakane M, Hitomi Y et al (2002) Association between aldehyde dehydrogenase gene polymorphisms and the phenomenon of field cancerization in patients with head and neck cancer. Carcinogenesis 23(10):1759–1765
44. Yokoyama A, Kato H, Yokoyama T et al (2002) Genetic polymorphisms of alcohol and aldehyde dehydrogenases and glutathione S-transferase M1 and drinking, smoking, and diet in Japanese men with esophageal squamous cell carcinoma. Carcinogenesis 23(11):1851–1859
45. Boonyaphiphat P, Thongsuksai P, Sriplung H et al (2002) Lifestyle habits and genetic susceptibility and the risk of esophageal cancer in the Thai population. Cancer Lett 186(2):193–199
46. Yokoyama A, Yokoyama T, Muramatsu T et al (2003) Macrocytosis, a new predictor for esophageal squamous cell carcinoma in Japanese alcoholic men. Carcinogenesis 24(11):1773–1778
47. Yokoyama T, Yokoyama A, Kato H et al (2003) Alcohol flushing, alcohol and aldehyde dehydrogenase genotypes, and risk for esophageal squamous cell carcinoma in Japanese men. Cancer Epidemiol Biomarkers Prev 12(11 Pt 1):1227–1233
48. Muto M, Takahashi M, Ohtsu A et al (2005) Risk of multiple squamous cell carcinomas both in the esophagus and the head and neck region. Carcinogenesis 26(5):1008–1012
49. Yokoyama A, Omori T, Yokoyama T et al (2005) Esophageal melanosis, an endoscopic finding associated with squamous cell neoplasms of the upper aerodigestive tract, and inactive aldehyde dehydrogenase-2 in alcoholic Japanese men. J Gastroenterol 40(7):676–684
50. Yang CX, Matsuo K, Ito H et al (2005) Esophageal cancer risk by ALDH2 and ADH2 polymorphisms and alcohol consumption: exploration of gene-environment and gene-gene interactions. Asian Pac J Cancer Prev 6(3):256–262
51. Wu CF, Wu DC, Hsu HK et al (2005) Relationship between genetic polymorphisms of alcohol and aldehyde dehydrogenases and esophageal squamous cell carcinoma risk in males. World J Gastroenterol 11(33):5103–5108
52. Yokoyama A, Kato H, Yokoyama T et al (2006) Esophageal squamous cell carcinoma and aldehyde dehydrogenase-2 genotypes in Japanese females. Alcohol Clin Exp Res 30(3):491–500
53. Cai L, You NC, Lu H et al (2006) Dietary selenium intake, aldehyde dehydrogenase-2 and X-ray repair cross-complementing 1 genetic polymorphisms, and the risk of esophageal squamous cell carcinoma. Cancer 106(11):2345–2354
54. Yokoyama A, Mizukami T, Omori T et al (2006) Melanosis and squamous cell neoplasms of the upper aerodigestive tract in Japanese alcoholic men. Cancer Sci 97(9):905–911
55. Yokoyama A, Omori T, Yokoyama T et al (2006) Risk of squamous cell carcinoma of the upper aerodigestive tract in cancer-free alcoholic Japanese men: an endoscopic follow-up study. Cancer Epidemiol Biomarkers Prev 15(11):2209–2215
56. Chen YJ, Chen C, Wu DC et al (2006) Interactive effects of lifetime alcohol consumption and alcohol and aldehyde dehydrogenase polymorphisms on esophageal cancer risks. Int J Cancer 119(12):2827–2831
57. Hashimoto T, Uchida K, Okayama N et al (2006) ALDH2 1510 G/A (Glu487Lys) polymorphism interaction with age in head and neck squamous cell carcinoma. Tumour Biol 27(6):334–338
58. Yokoyama A, Yokoyama T, Omori T et al (2007) Helicobacter pylori, chronic atrophic gastritis, inactive aldehyde dehydrogenase-2, macrocytosis and multiple upper aerodigestive tract cancers and the risk for gastric cancer in alcoholic Japanese men. J Gastroenterol Hepatol 22(2):210–217
59. Asakage T, Yokoyama A, Haneda T et al (2007) Genetic polymorphisms of alcohol and aldehyde dehydrogenases, and drinking, smoking and diet in Japanese men with oral and pharyngeal squamous cell carcinoma. Carcinogenesis 28(4):865–874
60. Hiraki A, Matsuo K, Wakai K et al (2007) Gene-gene and gene-environment interactions between alcohol drinking habit and polymorphisms in alcohol-metabolizing enzyme genes and the risk of head and neck cancer in Japan. Cancer Sci 98(7):1087–1091

61. Yang SJ, Wang HY, Li XQ et al (2007) Genetic polymorphisms of ADH2 and ALDH2 association with esophageal cancer risk in southwest China. World J Gastroenterol 13(43):5760–5764
62. Li DP, Dandara C, Walther G et al (2008) Genetic polymorphisms of alcohol metabolising enzymes: their role in susceptibility to oesophageal cancer. Clin Chem Lab Med 46(3):323–328. doi:10.1515/CCLM.2008.073
63. Guo YM, Wang Q, Liu YZ et al (2008) Genetic polymorphisms in cytochrome P4502E1, alcohol and aldehyde dehydrogenases and the risk of esophageal squamous cell carcinoma in Gansu Chinese males. World J Gastroenterol 14(9):1444–1449
64. Lee CH, Lee JM, Wu DC et al (2008) Carcinogenetic impact of ADH1B and ALDH2 genes on squamous cell carcinoma risk of the esophagus with regard to the consumption of alcohol, tobacco and betel quid. Int J Cancer 122(6):1347–1356
65. Yokoyama A, Omori T, Yokoyama T et al (2008) Risk of metachronous squamous cell carcinoma in the upper aerodigestive tract of Japanese alcoholic men with esophageal squamous cell carcinoma: a long-term endoscopic follow-up study. Cancer Sci 99(6):1164–1171. doi:10.1111/j.1349-7006.2008.00807
66. Ding JH, Li SP, Cao HX et al (2009) Polymorphisms of alcohol dehydrogenase-2 and aldehyde dehydrogenase-2 and esophageal cancer risk in Southeast Chinese males. World J Gastroenterol 15(19):2395–2400
67. Lee CH, Wu DC, Wu IC et al (2009) Genetic modulation of ADH1B and ALDH2 polymorphisms with regard to alcohol and tobacco consumption for younger aged esophageal squamous cell carcinoma diagnosis. Int J Cancer 125(5):1134–1142. doi:10.1002/ijc.24357
68. Cui R, Kamatani Y, Takahashi A et al (2009) Functional variants in ADH1B and ALDH2 coupled with alcohol and smoking synergistically enhance esophageal cancer risk. Gastroenterology 137(5):1768–1775. doi:10.1053/j.gastro.2009.07.070
69. Ding JH, Li SP, Cao HX et al (2010) Alcohol dehydrogenase-2 and aldehyde dehydrogenase-2 genotypes, alcohol drinking and the risk for esophageal cancer in a Chinese population. J Hum Genet 55(2):97–102. doi:10.1038/jhg.2009.129
70. Oze I, Matsuo K, Hosono S et al (2010) Comparison between self-reported facial flushing after alcohol consumption and ALDH2 Glu504Lys polymorphism for risk of upper aerodigestive tract cancer in a Japanese population. Cancer Sci 101(8):1875–1880. doi:10.1111/j.1349-7006.2010.01599
71. Oikawa T, Iijima K, Koike T et al (2010) Deficient aldehyde dehydrogenase 2 is associated with increased risk for esophageal squamous cell carcinoma in the presence of gastric hypochlorhydria. Scand J Gastroenterol 45(11):1338–1344. doi:10.3109/00365521.2010.495419
72. Tanaka F, Yamamoto K, Suzuki S et al (2010) Strong interaction between the effects of alcohol consumption and smoking on oesophageal squamous cell carcinoma among individuals with ADH1B and/or ALDH2 risk alleles. Gut 59(11):1457–1464. doi:10.1136/gut.2009.205724
73. Li QD, Li H, Wang MS et al (2011) Multi-susceptibility genes associated with the risk of the development stages of esophageal squamous cell cancer in Feicheng County. BMC Gastroenterol 11:74. doi:10.1186/1471-230X-11-74
74. Yokoyama A, Tanaka Y, Yokoyama T et al (2011) p53 protein accumulation, iodine-unstained lesions, and alcohol dehydrogenase-1B and aldehyde dehydrogenase-2 genotypes in Japanese alcoholic men with esophageal dysplasia. Cancer Lett 308(1):112–117. doi:10.1016/j.canlet.2011.04.020
75. Wang Y, Ji R, Wei X, Gu L et al (2011) Esophageal squamous cell carcinoma and ALDH2 and ADH1B polymorphisms in Chinese females. Asian Pac J Cancer Prev 12(8): 2065–2068
76. Matsuo K, Rossi M, Negri E et al (2012) Folate, alcohol, and aldehyde dehydrogenase 2 polymorphism and the risk of oral and pharyngeal cancer in Japanese. Eur J Cancer Prev 21(2):193–198. doi:10.1097/CEJ.0b013e32834c9be5
77. Yokoyama A, Hirota T, Omori T et al (2012) Development of squamous neoplasia in esophageal iodine-unstained lesions and the alcohol and aldehyde dehydrogenase genotypes of Japanese alcoholic men. Int J Cancer 130(12):2949–2960. doi:10.1002/ijc.26296

78. Katada C, Muto M, Nakayama M et al (2012) Risk of superficial squamous cell carcinoma developing in the head and neck region in patients with esophageal squamous cell carcinoma. Laryngoscope 122(6):1291–1296. doi:10.1002/lary.23249
79. Guo LK, Zhang CX, Guo XF et al (2012) Association of genetic polymorphisms of aldehyde dehydrogenase-2 and cytochrome P450 2E1-RsaI and alcohol consumption with oral squamous cell carcinoma. Zhongguo Yi Xue Ke Xue Yuan Xue Bao 34(4):390–395. doi:10.3881/j.issn.1000-503X.2012.04.015
80. Gu H, Gong D, Ding G et al (2012) A variant allele of ADH1B and ALDH2, is associated with the risk of esophageal cancer. Exp Ther Med 4(1):135–140
81. Chung CS, Lee YC, Liou JM et al (2014) Tag single nucleotide polymorphisms of alcohol-metabolizing enzymes modify the risk of upper aerodigestive tract cancers: HapMap database analysis. Dis Esophagus 27(5):493–503. doi:10.1111/j.1442-2050.2012.01437
82. Wu C, Kraft P, Zhai K et al (2012) Genome-wide association analyses of esophageal squamous cell carcinoma in Chinese identify multiple susceptibility loci and gene-environment interactions. Nat Genet 44(10):1090–1097. doi:10.1038/ng.2411
83. Wu M, Chang SC, Kampman E et al (2013) Single nucleotide polymorphisms of ADH1B, ADH1C and ALDH2 genes and esophageal cancer: a population-based case-control study in China. Int J Cancer 132(8):1868–1877. doi:10.1002/ijc.27803
84. Bye H, Prescott NJ, Matejcic M et al (2011) Population-specific genetic associations with oesophageal squamous cell carcinoma in South Africa. Carcinogenesis 32(12):1855–1861. doi:10.1093/carcin/bgr211
85. Tian D, Feng Z, Hanley NM et al (1998) Multifocal accumulation of p53 protein in esophageal carcinoma: evidence for field cancerization. Int J Cancer 78(5):568–575
86. Katoh T, Kaneko S, Kohshi K et al (1999) Genetic polymorphisms of tobacco- and alcohol-related metabolizing enzymes and oral cavity cancer. Int J Cancer 83(5):606–609
87. Ji YB, Tae K, Ahn TH et al (2011) Candidate gene polymorphisms for diabetes mellitus, cardiovascular disease and cancer are associated with longevity in Koreans. Exp Mol Med 41(11):772–781. doi:10.3858/emm.2009.41.11.083
88. Zhou YZ, Diao YT, Li H et al (2010) Association of genetic polymorphisms of aldehyde dehydrogenase-2 with esophageal squamous cell dysplasia. World J Gastroenterol 16(27):3445–3449
89. Gao Y, He Y, Xu J et al (2013) Genetic variants at 4q21, 4q23 and 12q24 are associated with esophageal squamous cell carcinoma risk in a Chinese population. Hum Genet 132(6):649–656. doi:10.1007/s00439-013-1276-5
90. Hashibe M, Boffetta P, Zaridze D et al (2006) Evidence for an important role of alcohol- and aldehyde-metabolizing genes in cancers of the upper aerodigestive tract. Cancer Epidemiol Biomarkers Prev 15(4):696–703
91. McKay JD, Truong T, Gaborieau V et al (2011) A genome-wide association study of upper aerodigestive tract cancers conducted within the INHANCE consortium. PLoS Genet 7(3):e1001333. doi:10.1371/journal.pgen.1001333
92. Hakenewerth AM, Millikan RC, Rusyn I et al (2011) Joint effects of alcohol consumption and polymorphisms in alcohol and oxidative stress metabolism genes on risk of head and neck cancer. Cancer Epidemiol Biomarkers Prev 20(11):2438–49. doi:10.1158/1055-9965.EPI-11-0649
93. Dickson PA, James MR, Heath AC et al (2006) Effects of variation at the ALDH2 locus on alcohol metabolism, sensitivity, consumption, and dependence in Europeans. Alcohol Clin Exp Res 30(7):1093–1100
94. Ma WJ, Lv GD, Zheng ST et al (2010) DNA polymorphism and risk of esophageal squamous cell carcinoma in a population of North Xinjiang, China. World J Gastroenterol 16(5):641–647
95. Brennan P, Lewis S, Hashibe M et al (2004) Pooled analysis of alcohol dehydrogenase genotypes and head and neck cancer: a HuGE review. Am J Epidemiol 159(1):1–16
96. Yokoyama A, Tsutsumi E, Imazeki H et al (2007) Contribution of the alcohol dehydrogenase-1B genotype and oral microorganisms to high salivary acetaldehyde concentrations in Japanese alcoholic men. Int J Cancer 121(5):1047–1054

97. Matsuo K, Oze I, Hosono S et al (2013) The aldehyde dehydrogenase 2 (ALDH2) Glu504Lys polymorphism interacts with alcohol drinking in the risk of stomach cancer. Carcinogenesis 34(7):1510–1515. doi:10.1093/carcin/bgt080
98. Nan HM, Park JW, Song YJ et al (2005) Kimchi and soybean pastes are risk factors of gastric cancer. World J Gastroenterol 11(21):3175–3181
99. Nan HM, Song YJ, Yun HY et al (2005) Effects of dietary intake and genetic factors on hypermethylation of the hMLH1 gene promoter in gastric cancer. World J Gastroenterol 11(25):3834–3841
100. Shin CM, Kim N, Cho SI et al (2011) Association between alcohol intake and risk for gastric cancer with regard to ALDH2 genotype in the Korean population. Int J Epidemiol 40(4):1047–1055. doi:10.1093/ije/dyr067
101. You WC, Hong JY, Zhang L et al (2005) Genetic polymorphisms of CYP2E1, GSTT1, GSTP1, GSTM1, ALDH2, and ODC and the risk of advanced precancerous gastric lesions in a Chinese population. Cancer Epidemiol Biomarkers Prev 14(2):451–458
102. Cao HX, Li SP, Wu JZ et al (2010) Alcohol dehydrogenase-2 and aldehyde dehydrogenase-2 genotypes, alcohol drinking and the risk for stomach cancer in Chinese males. Asian Pac J Cancer Prev 11(4):1073–1077
103. Zhang FF, Hou L, Terry MB et al (2007) Genetic polymorphisms in alcohol metabolism, alcohol intake and the risk of stomach cancer in Warsaw, Poland. Int J Cancer 121(9): 2060–2064
104. Duell EJ, Sala N, Travier N et al (2012) Genetic variation in alcohol dehydrogenase (ADH1A, ADH1B, ADH1C, ADH7) and aldehyde dehydrogenase (ALDH2), alcohol consumption and gastric cancer risk in the European Prospective Investigation into Cancer and Nutrition (EPIC) cohort. Carcinogenesis 33(2):361–367. doi:10.1093/carcin/bgr285
105. Chiang CP, Jao SW, Lee SP et al (2012) Expression pattern, ethanol-metabolizing activities, and cellular localization of alcohol and aldehyde dehydrogenases in human large bowel: association of the functional polymorphisms of ADH and ALDH genes with hemorrhoids and colorectal cancer. Alcohol 46(1):37–49. doi:10.1016/j.alcohol.2011.08.004
106. Matsuo K, Hamajima N, Hirai T et al (2002) Aldehyde dehydrogenase 2 (ALDH2) genotype affects rectal cancer susceptibility due to alcohol consumption. J Epidemiol 12(2):70–76
107. Murata M, Tagawa M, Watanabe S et al (1999) Genotype difference of aldehyde dehydrogenase 2 gene in alcohol drinkers influences the incidence of Japanese colorectal cancer patients. Jpn J Cancer Res 90(7):711–719
108. Yin G, Kono S, Toyomura K et al (2007) Alcohol dehydrogenase and aldehyde dehydrogenase polymorphisms and colorectal cancer: the Fukuoka Colorectal Cancer Study. Cancer Sci 98(8):1248–1253
109. Otani T, Iwasaki M, Hanaoka T et al (2005) Folate, vitamin B6, vitamin B12, and vitamin B2 intake, genetic polymorphisms of related enzymes, and risk of colorectal cancer in a hospital-based case-control study in Japan. Nutr Cancer 53(1):42–50
110. Miyasaka K, Hosoya H, Tanaka Y et al (2010) Association of aldehyde dehydrogenase 2 gene polymorphism with pancreatic cancer but not colon cancer. Geriatr Gerontol Int 10(Suppl 1):S120–6. doi:10.1111/j.1447-0594.2010.00616
111. Matsuo K, Wakai K, Hirose K et al (2006) A gene-gene interaction between ALDH2 Glu487Lys and ADH2 His47Arg polymorphisms regarding the risk of colorectal cancer in Japan. Carcinogenesis 27(5):1018–1023
112. Yang H, Zhou Y, Zhou Z et al (2009) A novel polymorphism rs1329149 of CYP2E1 and a known polymorphism rs671 of ALDH2 of alcohol metabolizing enzymes are associated with colorectal cancer in a southwestern Chinese population. Cancer Epidemiol Biomarkers Prev 18(9):2522–2527. doi:10.1158/1055-9965
113. Kuriki K, Hamajima N, Chiba H et al (2005) Relation of the CD36 gene A52C polymorphism to the risk of colorectal cancer among Japanese, with reference to with the aldehyde dehydrogenase 2 gene Glu487Lys polymorphism and drinking habit. Asian Pac J Cancer Prev 6(1):62–68
114. Gao CM, Takezaki T, Wu JZ et al (2008) Polymorphisms of alcohol dehydrogenase 2 and aldehyde dehydrogenase 2 and colorectal cancer risk in Chinese males. World J Gastroenterol 14(32):5078–5083

115. Landi S, Gemignani F, Moreno V et al (2005) A comprehensive analysis of phase I and phase II metabolism gene polymorphisms and risk of colorectal cancer. Pharmacogenet Genomics 15(8):535–546
116. Ferrari P, McKay JD, Jenab M et al (2012) Alcohol dehydrogenase and aldehyde dehydrogenase gene polymorphisms, alcohol intake and the risk of colorectal cancer in the European Prospective Investigation into Cancer and Nutrition study. Eur J Clin Nutr 66(12):1303–1308. doi:10.1038/ejcn.2012.173
117. Miyasaka K, Kawanami T, Shimokata H et al (2005) Inactive aldehyde dehydrogenase-2 increased the risk of pancreatic cancer among smokers in a Japanese male population. Pancreas 30(2):95–98
118. Kanda J, Matsuo K, Suzuki T et al (2009) Impact of alcohol consumption with polymorphisms in alcohol-metabolizing enzymes on pancreatic cancer risk in Japanese. Cancer Sci 100(2):296–302. doi:10.1111/j.1349-7006.2008.01044
119. Kato S, Tajiri T, Matsukura N et al (2003) Genetic polymorphisms of aldehyde dehydrogenase 2, cytochrome p450 2E1 for liver cancer risk in HCV antibody-positive Japanese patients and the variations of CYP2E1 mRNA expression levels in the liver due to its polymorphism. Scand J Gastroenterol 38(8):886–893
120. Tomoda T, Nouso K, Sakai A et al (2012) Genetic risk of hepatocellular carcinoma in patients with hepatitis C virus: a case control study. J Gastroenterol Hepatol 27(4):797–804. doi:10.1111/j.1440-1746.2011.06948
121. Sakamoto T, Hara M, Higaki Y et al (2006) Influence of alcohol consumption and gene polymorphisms of ADH2 and ALDH2 on hepatocellular carcinoma in a Japanese population. Int J Cancer 118(6):1501–1507
122. Munaka M, Kohshi K, Kawamoto T et al (2003) Genetic polymorphisms of tobacco- and alcohol-related metabolizing enzymes and the risk of hepatocellular carcinoma. J Cancer Res Clin Oncol 129(6):355–360
123. Ding J, Li S, Wu J et al (2008) Alcohol dehydrogenase-2 and aldehyde dehydrogenase-2 genotypes, alcohol drinking and the risk of primary hepatocellular carcinoma in a Chinese population. Asian Pac J Cancer Prev 9(1):31–35
124. Takeshita T, Yang X, Inoue Y et al (2000) Relationship between alcohol drinking, ADH2 and ALDH2 genotypes, and risk for hepatocellular carcinoma in Japanese. Cancer Lett 149(1–2):69–76
125. Koide T, Ohno T, Huang XE et al (2000) HBV/HCV infection, alcohol, tobacco and genetic polymorphisms for hepatocellular carcinoma in Nagoya, Japan. Asian Pac J Cancer Prev 1(3):239–245
126. Yu SZ, Huang XE, Koide T et al (2002) Hepatitis B and C viruses infection, lifestyle and genetic polymorphisms as risk factors for hepatocellular carcinoma in Haimen, China. Jpn J Cancer Res 93(12):1287–1292
127. Shibata A, Fukuda K, Nishiyori A et al (1998) A case-control study on male hepatocellular carcinoma based on hospital and community controls. J Epidemiol 8(1):1–5
128. Ohhira M, Fujimoto Y, Matsumoto A et al (1996) Hepatocellular carcinoma associated with alcoholic liver disease: a clinicopathological study and genetic polymorphism of aldehyde dehydrogenase 2. Alcohol Clin Exp Res 20(9 Suppl):378A–382A
129. Yamagishi Y, Horie Y, Kajihara M (2004) Hepatocellular carcinoma in heavy drinkers with negative markers for viral hepatitis. Hepatol Res 28(4):177–183
130. Minegishi Y, Tsukino H, Muto M et al (2007) Susceptibility to lung cancer and genetic polymorphisms in the alcohol metabolite-related enzymes alcohol dehydrogenase 3, aldehyde dehydrogenase 2, and cytochrome P450 2E1 in the Japanese population. Cancer 110(2):353–362
131. Eom SY, Zhang YW, Kim SH et al (2009) Influence of NQO1, ALDH2, and CYP2E1 genetic polymorphisms, smoking, and alcohol drinking on the risk of lung cancer in Koreans. Cancer Causes Control 20(2):137–145. doi:10.1007/s10552-008-9225-7
132. Park JY, Matsuo K, Suzuki T et al (2010) Impact of smoking on lung cancer risk is stronger in those with the homozygous aldehyde dehydrogenase 2 null allele in a Japanese population. Carcinogenesis 31(4):660–665. doi:10.1093/carcin/bgq021

133. Zienolddiny S, Campa D, Lind H et al (2008) A comprehensive analysis of phase I and phase II metabolism gene polymorphisms and risk of non-small cell lung cancer in smokers. Carcinogenesis 29(6):1164–1169. doi:10.1093/carcin/bgn020
134. Choi JY, Abel J, Neuhaus T et al (2003) Role of alcohol and genetic polymorphisms of CYP2E1 and ALDH2 in breast cancer development. Pharmacogenetics 13(2):67–72
135. Kawase T, Matsuo K, Hiraki A et al (2009) Interaction of the effects of alcohol drinking and polymorphisms in alcohol-metabolizing enzymes on the risk of female breast cancer in Japan. J Epidemiol 19(5):244–250
136. Sangrajrang S, Sato Y, Sakamoto H et al (2010) Genetic polymorphisms in folate and alcohol metabolism and breast cancer risk: a case-control study in Thai women. Breast Cancer Res Treat 123(3):885–893. doi:10.1007/s10549-010-0804-4
137. Ribas G, Milne RL, Gonzalez-Neira A (2008) Haplotype patterns in cancer-related genes with long-range linkage disequilibrium: no evidence of association with breast cancer or positive selection. Eur J Hum Genet 16(2):252–60

Chapter 4
Alcohol and Cancer: An Overview with Special Emphasis on the Role of Acetaldehyde and Cytochrome P450 2E1

Helmut K. Seitz and Sebastian Mueller

Abstract The mechanisms by which chronic alcohol consumption enhances carcinogenesis include acetaldehyde (AA) generated by alcohol dehydrogenase and reactive oxygen species (ROS) generated predominantly by cytochrome P450 2E1 (CYP2E1), but also by other factors during inflammation. In addition, ethanol also alters epigenetics by changing DNA and histone methylation and acetylation. A loss of retinoic acid due to a CYP2E1-related enhanced degradation results in enhanced cellular proliferation and decreased cell differentiation. Changes in cancer genes and in signaling pathways (MAPK, RAS, Rb, TGFβ, p53, PTEN, ECM, osteopontin, Wnt) may also contribute to ethanol-mediated mechanisms in carcinogenesis. Finally, immunosuppression may facilitate tumor spread. In the present review major emphasis is led on AA and ROS. While AA binds to proteins and DNA generating carcinogenic DNA adducts and inhibiting DNA repair and DNA methylation, ROS results in lipid peroxidation with the generation of lipid peroxidation products such as 4-hydoxynonenal which binds to DNA-forming highly carcinogenic exocyclic DNA adducts. ROS production correlates significantly with CYP2E1 in the liver but also in the esophagus, and its generation can be significantly reduced by the specific CYP2E1 inhibitor clomethiazole. Finally, CMZ also inhibits alcohol-mediated nitrosamine-induced hepatocarcinogenesis.

Keywords Acetaldehyde • Reactive oxygen species • Cytochrome P450 2E1 • Retinoic acid • Epigenetics • Lipid peroxidation • 4-Hydroxynonenal • Clomethiazole • DNA adducts

H.K. Seitz, M.D., A.G.A.F. (✉)
Centre of Alcohol Research, University of Heidelberg, Germany and Salem Medical Centre, Heidelberg, Germany
e-mail: helmut_karl.seitz@urz.uni-heidelberg.de

S. Mueller
Centre of Alcohol Research, University of Heidelberg AND Department of Medicine, Salem Medical Centre, Zeppelinstr. 11-33, 69121 Heidelberg, Germany

V. Vasiliou et al. (eds.), *Biological Basis of Alcohol-Induced Cancer*, Advances in Experimental Medicine and Biology 815,
DOI 10.1007/978-3-319-09614-8_4

4.1 Introduction

4.1.1 Historic Background

The fact that alcohol is a risk factor for certain types of cancer is well established. The first report about such an association was published in Paris in 1910 by French pathologists about the significant correlation between heavy drinking (absinth), smoking, and esophageal cancer [1]. In the 1960s and 1970s, innumerous epidemiological studies have been performed demonstrating that chronic alcohol consumption is a risk factor for cancer of the upper aerodigestive tract (oropharynx, larynx, esophagus) and of the liver (when cirrhosis is present). Due to these new aspects, an alcohol and cancer workshop sponsored by the Division of Cancer Control and Rehabilitation of the National Cancer Institute and the Division of Extramural Research of the National Institute of Alcohol Abuse and Alcoholism has been held at the National Institute of Health (NIH) in Bethesda, Maryland, October 23–24, 1978.

In addition, in the 1980s and 1990s, two other targets of alcohol-mediated cancer occurred, namely, the colorectum and the female breast. Thus, a second workshop has then been held at the National Institute of Alcohol and Alcohol abuse in 2004 especially dealing with mechanisms of alcohol-mediated cancer. In 2007 the International Agency of Research on Cancer in Lyon (France) published the result of a one-week expert conference on alcohol and cancer. The experts came to the conclusion that alcoholic beverages are carcinogenic to humans and that the occurrence of malignant tumors of the oral cavity, pharynx, larynx, esophagus, liver, colorectum, and female breast is causally related to alcohol consumption. Finally, they stated that there is a substantial mechanistic evidence in humans deficient in *aldehyde dehydrogenase* that acetaldehyde derived from ethanol metabolism causes malignant esophageal tumors [2].

In 2010 a satellite congress of the European Society of Biomedical Research on Alcoholism (ESBRA) on alcohol and cancer was held at the German Cancer Research Centre (DKFZ) in Heidelberg, Germany. This was the first congress sponsored by ESBRA. One year later the NIAAA again focused on the topic of alcohol and cancer in a two-day meeting elaborating major fields of mechanisms to explain the carcinogenic effect of alcohol. This meeting was finally published in a book edited by Drs. S. Zakhari, V. Vasiliou, and M. Guo [3]. The second International Congress on Alcohol and Cancer in Breckenridge, Colorado, has been predominantly focused on new mechanistic aspects in alcohol-mediated carcinogenesis.

4.1.2 Epidemiology and Animal Experiments

Epidemiology has clearly identified alcohol as a carcinogen for the targets mentioned above. In addition, experimental studies in animals have shown that alcohol indeed is a carcinogen. When given to B6C3F1 mice for more than a hundred weeks as a 2.5 % and 5 % solution in drinking water, there was a significant dose-related trend to more hepatocellular cancers and hepatocellular adenomas in these animals [4]. When given as 10 of 15 % in drinking water for 25 months to ICA mice, 25 % more animals with papillary and medullary adenocarcinomas of the breast were detected [5]. When given as 15 and 20 % of alcohol in drinking water to C57/B6 APC min mice for 10 weeks, more intestinal tumors and more tumors in the distal small intestine were found as compared to controls [6], and it has to be mentioned that this strain of mice represents a genetic modal that resembles that of FAP in humans. Finally, when given as 10 % solution in drinking water livelong to Sprague-Dawley rats, more tumors of the oral cavity, the lips, the tongue, and the forestomach were found [7]. More recently, also liver tumors were detected after alcohol consumption without any carcinogen given [8]. In this context it is noteworthy that innumerous animal experiments have been performed with various procarcinogens and carcinogens administered to the animals with and without alcohol. The majority of these experiments found that animals treated with alcohol develop more tumors and mostly at a more rapid time interval [9].

4.2 Specific Mechanisms of Ethanol-Mediated Carcinogenesis

Table 4.1 summarizes various mechanisms involved in alcohol-associated carcinogenesis. In the present paper only acetaldehyde (AA) as a carcinogen as well as oxidative stress will be discussed, and Figure 4.1 illustrates the role of these important factors in ethanol-mediated carcinogenesis. In addition, a few comments will be made on retinoids. For other aspects it is referred to further review articles.

4.2.1 Acetaldehyde as a Carcinogen

4.2.1.1 Mechanisms

AA produced by various alcohol dehydrogenases (ADHs) is a toxin and carcinogen which rapidly binds to proteins and DNA. It is degradated by acetaldehyde dehydrogenases (ALDHs) to acetate, which is not toxic. AA has three major modes of action with respect to carcinogenesis:

1. It forms with DNA carcinogenic DNA adducts.
2. It inhibits DNA repair.
3. It has significant effects on epigenetics (DNA methylation).

Ad (1) AA forms DNA adducts leading to N2-ethyl-2′-deoxyguanosine (N2-Et-dG) which is rather a marker for chronic alcohol consumption than a carcinogenic lesion. On the other hand, data by Dr. Brooks show clearly that

Table 4.1 Specific mechanisms of alcohol-mediated carcinogenesis

Carcinogenesis
1. Acetaldehyde
2. Oxidative stress: induction of cytochrome P450 2E1 and reactive oxygen species
3. Retinoids
4. Epigenetic changes
5. Cancer genes and signaling pathways
Mitogenic signals: MAPK, RAS
Insensitivity to antigrowth signals: Rb and cell cycle control, TGFβ
Apoptosis: p53, PTEN
Angiogenesis: VEGF
Metastasis: cell adhesion, ECM, osteopontin
Developmental signaling pathways
6. Inflammation
7. Immunosuppression
8. Organ-specific mechanisms: cirrhosis, gastroesophageal reflux
Disease, estrogens

MAPK mitogen-activated protein kinase, *Rb* retinoblastoma protein, *TGFβ* transforming growth factor ß, *p53* tumor protein 53, *PTEN* phosphatase and tensin homolog, *VEGF* vascular endothelial growth factor, *ECM* extracellular matrix

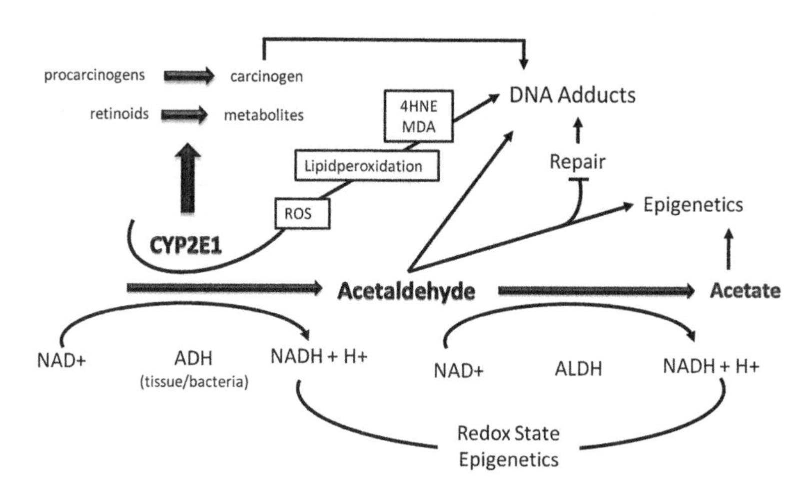

Fig. 4.1 The role of cytochrome P450 2E1 (CYP2E1) and acetaldehyde in ethanol-mediated carcinogenesis. Ethanol is metabolized via alcohol dehydrogenase (ADH) and CYP2E1 to acetaldehyde, and acetaldehyde is further metabolized via acetaldehyde dehydrogenase (ALDH) to acetate. Through the ADH and ALDH reaction reducing equivalence in the form of NADH are generated influencing the redox state of cell with major influence of epigenetics. Acetaldehyde by itself forms carcinogenic DNA adducts, inhibits DNA repair, and leads to epigenetic changes due to a decreased availability of methyl groups resulting in hypomethylation of DNA and histones. CYP2E1 is important in the metabolism of retinoids to their polar metabolites and of procarcinogens to their ultimate carcinogens. The loss of retinol and retinoic acids leads to dedifferentiation and hyperproliferation. In addition, ethanol metabolism via CYP2E1 results also in the generation of reactive oxygen species (ROS) with lipid peroxidation and the occurrence of lipid peroxidation products such 4-hydroxynonenal (4HNE) and malondialdehyde (MDA) which form highly carcinogenic exocyclic etheno DNA adducts. Finally, acetate may influence epigenetics due to hyperacetylation of histones

AA under basic conditions and in the presence of basic amino acids, histones and polyamines, generates 1,N2-propano-2′-deoxyguanosine (PDG) which is a carcinogenic lesion [10]. This will be discussed by Dr. Brooks in a separate article of this book.

Ad (2) AA inhibits DNA repair and has an inhibitory effect on 3N-A-DNA-G and on MMST-DNA-G HOGG1. In addition, AA binds to glutathione and inhibits, therefore, the antioxidative defense system (AODS) which is responsible for the detoxification of reactive oxygen species (ROS) and reactive nitrogen species (RNS). Furthermore, AA stimulates NFκB which inhibits apoptosis, a major feature of carcinogenesis. AA can either act directly or via interleukin 6 (Il6) or MCL2 [11, 12].

Ad (3) Methylation of DNA and histones is an important mechanism in epigenetics. Alcohol interferes with methyl transfer at various sites. First of all, ethanol leads to a decreased intake of folic acid [13]. Folic acid is an important factor in the methylation of homocysteine to become methionine. Deficiency in folic acid leads to an accumulation on homocysteine and to diminished generation of methionine. Furthermore, AA inhibits methyl adenosine transferase 1 (MAT 1) with the result of a reduced production of S-adenosyl methionine (SAMe), the final precursor of methyl transfer. Subsequently alcohol also inhibits DNA methyltransferase. As a side effect alcohol also decreases vitamin B6 which is involved in these processes. In conclusion, there is a diminished production of SAMe with a diminished methyl transfer and diminished methylation of DNA and histones associated with an accumulation of homocysteine, an important component to enhance endoplasmic reticulum stress [14]. In this context it has to be pointed out that in addition to DNA and histone hypomethylation due to the mechanisms discussed, a hyperacetylation of histones also occurs due to two factors: (1) an increased production of acetate following ethanol oxidation and (2) a change in the hepatic redox state with an increase in NADH and a decrease in NAD again due to ethanol oxidation. These are prerequisites to inhibit the activity of histone deacetylase (HDAC) [15].

4.2.1.2 Genetic Aspects of Acetaldehyde Accumulation

Genetic linkage studies with alcoholics have provided strong support for the assumption that AA plays a central role in alcohol-associated carcinogenesis. These studies found that individuals who accumulate AA because they carry certain alleles of the genes encoding ADH or ALDH have an increased cancer risk [16, 17]. For both the ADH1B and the ADH1C genes, several alleles exist that result in differences in the activity of the ADH molecules that they encode. For example, the ADH1B*2 allele encodes an enzyme that is approximately 40 times more active than the enzyme encoded by the ADH1B*1 allele. Similarly, the enzyme encoded by the ADH1C*1 allele is 2.5 times more active than the enzyme encoded by the ADH1C*2 allele [18]. Individuals who carry the highly active ADH1B*2 allele rapidly convert ethanol to AA. This leads to AA accumulation following alcohol consumption and results in toxic side effects, such as a flushing syndrome with sweating, accelerated heart rate, nausea, and vomiting. These adverse symptoms exert a protective effect against acute and chronic alcohol consumption. The ADH1B*2 allele is rarely found in Caucasians but occurs more frequently in Asian populations. It has recently been demonstrated that also the low activity ADH1B*1/1* genotype associates with enhanced exposure to AA through saliva. This might be due to lower systemic elimination rate of ethanol from the body which results in prolonged exposure to acetaldehyde produced by oral microbes [19]. There is also strong evidence that the combination of ALDH2 deficiency and slow ADH1B associates with the highest risk for esophageal cancer especially among heavy drinkers [20, 21]. The effects of the different ADH1C alleles on alcohol metabolism and, consequently, on drinking levels and alcohol-related carcinogenesis are more subtle. They can be studied best in Caucasian populations in which the highly active ADH1B*2 allele is rare. Studies on the relationship between ADH1C alleles and cancer occurrence in Caucasians have led to contradictory results which have been discussed elsewhere [22]. Harty and colleagues [23] were the first who compared the risk of oral cancer associated with various alcohol consumption levels in individuals homozygous for the more active ADH1C*1 allele with the risk in heterozygotes who carried only one copy of this allele or were homozygous for the less active ADH1C*2 allele. The study found that individuals who consumed eight or more drinks per day and were homozygous for the more active ADH1C*1 allele had a 40-fold increased risk for oral cancer compared with nondrinkers. In contrast, people who consumed the same amount of alcohol but who were heterozygous or homozygous for the less active ADH1C* allele had only a four- to sevenfold increased risk compared with nondrinkers.

We determined ADH1C polymorphisms in more than 400 heavy drinkers with daily alcohol intake of more than 60 g and various cancers of the upper aerodigestive tract, liver, and breast [24–27]. Cases of cancer patients were compared with carefully matched control patients with alcohol-related diseases (e.g., cirrhosis of the liver, pancreatitis, and alcohol dependence) but no cancer. Cancer patients and control subjects were of similar age and had similar histories of alcohol consumption and cigarette smoking. In this study, significantly more patients with alcohol-related cancers either had at least one ADH1C*1 allele or were homozygous for ADH1C*1 than did patients with other alcohol-related diseases. Statistical analyses determined a significant association between ADH1C*1 allele frequency and rate of homozygosity and an increased risk for alcohol-related cancer ($P<0.001$). Finally, individuals homozygous for ADH1C*1 had a relative risk of developing esophageal, liver, and head and neck cancers of 2.9, 3.6, as well as 2.2, respectively, compared with people homozygous for ADH1C*2. We also found that individuals who are homozygous for the ADH1C*1 allele had significantly higher AA levels in their saliva than did heterozygous individuals or individuals who are homozygous for the ADH1C*2 allele [24]. Saliva rinses the mucosa of the upper aerodigestive tract, and any AA in the saliva may be taken up by mucosal cells. Moreover, mucosal cells display little ALDH2 enzyme activity and therefore cannot efficiently detoxify AA. As a result, AA may bind to proteins and DNA in the mucosal cells and may initiate carcinogenesis. The hypothesis that AA in the saliva contributes to tumor development is supported by the observation that AA-fed rats with intact salivary glands showed excessive proliferation of the upper gastrointestinal mucosa [28], similar to the changes observed following chronic alcohol consumption [29]. When the glands were surgically removed (i.e., when the animals no longer produced saliva), however, this excessive cell proliferation disappeared [30].

In conclusion, the evidence for the involvement for AA in ethanol-associated cancer can be summarized as follows:

1. High AA levels occur in the saliva and in the colon following alcohol consumption.
2. AA is a toxin which binds to proteins leading to structural and functional alterations of the cell with epigenetic (disturbed DNA methylation and DNA reparation) and genetic effects. AA forms mutagenic and carcinogenic DNA adducts especially in hyperregenerative tissues.
3. Elevated AA levels in experimental carcinogenesis accelerate cancer development in the colon.
4. Data in humans demonstrate high risk for upper aerodigestive tract and colorectal cancer in individuals who accumulate AA due to decrease detoxification (ALDH2 deficiency) or increased production (ADH1c1 homozygosity).

4.2.2 Alcohol and Oxidative Stress

4.2.2.1 Mechanisms of ROS and Cytochrome P450 2E1 in Alcohol-Mediated Carcinogenesis

Various factors contributed to oxidative stress in alcoholic liver disease (ALD) including inflammation and the invasion of neutrophils and macrophages. However, one major factor is the induction of cytochrome P450 2E1 (CYP2E1) after chronic alcohol consumption. This induction varies *interindividually* and occurs already at an alcohol dose of 40 g per day as well as already at a time interval of one week [31]. As longer as alcohol is consumed as higher is the induction. However, it is important to note that some individuals have no CYP2E1 induction at all. It is believed that individuals who have a strongly induced CYP2E1 status have an increased risk for ALD due to the production of ROS and their consequences. After alcohol withdrawal CYP2E1 disappears rapidly within days [31]. Approximately 10 to 15 % of alcohol oxidation occurs via CYP2E1. However, in the induced state, more than 30 % may be metabolized via CYP2E1. In this situation ROS occurs. ROS may lead to lipid peroxidation with a generation of lipid peroxidation products such as malondialdehyde and 4-hydroxynonenal (4HNE) [11].

Furthermore, CYP2E1 is involved in the activation of various xenobiotics and procarcinogens to their ultimate carcinogenic intermediates which is important in situations where carcinogens are consumed together with alcohol [11]. For example, aflatoxins occur in certain food products contaminated with fungi. These aflatoxins are activated through CYP2E1 which is induced after alcohol consumption [32]. CYP2E1 is also involved in drug metabolism including paracetamol and isoniazid. In this context it is referred to other review articles [33].

Finally, CYP2E1 catalyzes the metabolism of retinol and retinoid acid (RA) leading to a decrease in RA concentrations associated with loss of cell differentiation and cellular hyperregeneration [34]. These low levels in RA lead to a decrease of the RXR and RAR receptor as well as to a decrease in MKP-1 and an increase of JNK. At the same time CYP2E1 leads to oxidative stress which by itself increases JNK leading to an increased phosphorylation of c-fos and c-jun, the proteins of AP1 [35]. Both increased AP1 expression and decreased RXR and RAR receptor regulate cell proliferation and apoptosis and lead to an increased cellular proliferation which stimulates carcinogenesis. On the other hand, if CYP2E1 is inhibited by CMZ, oxidative stress decreases, RA increases, and carcinogenesis is prevented [36]. It has been shown that the supplementation of vitamin A or RA is not helpful in preventing carcinogenesis since both compounds in the presence of an induced CYP2E1 are metabolized to polar retinoic metabolites which have apoptotic properties [37].

4-HNE is well known to form exocyclic ethanol DNA adducts with DNA which are highly carcinogenic. It has been shown in CYP2E1 knockout mice that carcinogenic DNA lesions occur more frequently [38]. In recent experiements we could show that in CYP2E1 overexpressing HepG2 cells an increased exocyclic ethanol adduct formation with increasing concentrations of alcohol exists an increased exocyclic ethanol adduct formation with increasing concentrations of alcohol [39]. The occurrence of these adducts could be diminished significantly in the presence of clomethiazole (CMZ), a specific CYP2E1 inhibitor. These cell experiments could be extended to the human situation [39]. In biopsies from patients with ALD a significant correlation between CYP2E1, 4-HNE and exocyclic ethanol adducts of adenine and cytosine was found. In a recent nonpublished study, we have examined 89 patients with various stages of non-cirrhotic ALD, and we clearly found that CYP2E1 was not associated with the amount of alcohol consumed which is not surprising, keeping in mind the fact that the induction of CYP2E1 is interindividually different. However, there was a significant correlation between etheno-DNA adduct formation and CYP2E1 induction (Seitz, personal communication).

CYP2E1 also plays a role in DNA lesions in esophageal biopsies from patients with esophageal cancer after chronic alcohol consumption. We could show that CYP2E1 correlates with exocyclic etheno-DNA adducts in esophageal tissue and in contrast to the liver with the amount of alcohol consumption [40]. It has to be mentioned that drinking and smoking enhance esophageal cell proliferation significantly which is a risk situation in carcinogenesis.

4.2.2.2 The Role of CYP2E1 in Experimental Carcinogenesis

It has been shown that CYP2E1 is critical for the accumulation of apurinic and apyrimidinic DNA sites in CYP2E1 knockout mice as compared to wild-type mice [38]. This was the first hint that CYP2E1 may be involved in carcinogenesis directly. A study by Lu et al. showed that the administration of CMZ, a specific CYP2E1 inhibitor, to an alcohol diet inhibits both hepatic ROS generation and hepatic nitrogen species produced by alcohol [41]. To further study this effect, we applied alcohol in a carcinogenesis model. Sprague-Dawley rats received one single dose of dimethylnitrosamine (20 mg/kg bwt) to induce hepatic tumors. The animals received Lieber-DeCarli diets for one month and for ten months either with or without CMZ. After 1 month CYP2E1 was significantly induced by ethanol but did not increase in the presence of CMZ. In addition, alcohol increased cell proliferation, NFκB, and TNFα mRNA and decreased hepatic apoptosis. None of the effects have been observed when CMZ was added to the alcohol diet [42, 43]. In addition, the number of p-GST-positive hepatic foci induced by the nitrosamine was found to be significantly increased in the ethanol-fed rats but not in the presence of CMZ.

After 10 months various histological features have been observed including hepatic amphophilic foci of cellular alteration, nodular regenerative hyperplasia, and hepatocellular adenoma. Four out of five animals receiving alcohol in addition to dimethylnitrosamine revealed adenomas after 10 months, while none of six animals receiving ethanol with CMZ had adenoma. CYP2E1 was found to be significantly increased following the alcohol diet but not when CMZ was administered simultaneously [43]. Most recently, Tsuchishima et al. demonstrated also that chronic ingestion of ethanol induces hepatocellular carcinoma in mice without any carcinogen and that this was associated with the induction of CYP2E1 [8].

4.3 CYP2E1 and Its Role in Nonalcoholic Fatty Liver Diseases

It is well known that CYP2E1 is also induced in nonalcoholic fatty liver disease (NAFLD) [44] due to either free fatty acids or acetone. Dey and Cederbaum have clearly shown that the induction of cytochrome P450 2E1 promotes liver injury in ob/ob mice, an experimental model for NASH [45]. We studied the effect of alcohol on top of a dietary-induced fatty liver [46]. Sprague-Dawley rats received a 71 % high-fat diet to induce NASH for 4 weeks. Thereafter, the animals were divided into two groups: group I received a Lieber-DeCarli diet with 17 % ethanol, while group II received carbohydrates. The ethanol content of the diet was less than half of a regular Lieber-DeCarli diet. Both the high-fat diet with or without ethanol increased hepatic oxidative stress, hepatic Bax protein, hepatic TNF-α mRNA, and hepatic TNFR1 mRNA and decreased hepatic Bcl-2 protein as compared to controls. Hepatic apoptosis, hepatic Fas mRNA, and hepatic FasL mRNA were found to be further significantly increased in the ethanol-fed group as compared to the high-fat group. Histologically the liver of the animals fed with high-fat diet with alcohol showed an increase in inflammatory foci. The conclusion from this study is that moderate alcohol consumption aggravates high-fat diet-induced steatohepatitis in rats.

In addition, we were also using obese Zucker rats (fa/fa) and their lean littermates as a genetic model for NASH. These animals are leptin deficient and insulin resistant. A significant increase in hepatic CYP2E1 was shown in obese animals as compared to lean animals and furthermore in obese animals after ethanol administration as compared to lean animals following alcohol ingestion. This increase in CYP2E1 was paralleled by the generation of exocyclic etheno DNA adducts [39].

Most recently we investigated CYP2E1 and exocyclic etheno DNA adducts in children and adolescents with NASH. In 3 out of 21 liver biopsies from these children, we found significant levels of etheno-DNA adducts. These adduct levels did not correlate with CYP2E1 which is not surprising since in NASH additional inflammatory processes take place which may lead to exocyclic etheno adduct formation. In this context it is noteworthy that Ascha and coworkers [47] identified alcohol as an important risk factor for hepatocellular cancer in NASH patients. The risk ratio was found to be 3.6 for any alcohol level consumed. Therefore, the conclusion was stated in the American Journal of Gastroenterology that "until further data from rigorous prospective studies become available people with NAFLD should avoid alcohol of any type or amount" [48].

The role of CYP2E1 in ethanol-mediated carcinogenesis can be summarized as follows [49]:

The *induction of CYP2E1* by chronic ethanol consumption results in:

1. An enhanced metabolism of drugs and xenobiotics resulting in either low blood drug levels or increased drug toxicity
2. An increased activation of various dietary and tobacco procarcinogens to their ultimate carcinogenic metabolites
3. An enhanced metabolism of retinol and retinoic acid associated with the activation of the AP-1 gene resulting in hyperregeneration
4. The generation of ROS with lipid peroxidation and the occurrence of highly carcinogenic DNA lesions in man
5. An enhancement of liver pathology induced by chronic ethanol ingestion (hepatic steatosis and fibrosis)

The *inhibition of CYP2E1* by chlormethiazole results in:

1. An improvement of experimental alcoholic liver disease
2. A significant reduction of carcinogenic DNA lesions in cell cultures and animal experiments induced by ethanol
3. A normalization of retinol and retinoic acid levels and thus cell cycle behavior
4. An inhibition of nitrosamine-induced hepatic carcinogenesis in rats

Acknowledgment Original studies were supported by the Dietmar Hopp and Manfred Lautenschläger Foundations.

References

1. Lamu L (1910) Etude de statistique clinique de 131 cas de cancer de l'oesophage et due cardia. Arch Mal Appar Dig Mal Nutr 4:451–456
2. Baan R, Straif K, Grosse Y et al (2007) Carcinogenicity of alcoholic beverages. Lancet Oncol 8:292–293

3. Zakhari S, Vasiliou V, Max Guo Q (eds) (2011) Alcohol and cancer. Springer, New York
4. Beland FA, Benson RW, Mellick PW et al (2005) Effect of ethanol on the tumorigenicity of urethane (ethyl carbamate) in B6C3F1 mice. Food Chem Toxicol 43:1–19
5. Watabiki T, Okii Y, Tokiyasu T et al (2000) Long-term ethanol consumption in ICR mice causes mammary tumor in females and liver fibrosis in males. Alcohol Clin Exp Res 24:117S–122S
6. Roy HK, Gulizia JM, Karolski WJ et al (2002) Ethanol promotes intestinal tumorigenesis in the MIN mouse. Multiple intestinal neoplasia. Cancer Epidemiol Biomarkers Prev 11:1499–1502
7. Soffritti M, Belpoggi F, Cevolani D et al (2002) Results of long-term experimental studies on the carcinogenicity of methyl alcohol and ethyl alcohol in rats. Ann N Y Acad Sci 982:46–69
8. Tsuchishima M, George J, Shiroeda H et al (2013) Chronic ingestion of ethanol induces hepatocellular carcinoma in mice without additional hepatic insult. Dig Dis Sci 58:1923–1933
9. WHO (2010) IARC monographs on the evaluation of carcinogenic risks to humans, vol 96, Alcohol consumption and ethyl carbamate. IARC, France
10. Theruvathu JA, Jaruga P, Nath RG et al (2005) Polyamines stimulate the formation of mutagenic 1, N2-propanodeoxyguanosine adducts from acetaldehyde. Nucleic Acids Res 33:3513–3520
11. Seitz HK, Stickel F (2007) Molecular mechanisms of alcohol-mediated carcinogenesis. Nat Rev Cancer 7:599–612
12. Seitz HK, Stickel F (2006) Risk factors and mechanisms of hepatocarcinogenesis with special emphasis on alcohol and oxidative stress. Biol Chem 387:349–360
13. Seitz HK, Mueller S (2013) Alcohol: metabolism, toxicity and its impact on nutrition. In: Dulbecco & Abelson (eds) Encyclopedia of Human Biology, 3rd Edition, Elsevier (in press)
14. Stickel F, Herold C, Seitz HK et al (2006) Alcohol and methyl transfer: Implications for alcohol related hepatocarcinogenesis. In: Ali S, Friedman SL, Mann DA (eds) Liver diseases: biochemical mechanisms and new therapeutic insights. Science, Enfield, pp 45–58
15. Kirpich I, Ghare S, Zhang J et al (2012) Binge alcohol-induced microvesicular liver steatosis and injury are associated with down regulation of hepatic Hdac 1, 7,9,10,11 and up regulation of Hdac 3. Alcohol Clin Exp Res 36:1578–1586
16. Seitz HK, Matsuzaki S, Yokoyama A et al (2001) Alcohol and cancer. Alcohol Clin Exp Res 25:137S–143S
17. Yokoyama A, Muramatsu T, Ohmori T et al (1998) Alcohol-related cancers and aldehyde dehydrogenase-2 in Japanese alcoholics. Carcinogenesis 19:1383–1387
18. Bosron WF, Ehrig T, Li TK (1993) Genetic factors in alcohol metabolism and alcoholism. Semin Liver Dis 13:126–135
19. Yokoyama A, Tsutsumi E, Imazeki H et al (2007) Contribution of the alcohol dehydro-genase-1B genotype and oral microorganisms to high salivary acetaldehyde concentrations in Japanese alcoholic men. Int J Cancer 121:1047–1054
20. Lee CH, Lee JM, Wu DC et al (2008) Carcinogenetic impact of ADH1B and ALDH2 genes on squamous cell carcinoma risk of the esophagus with regard to the consumption of alcohol, tobacco and betel quid. Int J Cancer 122:1347–1356
21. Yang SJ, Wang HY, Li XQ et al (2007) Genetic polymorphisms of ADH2 and ALDH2 association with esophageal cancer risk in southwest China. World J Gastroenterol 13:5760–5764
22. Seitz HK, Stickel F (2009) Acetaldehyde as an underestimated risk factor for cancer development: role of genetics in ethanol metabolism. Genes Nutr 5:121–128
23. Harty LC, Caporaso NE, Hayes RB et al (1997) Alcohol dehydrogenase 3 genotype and risk of oral cavity and pharyngeal cancers. J Natl Cancer Inst 89:1698–1705
24. Visapaa JP, Gotte K, Benesova M et al (2004) Increased cancer risk in heavy drinkers with the alcohol dehydrogenase 1C*1 allele, possibly due to salivary acetaldehyde. Gut 53:871–876
25. Homann N, Stickel F, Konig IR et al (2006) Alcohol dehydrogenase 1C*1 allele is a genetic marker for alcohol-associated cancer in heavy drinkers. Int J Cancer 118:1998–2002
26. Homann N, Konig IR, Marks M et al (2009) Alcohol and colorectal cancer: the role of alcohol dehydrogenase 1C polymorphism. Alcohol Clin Exp Res 33:551–556
27. Coutelle C, Höhn B, Benesova M et al (2004) Risk factors in alcohol-associated breast cancer: alcohol dehydrogenase polymorphisms and estrogens. Int J Oncol 25:1127–1132

28. Homann N, Karkkainen P, Koivisto T et al (1997) Effects of acetaldehyde on cell regeneration and differentiation of the upper gastrointestinal tract mucosa. J Natl Cancer Inst 89: 1692–1697
29. Simanowski UA, Suter P, Stickel F et al (1993) Esophageal epithelial hyperproliferation following long-term alcohol consumption in rats: effects of age and salivary gland function. J Natl Cancer Inst 85:2030–2033
30. Maier H, Weidauer H, Zoller J et al (1994) Effect of chronic alcohol consumption on the morphology of the oral mucosa. Alcohol Clin Exp Res 18:387–391
31. Oneta CM, Lieber CS, Li J et al (2002) Dynamics of cytochrome P4502E1 activity in man: induction by ethanol and disappearance during withdrawal phase. J Hepatol 36:47–52
32. Seitz HK, Osswald B (1992) Effect of ethanol on procarcinogen activation. In: Watson R (ed) Alcohol and cancer. CRC, Boca Raton, pp 55–72
33. Lieber CS (1997) Cytochrome P-4502E1: its physiological and pathological role. Physiol Rev 77:517–544
34. Liu C, Russell RM, Seitz HK et al (2001) Ethanol enhances retinoic acid metabolism into polar metabolites in rat liver via induction of cytochrome P4502E1. Gastroenterology 120:179–189
35. Wang XD, Liu C, Chung J et al (1998) Chronic alcohol intake reduces retinoic acid concentration and enhances AP-1 (c-Jun and c-Fos) expression in rat liver. Hepatology 28:744–750
36. Wang XD, Seitz HK (2004) Alcohol and retinoid Interaction. In: Watson R (ed) Nutrition and alcohol: linking nutrient interactions and dietary intake. CRC, Boca Raton, pp 313–321
37. Dan Z, Popov Y, Patsenker E et al (2005) Hepatotoxicity of alcohol-induced polar retinol metabolites involves apoptosis via loss of mitochondrial membrane potential. FASEB J 19:845–847
38. Bradford BU, Kono H, Isayama F, Kosyk O (2005) Cytochrome P450 CYP2E1, but not nicotinamide adenine dinucleotide phosphate oxidase, is required for ethanol-induced oxidative DNA damage in rodent liver. Hepatology 41:336–344
39. Wang Y, Millonig G, Nair J et al (2009) Ethanol-induced cytochrome P4502E1 causes carcinogenic etheno-DNA lesions in alcoholic liver disease. Hepatology 50:453–461
40. Millonig G, Wang Y, Homann N et al (2011) Ethanol-mediated carcinogenesis in the human esophagus implicates CYP2E1 induction and the generation of carcinogenic DNA-lesions. Int J Cancer 128:533–540
41. Lu Y, Zhuge J, Wang X et al (2008) Cytochrome P450 2E1 contributes to ethanol-induced fatty liver in mice. Hepatology 47:1483–1494
42. Chavez PR, Lian F, Chung J, Liu C, Paiva SA, Seitz HK et al (2011) Long-term ethanol consumption promotes hepatic tumorigenesis but impairs normal hepatocyte proliferation in rats. J Nutr 141:1049–1055
43. Ye Q, Lian F, Chavez PR, Chung J, Ling W, Qin H et al (2012) Cytochrome P450 2E1 inhibition prevents hepatic carcinogenesis induced by diethylnitrosamine in alcohol-fed rats. Hepatobiliary Surg Nutr 1:5–18
44. Weltman MD, Farrell GC, Hall P et al (1998) Hepatic cytochrome P450 2E1 is increased in patients with nonalcoholic steatohepatitis. Hepatology 27:128–133
45. Dey A, Cederbaum AI (2007) Induction of cytochrome P450 2E1 promotes liver injury in ob/ob mice. Hepatology 45:1355–1365
46. Wang Y, Seitz H, Wang X (2010) Moderate alcohol consumption aggravates high-fat diet induced steatohepatitis in rats. Alcohol Clin Exp Res 34:567–573
47. Ascha MS, Hanouneh IA, Lopez R, Tamimi TA, Feldstein AF, Zein NN (2010) The incidence and risk factors of hepatocellular carcinoma in patients with nonalcoholic steatohepatitis. Hepatology 51:1972–1978
48. Liangpunsakul S, Chalasani N (2012) What should we recommend to our patients with NAFLD regarding alcohol use? Am J Gastroenterol 107:976–978
49. Seitz HK, Wang XD (2013) The role of cytochrome P450 2E1 in ethanol-mediated carcinogenesis. Subcell Biochem 67:131–143

Chapter 5
Implications of Acetaldehyde-Derived DNA Adducts for Understanding Alcohol-Related Carcinogenesis

Silvia Balbo and Philip J. Brooks

Abstract Among various potential mechanisms that could explain alcohol carcinogenicity, the metabolism of ethanol to acetaldehyde represents an obvious possible mechanism, at least in some tissues. The fundamental principle of genotoxic carcinogenesis is the formation of mutagenic DNA adducts in proliferating cells. If not repaired, these adducts can result in mutations during DNA replication, which are passed on to cells during mitosis. Consistent with a genotoxic mechanism, acetaldehyde does react with DNA to form a variety of different types of DNA adducts. In this chapter we will focus more specifically on N^2-ethylidene-deoxyguanosine (N^2-ethylidene-dG), the major DNA adduct formed from the reaction of acetaldehyde with DNA and specifically highlight recent data on the measurement of this DNA adduct in the human body after alcohol exposure. Because results are of particular biological relevance for alcohol-related cancer of the upper aerodigestive tract (UADT), we will also discuss the histology and cytology of the UADT, with the goal of placing the adduct data in the relevant cellular context for mechanistic interpretation. Furthermore, we will discuss the sources and concentrations of acetaldehyde and ethanol in different cell types during alcohol consumption in humans. Finally, in the last part of the chapter, we will critically evaluate the concept of carcinogenic levels of acetaldehyde, which has been raised in the literature, and discuss how data from acetaldehyde genotoxicity are and can be utilized in physiologically based models to evaluate exposure risk.

Keywords Acetaldehyde • DNA adducts • N^2-ethyldeoxyguanosine • Upper aerodigestive tract cancers

S. Balbo (✉)
Masonic Cancer Center, University of Minnesota, Minneapolis, MN 55455, USA
e-mail: balbo006@umn.edu

P.J. Brooks
Division of Metabolism and Health Effects, and Laboratory of Neurogenetics, National Institute on Alcohol Abuse and Alcoholism, Bethesda, MD 20892-9304, USA
e-mail: pjbrooks@mail.nih.gov

V. Vasiliou et al. (eds.), *Biological Basis of Alcohol-Induced Cancer*,
Advances in Experimental Medicine and Biology 815,
DOI 10.1007/978-3-319-09614-8_5

5.1 Introduction

The designation of alcohol as carcinogenic to humans (Group 1) by the International Agency for Research on Cancer (IARC) represented an important change in our understanding of the health effects of alcohol consumption [1]. While previous IARC working groups had classified the carcinogenicity of alcoholic beverages, they left open the possibility that the carcinogenic effects resulted from contaminants in the alcoholic beverages, rather than alcohol itself. Thus, the important question of the carcinogenicity of alcohol per se was not definitively addressed. The notable aspect of the 2007 working group meeting was that alcohol (ethanol) itself was identified as carcinogenic to humans [2]. This classification therefore allows the scientific community to focus on the mechanistic question of how alcohol in alcoholic beverages increases the risk of cancers at certain sites in the body. Given the diversity of target tissues for alcohol-related carcinogenicity (liver, female breast, colorectum, upper aerodigestive tract), it is possible, and indeed likely, that different mechanisms are involved at different target tissues.

An obvious possible mechanism for the carcinogenicity of alcohol, at least in some tissues, involves the metabolism of ethanol to acetaldehyde. Redressing an oversight from the 2007 monograph, the 2009 IARC working group concluded that "acetaldehyde associated with the consumption of alcoholic beverages is carcinogenic to humans (Group 1)" [3]. This conclusion was based in a large part on the dramatically elevated risk of esophageal cancer from alcohol drinking in individuals who are unable to metabolize acetaldehyde due to a genetic variant in ALDH2 [4–6]. Based on these and other data, the strongest evidence for a causative role for acetaldehyde is for alcohol-related cancers of the UADT. The UADT includes the oral cavity, larynx, pharynx, and esophagus.

It is worth emphasizing here that the IARC Group 1 classification specifically applies to acetaldehyde *associated with the consumption of alcoholic beverages*. Acetaldehyde alone remains classified as Group 2b, possibly carcinogenic to humans. We will return to this topic in the last part of this chapter focusing on carcinogenic levels of acetaldehyde.

Broadly speaking, there are two mechanistically different types of carcinogens: genotoxic and non-genotoxic [7]. Genotoxic carcinogens react directly to chemically modify the DNA, resulting in the increased rate of mutagenesis and therefore increased rate of carcinogenesis. Well-known examples of genotoxic carcinogens are ultraviolet light, components of cigarette smoke, and aflatoxin. In contrast, non-genotoxic carcinogens increase the risk of cancer by mechanisms that do not involve direct DNA damage. Examples of non-genotoxic carcinogenic mechanisms include inflammation, which can result in DNA damage from inflammatory mediators, and hormone-like effects. The two mechanisms are not mutually exclusive. Notably, the IARC carcinogen classifications encompass both genotoxic and non-genotoxic agents. As stated in the preamble to the IARC *Monographs:* "… an agent is termed 'carcinogenic' if it is capable of increasing the incidence of malignant neoplasms, reducing their latency, or increasing their severity or multiplicity." This broad and mechanism-independent aspect of the IARC classification system is intentional;

"The aim of the *Monographs* has been, from their inception, to evaluate evidence of carcinogenicity at any stage in the carcinogenesis process, independently of the underlying mechanisms."

From a mechanistic standpoint, genotoxic and non-genotoxic mechanisms may have different time courses and other significant differences with practical implications for risk assessment and disease prevention (e.g., linear risk extrapolation versus thresholds for exposure; see [7]). Therefore, the main focus of this chapter will be on efforts made and strategies developed to assess the role of direct genotoxicity in the carcinogenic effect of acetaldehyde. Consistent with a genotoxic mechanism, acetaldehyde can react with DNA to form a variety of different types of DNA adducts. Since the general topic of acetaldehyde-DNA adducts was covered in a recent review [8], here we will focus more specifically on the major DNA adduct formed from the reaction of acetaldehyde with DNA, N^2-ethylidene-deoxyguanosine (N^2-ethylidene-dG) and highlight recent data on the measurement of this DNA adduct in the human body after alcohol exposure. Because the results are of particular biological relevance for alcohol-related cancer of the UADT, we will also focus on the histology and cytology of these target tissues, with the goal of placing the adduct data in the relevant cellular context for mechanistic interpretation. We also discuss the sources and concentrations of acetaldehyde and ethanol in different cell types during alcohol consumption in humans. Finally, we will critically evaluate the concept of carcinogenic levels of acetaldehyde, which has been raised in the literature [9–11], and discuss how data from acetaldehyde genotoxicity are utilized to identify exposure risk.

5.2 DNA Adducts from Acetaldehyde and Alcohol

Acetaldehyde's genotoxic effect is attributable to its reactivity. The electrophilic nature of its carbonyl carbon results in reactions with DNA, generating DNA adducts [12]. The main reactions occur with deoxyguanosine (dG) followed by deoxyadenosine (dA) and then deoxycytosine (dC) [6, 13]. The binding of acetaldehyde to these nucleosides leads principally to the formation of a Schiff base on the exocyclic amino groups. The resulting imines are unstable at room temperature and neutral pH. However, these compounds can be stabilized using reducing agents, ultimately resulting in ethyl-adducts which are then easier to detect and to quantify.

The most abundant and well-studied acetaldehyde-DNA adduct is N^2-ethylidene-dG which can be stabilized by reduction to N^2-ethyl-dG. These adducts are illustrated in Fig. 5.1.

The instability of N^2-ethylidene-dG prevents direct investigation of its biological properties. In contrast, N^2-ethyl-dG is stable in aqueous solution, as well as under the conditions used for automated oligonucleotide synthesis. Therefore, most of the experimental data available for the biological effects of N^2-ethylidene-dG are inferred from experiments using N^2-ethyl-dG as a stable analog. This is a common approach in the field of DNA damage and mutagenesis. For example, abasic sites in DNA,

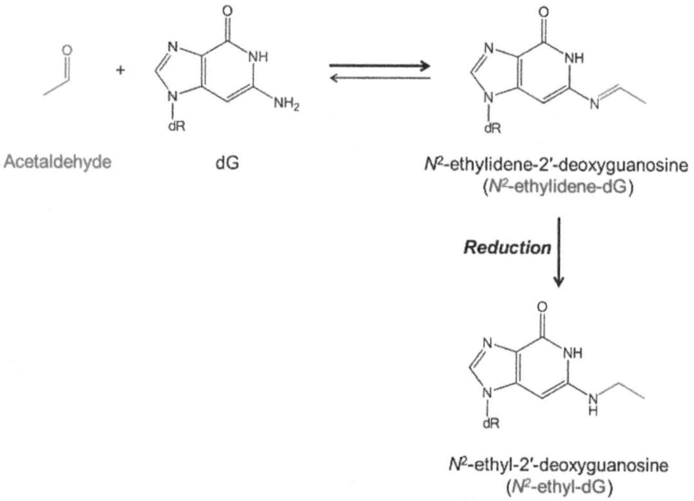

Fig. 5.1 The reaction of acetaldehyde with deoxyguanosine results in the formation of *N2-ethylidendene-dG*; this unstable Schiff base can be converted through a reduction step to the more stable form: N^2-ethyl-dG (dR = 2′-deoxyribose)

which result from depurination, are one of the most common forms of endogenous DNA damage [14]. Because authentic abasic sites are unstable, much of the information we have about their biological effects is derived from studies of tetrahydrofuran as a structural analog [15].

Studies in vitro indicate that N^2-ethyl-dG does not significantly inhibit the replicative DNA polymerase delta [16]. The effects of the lesion on the other major replicative DNA polymerase, epsilon, have not been directly assessed. However, in vivo studies in mammalian cells indicate that the lesion does block replication but is weakly mutagenic, causing primarily −1 frameshift deletion mutations [17–19]. In light of the discussion above, however, it is important to carefully evaluate the limitations of N^2-ethyl-dG as a structural analog.

Figure 5.2 shows energy-minimized models of an N^2-ethylidene-dG paired with dC, an N^2-ethyl-dG paired with dC, and an unmodified dG paired with dC. At first glance, all three models appear similar. The ethyl/ethylidene moiety can be accommodated in the minor groove, with no structural impediment to the guanosine base forming Watson–Crick type H bonds with the appropriate atoms on dC. However, one notable difference is that while dG and N^2-ethyl-dG each form three H bonds with dC, N^2-ethylidene-dG can only form two H bonds. The missing H atom is due to the presence of a double bond between the nitrogen atom and the carbon from acetaldehyde. Viewed from this perspective, the formation of N^2-ethylidene-dG essentially results in G:C base pair with the stability of an A:T base pair.

Studies of frameshift mutagenesis in experimental systems have documented that runs of A:T base pairs are prone to frameshift mutations resulting from a template dislocation mechanism [20]. This observation has generally been explained

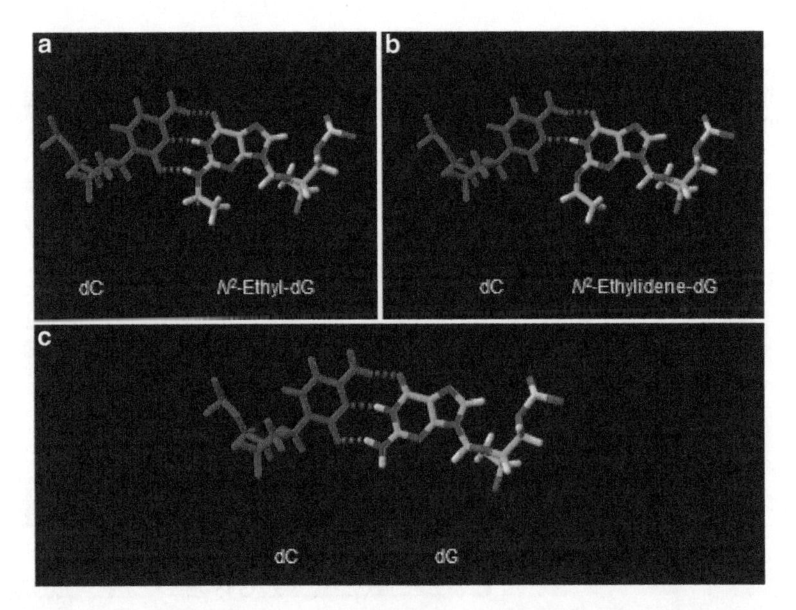

Fig. 5.2 Energy-minimized models showing the difference in base pairing of N^2-ethylidene-dG and N^2-ethyl-dG compared to deoxyguanosine. *Panel A*: the model shows N^2-ethylidene-dG paired with deoxycytosine (dC). *Panel B*: the model shows N^2-ethyl-dG paired with dC. *Panel C*: the model shows an unmodified dG paired with dC

by the reduced stability of A:T base pairs, which increases the probability of primer-template misalignment resulting in frameshift mutations. Adapting this model to N^2-ethylidene-dG in a run of G:C base pairs, an analogous mechanism for frameshift mutagenesis can be hypothesized (see Fig. 5.3). The important point is that it would not be possible to test this hypothesis using N^2-ethyl-dG as a model, because N^2-ethyl-dG forms three H bonds with dC (as shown in panel A of Fig. 5.2). However, this hypothesis could be tested in vivo, perhaps using a yeast strain with run of G:C base pairs in a mutational reporter gene. It should be pointed out, however, that runs of G:C base pairs are generally refractory to frameshift mutagenesis due to their inherent stability.

5.2.1 Other Acetaldehyde-DNA Adducts

In addition to N^2-ethylidene-dG, the most well-studied acetaldehyde-related DNA adducts are the crotonaldehyde-derived propano-dG (CrPdGs) adducts [21]. The condensation of two molecules of acetaldehyde can also produce a reactive electrophile, 3-hydroxybutanal (crotonaldehyde), which can also form a Schiff base on the same amino group of dG. These CrPdG adducts can have multiple biologic effects as a result of their ability to undergo a ring opening reaction. Ring opening yields another aldehyde moiety which can react with proteins to form DNA-protein

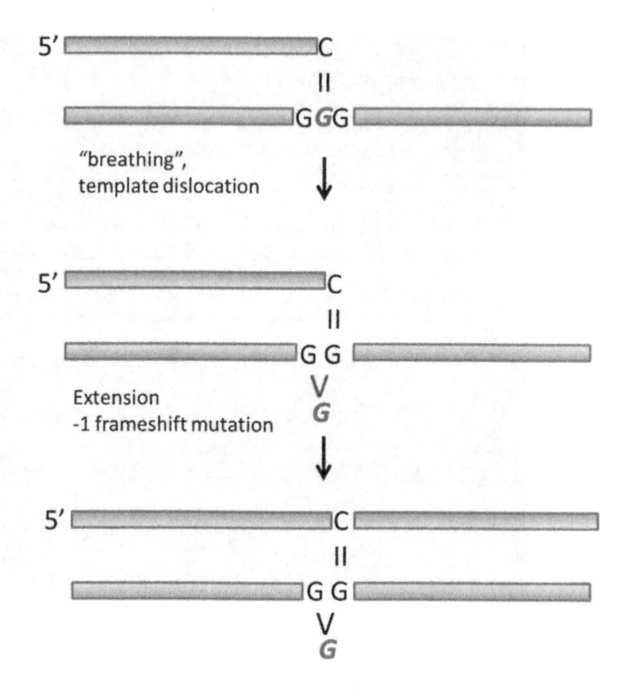

Fig. 5.3 Mechanisms of frameshift mutagenesis hypothesized for N^2-ethylidene-dG

cross-links or (in some sequence contexts) with deoxyguanosine on the opposite strand to form DNA-interstrand cross-links. The biological effects of these adducts have been reviewed recently [8].

It is also worth mentioning the early studies of Fraenkel-Conrat and colleagues, who showed that ethanol and acetaldehyde in combination could react with DNA bases to generate mixed acetal DNA adducts [22]. However, these adducts were very unstable, at least under the in vitro conditions investigated. As such, the biologic significance of these adducts, if any, is currently unclear.

In summary, acetaldehyde has been shown to form several DNA adducts, including N^2-ethylidene-dG and the CrPdG adducts. Under in vivo conditions, which are of the most direct relevance to human carcinogenesis, N^2-ethylidene-dG has been the most well studied, and therefore, we review these studies in the following paragraphs.

5.3 N^2-Ethylidene-dG as a Biomarker of DNA Damage Resulting from Acetaldehyde Derived from Ethanol

As mentioned above, the major reaction of acetaldehyde with DNA occurs on the exocyclic amino group of guanine forming N^2-ethylidene-dG. This adduct is stable in DNA but it easily breaks down when released as a nucleoside.

Fang et al. were the first to report the detection of this adduct, as its reduced form N^2-ethyl-dG, in leukocyte DNA of alcoholics. In their work a ^{32}P-postlabelling method was used for the adduct quantitation. Only samples from heavy drinkers showed detectable amounts of N^2-ethyl-dG likely resulting from the reduction of N^2-ethylidene-dG by endogenous reducing agents such as ascorbic acid and glutathione [23].

In order to avoid the degradation of N^2-ethylidene-dG during DNA hydrolysis and increase the sensitivity of the method, a new approach was developed by Wang et al. [24]. A reducing agent, $NaBH_3CN$, was introduced prior to DNA hydrolysis. Additionally, a stable isotope dilution method was used for the quantitation of the DNA adduct by liquid-chromatography-electrospray ionization-tandem mass spectrometry-selected reaction monitoring (LC-ESI-MS/MS-SRM). The use of this new method allowed a selective detection of N^2-ethyl-dG and an accurate quantitation. By reducing the degradation of the adduct through its conversion into a more stable compound, N^2-ethyl-dG was detected in DNA from all samples analyzed. This resulted in an increased sensitivity which allowed for the detection of lower levels of the DNA adduct, expanding its application beyond the quantitation in samples from heavy drinkers, and set the stage for the broader use of this adduct as a marker for acetaldehyde-induced DNA damage. Since then, N^2-ethyl-dG has been measured in DNA from various samples, for the investigation of the effects on DNA of acetaldehyde from different sources [25, 26].

N^2-Ethyl-dG has been used successfully to measure ethanol-induced DNA damage in HeLa cells expressing ADH1B, which corresponded to the activation of the Fanconi anemia-breast cancer susceptibility (FA-BRCA) DNA damage response network [27]. In a different study, the same DNA adduct was quantified to investigate ethanol-induced DNA damage in the brain of ethanol-treated mice. Higher levels of N^2-ethyl-dG were observed in brain DNA from mice exposed chronically and acutely to ethanol compared to controls [28]. These examples demonstrate that N^2-ethyl-dG is an extremely valuable tool for the investigation of DNA damage associated with acetaldehyde exposure from alcohol consumption and thus for the investigation of alcohol-related mechanisms of carcinogenesis.

The measurement of levels of N^2-ethyl-dG has indeed been crucial in studies focusing on the investigation of effects of alcohol exposure in ALDH2 deficiency. Several studies have used this adduct to detect the DNA damage induced by ethanol exposure in wild-type and *Aldh2* knockout mice, used as a model for *ALDH2* deficiency in humans. Increased levels of the adduct have been found in the liver, esophagus, tongue, and submandibular gland DNA of *Aldh2* knockout mice exposed to ethanol [29, 30]. These findings, together with the results from a study showing increased levels of N^2-ethyl-dG in peripheral blood of ALDH2-deficient alcoholics [31], contributed substantially to the evidence supporting an acetaldehyde-mediated mechanism in alcohol-related carcinogenesis. Together with the epidemiological data showing a dramatic increase of risk for esophageal and head and neck cancers in ALDH2-deficient drinkers, the results from the DNA damage studies contributed to the classification of acetaldehyde related to alcohol consumption as carcinogenic to humans (Group 1) by the IARC [3].

5.3.1 Experimental Studies of Acetaldehyde-DNA Adduct Formation from Alcohol Drinking in Humans

The studies described above clearly demonstrate that increased levels of acetaldehyde-DNA adducts can be observed in animals exposed to ethanol and in human alcohol abusers. However, the studies do not address the minimal amount of ethanol exposure necessary to increase DNA adduct levels or the time course or persistence of the adducts. Moreover, previous studies have not specifically investigated acetaldehyde-DNA adduct formation in humans in a known target tissue for alcohol-related carcinogenesis. Overall, little is known about the formation and lifetime of DNA adducts in the human body. Experimental studies have shown that with constant dosing a steady state concentration of DNA adducts will occur, where the number of new adducts formed equals the number of adducts lost due to repair or instability. However, repair processes vary depending on the cell type and remove different adducts with various efficiencies. Consequently, the lifetime of DNA adducts in vivo can be highly variable according to the tissue or cell type in which they are formed [32]. For instance, easily accessible surrogate tissues such as buccal cells or peripheral blood cells have very different lifetimes. In particular peripheral blood white cells include large cell subpopulations with major differences in lifespan. Lymphocytes are long-lived cells with a life span up to several years, while neutrophils are extremely short-lived cells with a life span of 2–3 days. The quantitation of N^2-ethylidene-dG in these cell types could potentially reflect very different exposure effects and provide very different information on the formation, accumulation, and elimination of the DNA adduct. Therefore, to address these important questions, Balbo et al. performed a biomonitoring study on human subjects before and after consumption of increasing amounts of ethanol [33, 34].

Ten healthy volunteers were recruited. Subjects were required to refrain from any alcohol consumption other than that administered for the study, starting from 1 week prior to the beginning of the experiment and throughout its entire duration. Three increasing alcohol doses were administered during the experiment, one dose a week, starting from the lowest. The alcohol dose administered was selected taking into account gender and weight in order to target specific blood alcohol levels, all below intoxication (defined as a blood alcohol level of 0.08 % [35]). The 3 doses selected for the study can roughly be described as corresponding to 1, 2, and 3 vodka drinks per subject. Overall, the alcohol doses had an ethanol content that ranged between 20 and 50 g which corresponded to an ethanol concentration in the drink ranging between 1.5 and 2.5 M. These concentrations ultimately resulted in a final blood alcohol concentration in the range of 0.01–0.02 M.

Levels of N^2-ethyl-dG were measured in DNA isolated from oral cells collected with a nonalcoholic mouthwash and from white blood cells. Granulocytes and lymphocytes were isolated from the blood to test potential effects of different cellular life span and repair mechanisms. A sample was collected before drinking to establish a baseline and then at several time points after exposure to each dose (2, 4, 6, 24, 48, and 120 h) to assess the kinetics of adduct formation and disappearance.

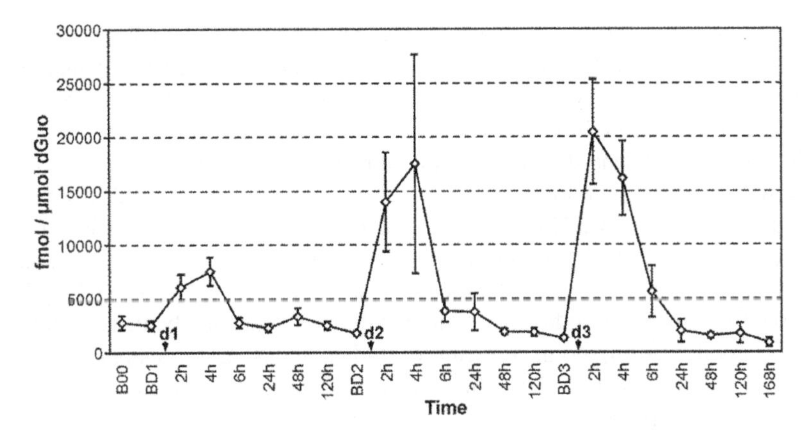

Fig. 5.4 Summary of the results obtained from a study investigating the effects on oral cell DNA of consumption of increasing doses of alcohol. The study was performed on samples collected form ten subjects who abstained from drinking any alcoholic beverage other than the doses provided over the entire duration of the study. The graph reports the mean levels of N^2-ethyl-dG (fmol/μmol dG) measured in oral cell DNA at various intervals before and after three increasing alcohol doses. The first time point reported on the *left* (B00) refers to 1 week before consumption of the first alcohol dose. Starting from this time point, participants began to abstain from consuming any alcoholic beverage. The next time point (BD1) refers to the baseline level detected 1 week later, 1 h before consumption of the first dose (d1, lowest). Subsequently, the graph shows the levels of N^2-ethyl-dG measured at the various time points considered after each dose (2–120 h). The DNA adduct levels were measured at the same time points before and after exposure to the next two doses (d2, intermediate, and d3, highest). Levels of the adduct increased 2 h after exposure even after consumption of the lowest dose and returned to baseline 24 h after exposure. A clear dose–response effect of alcohol on N^2-ethyl-dG levels was found. The baseline time points measured 1 h before the dose (BD1–BD3) are 7 days apart. Values are means and SEs. Data from reference [33, 34]

Considering the results from the oral cavity first, as shown in Fig. 5.4, increased levels of N^2-ethyl-dG were already detected at 2 h after alcohol consumption and reached a peak between 2 and 6 h. Interestingly, adduct levels had returned to baseline after 24 h, indicative of either DNA repair or cell turnover. We will return to this point in the discussion of the histology of the oral epithelia (discussed in detail below). Most importantly, peak adduct levels showed a clear dose–response relationship to the amount of alcohol consumed.

A different pattern of adduct formation was observed in white blood cells after alcohol drinking. Quantitation of the DNA adduct in granulocytes and lymphocytes did not show a major difference between the two cell types. An increase after the alcohol doses was detected, but in contrast to the oral cavity, no clear dose response was observed. Additionally, the high baseline levels and the high intra- and interindividual variability did not allow the clear identification of an effect directly attributable to the alcohol dose.

To our knowledge, this is the first study to investigate the effects of alcohol consumption on the time course of DNA adduct formation in healthy volunteers. All previous published studies on N^2-ethyl-dG levels in humans were done on heavy drinkers or alcoholics [23, 29]. Furthermore, no information on the persistence of this specific modification was reported. These results clearly demonstrate that even a single drink of alcohol results in a significant and dose-dependent increase in acetaldehyde-DNA adducts in cells in the human oral cavity, a known target tissue for alcohol-related carcinogenesis.

While N^2-ethylidene-dG is the major adduct formed after reaction of acetaldehyde with DNA, as mentioned above, several other adducts can result from acetaldehyde reactions with DNA, although generally they are formed in lower yield. Because of its high levels, N^2-ethylidene-dG is easier to detect and measure and thus could be considered as a general indicator of exposure of DNA to acetaldehyde and a proxy for detection of other DNA modifications. Consequently, these results provide a good starting point for future studies focusing on mapping multiple acetaldehyde-derived DNA modifications and investigating their potential role in the carcinogenic process.

These findings demonstrate the utility of oral cell DNA for the investigation of the role of alcohol-related DNA adducts in head and neck carcinogenesis. These observations support the hypothesis that alcohol drinking increases the risk of oral cancer via a mechanism involving a genotoxic effect of acetaldehyde. Before exploring this question in more detail, however, it is necessary to put the results from Balbo et al. into the relevant cellular context. For this purpose, below we briefly review the anatomy and histology of the oral cavity and esophagus.

5.3.2 Anatomical Considerations

A fundamental aspect of genotoxic carcinogenicity is the formation of mutagenic DNA adducts in proliferating cells. If left unrepaired, these adducts can result in mutations during DNA replication, which are passed on to daughter cells during mitosis. DNA adducts that form in terminally differentiated cells (G0) do not directly contribute to carcinogenesis, because of the absence of DNA replication. As such, it follows that a key issue for interpreting the relationship between DNA adduct formation and carcinogenesis is the cell type in which the DNA adduct formed.

The oral cavity and esophagus are both squamous epithelial tissues, in which cells in the upper layers are continuously replaced by new cells generated in the lower layers (for review see [36]). A schematic representation of these tissues is shown in Fig. 5.5. As indicated in the figure, proliferating cells (i.e., cells that replicate their DNA) are located in the basal and suprabasal layers of the epithelium. After differentiation, cells move up toward the surface of the epithelial layer. During transit, these cells flatten out (desquamate) and are ultimately sloughed off into the saliva or lumen of the esophagus. The time for newly generated cells in the oral cavity to transit from the basal layer through the surface of the (non-cornified) epithelial layer has been estimated at roughly 4 days (in rabbits) [22, 37]. As the DNA labeling index of cells in the basal layers of humans and rabbits is similar, the 4-day transit time is likely to be a reasonable estimate for humans as well.

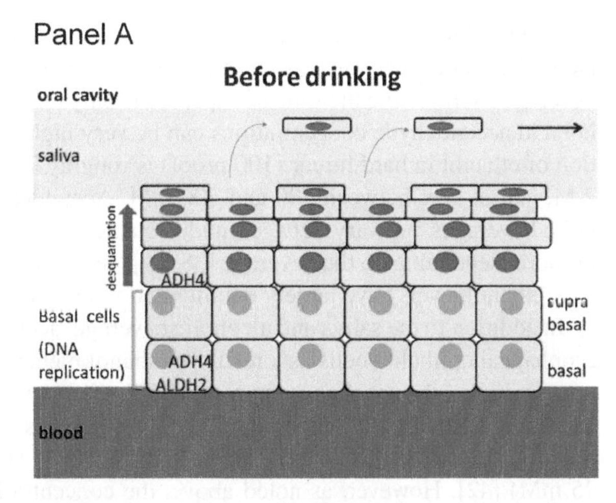

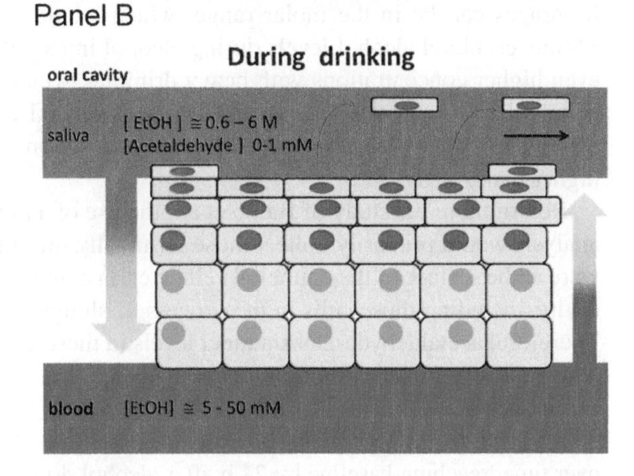

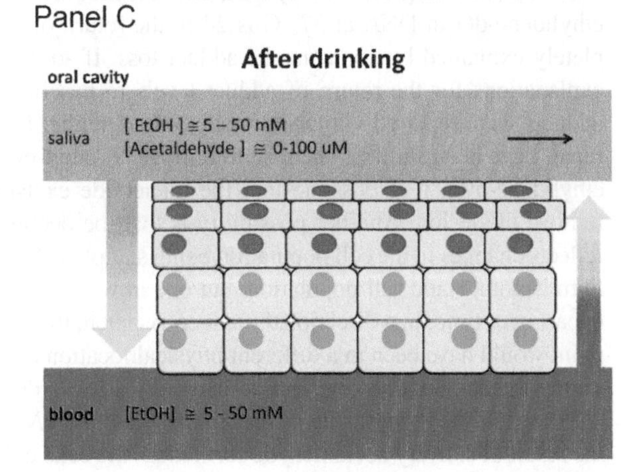

Fig. 5.5 Schematic representation of the squamous epithelial tissue of the oral cavity before, during, and after alcohol drinking. *Panel A* shows the various cell layers: cells in the upper layers are continuously replaced by new cells generated in the lower layers, the basal and suprabasal layers of the epithelium, where proliferating cells (i.e., cells that replicate their DNA) are located. After differentiation, cells move up toward the surface of the epithelial layer. During transit, these cells flatten out (desquamate) and are ultimately sloughed off into the saliva. *Panel B* shows the levels of exposure to alcohol and to acetaldehyde of the various levels of the epithelium when drinking alcohol. In addition to high concentrations of ethanol and acetaldehyde diffusing from the epithelial surface downward into the deeper cell layers (*brown arrow*), some alcohol reaches the blood stream from where it can diffuse into epithelial cells (*red arrow*), allowing metabolism to acetaldehyde in situ. Acetaldehyde levels in the blood are very low and considered negligible for this model. *Panel C* illustrates the levels of exposure to ethanol and acetaldehyde of the various layers of the epithelium after alcohol drinking when the ethanol concentration levels between the saliva and the blood stream reach the equilibrium

When drinking alcohol, the oral cavity and esophagus are transiently exposed to alcohol and acetaldehyde concentrations that are essentially the same as those in the beverage itself. While the time of exposure is only on the order of seconds, the alcohol and acetaldehyde concentrations can be very high. For example, the concentration of ethanol in hard liquor (100 proof) is roughly 7–8 M, in wine approximately 2 M, and in beer between 500 and 700 mM. Acetaldehyde concentrations in alcoholic beverages can vary between undetectable levels (vodka) to roughly 200 mM or more, depending on the beverage [38, 39]. By 30 min after alcohol drinking, salivary ethanol levels have largely equilibrated with blood levels [40, 41].

In addition to the saliva and alcoholic beverage, acetaldehyde is generated within esophageal epithelial cells as a result of ethanol metabolism. Cells of the upper GI tract including the esophagus express ADH7, as opposed to ADH1 proteins that are expressed in the liver. Compared to ADH1, which has a low Km for ethanol oxidation (on the order of 1 mM), the Km of ADH7 is substantially higher (around 25 mM) [42]. However, as noted above, the concentration of ethanol in alcoholic beverages can be in the molar range, which would clearly saturate the enzyme. Moreover, blood alcohol levels during alcohol intoxication could reach 25 mM, or even higher concentrations with heavy drinking, which would result in substantial metabolism to acetaldehyde by ADH7 in epithelial cells. Notably, the catalytic activity of ADH7 (K_{cat}) for acetaldehyde production is 1–2 orders of magnitude higher than ADH1 [42].

Returning to the study of Balbo et al., the use of a mouthwash to obtain cells for analysis would primarily collect those terminally differentiated epithelial cells that were at the surface of the epithelial cell layer, in contact with the contents of the oral cavity, including those cells in the process of sloughing off (see Fig. 5.5 panel A). Therefore, acetaldehyde-DNA adduct levels in these cells does not directly monitor adduct levels in the proliferating cell layers that are of greatest relevance to carcinogenesis.

An important aspect of the Balbo et al. findings is the reduction in adduct levels over time, reaching baseline by 24 h after alcohol drinking. Since the half-life of N^2-ethylidene-dG in DNA at 37 °C is 24 h, the return of adduct levels cannot be completely explained by spontaneous adduct loss. If so, there are at least two possible explanations for the return of adduct levels to baseline. One possibility is that the adducts were repaired via DNA repair. While neither base excision repair nor direct repair have been shown to be able to remove N^2-ethyl-dG (used as a surrogate for N^2-ethylidene-dG), it is possible that the nucleotide excision repair mechanism could remove the lesion. Another possibility is that the decline in adduct levels over time reflects changes in the cell population being sampled. As shown in Fig. 5.5, the epithelium is not a static cell population, but one in which cells are born, differentiate, and depart over time, in a directional manner. As such, the cells collected at the 24 h time point would have been in a different physical location relative to the epithelial surface during the alcohol drinking and immediately afterwards, when salivary acetaldehyde levels would be highest. Therefore, to the extent that N^2-ethylidene-dG adducts were the result of salivary acetaldehyde formation, the cells collected at the 24 h time point may have been at least partially protected from DNA damage by the overlying cells.

The two possibilities are not mutually exclusive, and both could be assessed experimentally. The role of NER could be readily tested by exposing normal and NER-deficient human cells to acetaldehyde, then assaying the disappearance of N^2-ethylidene-dG adducts over time in the two cell types. Assessing the kinetics of adduct formation in different cell layers of the oral epithelium in vivo is more technically challenging. Theoretically, antibodies against acetaldehyde-DNA adducts could be developed, which might be useful for a semiquantitative assay of adduct levels in different cell types with human biopsies. However, the sensitivity and specificity of such antibodies are difficult to ensure. Alternatively, the proliferating cell layer could be dissected out for adduct analysis (e.g., by laser capture microdissection), but the amount of DNA would be insufficient for analysis by mass spectrometry given the sensitivity of the technique currently. Conceivably, cell types of interest could be separated in bulk using cell-sorting techniques (e.g., [43]). However, this would require amounts of tissue that could not be obtained from humans.

Regarding animal models, it is important to note here that there is a major species difference in the K_m of ethanol oxidation by ADH7 between humans, mice, and rats. Specifically, while as noted above the Km of the human ADH7 for ethanol is approximately 25 mM [42], that of the mouse ADH7 homolog is roughly 200 mM [44], and for the rat is >2 M [45]. Therefore, to the extent that ethanol metabolism in oral or esophageal epithelial cells plays an important role in alcohol-related acetaldehyde-DNA adduct formation, the major difference in ADH7 for ethanol metabolism is a significant limitation of either rodent species as an animal model of humans.

5.4 Why Are ALDH2-Deficient Individuals at Such Elevated Risk of Esophageal Cancer from Drinking Alcohol?

The dramatically elevated risk of esophageal cancer in individuals with deficient ALDH2 is well known [46, 47]. However, the question of why these individuals are at such elevated risk of esophageal cancer, as opposed to other types of cancers, is not well understood. It has been shown that ALDH2-deficient individuals experience higher levels of salivary acetaldehyde from ethanol [48, 49] Importantly, this difference is only detectable when ethanol is actually ingested [50]. However, it is not clear that the difference in acetaldehyde concentrations in the saliva during alcohol drinking can fully explain the dramatically elevated esophageal cancer risk. In this context, a crucially important question concerns the localization of the ALDH2 enzyme in cells of the esophageal epithelium. Yin et al. [51] reported that ALDH2 activity in homogenates of the human esophageal mucosa, as measured on agarose gels, was barely detectable. Using standard enzymes assays, "low K_m ALDH" (assayed at 200 μM acetaldehyde) was reported to be less than 10 % of the high K_m form. However, these studies did not completely exclude the possibility of ALDH2 expression in a population of cells in the esophagus.

In contrast to biochemical studies, immunohistochemical studies of ALDH2 in the human esophagus [50, 52] did in fact detect ALDH2 staining in a subset of cells, specifically cells localized to the basal cell layer where proliferating cells are found. The ALDH2 antibody used in this work was developed by Weiner and colleagues [53] and had been validated using tissues from knockout mice lacking Aldh2. In support of these findings, data available on a public database shows that ALDH2 mRNA can be detected in the human esophagus (http://www.ncbi.nlm.nih.gov/geoprofiles/83899354) [54]. Interestingly, the staining intensity varied depending upon the drinking history of the tissue donors: the strongest staining was more often observed in samples from individuals with a history of alcohol drinking. These observations raise the intriguing possibility that ALDH2 expression may be inducible by heavy alcohol drinking in the human esophagus. Consistent with these observations and the idea that ALDH2 can be inducible, ALDH2 expression can be increased by low pH in a human esophageal cell line (http://www.ncbi.nlm.nih.gov/geoprofiles/11619953) [55].

Taken together, these data indicate that ALDH2 is expressed in proliferating cells of the human esophagus, where it could play a role in protecting genomic DNA against acetaldehyde generated from ethanol metabolism by ADH7 in situ. If so, then the absence of this activity in proliferating esophageal cells of ALDH2-deficient alcoholics would provide a compelling explanation for the dramatically elevated risk of esophageal cancer from high levels of alcohol drinking in this population.

One final point to be made here is that the proliferating cell layer of the oral and esophageal epithelium is not static. Under normal conditions, cell proliferation and differentiation are balanced to maintain the structure and function of the epithelium. However, in response to wounding or damage, the balance between proliferation and differentiation can shift to regenerate the damaged tissue [56]. An important study by Salaspuro and colleagues [57] in fact demonstrated that chronic exposure of rats to acetaldehyde in the drinking water does increase the size of the proliferating cell compartment in oral and gastrointestinal epithelia, as measured by the thickness of the epithelial layer and number and depth of cells staining for the proliferation maker Ki67. While the concentration of acetaldehyde in the drinking water used in that work (120 mM) is far in excess of what would be considered clinically relevant during alcohol drinking in humans, as noted by the authors, acetaldehyde is highly volatile (boiling point 24 °C) and would therefore likely diffuse from the bottles, reducing the actual concentration. Also, the bottles were only changed every three days. Therefore, the actual acetaldehyde concentration that tissues were exposed to from the drinking water was likely to be far less than 120 mM, especially by the third day after bottle change. Also, as shown in Fig. 5.5b, during alcohol drinking in humans, acetaldehyde can be generated intracellularly from ethanol metabolism at high blood ethanol concentrations, and these intracellular levels could be quite high, especially in ALDH2-deficient individuals. This issue could be addressed in the laboratory using cells expressing human ADH7, exposed to different ethanol concentrations spanning the range that could be generated in the blood during heavy alcohol drinking in humans.

The expansion of the proliferative cell compartment and hyperregeneration as a result of the toxic or damaging effects of acetaldehyde derived from ethanol metabolism (see [58, 59]) would be examples of non-genotoxic mechanisms for acetaldehyde-related carcinogenesis. However, these mechanisms could synergize with genotoxicity, in that more proliferating cells with mutagenic DNA adducts essentially expand the target tissue for carcinogenesis. It is therefore likely that acetaldehyde acts as both a genotoxic and non-genotoxic carcinogens in the human UADT.

5.5 Is There a "Carcinogenic" Level of Acetaldehyde?

As noted in the Introduction, the 2009 IARC classification specifically identifies acetaldehyde associated with the consumption of alcoholic beverages as a Group 1 carcinogen. Acetaldehyde alone remains classified in Group 2B (possibly carcinogenic to humans). However, the concept of a "carcinogenic level of acetaldehyde" (variously described as between 50 and 100 μM) has entered the literature [11, 39, 60]. As this term is based upon studies of DNA adduct formation, and genotoxic end points, we would like to briefly discuss its derivation and implications.

One of the studies cited in support of this concept was the work of Theruvathu et al. [61] (one of us, P.J.B., was an author on that paper). This work was intended to address a specific mechanistic question, which was whether polyamines could stimulate the formation of CrPdG adducts in DNA during exposure to acetaldehyde. In that study, Theruvathu et al. incubated purified genomic DNA with increasing concentrations of acetaldehyde, with or without a physiologically relevant concentration of polyamine, at 37 °C for 24 h [61]. While the acetaldehyde concentrations used were within the range that could plausibly occur in the human body during alcohol drinking, the work was not intended to be a basis for establishing a mutagenic or carcinogenic level of acetaldehyde. DNA adduct formation in living cells exposed to acetaldehyde is likely to be much lower than when pure DNA is exposed, due to reactions of acetaldehyde with other cellular molecules, as well as ongoing DNA repair. In fact, it would be important and relevant to assess CrPdG adduct levels in human cells exposed to different acetaldehyde concentrations for different periods of time. Highly sensitive and specific assays for CrPdG adducts in cellular DNA have been developed and could be used utilized for this purpose [31].

Another study that has been cited to support the carcinogenic level of acetaldehyde is the work of Obe and Ristow [62], who investigated the effect of acetaldehyde on sister chromatid exchanges (SCE) in mammalian cells. The lowest concentration that was shown to increase SCE levels in that study was roughly 90 μM. While an increase in SCEs is evidence of genotoxicity, SCEs are not mutations, and their relevance for predicting cancer risk is controversial [63].

A more important issue that is not explicitly considered when discussing carcinogenic levels of acetaldehyde is the histology of the oral cavity. As we have illustrated in Fig. 5.5, the proliferating cells in the oral cavity that are of most relevance to oral cancer lie several layers below the epithelial surface. For this reason, the concentration of acetaldehyde that proliferating cells in the oral epithelium are actually exposed to will be much lower than the levels in saliva, due to the reaction of acetaldehyde with components of the overlying, terminally differentiated cells.

Estimating a carcinogenic acetaldehyde concentration in the oral cavity based upon an acetaldehyde concentration that increases SCEs in monolayer cells in tissue culture fails to take this protective cell layer effect into account.

It is in fact possible to develop valid models of the effect of acetaldehyde on different epithelial tissues, which could be relevant to understand carcinogenic mechanisms, and to provide a basis for risk assessment. Templates for such an approach are physiologically based pharmacokinetic models for vinyl acetate toxicity and carcinogenesis in the oral and nasal epithelium [63]. Importantly, vinyl acetate is metabolized to acetaldehyde and acetic acid in epithelial cells, and therefore, some of the information that was used in developing the vinyl acetate models may be directly applicable to acetaldehyde carcinogenicity as well. As in the case of vinyl acetate, different models should be developed for different tissues, taking into account the expected concentrations of ethanol and acetaldehyde, as well as the time course of exposure, and considering the unique biology of different tissues (e.g., oral cavity versus esophagus versus colorectum).

5.6 Summary and Conclusions

The ability of acetaldehyde to form covalent adducts with DNA is consistent with a genotoxic mechanism for alcohol-related carcinogenicity, but does not prove such a mechanism, nor does it rule out non-genotoxic effects. More data on the kinetics of acetaldehyde-DNA adduct formation and repair in specific cell types in target tissues for alcohol-related carcinogenicity will be important to fully understand carcinogenic mechanisms. Data from humans is of the most direct relevance to human carcinogenesis, and whole genome sequencing of human tumors may provide evidence of genetic signatures corresponding to specific DNA adducts (see [8]). However, given the ethical and practical limitations of collecting relevant tissues from human, data from well-designed animal studies, taking into account relevant species differences as noted above, will also provide important mechanistic insights of relevance not only for alcohol-related carcinogenesis but also to the broader question of risk assessment for acetaldehyde alone.

Acknowledgment We thank Cheryl Marietta for proofreading the manuscript. Siliva Balbo's research is supported by ES-11297 from the National Institute of Environmental Health Sciences.

References

1. International Agency for Research on Cancer (1988) IARC monographs on the evaluation of carcinogenic risks to humans, vol 44. Lyon, IARC, pp 101–152
2. International Agency for Research on Cancer (2010) IARC monographs on the evaluation of carcinogenic risks to humans. Lyon, IARC
3. Secretan B, Straif K, Baan R, Grosse Y, El Ghissassi F, Bouvard V, Benbrahim-Tallaa L, Guha N, Freeman C, Galichet L, Cogliano V (2009) Lancet Oncol 10:1033–1034

4. Yokoyama A, Muramatsu T, Ohmori T, Yokoyama T, Okuyama K, Takahashi H, Hasegawa Y, Higuchi S, Maruyama K, Shirakura K, Ishii H (1998) Carcinogenesis 19:1383–1387
5. Yokoyama A, Muramatsu T, Omori T, Yokoyama T, Matsushita S, Higuchi S, Maruyama K, Ishii H (2001) Carcinogenesis 22:433–439
6. Yokoyama A, Omori T (2003) Jpn J Clin Oncol 33:111–121
7. Hernandez LG, van Steeg H, Luijten M, van Benthem J (2009) Mutat Res 682:94–109
8. Brooks PJ, Zakhari S (2014) Environ Mol Mutagen 55:77–91
9. Lachenmeier DW, Sarsh B, Rehm J (2009) Alcohol Alcohol 44:93–102
10. Lachenmeier DW, Monakhova YB (2011) J Exp Clin Cancer Res 30:3
11. Linderborg K, Salaspuro M, Vakevainen S (2011) Food Chem Toxicol 49:2103–2106
12. Wang M, McIntee EJ, Cheng G, Shi Y, Villalta PW, Hecht SS (2000) Chem Res Toxicol 13:1149–1157
13. Vaca CE, Fang JL, Schweda EK (1995) Chem Biol Interact 98:51–67
14. Lindahl T (1993) Nature 362:709–715
15. Efrati E, Tocco G, Eritja R, Wilson SH, Goodman MFJ (1997) Biol Chem 272:2559–2569
16. Choi JY, Guengerich FPJ (2005) Mol Biol 352:72–90
17. Upton DC, Wang XY, Blans P, Perrino FW, Fishbein JC, Akman SA (2006) Chem Res Toxicol 19:960–967
18. Pence MG, Blans P, Zink CN, Hollis T, Fishbein JC, Perrino FWJ (2009) Biol Chem 284:1732–1740
19. Perrino FW, Blans P, Harvey S, Gelhaus SL, McGrath C, Akman SA, Jenkins GS, LaCourse WR, Fishbein JC (2003) Chem Res Toxicol 16:1616–1623
20. Montgomery SB, Goode DL, Kvikstad E, Albers CA, Zhang ZD, Mu XJ, Ananda G, Howie B, Karczewski KJ, Smith KS, Anaya V, Richardson R, Davis J, MacArthur DG, Sidow A, Duret L, Gerstein M, Makova KD, Marchini J, McVean G, Lunter G (2013) Genome Res 23:749–761
21. Brooks PJ, Theruvathu JA (2005) Alcohol 35:187–193
22. Fraenkel-Conrat H, Singer B (1988) Proc Natl Acad Sci U S A 85:3758–3761
23. Fang JL, Vaca CE (1997) Carcinogenesis 18:627–632
24. Wang MY, Yu NX, Chen L, Villalta PW, Hochalter JB, Hecht SS (2006) Chem Res Toxicol 19:319–324
25. Chen L, Wang MY, Villalta PW, Luo XH, Feuer R, Jensen J, Hatsukami DK, Hecht SS (2007) Chem Res Toxicol 20:108–113
26. Singh R, Sandhu J, Kaur B, Juren T, Steward WP, Segerback D, Farmer PB (2009) Chem Res Toxicol 22:1181–1188
27. Abraham J, Balbo S, Crabb D, Brooks PJ (2011) Alcohol Clin Exp Res 35:2113–2120
28. Fowler AK, Hewetson A, Agrawal RG, Dagda M, Dagda R, Moaddel R, Balbo S, Sanghvi M, Chen Y, Hogue RJ, Bergeson SE, Henderson GI, Kruman IIJ (2012) Biol Chem 287:43533–43542
29. Matsuda T, Matsumoto A, Uchida M, Kanaly RA, Misaki K, Shibutani S, Kawamoto T, Kitagawa K, Nakayama KI, Tomokuni K, Ichiba M (2007) Carcinogenesis 28:2363–2366
30. Yu HS, Oyama T, Matsuda T, Isse T, Yamaguchi T, Tanaka M, Tsuji M, Kawamoto T (2012) Biomarkers 17:269–274
31. Matsuda T, Yabushita H, Kanaly RA, Shibutani S, Yokoyama A (2006) Chem Res Toxicol 19:1374–1378
32. Swenberg JA (2004) Toxicological considerations in the application and interpretation of DNA adducts in epidemiological studies. In: Buffler P, Rice J, Baan R, Bird M, Boffetta P (eds) Mechanisms of carcinogenesis: contributions of molecular epidemiology, vol 157. IARC, Lyon, pp 237–247
33. Balbo S, Meng L, Bliss RL, Jensen JA, Hatsukami DK, Hecht SS (2012) Cancer Epidemiol Biomarkers Prev 21:601–608
34. Balbo S, Meng L, Bliss RL, Jensen JA, Hatsukami DK, Hecht SS (2012) Mutagenesis 27:485–490
35. Fisher HR, Simpson RI, Kapur BM (1987) Can J Public Health 78:300–304
36. Squier CA (2001) J Natl Cancer Inst Monogr 7–15

37. Gerson S, Fry RJ, Kisieleski WE, Sallese ARJ (1980) J Investig Dermatol 74:192–196
38. Boffetta P, Kaihovaara P, Rudnai P, Znaor A, Lissowska J, Swiatkowska B, Mates D, Pandics T, Salaspuro M (2011) Eur J Cancer Prev 20:526–529
39. Lachenmeier DW, Sohnius EM (2008) Food Chem Toxicol 46:2903–2911
40. Gubala W, Zuba D (2003) Pol J Pharmacol 55:639–644
41. Jones AW (1979) Clin Exp Pharmacol Physiol 6:53–59
42. Kedishvili NY, Bosron WF, Stone CL, Hurley TD, Peggs CF, Thomasson HR, Popov KM, Carr LG, Edenberg HJ, Li TKJ (1995) Biol Chem 270:3625–3630
43. Balbo S, James-Yi S, Johnson CS, O'Sullivan MG, Stepanov I, Wang M, Bandyopadhyay D, Kassie F, Carmella S, Upadhyaya P, Hecht SS (2013) Carcinogenesis 34:2178–2183
44. Algar EM, Seeley TL, Holmes RS (1983) Eur J Biochem 137:139–147
45. Farres J, Moreno A, Crosas B, Peralba JM, Allali-Hassani A, Hjelmqvist L, Jornvall H, Pares X (1994) Eur J Biochem 224:549–557
46. Brooks PJ, Enoch MA, Goldman D, Li TK, Yokoyama A (2009) PLoS Med 6:3
47. Yokoyama A, Muramatsu T, Ohmori T, Higuchi S, Hayashida M, Ishii H (1996) Cancer Epidemiol Biomarkers Prev 5:99–102
48. Vakevainen S, Tillonen J, Agarwal DP, Srivastava N, Salaspuro M (2000) Alcohol Clin Exp Res 24:873–877
49. Yokoyama A, Tsutsumi E, Imazeki H, Suwa Y, Nakamura C, Yokoyama T (2010) Alcohol Clin Exp Res 34:1246–1256
50. Helminen A, Vakevainen S, Salaspuro M (2013) PLoS One 8:e74418
51. Chao YC, Liou SR, Tsai SF, Yin SJ (1993) Proc Natl Sci Counc Repub China B Life Sci 17:98–102
52. Morita M, Oyama T, Kagawa N, Nakata S, Ono K, Sugaya M, Uramoto H, Yoshimatsu T, Hanagiri T, Sugio K, Kakeji Y, Yasumoto K (2005) Front Biosci 10:2319–2324
53. Zheng CF, Wang TT, Weiner H (1993) Alcohol Clin Exp Res 17:828–831
54. http://www.ncbi.nlm.nih.gov/geoprofiles/83899354. 2013.
55. http://www.ncbi.nlm.nih.gov/geoprofiles/11619953. 2013.
56. Doupe DP, Alcolea MP, Roshan A, Zhang G, Klein AM, Simons BD, Jones PH (2012) Science 337:1091–1093
57. Homann N, Karkkainen P, Koivisto T, Nosova T, Jokelainen K, Salaspuro MJ (1997) Natl Cancer Inst 89:1692–1697
58. Simanowski UA, Suter P, Stickel F, Maier H, Waldherr R, Smith D, Russell RM, Seitz HKJ (1993) Natl Cancer Inst 85:2030–2033
59. Simanowski UA, Stickel F, Maier H, Gartner U, Seitz HK (1995) Alcohol 12:111–115
60. Salaspuro M (2007) Novartis Found Symp 285:80–89
61. Theruvathu JA, Jaruga P, Nath RG, Dizdaroglu M, Brooks PJ (2005) Nucleic Acids Res 33:3513–3520
62. Ristow H, Obe G (1978) Mutat Res 58:115–119
63. Bogdanffy MS, Sarangapani R, Plowchalk DR, Jarabek A, Andersen ME (1999) Toxicol Sci 51:19–35

Chapter 6
The Role of Iron in Alcohol-Mediated Hepatocarcinogenesis

Sebastian Mueller and Vanessa Rausch

Abstract Alcoholic liver disease (ALD) is the major liver disease in the developed world and characterized by hepatic iron overload in ca. 50 % of all patients. This iron overload is an independent factor of disease progression, hepatocellular carcinoma and it determines survival. Since simple phlebotomy does not allow the efficient removal of excess iron in ALD, a better understanding of the underlying mechanisms is urgently needed to identify novel targeted treatment strategies. This review summarizes the present knowledge on iron overload in patients with ALD.

Although multiple sides of the cellular and systemic iron homeostasis may be affected during alcohol consumption, most studies have focused on potential hepatic causes. However, it should not be overlooked that more than 90 % of the major iron pool, the hemoglobin-associated iron, is efficiently recycled within the human body and it is also strongly affected by alcohol. The few available studies suggest various molecular mechanisms that involve iron regulatory protein (IRP1), transferrin receptor 1 (TfR1), and the systemic iron master switch hepcidin, but not classical mutations of the HFE gene. Notably, reactive oxygen species (ROS), namely, hydrogen peroxide (H_2O_2), are powerful modulators of these iron-steering proteins. For instance, depending on the level, H_2O_2 may both strongly suppress and induce the expression of hepcidin that could partly explain the anemia and iron overload observed in these patients. More studies with appropriate ROS models such as the novel GOX/CAT system are required to unravel the mechanisms of iron overload in ALD to consequently identify molecular-targeted therapies in the future.

Keywords Alcoholic liver disease • Alcohol • Carcinogenesis • Iron • HFE • Hemochromatosis • Hepatocellular carcinoma • Reactive oxygen species

S. Mueller (✉) • V. Rausch
Department of Internal Medicine, Salem Medical Center and Center for Alcohol Research, University of Heidelberg, Zeppelinstraße 11-33, 69121 Heidelberg, Germany
e-mail: sebastian.mueller@urz.uni-heidelberg.de

V. Vasiliou et al. (eds.), *Biological Basis of Alcohol-Induced Cancer*,
Advances in Experimental Medicine and Biology 815,
DOI 10.1007/978-3-319-09614-8_6

Abbreviations

ALAS	Aminolevulinic acid synthase
ALD	Alcoholic liver disease
DMT1	Divalent metal transporter 1
FP	Ferroportin
GGT	Gamma-glutamyltransferase
HCC	Hepatocellular carcinoma
HCV	Hepatitis C virus
HFE	Gene that is mutated in hereditary hemochromatosis
HH	Hereditary hemochromatosis
HO-1	Heme oxygenase 1
IRE	Iron-responsive element
IRP	Iron regulatory protein
ISC	Iron sulfur cluster
MPO	Myeloperoxidase
NO	Nitrogen oxide
RNS	Reactive nitrogen species
ROS	Reactive oxygen species
SOCS3	Suppressor of cytokine signaling 3
STAT3	Signal transducer and activator of transcription 3
TfR	Transferrin receptor
TNF	α Tumor necrosis factor α
YLL	Years of life loss

6.1 Introduction

Carcinogenesis by alcohol is complex and still poorly understood for many reasons [1]. One major reason is due to the fact that it is rather the metabolites and side products of alcohol turnover rather than direct actions of alcohol itself (see Fig. 6.1). Indeed, alcohol is rapidly converted into more reactive, toxic, or even carcinogenic compounds such as acetaldehyde or reactive oxygen species (ROS) via nonenzymatic or enzymatic reactions. The production of these metabolites during

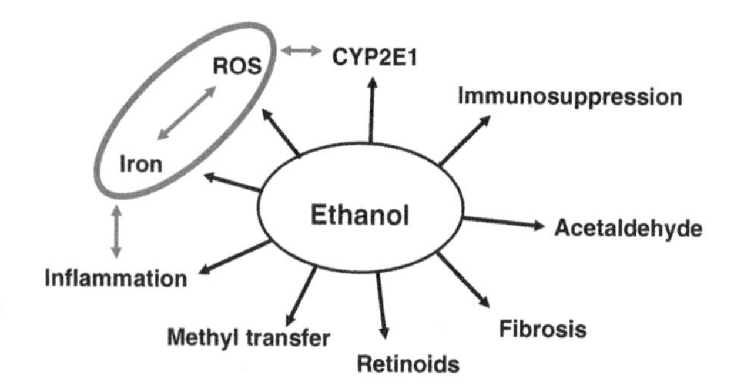

Fig. 6.1 Iron accumulation and ROS are important factors in the alcohol-mediated hepatocarcinogenesis

alcohol consumption is complicated since it is dependent on many additional factors that include the genetic background or the tissue-specific induction of detoxifying enzymes [2]. The complete understanding of alcohol-mediated carcinogenesis is also hampered by the fact that side products such as ROS are very difficult to measure and appropriate models to study ROS and ROS-mediated carcinogenesis are still lacking [3]. Finally, commonly used animal models such as rodents significantly differ from humans with regard to alcohol sensitivity and metabolism and consequently with regard to cancer development. Although iron and its accumulation in the liver during alcohol consumption has been appreciated for a long time, a better understanding of the underlying molecular mechanisms has been only gained in the last decade with the discoveries of key regulatory molecules such as hepcidin and the HFE gene. This overview aims to summarize the actual knowledge of iron toxicity and accumulation in ALD, which is considered an important single factor for the development of liver cancer and overall survival.

6.2 Health Statistics of ALD

Chronic alcohol consumption is one of the major risk factors worldwide, significantly affecting both mortality and years of life loss (YLL) [4]. Ca. 5 % of the Western world show risky alcohol consumption and some countries such as China reach a yearly regional increase of alcohol consumption of 400 % [5]. The liver is the major target organ of alcohol consumption, although alcohol may affect many other organ systems such as the heart, nervous system, pancreas, breast, and others. According to the recently published "burden of disease study 2010," liver cirrhosis and liver cancer are ranked at position 12 and 16 in the actual death statistics [6]. In 2010 ca. one million people died from liver cirrhosis with one third directly attributable to alcohol. This is a considerable number when comparing with coronary heart disease with about ten million deaths and ranking as the leading cause of mortality in the global death statistics. In Central Europe, liver cirrhosis even ranks at the fourth position of YLL and HCC is now the most common fatal complication of patients with alcoholic liver cirrhosis. Moreover, it has the second fastest increase of all tumors worldwide after kidney tumors, and alcohol-associated HCC ranks on third position after HCC caused by viral hepatitis B and C infections. These epidemiological data explain the high interest in gaining a better understanding of molecular mechanisms of alcohol-driven hepatocarcinogenesis to identify novel target approaches for future therapies.

6.3 Prevalence of Iron Overload in ALD

A significant iron deposition in patients with chronic alcohol consumption has been appreciated for a long time [7] (see Fig. 6.2). Considerable confusion still exists today with genetic hemochromatosis, which is characterized by a typical and severe hepatocellular iron overload and shows other classical symptoms such as arthritis,

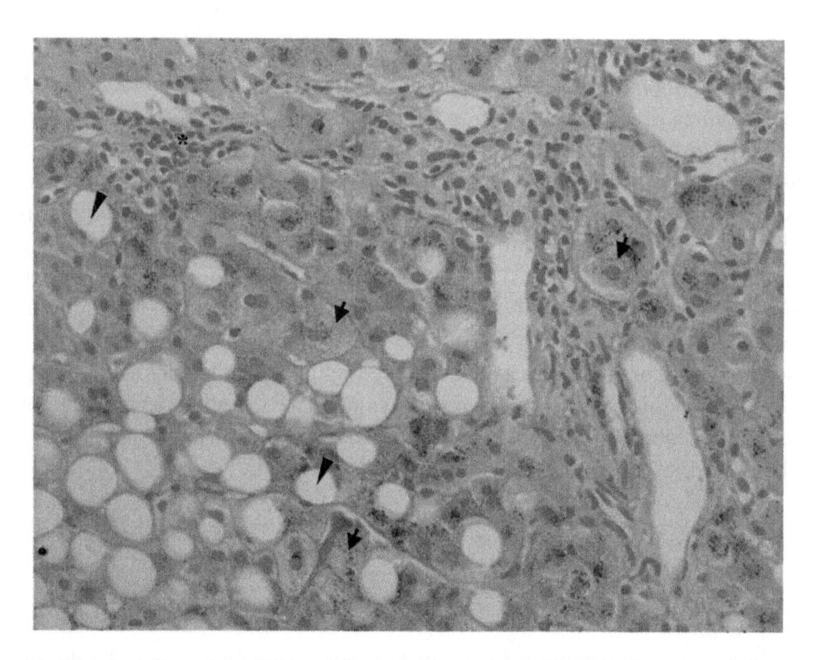

Fig. 6.2 Histological section of a liver biopsy from a patient with ALD. Staining of iron-loaded cells by Prussian blue (*black arrows* = intracellular inclusions of iron in hepatocytes) in the liver of ALD patient. *Asterisk* = inflammatory foci, *arrow head* = macrovesicular steatosis (fat droplet with displaced nucleus). In this patient, a predominant hepatocellular iron deposition is seen while Kupffer cells are not stained for iron

skin pigmentation, and diabetes mellitus. The genetic nature of hereditary hemochromatosis was already clear when the first monographs on hemochromatosis in 1935 by Sheldon and the extensive review by Finch et al. in 1955 [8, 9] appeared. However, it was also noted [8] that patients with hemochromatosis showed a high prevalence of alcoholism of ca. 30 %. Although the introduction of percutaneous needle biopsy of the liver in the 1950s added a new tool for exact quantitation of hepatic iron deposition, it further led to considerable confusion since histological stainable iron deposits are common in cirrhotic livers [10, 11]. This is especially the case in alcoholic subjects and some investigators have even equated mild to moderate alcoholic cirrhosis with hemochromatosis. Notably, it is still not clear in what form the iron is actually deposited and in what form the iron is toxic (ferritin, hemosiderin, or labile iron pool [12, 13]). When analyzing the Finch data for differences between alcoholic and nonalcoholic patients, Powell noted that laboratory signs of hepatocellular damage showed a significant higher prevalence in the alcoholic group (serum transaminase elevation 35.5 % vs. 15.4 %). Interestingly, most large studies seem to indicate that mutations in the HFE gene [14] are not responsible for the iron accumulation [15–17] in ALD patients. Moreover, since hereditary hemochromatosis shows a very weak clinical penetration of about 10 %, it still remains open what exact role alcohol plays in patients with hemochromatosis and whether it is an important but underestimated disease modifier.

Important studies in the 1980s focused on pathologic abnormalities of available serum iron parameters. For instance, Chapman et al. determined the hepatic iron content in 60 alcoholics with liver disease and in 15 patients with untreated idiopathic hemochromatosis [18]. A significantly higher mean liver iron concentration was found in alcoholics as compared to controls (156.4 vs. 53 μg per 100 mg dry weight). However, liver iron was much higher in the hemochromatosis group (2,094.5 μg). In addition, no relationship between liver iron concentration and the amount of alcohol consumed was noted. Interestingly, Chapman also recognized that the serum ferritin concentrations reflected well the iron overload in patients with hemochromatosis and mild alcoholic liver disease while no association was found with serum ferritin in patients with severe alcoholic liver disease. He also concluded that serum iron and transferrin saturation were of little value in assessing iron status in patients with ALD. In 1994, Bell studied 312 patients with different liver diseases and, after a careful assessment of fasting serum iron, transferrin, and ferritin levels, found that serum ferritin is more frequently elevated in ALD than in other liver diseases [19]. In addition, measurement of ferritin should be postponed until the patients are abstaining because he noted that serum ferritin decreased from 1,483 to 388 μg/L after 1.5–6 weeks of abstaining from alcohol. The author claimed that most patients with increased serum ferritin had normal transferrin saturation. Again, ferritin levels higher than 1,000 μg/L were observed in 100 % of patients with hemochromatosis but only in 11 % in patients with ALD. Only 7.7 % of patients with other liver diseases had such high ferritin concentrations. It should be noted, however, that 15.2 % of ALD patients have increased transferrin saturation. In a recent study from Japan, excess iron accumulation was found in 22 hepatic tissues with alcoholic liver diseases, but not in any normal hepatic tissues [20].

We have recently characterized a large cohort at the Heidelberg Salem Medical Center using biopsy, but also various newer noninvasive tests including transient elastography. Selected data [21] from a cohort of 235 patients with ALD, of whom 86 had underwent a liver biopsy, with detailed iron characterization regarding intensity and location (macrophages vs. hepatocytes), are shown in Table 6.1. It can be summarized that ca. 50 % of patients with ALD have significant pathological iron deposition. Interestingly, iron deposits are found both at the same rate in macrophages and hepatocytes. In addition, there is a mixture of patient groups with only macrophages, only hepatocytes, and mixed iron deposition. It should be clearly noted that this iron deposition depends on fibrosis stage, as has already been seen many decades before. However, with new noninvasive technologies to better quantitate the stage of fibrosis (e.g., transient elastography), more quantitative associations can be shown (see Table 6.2). It should be also noted that the Heidelberger alcohol cohort shows a rather representative fibrosis distribution with about 60 % without any fibrosis, 10 % with fibrosis stages F1 and F2, 10 % fibrosis stage F3, and 15 % fibrosis stage F4. According to our data, the mean histological iron score (0–3) is 1 in patients with advanced fibrosis stage F3 and F4. About 2/3 of these patients with advanced fibrosis stage have elevated ferritin levels higher 400 μg/mL and an increased transferrin saturation higher in 45 %. This is especially important to consider, since ferritin levels and transferrin saturation are commonly used to screen patients with

Table 6.1 Location and intensity of histological iron deposits in the ALD cohort from the Heidelberg Salem Medical Center

Parameter	Units	Mean	Fibrosis stage			Elevated (%)
			F0–2	F3	F4	
Ferritin (ng/mL)	m (30–400)	567.2	678.2	1,062.0	509.2	
Transferrin (g/L)	(2–3.6)	2.5	2.3	2.4	1.9	
Serum iron (ug/dL)	(59–158)	157.0	115.1	90.6	83.2	
Cirrhosis (ultrasound)		18.75 %	6.52 %	25.00 %	64.71 %	
Liver stiffness (kPa)	kPa	17.4	7.1	23.4	51.1	
Iron (macrophages)	0–3	0.67	0.59	0.88	0.63	45.24
Iron (hepatocytes)	0–3	0.61	0.57	1.00	0.38	42.86
Steatosis (Kleiner score)	0–3	1.73	1.96	1.88	1.76	89.41
Lobular inflammation	0–3	1.15	1.24	1.59	1.35	90.59
Ballooning	0–2	0.72	0.67	0.94	1.00	63.53
Mallory–Denk bodies	0–1	0.26	0.16	0.35	0.47	25.88
Cirrhosis (Kleiner)	0–4	2.20	1.59	3.00	4.00	88.68

Routine laboratory tests, liver stiffness measurements (Fibroscan), and histological assessment of liver biopsies in the context of fibrosis grading are shown ($n=235$)

Table 6.2 Typical routine laboratory tests and ultrasound parameters in the actual Heidelberg cohort of ALD patients

Parameter	All (%)	No or low fibrosis (F0–2) (%)	advanced fibrosis (F3–4) (%)
Laboratory			
GGT >60 IU/L	75.0	61.9	95.2
GOT/GPT >1	81.3	77.1	92.9
GOT >50 IU/L	60.1	51.4	88.1
GPT >50 IU/L	46.5	41.6	66.7
Ferritin >400 ng/mL	37.3	28.6	66.7
Transferrin saturation >45 % (%)	36.1	27.8	55.3
Bilirubin >1.3 mg/dL	15.8	5.7	26.2
Ultrasound			
Sonographic signs of cirrhosis (%)	19.6	1.0	40.0
Splenomegaly (>11 cm) (%)	11.4	0.0	15.0
Ascites (%)	25.7	17.6	32.3

hereditary hemochromatosis. In the subanalysis (Mueller et al. 2013, unpublished data), we could show that the serum ferritin levels were highly associated with GGT, transaminases levels, transferrin saturation, liver stiffness, and histological iron deposition in hepatocytes and to a lesser extent to iron in macrophages. Longitudinal analysis of serum ferritin levels during abstaining from alcohol showed that concentrations decreased by 50 % within 2.1 days. Importantly, a careful comparison between patients with alcoholic liver disease and patients with nonalcoholic liver disease (NALD) in 30 cases matched with regard to gender, age, and fibrosis stage showed significant differences between the two cohorts with regard to iron-related

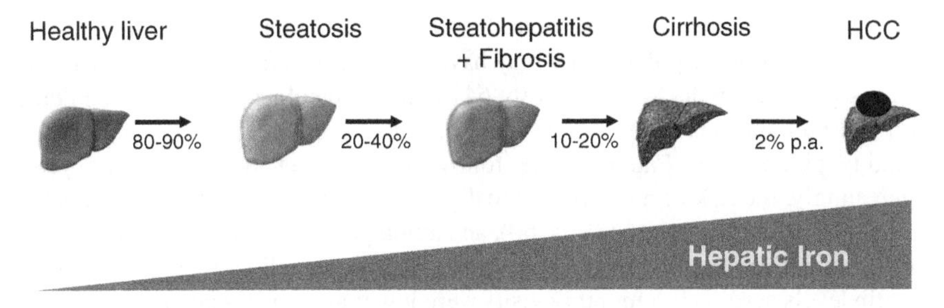

Fig. 6.3 Natural course of ALD ultimately leading to liver cancer

proteins. So ALD patients showed lower transferrin levels and lower erythrocyte counts and hemoglobin levels but significantly higher ferritin level [22].

Taken together, patients with ALD show a significantly increased histological iron deposition in about half of the population that increases with disease progression toward cirrhosis (see Fig. 6.3). This iron deposition is reflected by serum ferritin levels, but unfortunately overlayered by additional inflammatory conditions. In fact, according to our analysis, serum ferritin levels reflect hepatocyte iron deposition more closely than macrophage iron deposition in ALD. Moreover and often overlooked, a significant proportion of patients show elevated levels of transferrin saturation and ferritin levels that can be easily mistaken for hemochromatosis.

6.4 Iron Toxicity and Carcinogenesis

Iron toxicity has been well known for many decades and usually attributed to Fenton chemistry, first described in 1896 by Fenton and referring to the reaction of reduced ferrous iron with H_2O_2 to ferric iron and hydroxyl radicals [23]. These radicals are highly and universally reactive to all cellular compounds including DNA, proteins, lipids, and carbohydrates. They cause oxidative damage to DNA, including base modifications and DNA strand breaks, and may change the structure of proteins and lipids, eventually leading to mutagenesis.

Although, the direct chemical nature of the iron toxicity in Fenton's chemistry has been recently questioned by Saran et al. [24], it is generally assumed that at least Fenton-like reactions contribute to the toxicity of iron. For this reason, we will not further differentiate both reactions in this review. In vitro studies have shown that liver viability decreases in the presence of high amounts of iron and that the depletion of iron by iron chelators can protect cells from peroxide toxicity. Hepatic iron overload and the risk for liver cancer have also been extensively studied using in vivo experiments in the past. In patients with hereditary hemochromatosis due to HFE mutations, the hepatic iron overload ultimately determines the overall survival [25]. In a large transplant registry study, Ko et al. showed that iron alone increased the risk of HCC by a factor of 2.2 [26]. Turlin et al. showed in 1995 that the iron-associated risk of HCC was not dependent on whether the patient had cirrhosis or not [27]. In an interesting study from the NIH, Zacharski et al. could demonstrate that iron depletion by phlebotomy may have a protective effect on general carcinogenesis [28]. He studied prospectively 641 controlled subjects without iron depletion and 636 subjects in an iron reduction group. Iron reduction was achieved by monthly phlebotomy. Patients were followed up for an average for 4.5 years. Interestingly, the risk of new visceral malignancies was lower in the iron reduction group than controlled group (38 vs. 60), and among patients with new cancers, those in the iron reduction group had lower cancer-specific and all-cause mortality. Mean ferritin levels across all 6-monthly visits were lower among all patients who did not develop cancer. Notably, in a 1988 animal study, Hann et al. showed reduced tumor growth in iron-deficient mice [29], and Tsukamoto et al. in 1995 showed in rodents that a mixture of high fat diet and alcohol together with supplemented iron led to pronounced cirrhosis [30]. In a meta-analysis by D'Amico et al., it was concluded that liver iron is one among several independent prognostic factors of survival in liver cirrhosis [31]. Ganne-Carrie showed in 229 patients with ALD or hepatitis C virus-related cirrhosis that iron is the best parameter for predicting mortality in ALD patients ($p=0.007$) followed over 57 months [32]. This study found no prognostic significance of hepatic iron for HCV.

Taken together, there seems to be enough evidence from human and animal studies to suggest that hepatic iron overload leads to hepatocellular damage. This in turn leads to increased fibrosis progression and risk for hepatocarcinogenesis, and ultimately iron seems to be an important prognostic factor for overall survival in patients with ALD. The most evident and generally accepted mechanisms of iron carcinogenesis include Fenton-like reactions that lead to the highly aggressive hydroxyl radicals.

6.5 Mechanism of Iron Overload

Enormous progress has been made in the last two decades to better understand the molecular mechanisms of iron regulation and homeostasis both at the cellular and systemic level [33–35]. These milestones include the discovery of iron regulatory proteins (IRP) and their posttranscriptional regulation of cellular iron homeostasis in the late 1980s and early 1990s as well as identification of the systemic master switch peptide hepcidin about 10 years ago [36]. Despite this progress, numerous aspects of iron regulation still remain unexplained in humans. Apart from the diagnosis of rare and common variants of hereditary hemochromatosis, the progress has led to significant improvements in the diagnosis and therapy of iron overload. The discovery of the HFE gene that causes the majority of hereditary hemochromatosis forms has clearly helped to resolve the role of hemochromatosis in alcoholics [14]. Interestingly, most large studies seem to indicate that mutations in the HFE gene [14] are not responsible for the iron accumulation in patients with ALD [15–17]. These studies thus seem to rule out HFE as important genetic disease modifier. In contrast, as briefly discussed above, alcohol could be an important disease modifier in patients with hemochromatosis, but typically with weak clinical penetration.

In addition, at the systemic level, ALD appears rather complex since alcohol affects not only the liver but also many other organ systems including the bone marrow and the immune system so that iron regulation obviously is altered at many different levels and locations throughout the body. In the following, a brief description of the cellular and systemic iron homeostasis is given and the *status quo* of molecular findings in patients with ALD is discussed.

6.6 Systemic Regulation

The majority of body iron, roughly 2 g in humans, is distributed in the oxygen carrier hemoglobin of red blood cells and developing erythroid cells (Fig. 6.4). Excess iron is usually stored in the liver, which normally contains about 1 g of iron, predominantly in the form of ferritin [34]. Ferritin is a protein composed of 24 subunits with a ferrihydrite core where 4500 iron atoms can be stored in cage-like structures. It consists of ferritin H chain (heavy or heart) and ferritin L chain (liver or light) subunits and can be regulated by IRPs [37]. Other significant amounts of iron are found in macrophages (0.6 g) and in the myoglobin of muscles (0.3 g). It is generally believed that mammals lose iron from regular sloughing of the mucosa and skin cells or during bleeding and do not possess any mechanisms for iron excretion from the

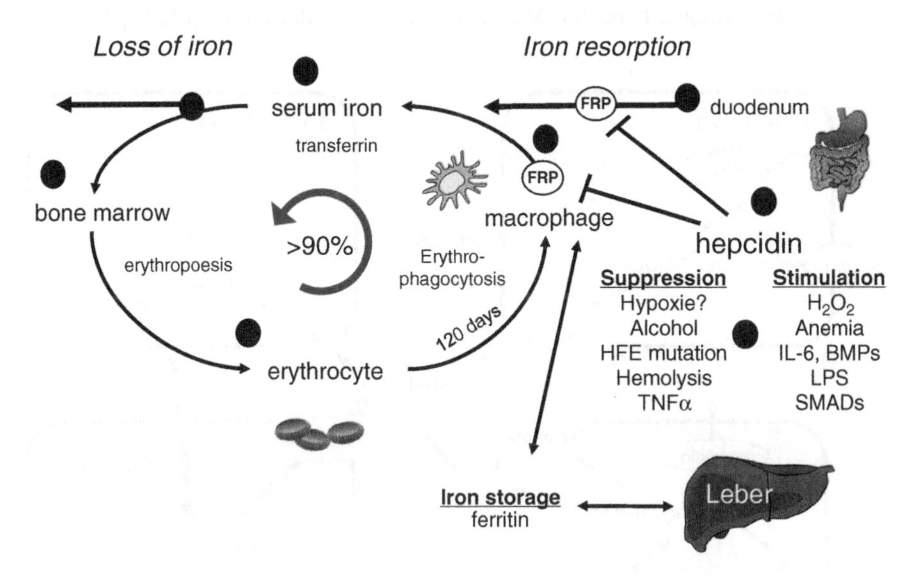

Fig. 6.4 Systemic iron homeostasis and utilization in the body. Black circles indicate potential target sites of iron regulation affected by chronic or acute alcohol consumption. Dietary iron is absorbed in the duodenum and bound to transferrin. Iron is then mainly delivered to the bone marrow for erythropoiesis while senescent erythrocytes are phagocytosed by macrophages. This efficient mechanism recycles >90 % of all iron for new heme synthesis. Excess iron is stored in hepatic ferritin or temporarily in macrophage ferritin. Regulation of iron metabolism by hepcidin and upstream factors are also shown

body [33]. Therefore, the balance is maintained by the tight control of dietary iron absorption at the brush border of enterocytes in the proximal duodenum (Fig. 6.4). Dietary iron uptake involves the reduction of ferric iron to ferrous iron in the intestinal lumen by ferric reductases, such as duodenal cytochrome B (Dcytb), and subsequent transport of iron via the apical membrane of enterocytes by divalent metal transporters (such as DMT1 encoded by SCL11A2 (solute carrier family 11, member 2) gene). Iron can also be transported across the enteral membrane in the form of heme by unknown mechanisms and the iron is then released inside enterocytes via heme oxygenase 1 (HO-1). Cytosolic iron is exported by the basolateral iron transporter ferroportin (FP-1 or SLC40A1) from the enterocytes to the blood compartment and then bound to transferrin, the major abundant iron-carrying protein within serum. Before binding, ferrous iron is oxidized by ferroxidases (e.g., hephaestin/ceruloplasmin) to the ferric form. Iron is then distributed within the body and used in various pathways, but mainly utilized in the bone marrow for the synthesis of new heme.

About 1–2 mg per day of iron is absorbed to keep a stable iron balance. Of note, iron undergoes an efficient recycling. Senescent erythrocytes that have a life span of 120 days are sequestered by macrophages and the iron is reused for new heme synthesis. This recycling machinery accounts for about 90 % of newly synthesized hemoglobin. It is important to note that the iron export pump, ferroportin, is not only found on enterocytes but also on macrophages and hepatocytes (Fig. 6.5). The ferroportin-mediated efflux of ferrous iron from enterocytes and macrophages into the serum is critical for systemic iron homeostasis and mainly controlled via the liver-secreted 25 AS peptide hormone, hepcidin. Mechanistically, hepcidin binds to ferroportin and

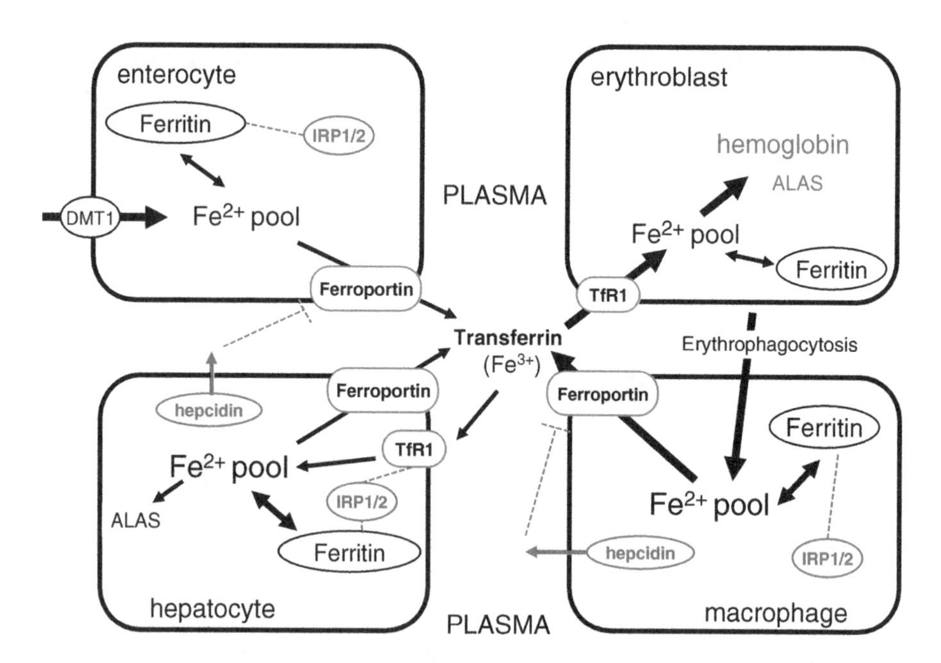

Fig. 6.5 Interaction of systemic and cellular iron homeostasis at four major sites: enterocyte, hepatocyte, erythroblast, and macrophage

promotes its phosphorylation, internalization, and lysosomal degradation [38, 39]. Hepcidin is primarily expressed in hepatocytes as a precursor pro-peptide, although other locations of secretion have been described, such as macrophages [40] (Fig. 6.5) and to a lesser extent cardiomyocytes. Hepcidin efficiently blocks ferroportin, which leads to accumulation of iron within macrophages and blocks the iron absorption via enterocytes (Fig. 6.4).

An important evolutionary conserved mechanism to induce hepcidin is an infection or inflammatory state. Via cytokines, namely, IL-6, but also microbial molecules (such as lipopolysaccharide), hepcidin is strongly induced leading to a rapid decrease of serum iron, which is thought to function as an antibacterial defense mechanism. More recently, another important inflammatory cofactor, H_2O_2, has been identified as potent inducer of hepcidin. In contrast, hepcidin levels seem to be suppressed in patients with genetic hemochromatosis, leading to increased uptake of iron via the duodenum and increased release of iron through macrophages. Cytokine-mediated induction of hepcidin is thought to be the reason for anemia of chronic disease [41], while the disruption of hepcidin is generally associated with the systemic iron overload (e.g., genetic hemochromatosis). Despite the progress and the discovery of various upstream regulators of hepcidin (such as C/EBPα; BMP6; SMAD 1, 5, 8, and 4; TMPRSS6; IL-6; CREBH; CHOP; and TLR4), an overall and conclusive understanding of the regulatory network with respect to the control of iron is not yet completely understood. Hitherto unexplained features of hepcidin regulation include:

1. Expression of hepcidin in cells other than hepatocytes
2. The nature of co-expressing ferroportin in hepatocytes and macrophages
3. The experimental and clinical finding that hepcidin responds differentially to iron overload in vitro and in vivo
4. The controversial findings of hypoxia-mediated regulation of hepcidin
5. The controversial and conflicting findings of the response of hepcidin toward ROS

At least to the latter point we could recently identify a partial explanation [42]. Thus, the central ROS, H_2O_2 induces hepcidin in hepatocytes independent of IL-6 when exposed in continuous manner. Bolus treatments, however, which reflect an artificially high H_2O_2 exposure blocked hepcidin expression [43]. They are even toxic at H_2O_2 concentrations higher than 50 μmol and hepcidin suppression at such conditions was due to unspecific inhibition of the mRNA transcription machinery [43].

6.7 Cellular Iron Regulation

Developing erythrocytes, as well as most other cell types, require iron from plasma transferrin. Iron is loaded onto transferrin with a capacity of two atoms of ferric iron per molecule. Transferrin binds with a very high affinity to cell surface transferrin receptor 1 (TfR1) ubiquitously expressed and with much lower affinity to transferrin receptor 2 (TfR2) sharing 45 % homology to TfR1 [44, 45] (see Fig. 6.6).

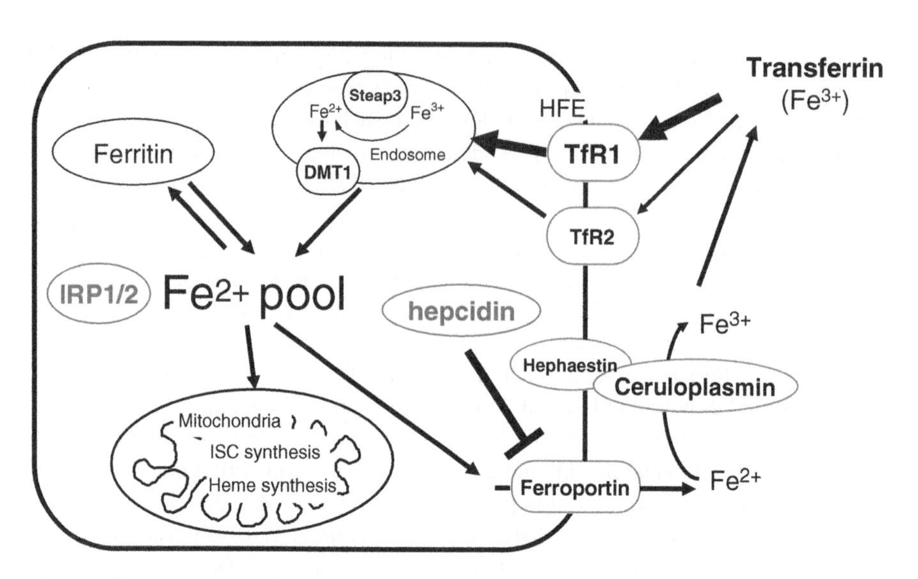

Fig. 6.6 Cellular iron metabolism. Iron-loaded holo transferrin binds to its cell surface receptor (high affinity TfR1, low affinity TfR2) and the Tf-TfR complex is endocytosed. In the endosome, iron is released and reduced by Steap3 and transported across the endosomal membrane by DMT1. Internalized iron is directed to mitochondria for heme or ISC synthesis and excess iron is stored in ferritin. Cytosolic iron is exported by the iron transporter ferroportin to the blood compartment and then again bound to transferrin. Regulatory mechanisms by hepcidin or IRPs are shown in *red*

Homology aside, TfR2 exhibits different properties from TfR1. TfR2 is only expressed in liver and to a lesser extent in erythroid, spleen, lung, and muscle cells. Also, its mRNA levels are not directly regulated by cellular iron content. Its expression is higher in iron-replete conditions, suggesting a role as transferrin-iron sensor candidate rather than acting primary in iron uptake [46]. The TF-TfR1 complex is endocytosed via clathrin-coated pits. The release of ferrous iron into the intracellular compartment requires acidification of the endosome using a proton pump and the reduction of ferric iron to the ferrous form via Steap3. DMT1 transports the ferrous iron across the endosomal membrane into the cytosol where it can be stored in ferritin complexes in non-erythroid cells or incorporated into hemoglobin in erythroid cells [47] (see Fig. 6.6). Iron is then either reused for various synthesis pathways such as intracellular heme synthesis and iron cluster protein synthesis or transported into the major iron storage protein ferritin.

The expression of DMT1 and ferritin is coordinately posttranscriptionally regulated by binding of trans-acting iron-responsive proteins (IRP1 and IRP2) to iron-responsive elements (IRE) in the untranslated regions (UTRs) of their respective mRNAs [48–50]. In iron-starved cells IRPs bind with high affinity to cognate IREs (see Fig. 6.7a, b). IREs are evolutionary conserved hairpin structures of about 30 nucleotides with characteristic sequences. The effect of IRP binding to IREs is dependent on their position. TfR1 mRNA contains five IREs within its long 3′ UTR that stabilizes and protects the transcript from degradation leading to protein upregulation,

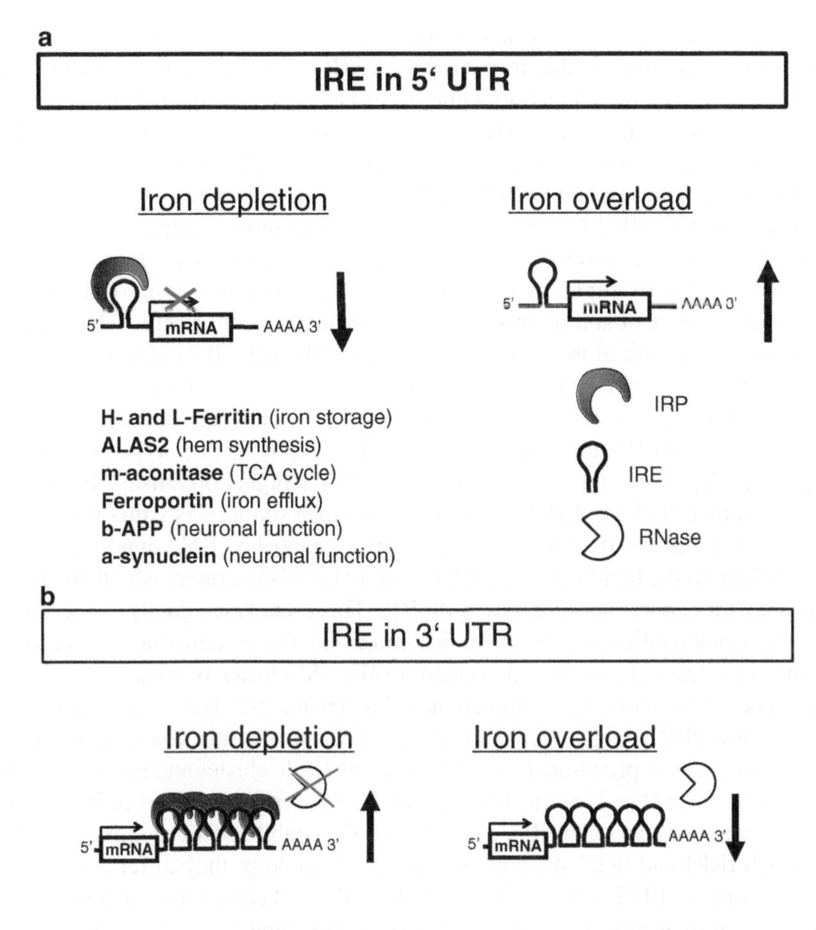

Fig. 6.7 Posttranscriptional regulation of iron-responsive element (IRE)-containing mRNAs either in the A) 5'UTR or B) 3′ UTR by iron regulatory proteins (IRPs). Binding of IRP in the 3′ UTR modulates TfR1 and DMT1 mRNA stability therefore influencing iron uptake and transport and renders translation of mRNAs with IREs in the 5′ UTR encoding H- and L-ferritins, ALAS2, m-aconitase, ferroportin, β-APP, and α-synuclein, which control iron storage, heme synthesis, energy homeostasis, iron efflux, and neurological functions, respectively

but other mRNAs, for example, mRNAs encoding H- and L-ferritins, contain a single IRE in their 5′ UTRs where binding results in decreased protein translation by steric blockade. As a result increased TfR1 levels stimulated acquisition of iron from plasma transferrin to counteract iron deficiency. The inhibition of ferritin synthesis leads to decreased abundance of this protein as iron storage becomes obsolete under these conditions. Conversely, in cells with high iron content, both IRP1 and IRP2 become unavailable for IRE binding lowering TfR1mRNA degradation and ferritin mRNA translation. Thus, when iron supply exceeds cellular means, the IRE-IRP switch minimizes further iron uptake via TfR1 and favors the storage of excess iron in newly synthesized ferritin. Other IREs have been discovered in the genes of ALAS2, mitochondrial aconitase, ferroportin, HIF2α, β-APP, and α-synuclein, which in turn control iron storage and erythroid iron utilization, energy homeostasis, hypoxia responses, and neurological pathways, respectively.

Although, the IRE-IRP network allows an autonomous independent control of iron homeostasis for individual cells, it can be overwritten by additional control mechanisms. For example, TfR1 expression is regulated transcriptionally by erythroid active element and its promoter in erythroid progenitor cells, which take up an enormous amount of iron for heme synthesis [51]. IRE-IRP independent regulation has been described in details elsewhere [52, 53]. An extensive phylogenetic analysis confirmed the IRE-containing mRNAs are exclusively found in metazoans [54]. The ferritin IRE motive may represent the exceptional prototype that was subsequently adapted during evolution by other genes in higher organisms. It should be also noted that IRP1 and IRP2 do not share sequence similarities with known RNA-binding proteins and do not contain any other established RNA-binding motives. IRPs belong to the family of iron sulfur cluster (ISC) isomerases, which are homologous to mitochondrial aconitase [48–50]. These enzymes catalyze the stereo-specific isomerization of citrate to isocitrate via the intermediate cis-aconitate during the citric acid cycle and contain a (4Fe-4S) cluster in its active side. ISC IRP1 assembles analogous to mitochondrial aconitase ISC. However, in contrast to m-aconitase, IRP1 only retains its ISC and its aconitase function in iron-repleted cells. Under iron deprivation, holo-IRP1 loses its labile cluster and is converted into apo-IRP1, which then binds to IRE-containing transcripts involved in iron uptake, storage, and transport. The ISC of IRP1 is also stabilized by hypoxia [55, 56]. Although IRP1 and IRP2 share 64 % sequence homology, they differ considerably in some aspects. IRP2 is neither assembled in (4Fe-4S) cluster nor retains aconitase activity. Consequently, IRP2 only exhibits an IRE-binding activity and does not have any enzymatic function. IRP2 is only regulated by its stability and is newly synthesized in response to low iron. Iron depletion or hypoxia leads to IRP2 stabilization [48, 50], whereas in iron-repleted cells, IRP2 becomes destabilized and undergoes rapid ubiquitination and degradation via the proteasome. More recently, an E3 ligase complex (SKP1-Cul1-FBXL5) was described to be responsible for iron-mediated IRP2 degradation [57, 58].

6.8 Redox Regulation of Iron Metabolism

Both IRP1 and IRP2 are sensitive to ROS and RNS (reactive nitrogen species) (reviewed in [59, 60]). This is evident through the redox regulation of IRP1 through its ISC. Exposure of cells to micromolar concentrations of ROS and RNS (especially NO) also leads to the destabilization of the IRP1-ISC complex with subsequent induction of IRE-binding activity via an incompletely characterized mechanism. This response can be antagonized by MPO-derived hypochlorite [61].

In vitro studies showed that ROS and RNS remove iron of the ISC and convert it to a nonfunctional cluster. IRP also responds to NO, which likewise induces IRE binding at the expense of its aconitase activity [62]. However, ROS and RNS may also modify potential cluster-destabilizing factors rather than the IRP1-ISC complex itself. Furthermore, some discrepancies are found in the literature about the regulation of IRP2 and NO. NO may either positively or negatively regulate IRP2, which could be explained by differential effects of the numerous NO species. Taken together, the redox regulation of IRE-IRP directly links the iron metabolism to inflammatory processes, hypoxia, and oxidative stress.

6.9 Non-hepatic Causes of Alcohol-Mediated Iron Overload

Although the liver is the major target organ of alcohol consumption, leading to the formation of steatosis in 90 %, inflammation (hepatitis) in 30 %, and cirrhosis in ca. 15 % (see also Fig. 6.3), various other target sites of iron homeostasis are affected (Fig. 6.4, black circles). Such potential sites include the duodenum, which is typically inflamed during chronic alcohol exposure. Chronic alcohol consumption may further affect the binding of iron to transferrin and the generation of newly formed erythrocytes via interference with vitamin B12 and folate metabolism, with secondary effects on iron utilization. Of course, overall iron metabolism can be affected in ALD patients simply by bleeding due to inflammatory conditions or ulcers in the upper gastrointestinal tract or due to complications of liver cirrhosis such as hypertensive gastropathy and esophageal varices. The stability of erythrocytes is also strongly affected by chronic alcohol consumption, leading to an earlier sequestration by macrophages as well as the release of iron from macrophages.

6.9.1 Alcohol and Hematopoietic System

Alcoholics are often found to have an abnormal hemogram because of direct damaging effects on hematopoietic cells and pathologic consequences on hematopoiesis. Until recently, the frequent coexisting formation of infections in liver diseases and malnutrition have been thought to be primary the cause of these changes. However, nowadays it is appreciated that alcohol alone is capable of producing several types of hematologic abnormalities such as diseases of red blood cells, white blood cells, and platelets. Red blood cells are pathologically affected by alcohol in nearly every stage of their life cycle [63]. Alcohol impairs red blood cell production in the bone marrow, which also has dramatic effects on erythrocyte maturation, delivery, and life span. Alcohol is also a potent folic acid antagonist. It blocks its absorption in the jejunum and decreases serum folate levels. Folic acid deficiency may be a key hematologic abnormality in ALD and the resulting anemia is almost invariably secondary to alcoholism and may therefore be the most common type of anemia seen in hospitals. Chronic alcohol abuse also leads to altered vitamin B6 (pyridoxine, pyridoxal, and pyridoxamine) metabolism often resulting in deficiency of the coenzyme pyridoxal phosphate, which is important for heme biosynthesis as described in a study by Hines et al. [64]. Lumeng and Li reported an alternative mechanism: acetaldehyde reduces the synthesis of intracellular pyridoxal phosphate by increasing membrane-bound phosphatases resulting in accelerated degradation of pyridoxal phosphate within erythrocytes [65]. Erythrocyte abnormalities are also due to alcohol and alcohol-mediated changes of the iron metabolism. Thus, a sideroblastic hypochromic anemia was found in alcoholic patients in whom the red blood cells appear similar to that seen in iron deficiencies. Further effects on red blood cell survival are seen as an effect of alcohol by producing several types of the hemolytic syndrome [66]. Chronic alcohol ingestion is also associated with a diminished granulocyte precursor pool in the bone marrow and functional disturbances that may lead to neutrocytopenia, brought into connection with the high susceptibility to infections of alcoholics [67]. Also dysfunctional monocyte-macrophages and lymphocytes may undoubtedly contribute to the predisposition to infections seen in ALD patients. Platelets are also affected by alcohol withdrawal, as thrombocytopenia has been frequently observed in ALD patients. Here, dietary folic acid deficiency is questionable as the underlying reason, but a direct effect of alcohol on platelet production and reduced survival may be the case [68]. All these changes and hematopoietic disorders seem to be reversible and return to normal levels after alcohol withdrawal, although, for example, an enlarged erythrocyte volume (MCV) can be detected still years after abstaining from alcohol.

6.9.2 Alcohol and Hemolysis

An increased sequestration and degradation of erythrocytes is observed in the spleen mainly due to alcohol-induced alterations to red blood cells. Typically observed morphological changes related to membrane lipids due to peroxidation of fatty acids [69] include stomatocytes, knizocytes (also called bridge cells or target cells), acanthoid cells, and irregularly spiculated cells [70]. Hemolysis often ultimately leads to an enlarged spleen (splenomegaly) rather than cirrhosis. Therefore, ALD patients often have elevated levels of heme oxygenases (HO-1) and indirect bilirubin (called hyperbilirubinemia), as a breakdown product from heme catabolism (unpublished observations). Disruption of the hepato-protective glutathione metabolism [71–73] may be another important reason contributing to hemolysis. A loss of reduced glutathione could sensitize cells for oxidative damage by iron overload or infections often coexisting in ALD patients.

6.9.3 Alcohol and Nutrition

Excessive and chronic alcohol consumption frequently causes malnutrition and vitamin deficiency. This malnutrition is often due to poor dietary intake because energy supply is to 50 % replaced by alcohol, therefore leading to inadequate calorie intake or periods without any food intake [74]. The status of malnutrition often correlates with the severity of ALD. Furthermore, gastrointestinal disturbances, e.g., maldigestion and malabsorption of nutrients, have also been observed [75]. These disturbances include dysgeusia (disorder of the sense of taste), anorexia, nausea, and early satiety [76]. Intestinal malabsorption and pancreatic dysfunction are seen in alcoholic patients with cirrhosis and in alcoholic patients with minimal and low liver disease. The substances that have been shown to be a malabsorbed are D-xylose, thiamine, folic acid, and lipid [77]. Deficiencies in nutrients like zinc, magnesium, and vitamin A, among many others, are also very common in these patients [78]. Therefore, additional sufficient nutritional support is an important therapeutic strategy [79].

6.10 Hepatic Causes of Iron Overload in ALD

The complete picture of the molecular changes in iron metabolism during chronic alcohol consumption is not yet completely understood (Fig. 6.8). There are various studies that focused on the central systemic master switch hepcidin. Both acute and chronic alcohol exposure seem to suppress hepcidin expression in the liver and in sera from patients with ALD and also pro-hepcidin levels are reduced. Ohtake and coworkers demonstrated that serum hepcidin is decreased in patients with ALD [80]. Most notably, Harrison-Findik et al. showed in their elegant study using an alcohol mouse model that the expression of hepatic hepcidin is rapidly suppressed upon exposure to alcohol [81]. 4-Methylpyrazole, a competitive inhibitor of

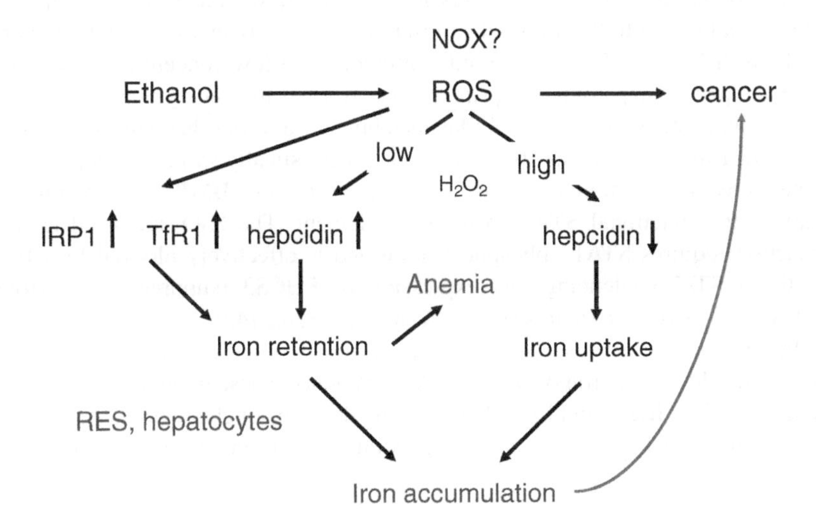

Fig. 6.8 Present understanding of systemic and cellular changes of iron homeostasis in ALD ultimately leading to cancer

alcohol-metabolizing enzymes (e.g., alcohol dehydrogenases), abolished this effect. However, ethanol did not alter the expression of TfR1 and ferritin or the activation of iron regulatory mRNA-binding proteins IRP1 and IRP2. Mice maintained on 10–20 % ethanol for 7 days displayed a downregulation of liver hepcidin without changes in liver triglycerides or histology. This was accompanied by elevated DMT1 and ferroportin expression in the duodenum [81]. Ethanol downregulated hepcidin promoter activity and the DNA-binding activity of the CCAT/NH-binding protein α (C/EBPα) but not β [82]. The same author showed later that the effect of alcohol and hepcidin was independent of Kupffer cell activation and TNFα signaling [83]. Conversely, there are observational studies showing that the hepatic expression of TfR1 is increased under conditions of alcohol consumption. In 2002 Suzuki et al. showed that TfR1 is increased in patients with ALD and that abstinence from drinking decreased TfR1 [20]. Furthermore, ethanol increased transferrin-bound iron uptake into hepatocytes, possibly due to ethanol-induced TfR1 expression and partly mediated by activation of IRP1 [84, 85]. TfR1 upregulation may be directly related to translational affects by ROS and not due to IRP1 [86].

Notably, we could not find differences in the expression of various iron-related mRNAs in a preliminary microarray study comparing iron overload and normal ALD patients (Mueller, S. et al., unpublished). We also noted a high variability in the expression of these mRNAs, implicating some caution in the interpretation of liver samples from ALD patients. It seems that ALD lesions are quite heterogeneously scattered throughout the liver. Therefore, it remains an open question whether the effects observed in an acute alcohol-exposure model in rodents are directly related to alcohol and hepcidin. Additional indirect conditions on the perfusion of the splanchnic system could also be involved. Two recent publications have tried to further elucidate the mechanisms of alcohol-mediated hepcidin expression [43, 87]. Both studies showed that bolus H_2O_2 treatments, a major ROS insult, could drastically suppress hepcidin expression. However, we recently found that this effect is rather due to the artificial exposure to high H_2O_2 levels (>50 μM) than peroxide itself [42]. If H_2O_2 is released continuously at low concentrations (between 0.3 and 6 μM) to hepatoma and primary liver cells, as they are typically released by inflammatory cells or during cellular metabolism, a strong hepcidin upregulation rather than suppression was observed. H_2O_2 synergistically stimulates hepcidin promoter activity in combination with recombinant IL-6 or BMP-6 in a manner that requires the functional STAT3-responsive element. The H_2O_2-mediated hepcidin induction requires STAT3 phosphorylation and is effectively blocked by siRNA-mediated STAT3 silencing, overexpression of SOCS3 (suppressor of cytokine signaling 3), and antioxidants such as *N*-acetylcysteine [42].

In summary, numerous rodent and human studies suggest the involvement of key iron molecules in the iron overload of ALD patients. These molecules may include hepcidin, TfR1, IRP1, and ROS. However, no definite conclusions can be made due to methodological challenges and complex interactions among these molecules.

6.11 Therapy of ALD in the Context of Iron Overload

Based on our preliminary knowledge of underlying mechanisms of iron disturbances in patients with ALD, no treatment options are readily available. The idea of iron depletion by phlebotomy has been controversially discussed in patients with HCV that equally develop iron overload; however, phlebotomy is highly propagated in certain countries, such as Japan. There has been one trial initiated in France in 2012 and titled, "phlebotomy in risk of HCC in patients with compensated alcoholic cirrhosis" (Tirrox). Unfortunately, this study has stopped recruiting patients due to delayed recruitment compared to that anticipated (REF: ClinicalTrials.gov NCT01342705). For the moment it can be regularly perceived that iron overload is an important factor in the progression of ALD, also in combination with other liver diseases, such as HCV. It is an important challenge for the future to clearly identify the underlying molecular mechanisms and to develop novel targeted therapeutic strategies and to improve the early detection of hepatic iron overload in alcoholic liver disease (e.g., measurement of liver iron by susceptometry or hepcidin-ELISA for diagnostic purposes). Future studies will continue with the goal of preventing fibrosis progression and to prevent HCC development, thereby improving overall survival by targeted therapeutic approaches that may include a pharmacological intervention leading to stimulation of hepcidin.

6.12 Summary

Presently, the pathological progression regarding iron metabolism in patients with chronic alcohol consumption is not completely and conclusively understood. First findings on human subjects and animal models indicate that the systemic iron master switch hepcidin is suppressed. Hepcidin suppression likely contributes to the hepatic iron overload observed in these patients. On the other side, it seems not to be clear whether ROS play an important role in mediating this hepcidin suppression. Although, ROS are assuredly involved in alcohol metabolism, specific levels have never really been detected inside hepatic cells. Additionally, the suppression of hepcidin has only been shown in bolus treatment experiments that do not recapitulate the physiological condition. Contrary, low steady state levels of peroxide are able to induce hepcidin. Unfortunately, the clinical characterization of iron status in patients with ALD is complicated and heterogeneous. It is well known that these patients have low hemoglobin levels, which cannot be solely attributed to bleeding events. One mechanistic explanation includes an ROS-mediated induction of hepcidin. Furthermore, patients with ALD display increased transferrin saturation and ferritin levels, which are not only observed in cirrhotic patients and are clearly related to iron deposition found in macrophages and also in hepatocytes. This means that iron accumulation in the liver of ALD patients is not solely due to inflammatory events mediated by macrophages. The incomplete characterization of the iron physiology itself further aggravates the studies on ALD patients. For instance, it is not clear why in vivo and in vitro findings differ with regard to hepcidin responses toward iron exposure. Under in vivo conditions, iron loading leads to induction of hepcidin. The contrary is the case if liver cells in vitro were exposed to iron, which leads to subsequent downregulation of hepcidin. Therefore, it appears that the complicated network of ferroportin expression and potential systemic and paracrine effects of hepcidin are not yet fully understood. Presently, it should be well accepted that ROS are involved with and are responsible for iron disturbance in ALD. Additionally, low ROS levels lead to IRP1 and TfR1 upregulation that cause iron uptake and hepcidin upregulation that further results in iron retention in hepatocytes. Higher ROS levels that may destroy and damage hepatocytes can locally dramatically reduce and suppress hepcidin levels, what then could result in an "overflow" of iron and lead to further iron accumulation. It is quite conceivable that the presence of increased ROS (oxidative stress) and iron would dramatically increase Fenton-like reactions mentioned above, which then can be a highly cancerogenic environment. In summary, potential mechanisms of iron overload in ALD include:

1. Suppression of hepcidin via impaired redox signaling (role of NADPH oxidases) and overproduction of ROS
2. Impaired iron recycling (hemolysis)
3. Ineffective erythropoiesis
4. Retinoid signaling
5. Other iron proteins, e.g., TfR1 and IRP1
6. Interference of molecules with iron chelating and antioxidant property modifying iron absorption in the small intestine [88]

Acknowledgment This work was supported by the Dietmar-Hopp-Stiftung and the Manfred-Lautenschläger-Stiftung.

References

1. Seitz HK, Stickel F (2007) Molecular mechanisms of alcohol-mediated carcinogenesis. Nat Rev Cancer 7:599–612
2. Seitz HK, Mueller S (2011) Ethanol metabolism and its consequences. In: Anzenbacher P, Zanger U (eds) Metabolism of Drugs and Xenobiotics. Weinheim, Wiley
3. Mueller S, Millonig G, Seitz HK, Waite GN (2012) Chemiluminescence detection of hydrogen peroxide. In: Schipper H, Pantopoulos K (eds) Principles of Free Radical Biomedicine. Nova Publishers, New York
4. Mueller S, Millonig G, Seitz HK (2009) Alcoholic liver disease and hepatitis C: a frequently underestimated combination. World J Gastroenterol 15:3462–3471
5. Cochrane J, Chen H, Conigrave KM, Hao W (2003) Alcohol use in China. Alcohol Alcohol 38:537–542
6. Lozano R, Naghavi M, Foreman K, Lim S, Shibuya K, Aboyans V, Abraham J, Adair T, Aggarwal R, Ahn SY, Alvarado M, Anderson HR, Anderson LM, Andrews KG, Atkinson C, Baddour LM, Barker-Collo S, Bartels DH, Bell ML, Benjamin EJ, Bennett D, Bhalla K, Bikbov B, Bin Abdulhak A, Birbeck G, Blyth F, Bolliger I, Boufous S, Bucello C, Burch M,

Burney P, Carapetis J, Chen H, Chou D, Chugh SS, Coffeng LE, Colan SD, Colquhoun S, Colson KE, Condon J, Connor MD, Cooper LT, Corriere M, Cortinovis M, de Vaccaro KC, Couser W, Cowie BC, Criqui MH, Cross M, Dabhadkar KC, Dahodwala N, De Leo D, Degenhardt L, Delossantos A, Denenberg J, Des Jarlais DC, Dharmaratne SD, Dorsey ER, Driscoll T, Duber H, Ebel B, Erwin PJ, Espindola P, Ezzati M, Feigin V, Flaxman AD, Forouzanfar MH, Fowkes FG, Franklin R, Fransen M, Freeman MK, Gabriel SE, Gakidou E, Gaspari F, Gillum RF, Gonzalez-Medina D, Halasa YA, Haring D, Harrison JE, Havmoeller R, Hay RJ, Hoen B, Hotez PJ, Hoy D, Jacobsen KH, James SL, Jasrasaria R, Jayaraman S, Johns N, Karthikeyan G, Kassebaum N, Keren A, Khoo JP, Knowlton LM, Kobusingye O, Koranteng A, Krishnamurthi R, Lipnick M, Lipshultz SE, Ohno SL et al (2012) Global and regional mortality from 235 causes of death for 20 age groups in 1990 and 2010: a systematic analysis for the Global Burden of Disease Study 2010. Lancet 380:2095–2128
7. Powell LW (1975) The role of alcoholism in hepatic storage disease. In: Seixas FA, Williams K, Seggleston S (eds) Medical consequences of alcoholism. New York Academy of Sciences, New York
8. Finch SC, Finch CA (1955) Idiopathic hemochromatosis, an iron storage disease. A iron metabolism in hemochromatosis. Medicine (Baltimore) 34:381–430
9. Sheldon JH (1935) Haemochromatosis. Oxford University Press, London
10. Bell ET (1955) The relation of portal cirrhosis to hemochromatosis and to diabetes mellitus. Diabetes 4:435–446
11. Conrad ME, Berman A Jr, Crosby WH (1962) Iron kinetics in laennec's cirrhosis. Gastroenterology 43:385–390
12. Tavill AS, Qadri AM (2004) Alcohol and iron. Semin Liver Dis 24:317–325
13. Shoden A, Gabrio BW, Finch CA (1953) The relationship between ferritin and hemosiderin in rabbits and man. J Biol Chem 204:823–830
14. Feder JN, Gnirke A, Thomas W, Tsuchihashi Z, Ruddy DA, Basava A, Dormishian F, Domingo R Jr, Ellis MC, Fullan A, Hinton LM, Jones NL, Kimmel BE, Kronmal GS, Lauer P, Lee VK, Loeb DB, Mapa FA, McClelland E, Meyer NC, Mintier GA, Moeller N, Moore T, Morikang E, Wolff RK et al (1996) A novel MHC class I-like gene is mutated in patients with hereditary haemochromatosis. Nat Genet 13:399–408
15. Pascoe A, Kerlin P, Steadman C, Clouston A, Jones D, Powell L, Jazwinska E, Lynch S, Strong R (1999) Spur cell anaemia and hepatic iron stores in patients with alcoholic liver disease undergoing orthotopic liver transplantation. Gut 45:301–305
16. Sohda T, Takeyama Y, Irie M, Kamimura S, Shijo H (1999) Putative hemochromatosis gene mutations and alcoholic liver disease with iron overload in Japan. Alcohol Clin Exp Res 23:21S–23S
17. Dostalikova-Cimburova M, Kratka K, Stransky J, Putova I, Cieslarova B, Horak J (2012) Iron overload and HFE gene mutations in Czech patients with chronic liver diseases. Dis Markers 32:65–72
18. Chapman RW, Morgan MY, Laulicht M, Hoffbrand AV, Sherlock S (1982) Hepatic iron stores and markers of iron overload in alcoholics and patients with idiopathic hemochromatosis. Dig Dis Sci 27:909–916
19. Bell H, Skinningsrud A, Raknerud N, Try K (1994) Serum ferritin and transferrin saturation in patients with chronic alcoholic and non-alcoholic liver diseases. J Intern Med 236:315–322
20. Suzuki Y, Saito H, Suzuki M, Hosoki Y, Sakurai S, Fujimoto Y, Kohgo Y (2002) Up-regulation of transferrin receptor expression in hepatocytes by habitual alcohol drinking is implicated in hepatic iron overload in alcoholic liver disease. Alcohol Clin Exp Res 26:26S–31S
21. Mueller S (2013) Non-invasive assessment of patients with alcoholic liver disease. Clin Liver Dis 2:68–71
22. Mueller S (2013) The role of iron in alcoholic liver disease. Alcohol Alcohol 48:6
23. Fenton HJ (1894) Oxidation of tartaric acid in presence of iron. J Chem Soc Trans 65:899–910
24. Saran M, Michel C, Stettmaier K, Bors W (2000) Arguments against the significance of the Fenton reaction contributing to signal pathways under in vivo conditions. Free Radic Res 33:567–579

25. Niederau C, Fischer R, Purschel A, Stremmel W, Haussinger D, Strohmeyer G (1996) Long-term survival in patients with hereditary hemochromatosis. Gastroenterology 110:1107–1119
26. Ko C, Siddaiah N, Berger J, Gish R, Brandhagen D, Sterling RK, Cotler SJ, Fontana RJ, McCashland TM, Han SH, Gordon FD, Schilsky ML, Kowdley KV (2007) Prevalence of hepatic iron overload and association with hepatocellular cancer in end-stage liver disease: results from the National Hemochromatosis Transplant Registry. Liver Int 27:1394–1401
27. Turlin B, Juguet F, Moirand R, Le Quilleuc D, Loreal O, Campion JP, Launois B, Ramee MP, Brissot P, Deugnier Y (1995) Increased liver iron stores in patients with hepatocellular carcinoma developed on a noncirrhotic liver. Hepatology 22:446–450
28. Zacharski LR, Chow BK, Howes PS, Shamayeva G, Baron JA, Dalman RL, Malenka DJ, Ozaki CK, Lavori PW (2008) Decreased cancer risk after iron reduction in patients with peripheral arterial disease: results from a randomized trial. J Natl Cancer Inst 100:996–1002
29. Hann HW, Stahlhut MW, Blumberg BS (1988) Iron nutrition and tumor growth: decreased tumor growth in iron-deficient mice. Cancer Res 48:4168–4170
30. Tsukamoto H, Horne W, Kamimura S, Niemela O, Parkkila S, Yla-Herttuala S, Brittenham GM (1995) Experimental liver cirrhosis induced by alcohol and iron. J Clin Invest 96:620–630
31. D'Amico G, Garcia-Tsao G, Pagliaro L (2006) Natural history and prognostic indicators of survival in cirrhosis: a systematic review of 118 studies. J Hepatol 44:217–231
32. Ganne-Carrie N, Christidis C, Chastang C, Ziol M, Chapel F, Imbert-Bismut F, Trinchet JC, Guettier C, Beaugrand M (2000) Liver iron is predictive of death in alcoholic cirrhosis: a multivariate study of 229 consecutive patients with alcoholic and/or hepatitis C virus cirrhosis: a prospective follow up study. Gut 46:277–282
33. Wang J, Pantopoulos K (2011) Regulation of cellular iron metabolism. Biochem J 434:365–381
34. Andrews NC (1999) Disorders of iron metabolism. N Engl J Med 341:1986–1995
35. Ganz T (2007) Molecular control of iron transport. J Am Soc Nephrol 18:394–400
36. Nicolas G, Bennoun M, Devaux I, Beaumont C, Grandchamp B, Kahn A, Vaulont S (2001) Lack of hepcidin gene expression and severe tissue iron overload in upstream stimulatory factor 2 (USF2) knockout mice. Proc Natl Acad Sci U S A 98:8780–8785
37. Theil EC (2003) Ferritin: at the crossroads of iron and oxygen metabolism. J Nutr 133:1549S–1553S
38. Nemeth E, Ganz T (2009) The role of hepcidin in iron metabolism. Acta Haematol 122:78–86
39. Hentze MW, Muckenthaler MU, Galy B, Camaschella C (2010) Two to tango: regulation of mammalian iron metabolism. Cell 142:24–38
40. Sow FB, Florence WC, Satoskar AR, Schlesinger LS, Zwilling BS, Lafuse WP (2007) Expression and localization of hepcidin in macrophages: a role in host defense against tuberculosis. J Leukoc Biol 82:934–945
41. Weiss G, Goodnough LT (2005) Anemia of chronic disease. N Engl J Med 352:1011–1023
42. Millonig G, Ganzleben I, Peccerella T, Casanovas G, Brodziak-Jarosz L, Breitkopf-Heinlein K, Dick TP, Seitz HK, Muckenthaler MU, Mueller S (2012) Sustained submicromolar H2O2 levels induce hepcidin via signal transducer and activator of transcription 3 (STAT3). J Biol Chem 287:37472–37482
43. Miura K, Taura K, Kodama Y, Schnabl B, Brenner DA (2008) Hepatitis C virus-induced oxidative stress suppresses hepcidin expression through increased histone deacetylase activity. Hepatology 48:1420–1429
44. Ponka P, Beaumont C, Richardson DR (1998) Function and regulation of transferrin and ferritin. Semin Hematol 35:35–54
45. Kawabata H, Yang R, Hirama T, Vuong PT, Kawano S, Gombart AF, Koeffler HP (1999) Molecular cloning of transferrin receptor 2. A new member of the transferrin receptor-like family. J Biol Chem 274:20826–20832
46. Kawabata H, Nakamaki T, Ikonomi P, Smith RD, Germain RS, Koeffler HP (2001) Expression of transferrin receptor 2 in normal and neoplastic hematopoietic cells. Blood 98:2714–2719

47. Richardson DR, Lane DJ, Becker EM, Huang ML, Whitnall M, Suryo Rahmanto Y, Sheftel AD, Ponka P (2010) Mitochondrial iron trafficking and the integration of iron metabolism between the mitochondrion and cytosol. Proc Natl Acad Sci U S A 107:10775–10782
48. Rouault TA (2006) The role of iron regulatory proteins in mammalian iron homeostasis and disease. Nat Chem Biol 2:406–414
49. Recalcati S, Minotti G, Cairo G (2010) Iron regulatory proteins: from molecular mechanisms to drug development. Antioxid Redox Signal 13:1593–1616
50. Wallander ML, Leibold EA, Eisenstein RS (2006) Molecular control of vertebrate iron homeostasis by iron regulatory proteins. Biochim Biophys Acta 1763:668–689
51. Lok CN, Ponka P (2000) Identification of an erythroid active element in the transferrin receptor gene. J Biol Chem 275:24185–24190
52. Torti FM, Torti SV (2002) Regulation of ferritin genes and protein. Blood 99:3505–3516
53. Ponka P, Lok CN (1999) The transferrin receptor: role in health and disease. Int J Biochem Cell Biol 31:1111–1137
54. Piccinelli P, Samuelsson T (2007) Evolution of the iron-responsive element. RNA 13:952–966
55. Deck KM, Vasanthakumar A, Anderson SA, Goforth JB, Kennedy MC, Antholine WE, Eisenstein RS (2009) Evidence that phosphorylation of iron regulatory protein 1 at Serine 138 destabilizes the [4Fe-4S] cluster in cytosolic aconitase by enhancing 4Fe-3Fe cycling. J Biol Chem 284:12701–12709
56. Sanchez M, Galy B, Muckenthaler M, Hentze MW (2007) Iron-regulatory proteins limit hypoxia-inducible factor-2alpha expression in iron deficiency. Nat Struct Mol Biol 14:420–426
57. Vashisht AA, Zumbrennen KB, Huang X, Powers DN, Durazo A, Sun D, Bhaskaran N, Persson A, Uhlen M, Sangfelt O, Spruck C, Leibold EA, Wohlschlegel JA (2009) Control of iron homeostasis by an iron-regulated ubiquitin ligase. Science 326:718–721
58. Salahudeen AA, Thompson JW, Ruiz JC, Ma HW, Kinch LN, Li Q, Grishin NV, Bruick RK (2009) An E3 ligase possessing an iron-responsive hemerythrin domain is a regulator of iron homeostasis. Science 326:722–726
59. Fillebeen C, Pantopoulos K (2002) Redox control of iron regulatory proteins. Redox Rep 7:15–22
60. Mueller S (2005) Iron regulatory protein 1 as a sensor of reactive oxygen species. Biofactors 24:171–181
61. Mutze S, Hebling U, Stremmel W, Wang J, Arnhold J, Pantopoulos K, Mueller S (2003) Myeloperoxidase-derived hypochlorous acid antagonizes the oxidative stress-mediated activation of iron regulatory protein 1. J Biol Chem 278:40542–40549
62. Watts RN, Hawkins C, Ponka P, Richardson DR (2006) Nitrogen monoxide (NO)-mediated iron release from cells is linked to NO-induced glutathione efflux via multidrug resistance-associated protein 1. Proc Natl Acad Sci U S A 103:7670–7675
63. Hillman RS (1975) Alcohol and hematopoiesis. In: Seixas FA, Williams K, Seggleston S (eds) Medical consequences of alcoholism. New York Academy of Sciences, New York
64. Hines JD (1975) Hematologic abnormalities involving vitamin B6 and folate metabolism in alcoholic subjects. In: Seixas FA, Williams K, Seggleston S (eds) Medical consequences of alcoholism. New York Academy of Sciences, New York
65. Lumeng L, Li TK (1974) Vitamin B6 metabolism in chronic alcohol abuse. J Clin Invest 53:698
66. Herbert V, Tisman G (1975) Hematologic effects of alcohol. In: Seixas FA, Williams K, Seggleston S (eds) Medical consequences of alcoholism. New York Academy of Sciences, New York
67. Scharf R, Aul C (1988) Alcohol-induced disorders of the hematopoietic system. Z Gastroenterol 26(Suppl 3):75–83
68. Eichner ER (1973) The hematologic disorders of alcoholism. Am J Med 54:621–630
69. Lindenbaum J (1987) Hematologic complications of alcohol abuse. Semin Liver Dis 7:169–181

70. Niemela O, Parkkila S (2004) Alcoholic macrocytosis—is there a role for acetaldehyde and adducts? Addict Biol 9:3–10
71. Pitcher CS, Williams R (1963) Reduced red cell survival in jaundice and its relation to abnormal glutathione metabolism. Clin Sci 24:239–252
72. Krasnow SE, Walsh JR, Zimmerman HJ, Heller P (1957) Megaloblastic anemia in alcoholic cirrhosis. AMA Arch Intern Med 100:870–880
73. Maturu P, Reddy VD, Padmavathi P, Varadacharyulu N (2012) Ethanol induced adaptive changes in blood for the pathological and toxicological effects of chronic ethanol consumption in humans. Exp Toxicol Pathol 64:697–703
74. McClain CJ, Barve SS, Barve A, Marsano L (2011) Alcoholic liver disease and malnutrition. Alcohol Clin Exp Res 35:815–820
75. Mezey E (1975) Intestinal function in chronic alcoholism. In: Seixas FA, Williams K, Seggleston S (eds) Medical consequences of alcoholism. New York Academy of Sciences, New York
76. Madden AM, Bradbury W, Morgan MY (1997) Taste perception in cirrhosis: its relationship to circulating micronutrients and food preferences. Hepatology 26:40–48
77. Mezey E (1982) Liver disease and protein needs. Annu Rev Nutr 2:21–50
78. Hanje AJ, Fortune B, Song M, Hill D, McClain C (2006) The use of selected nutrition supplements and complementary and alternative medicine in liver disease. Nutr Clin Pract 21:255–272
79. Stickel F, Hoehn B, Schuppan D, Seitz HK (2003) Review article: Nutritional therapy in alcoholic liver disease. Aliment Pharmacol Ther 18:357–373
80. Ohtake T, Saito H, Hosoki Y, Inoue M, Miyoshi S, Suzuki Y, Fujimoto Y, Kohgo Y (2007) Hepcidin is down-regulated in alcohol loading. Alcohol Clin Exp Res 31:S2–S8
81. Harrison-Findik DD, Schafer D, Klein E, Timchenko NA, Kulaksiz H, Clemens D, Fein E, Andriopoulos B, Pantopoulos K, Gollan J (2006) Alcohol metabolism-mediated oxidative stress down-regulates hepcidin transcription and leads to increased duodenal iron transporter expression. J Biol Chem 281:22974–22982
82. Harrison-Findik DD, Klein E, Crist C, Evans J, Timchenko N, Gollan J (2007) Iron-mediated regulation of liver hepcidin expression in rats and mice is abolished by alcohol. Hepatology 46:1979–1985
83. Harrison-Findik DD, Klein E, Evans J, Gollan J (2009) Regulation of liver hepcidin expression by alcohol in vivo does not involve Kupffer cell activation or TNF-alpha signaling. Am J Physiol Gastrointest Liver Physiol 296:G112–G118
84. Shindo M, Torimoto Y, Saito H, Motomura W, Ikuta K, Sato K, Fujimoto Y, Kohgo Y (2006) Functional role of DMT1 in transferrin-independent iron uptake by human hepatocyte and hepatocellular carcinoma cell, HLF. Hepatol Res 35:152–162
85. Suzuki M, Fujimoto Y, Suzuki Y, Hosoki Y, Saito H, Nakayama K, Ohtake T, Kohgo Y (2004) Induction of transferrin receptor by ethanol in rat primary hepatocyte culture. Alcohol Clin Exp Res 28:98S–105S
86. Andriopoulos B, Hegedusch S, Mangin J, Hd R, Hebling U, Wang J, Pantopoulos K, Mueller S (2007) Sustained hydrogen peroxide induces iron uptake by transferrin receptor-1 independent of the iron regulatory protein/iron-responsive element network. J Biol Chem 282:20301–20308
87. Nishina S, Hino K, Korenaga M, Vecchi C, Pietrangelo A, Mizukami Y, Furutani T, Sakai A, Okuda M, Hidaka I, Okita K, Sakaida I (2008) Hepatitis C virus-induced reactive oxygen species raise hepatic iron level in mice by reducing hepcidin transcription. Gastroenterology 134:226–238
88. Ren Y, Deng F, Zhu H, Wan W, Ye J, Luo B (2011) Effect of epigallocatechin-3-gallate on iron overload in mice with alcoholic liver disease. Mol Biol Rep 38:879–886

Chapter 7
Alcoholic Cirrhosis and Hepatocellular Carcinoma

Felix Stickel

Abstract Hepatocellular carcinoma shows a rising incidence worldwide, and the largest burden of disease in Western countries derives from patients with alcoholic liver disease (ALD) and cirrhosis, the latter being the premier premalignant factor for HCC. The present chapter addresses key issues including the epidemiology of alcohol-associated HCC, and its link to other coexisting non-alcoholic liver diseases, and additional host and environmental risk factors including the underlying genetics. Also discussed are molecular mechanisms of alcohol-associated liver cancer evolution involving the mediators of alcohol toxicity and carcinogenicity, acetaldehyde and reactive oxygen species, as well as the recently described mutagenic adducts which these mediators form with DNA. Specifically, interference of alcohol with retinoids and cofactors of transmethylation processes are outlined. Information presented in this chapter illustrates that the development of HCC in the context of ALD is multifaceted and suggests several molecular targets for prevention and markers for the screening of risk groups.

Keywords Acetaldehyde • Alcohol dehydrogenase • Cytochrome P450 2E1 • Gene methylation • Liver cirrhosis • Non-alcoholic fatty liver disease • Retinoic acid • Viral hepatitis

7.1 Introduction

The incidence of hepatocellular cancer (HCC) is rising worldwide with more than 600,000 new cases per year (4 % of all cancers) resulting in HCC being the fifth most frequent cancer and third most frequent cause of cancer-related death only

F. Stickel, M.D. (✉)
Department of Clinical Research, University of Bern,
Murtenstrasse 35, CH-3010 Bern, Germany

Hepatology Unit, Clinic Beau-Site, Hirslanden, Bern, Germany
e-mail: Felix.stickel@ikp.unibe.ch

V. Vasiliou et al. (eds.), *Biological Basis of Alcohol-Induced Cancer*,
Advances in Experimental Medicine and Biology 815,
DOI 10.1007/978-3-319-09614-8_7

surpassed by cancers of the lungs and the stomach [1–4]. While chronic infections with hepatitis B and C viruses are the prime causes leading to cirrhosis and HCC in Southeast Asia and sub-Saharan Africa [5], persistent high alcohol consumption and progressive liver disease associated with the metabolic syndrome resulting in non-alcoholic steatohepatitis (NASH) are the leading and emerging causes in Western industrialized regions [6, 7]. The Global Burden of Disease Project of the WHO concludes that alcohol accounts for approximately 1.8 million annual deaths and one of the most significant diseases caused by chronic alcohol consumption is cancer [8]. In February 2007 an international group of specialist met at the International Agency for Research on Cancer (IARC) in Lyon, France, to evaluate the role of alcohol and its first metabolite acetaldehyde, as potential carcinogens. This working group concluded that the occurrence of malignant tumours of the oral cavity, pharynx, larynx, oesophagus, liver, colorectum and female breast is causally related to the consumption of alcoholic beverages [9]. Worldwide, a total of approximately 389,000 cases of cancer representing 3.6 % of all cancers derive from chronic alcohol consumption [10].

The intention of this chapter is to provide the reader with an overview on the topic without the claim of being either complete or even objective.

7.2 Epidemiology

In the USA, HCC is the fastest-growing cause of cancer-related death in men with incidence rates increasing more than twofold between 1985 and 2002 [11]. Incidence of HCC closely corresponds with its mortality with some 626,000 cases diagnosed each year and 598,000 deaths due to HCC. There are important differences between countries and regions in terms of HCC incidence. The latter can be as high as 99/100,000 in the Mongolian Republic and around 30–35/100,000 in East Asia and sub-Saharan Western Africa, whereas it can be only around 10–15/100,000 in Italy, Spain and Greece or even lower (1–5/100,000) in countries like France, Great Britain, Germany, Canada, Northern America and Scandinavia [12]. Over the last decades a gradually decreasing incidence of HCC in many high prevalence areas of the world has been observed, whereas it has nearly doubled in low prevalence regions such as the USA and Europe [13]. While the decline in Asia is predominantly the result of large-scale vaccination against hepatitis B virus infection and decreased exposure to dietary aflatoxins with improved handling of food, the increase in Western countries is caused by the rising incidence of progressively fibrosing viral hepatitis C and persistently high alcohol consumption. In an analysis of 1,605 patients diagnosed with HCC between 1993 and 1998, rates of HCC due to chronic hepatitis C infection increased threefold, while age-adjusted rates for HCC following chronic hepatitis B infection and alcohol abuse remained stable [14].

7.3 Pathophysiology

7.3.1 Tissue Remodelling as the Priming Condition for HCC

The vast majority of alcohol-associated HCCs develop in patients with established alcoholic cirrhosis, and alcohol-related HCC without cirrhosis is rare. However, case series have shown that the latter may occasionally occur [15–17]. Cirrhosis as a priming condition is not equally prone to HCC development across etiologies, and Fattovich et al. [2] calculated the 5-year incidence of HCC in alcoholic cirrhosis at 8 % which is somewhat lower than the figures observed in Asians with HCV infection (30 %) or hereditary hemochromatosis (21 %) (Fig. 7.1).

Obviously, structural and functional abnormalities pertinent to alcoholic cirrhotic transformation such as profound alterations of extracellular matrix composition, growth factor and cytokine milieu, disturbed angiogenesis, impaired immune function and reduced capacity of cirrhotic tissue to handle oxidative and/or toxic insults result in an environment that favours dedifferentiation and malignant growth.

Certain histological features typically seen in established HCC are already present, albeit to a lesser extent, in alcoholic cirrhosis and thereby support the assumption that some pathogenic hallmarks leading to cirrhosis precede those causing HCC [6]. Enzyme-altered foci and preneoplastic nodules are typically observed premalignant lesions which can also be induced in experimental rodent models of HCC [18]. Interestingly, Mallory body (MB) formation is high in HCC, and the incidence of HCC is significantly higher in cirrhosis with MBs than without leading to the hypothesis that MBs reflect an initial phenotypical alteration in the carcinogenic transformation of hepatocytes [19]. In addition, oval cells—pluripotent liver progenitor cells—are present in premalignant liver tissues HCC and adjacent tissues and evolve in response to long-term alcohol exposure [20].

HCC evolution is closely linked to chronic liver injury from various causes requiring cell/tissue regeneration for replacement of hepatocyte decay, but rarely develops

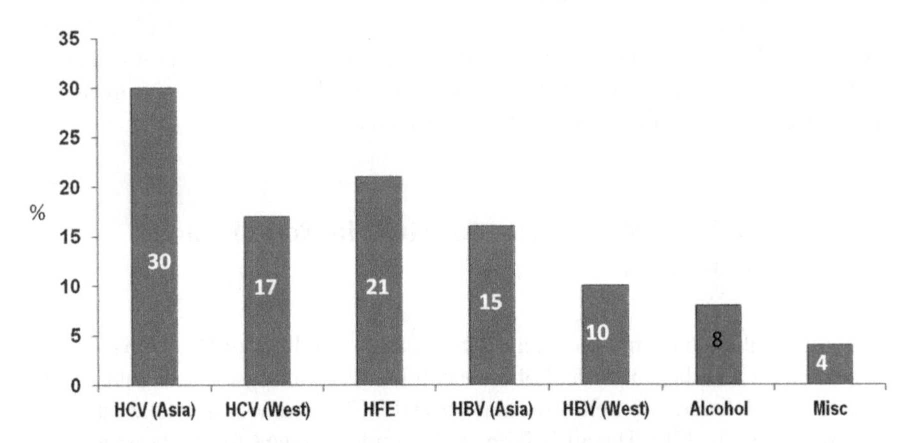

Fig. 7.1 Five-year cumulative incidence of HCC in different etiologies of liver cirrhosis (Data from Fattovich et al., Gastroenterology 2004;127:S35-50)

in healthy liver during physiological aging. One possible explanation for this tight correlation is that HCC development requires increased cell proliferation, leading to the stepwise accumulation of genetic hits allowing for dysplastic and eventually malignant transformation. The most common and unifying condition associated with hepatocarcinogenesis is cirrhosis which takes long to develop (20–40 years). As mentioned above, cirrhosis induces alterations of the microenvironment including altered cytokine secretion from activated hepatic stellate cells and portal fibroblasts, as well as proinflammatory stimuli from invading immune cells. Regarding the latter, molecular signals derived from proinflammatory tumour-necrosis factor α (TNFα) are considered pivotal in ALD [13]. Excessive alcohol consumption can lead to an increased portosystemic uptake of lipopolysaccharides/endotoxins from gram-negative gut bacteria which contribute to necroinflammation and fibrosis progression via various molecular mechanisms including TNFα and the CD14/toll-like receptor-4 complex to produce ROS via NADPH oxidase [21–23]. In fact, elevated TNFα levels and other proinflammatory cytokines are prominent features of ALD, resulting in hepatocyte proliferation or apoptotic/necrotic death, recruitment of inflammatory cells and tissue remodelling. Molecular responses are triggered upon binding of TNFα to its cellular receptors on hepatocytes and other liver cells leading to activation of adaptor protein 1 (AP-1; c-*jun*/c-*fos*), crosstalk with epidermal growth factor signalling and subsequently enhanced cell proliferation and potentially to apoptosis via caspase activation [24]. Beyond that, TNFα activates sphingomyelinase to increase intracellular ceramide which inhibits the mitochondrial electron transport chain. Consequently increased production of ROS promotes lipid peroxidation and apoptosis independently of caspases. However, increased oxidative stress also contributes to the activation of transcription nuclear factor κB which is instrumental for the initiation of cell survival mechanisms involving the upregulation of antiapoptotic proteins such as Bcl-2, manganese superoxide dismutase and nitric oxide synthase that can all protect mitochondrial integrity and function. Indeed, constitutive upregulation of nuclear factor κB expression has been convincingly demonstrated both in human and experimental ALD [25, 26] and is a typical feature in human HCC [27]. Hence, TNFα may dose-dependently activate cellular survival mechanisms or precipitate apoptosis and/or necrosis. This may provide an explanation why hepatocytes challenged by inflammatory insults below the threshold of cell death may acquire proliferative advantages and become susceptible to dedifferentiation triggered by carcinogens such as (alcohol-derived) acetaldehyde.

7.4 Alcohol as a Risk Factor for HCC in Non-alcoholic Liver Diseases

Chronic alcohol consumption is an established risk modifier of HCC development in patients with other, non-alcoholic concomitant liver diseases, particularly viral hepatitis [28], hereditary hemochromatosis (HH) [29] and non-alcoholic fatty liver disease (NAFLD) [30]. Hepatitis B and C infections account for the large majority of HCC cases in the developing world, whereas NAFLD along with the obesity epidemic is an increasing cause of HCC in Western countries. In these diseases, which render the liver susceptible to additional oncogenic insults, chronic alcohol consumption even at moderate levels could have a striking influence on the population-based risk of HCC risk [31].

7.4.1 Coexisting Chronic Viral Hepatitis

A large Italian case-control study including 464 subjects with HCC as cases and 824 subjects without liver diseases as controls demonstrated a synergism between alcohol drinking (>60 g/day) and hepatitis B or C virus infection regarding the risk of HCC with approximately a twofold increase in the odds ratio for each viral infection for drinkers [32] (Fig. 7.2).

Not surprisingly, the coexistence of two liver diseases (alcohol + chronic infection with hepatitis viruses) synergistically enhances the risk of liver disease progression and also HCC. However, the mechanisms leading to HCC evolution are imprecisely defined and may be distinct between the two types of viral hepatitis.

7.4.2 Hepatitis B

Data from human studies on the interaction between HBV infection and alcohol consumption are scarce. Marcellin and co-workers estimated the mortality related to HCV and HBV infections in France in 2001. They found that 95 % of patients with HCV infection who died had cirrhosis, and 33 % had HCC, similar to the

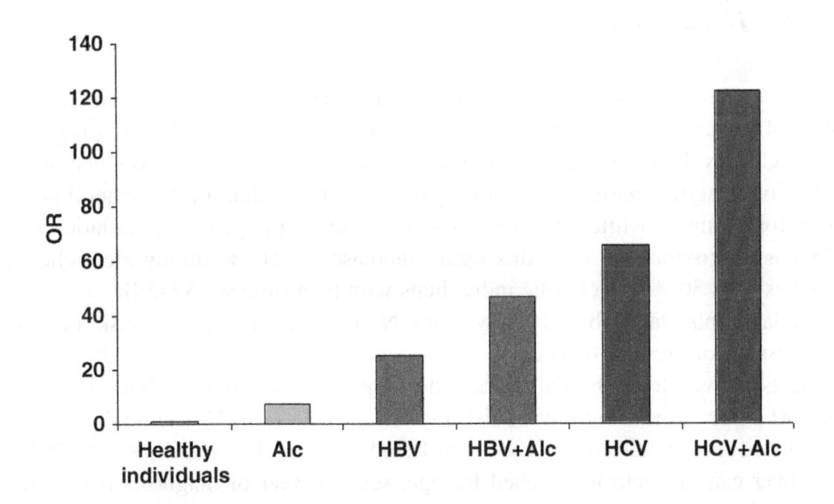

Fig. 7.2 Odds ratios (OR) for HCC in drinkers with/without chronic viral hepatitis B or C. Coexisting alcohol drinking doubles the risk of HCC in patients infected with hepatitis B or C virus (Data from Donato et al. Am J Epidemiol 2002;155:323–31)

figures in the HBV infection group in which 93 % of subjects had cirrhosis and 35 % had HCC. Overall, deaths related to either HBV or HCV infection occurred at an earlier age in patients with a history of excessive alcohol consumption [30]. In a similar study from Italy based on data from the Dionysos study, authors quantified the burden of chronic viral liver disease and the role of alcohol intake to morbidity and mortality in a population-based study. Here, ethanol intake was found to be an independent predictor of cirrhosis in subjects with chronic HCV infection and an independent predictor of death in subjects with either HCV or HBV infection [33]. Comparable data come from a prospective cohort study in Taiwan which demonstrated that habitual alcohol consumption can increase the risk of HCC in HBsAg-positive individuals with an odds ratio of 1.86 (1.32–2.61) [34]. A population-based cohort study from Korea found that in the subgroup of chronic HBV carriers, the HCC risk rose dose dependently with an alcohol intake of 50–99 g/day with a relative risk of 1.2 (95 % CI 1.0–1.5) and of 1.5 (95 % CI 1.2–2.0) for >100 g/day [35]. Whether this synergistic effect on the risk of HCC from alcohol and coexistent HBV infection is additive or exponential is not known.

Putative mechanisms are yet unknown, but may relate to a distinct pattern of epigenetic regulation of certain HCC-associated genes as evidenced by Lambert et al. who showed a high frequency of aberrant hypermethylation of specific genes (RASSF1A, GSTP1, CHRNA3 and DOK1) in HCCs compared to control cirrhotic or normal liver tissues [36]. An association between alcohol intake and hypomethylation of the methyl-guanine methyltransferase gene promoter was demonstrated, whereas HBV infection was linked to promoter hypermethylation of glutathione S-transferase, indicating that hypermethylation of the genes analysed in HCC tumours exhibits remarkably distinct patterns between associated risk factors.

7.4.3 Hepatitis C

Ample evidence exists demonstrating a clear synergistic effect between coexisting alcohol abuse and chronic HCV infection with regard to liver disease progression and HCC development. This is particularly important since the prevalence of HCV infection is significantly higher among heavy drinkers than in the general population; for example, while HCV antibody positivity in the general population of the USA is approximately 1 %, this figure increases to 16 % among alcoholics and reaches even 30 % in alcoholic individuals with liver disease [37–39].

A large population-based study from Northern Italy analysed risk factors of progression of chronic hepatitis C and development of HCC in anti-HCV-positive subjects derived from the Dionysos cohort and found a daily alcohol consumption of >90 g/day to be a significant risk factor for HCC [40]. Hassan and co-workers conducted a hospital-based, case-control study among 115 HCC patients and 230 non-liver cancer controls matched for age, sex and year of diagnosis [41]. Factors independently associated with HCC were chronic hepatitis B and C, alcohol consumption (>80 g/day) and type II diabetes. Significant synergism was observed between heavy alcohol consumption and chronic HCV infection (OR 53.9; 95 % CI 7.0–415.7) and diabetes mellitus (OR 9.9; 95 % CI, 2.5–39.3). The study underlines that heavy alcohol consumption contributes to the magnitude of HCC cases (32 %), whereas 22 %, 16 % and 20 % were explained by HCV, HBV and diabetes mellitus, respectively. Similar data have been gathered for Europe and Asia as well in which concomitant alcohol consumption in HCV-infected individuals increases the risk of HCC additively, if not exponentially [42, 43].

The underlying pathophysiology of this synergistic impact on HCC evolution is still not completely understood but may relate to joint effects of both alcohol and HCV on certain effects conveyed by HCV epitopes on key molecular events instrumental in hepatocarcinogenesis. In keeping, experimental evidence generated by Moriya and associates is highly suggestive of a direct oncogenic effect of the HCV core protein in mice [44]. In their study, the development of HCC in two independent lines of transgenic mice overexpressing the HCV core gene was reported. The same mice spontaneously developed steatosis early in life as a feature of chronic hepatitis C infection and of alcohol, but not more advanced liver disease (Fig. 7.3). The latter similarity allows for speculations with regard to coexisting alcohol abuse [45]: the downstream events triggered by the core protein are segregated into two components. One is the augmented production of oxidative stress along with the activation of scavenging system, including catalase and glutathione, in the putative preneoplastic stage with steatosis in the liver. Thus, oxidative stress production in the absence of inflammation by the core protein would partly contribute to the development of HCC. The generation of oxidative stress is estimated to originate from mitochondrial dysfunction in hepatocytes by HCV infection. Obviously,

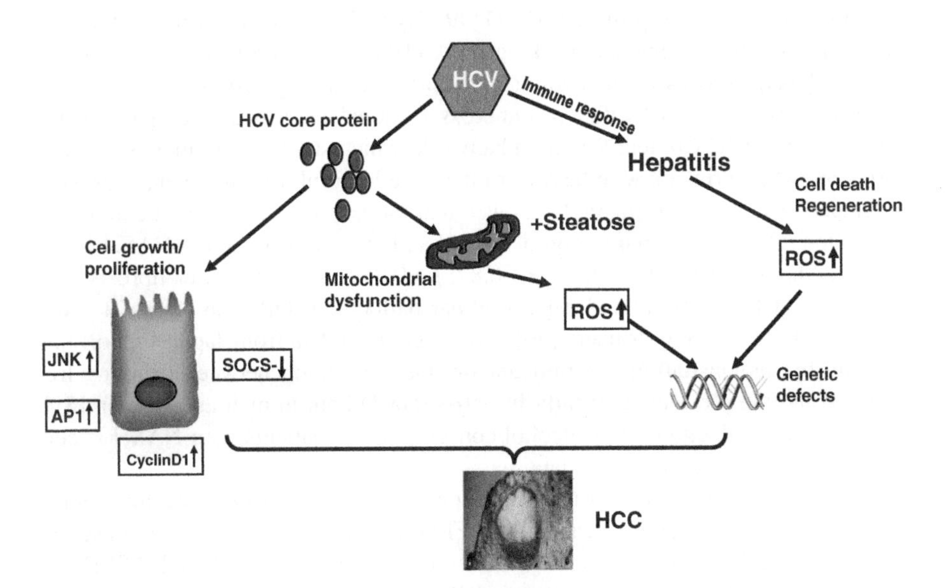

Fig. 7.3 HCV core protein promotes hepatocarcinogenesis indirectly via causing chronic inflammation and mitochondrial dysfunction. However, evidence exists that HCV may also directly cause malignant transformation through stimulating hyperproliferation, reduction of cytokine release and upregulation of cell survival mechanisms

oxidative stress from concomitant alcohol consumption would further intensify these HCV-related effects. The other component is the alteration of intracellular signalling cascade of mitogen-activated protein kinase (MAPK) and activating factor (AP)-1, leading to the activation of cell growth and proliferation. Notably, AP-1 upregulation is a key observation in alcohol-mediated liver cell regeneration via retinoic acid receptors, and MAP kinase cascades and their regulation by the phosphoinositide-3-kinase/Akt signalling cascade appear to be crucial in the onset of alcohol-mediated cell injury [46]. These mechanisms are also hallmarks of alcohol-associated hepatocarcinogenesis, and joint impact from both alcohol and HCV on these molecular targets represents a potential explanation for the increased incidence of HCC in HCV-infected alcoholics.

7.5 Non-alcoholic Fatty Liver Disease

Since hepatic histological changes in NASH as well as in alcoholic steatohepatitis (ASH) are strikingly similar, and regular alcohol consumption in obese subjects is rather the rule than the exception, it is not surprising that alcohol consumption may aggravate obesity-related liver disease and vice versa.

Accordingly, two studies from France demonstrated that alcohol consumption is a confirmed risk factor of steatosis and fibrosis progression in obese individuals [47, 48]. A more recent study from Sweden investigated whether low alcohol intake, consistent with the diagnosis of NAFLD, is associated with fibrosis progression in 71 subjects with biopsy-proven NAFLD [49]. By multivariate binary logistic regression analysis, heavy episodic drinking ($p<0.001$) and insulin resistance ($p<0.01$) were independently associated with significant fibrosis progression.

These results contrast with a recent cross-sectional analysis of adult participants in the NIH NASH Clinical Research Network, which studied only modest or non-drinkers, while drinkers with heavy or otherwise harmful drinking behaviour were excluded. This study surprisingly showed that modest drinkers (daily alcohol consumption <20 g) compared to non-drinkers had lower odds of having NASH (summary odds ratio 0.56, 95 % CI 0.39–0.84, $p=0.002$) and also lesser fibrosis (OR 0.56 95 % CI 0.41–0.77) and hepatocellular ballooning (OR 0.66 95 % CI 0.48–0.92) than lifetime non-drinkers [50]. Cross-sectional data from Japan support this finding by demonstrating a significant inverse correlation between drinking frequency and the prevalence of fatty liver ($p<0.001$) both in men and women [51]. Thus, a total embargo on any alcohol consumption in patients with NASH is currently not supported by human data.

However, very few data exist focussing on the interaction between features of the metabolic syndrome associated with NAFLD and concomitant alcohol consumption regarding HCC development. However, bearing in mind that both NAFLD and ALD are independent risk factors of HCC, coincidence of either etiology in an individual will likely increase the risk of HCC, particularly when considering that the mechanisms leading to HCC in either disease are remarkably similar (Fig. 7.4).

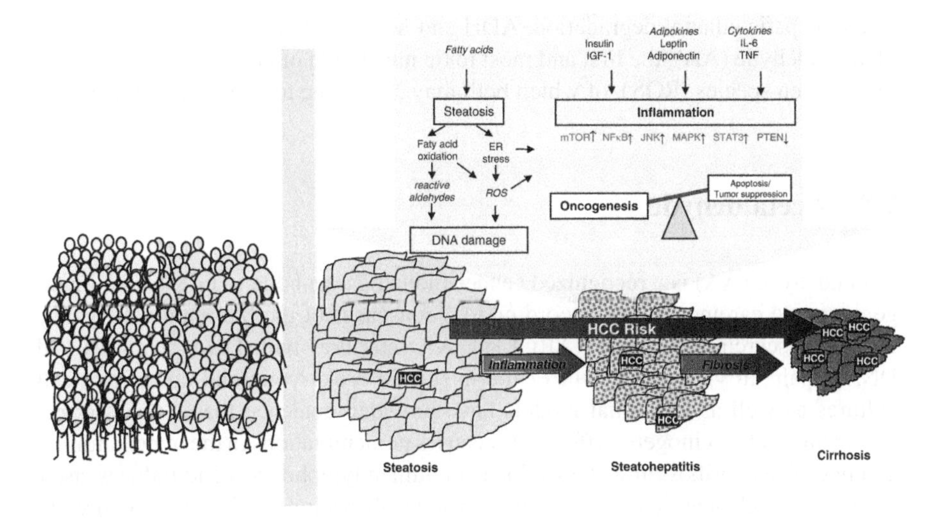

Fig. 7.4 Pathophysiology of alcoholic and non-alcoholic fatty liver disease shares many similarities which synergistically promote HCC development in subjects who harbour both risk factors as a result of their lifestyle (Adapted from Stickel F, Hellerbrand C. Gut 2010;59:1303–7)

7.6 Alcohol-Mediated Activation of Environmental Carcinogens

Alcoholics may be exposed to carcinogens or procarcinogens ingested along with alcoholic beverages which may contain nitrosamines, polycyclic hydrocarbons, asbestos fibres and fusel oils [52]. In addition, many alcoholics are smokers and epidemiological surveys have shown a hyperadditive effect of alcohol and smoking in increasing the risk of developing HCC [53]. Some of these procarcinogens are activated by cytochrome P450 2E1 (CYP2E1), which is induced by chronic ethanol consumption (see below). Thus, potent hepatocarcinogens, such as nitrosamines, aflatoxins as well as vinyl chloride, are substrates of CYP2E1 and undergo activation [54].

Importantly aflatoxin B1 can induce a mutation in codon 249 of the p53 tumour suppressor gene which is frequently found in human HCC [55]. Although animal experiments have been controversial as to whether ethanol enhances AFB_1-induced hepatocarcinogenesis, an epidemiological study on AFB_1 exposure demonstrated that even a moderate daily consumption of 24 g ethanol increases the risk of developing HCC induced by 4 μg of dietary AFB_1 by 35-fold [56].

7.7 Ethanol Metabolism and HCC

More than 90 % of ethanol is metabolized in the liver, either by alcohol dehydrogenase (ADH), the microsomal ethanol oxidizing system (MEOS) represented by CYP2E1 or to a much smaller extent by catalase. While catalase plays only a minor role in hepatic ethanol degradation, ADH and MEOS produce substantial amounts of acetaldehyde (AA), the first and most toxic metabolite of ethanol as well as reactive oxygen species (ROS), of which both may contribute to hepatocarcinogenesis.

7.8 Acetaldehyde

Acetaldehyde (AA) is a recognized cancer-inducing agent due to its toxicity, mutagenicity and carcinogenicity according to a consensus of the International Agency for the Research on Cancer (IARC) [9]. AA interferes with DNA synthesis and DNA repair. In vivo and in vitro experiments in prokaryotic and eukaryotic cell cultures as well as in animal models have provided evidence that AA has direct mutagenic and carcinogenic effects. AA causes point mutations in the hypoxanthine-guanine-phosphoribosyltransferase locus in human lymphocytes and induces sister chromatid exchanges and gross chromosomal aberrations [57–61]. AA directly inhibits O6-methylguanosyl transferase, an enzyme that repairs DNA adducts [62].

Most importantly, however, AA binds to DNA and forms stable adducts [61, 63–69]. Binding to DNA represents one mechanism by which AA could trigger replication errors and/or mutations in oncogenes and tumour suppressor genes. It has been shown that the major stable DNA adduct N2-ethyl-desoxyguanosine (N2-Et-dG) serves as a substrate of eukaryotic DNA polymerase. However, N2-Et-dG seems rather a marker for chronic ethanol consumption than a major risk lesion for cancer. In addition, another DNA adducts of AA, 1,N2-propano-desoxyguanosine (PdG) has been identified, especially in the presence of basic amino acids, histones and polyamines. While N2-Et-dG is non-mutagenic and may represent a marker of chronic alcohol ingestion, PdG has mutagenic properties.

Evidence of the causal role of AA in ethanol-mediated carcinogenesis comes from genetic candidate gene case-control studies in alcoholics. Individuals who accumulate AA due to polymorphisms and/or mutations in the genes coding for enzymes responsible for AA generation and degradation have been shown to have an increased cancer risk. Thus, individuals who drink alcohol and have a deficient aldehyde dehydrogenase (ALDH) such as 40 of the Asian population with increased AA levels after drinking also have a high risk for various cancers such as those of the upper aerodigestive tract and the colon [70]. Similarly individuals who produce more AA due to a rapid alcohol dehydrogenase (ADH1C1*1) also have an increased risk for these cancers including HCC [71].

7.9 Oxidative Stress

Oxidative stress by the formation of ROS such as superoxide anion and hydrogen peroxide is another hallmark of alcohol-mediated tissue injury. Several enzyme systems are capable to produce ROS, including CYP2E1, the mitochondrial respiratory chain and the cytosolic enzymes xanthine oxidase and aldehyde oxidase [72]. Ethanol-mediated ROS formation may be due to an increased electron leakage from the mitochondrial reparatory chain associated with the stimulation of reduced nicotinamide adenine dinucleotide (NADH) shuttling into mitochondria and to the interaction between N-acetylsphingosine (via TNF-α) and mitochondria. The induction of sphingomyelinase by TNF-α increases the levels of ceramide, an inhibitor of the activity of the mitochondrial electron transport chain, leading to increased mitochondrial production of ROS [73]. ROS can also be generated in alcoholic hepatitis with activated hepatic phagocytes [74]. Hepatic iron overload as observed in alcoholic liver disease increases ROS and finally nitric oxide production due to ethanol-mediated stimulation of inducible nitric oxide synthase results in the formation of highly reactive peroxynitrite [75].

Most important in ALD, however, is the production of ROS via alcohol-induced CYP2E1. This induction is an adaptive process and associated with an increased metabolism of ethanol to acetaldehyde and ROS. Most likely, increased CYP2E1 activity results from an inadequate degradation of CYP2E1 by the ubiquitin proteasome pathway impaired by alcohol. A significant increase in hepatic CYP2E1 activity may already occur following the ingestion of 40 g of ethanol daily for 1 week [76]. In animal experiments, the induction of CYP2E1 correlates with NAD phosphate (NADPH) oxidase activity, the generation of hydroxyethyl radicals, lipid peroxidation and the severity of hepatic damage, all of which could be prevented by the CYP2E1 inhibitor chlormethiazole [77]. In addition, premalignant DNA lesions were found more frequently in CYP2E1 wild-type mice compared to animals knocked out for CYP2E1 [78], and transgenic mice overexpressing CYP2E1 had more severe liver injury [79].

ROS produced by CYP2E1 results in lipid peroxidation. Various lipid peroxidation products including 4-hydroxynonenal may bind to purine and pyrimidine bases thereby forming exocyclic DNA adducts which can be mutagenic and carcinogenic [80, 81]. Biopsies from patients with various degrees and severities of alcoholic liver disease reveal a gradual increase of exocyclic DNA adducts along with the degree of alcoholic liver injury and a significant correlation with 4-HNE immunostaining and CYP2E1 induction [82]. This takes already place at the stage of alcoholic fatty liver [83]. Notably, the formation of etheno-DNA adducts can be inhibited by treatment with the CYP2E1 inhibitor chlormethiazole.

7.10 Alcohol and Altered DNA Methylation

Apart from genetic changes along with chronic alcoholism, i.e. mutations, DNA cross links or impaired DNA repair, chronic and acute alcohol intake may also affect epigenetic mechanisms of gene expression such as methylation of DNA. DNA methylation is an important determinant in controlling gene expression whereby hypermethylation has a silencing effect on genes and hypomethylation may lead to increased gene expression. And indeed, alcohol intercepts with these epigenetic mechanisms [84].

Alcohol interacts with absorption, storage, biologic transformation and excretion of compounds which are essential for methyl group transfer including folate, vitamin B6 and certain lipotropes. Especially, the production of S-adenosyl-L-METHIONINE (SAMe), the universal methyl group donor in methylation reactions, is impaired. Furthermore, ethanol compromises SAMe synthesis through inhibition of crucial enzymes involved in SAMe generation. This can impair the formation of endogenous antioxidants such as glutathione and also lead to reduced cellular membrane stability [85].

In addition, alcohol interacts with the methylation status of certain genes and thereby contributes to liver damage and tumour development. Accordingly, alcohol-induced depletion of lipotropes may cause hypomethylation of oncogenes leading to their activation. The decrease in methylation capacity caused by chronic alcohol consumption can therefore contribute to epigenetic alterations of genes involved in hepatocarcinogenesis.

7.11 Interaction of Alcohol with Retinoids

It has been shown for decades that chronic alcohol consumption lowers hepatic vitamin A levels, especially in advanced alcoholic liver disease [86]. Retinoic acid (RA) plays an important role in controlling cell growth differentiation, apoptosis and is of potential clinical interest in cancer prevention and treatment. Therefore, interaction of ethanol with RA metabolism has important implications on the evolution, prevention and treatment of alcohol-related cancers.

The mechanism of alcohol-associated decrease in retinol and RA has multiple causes (Fig. 7.5). Since ADH and ALDH share the common substrates ethanol and retinol as well as AA and retinal to form RA, an interaction at these enzyme sites is not surprising. It has been demonstrated that ethanol acts as a competitive inhibitor of retinol oxidation [87]. Beside the fact that ethanol competes with retinol for the binding site of ADH, there are other mechanisms explaining the decrease in RA. Since chronic ethanol consumption increases CYP2E1 activity, an enhanced catabolism of vitamin A and RA into polar metabolites due to an induction of cytochrome P-450 2E1 occurs [87]. Although a variety of cytochrome isoenzymes such CYP1A1, CYP2B4, CYP2C3, CYP2C7, CYP2E1 and CYP26 are involved in RA metabolism, CYP2E1 seems of major importance [87]. The involvement of CYP2E1 in the metabolism of RA was proven by the fact that the CYP2E1 inhibitor chlormethiazole can completely inhibit this degradation [88]. The prevention of reduced RA status in the liver of ethanol-fed rats by chlormethiazole treatment indicates that the breakdown of retinoids by microsomal CYP2E1 is a key mechanism for the ethanol-enhanced catabolism of retinoids in hepatic tissue after treatment with alcohol. Chronic ethanol consumption with low hepatic RA concentrations results in a downregulation of RA receptors and an up to eightfold increased expression of the AP-1 (c-*jun* and c-*fos*) transcriptional complex [87]. This explains hepatocellular hyperproliferation as AP-1 is a central complex downstream of various

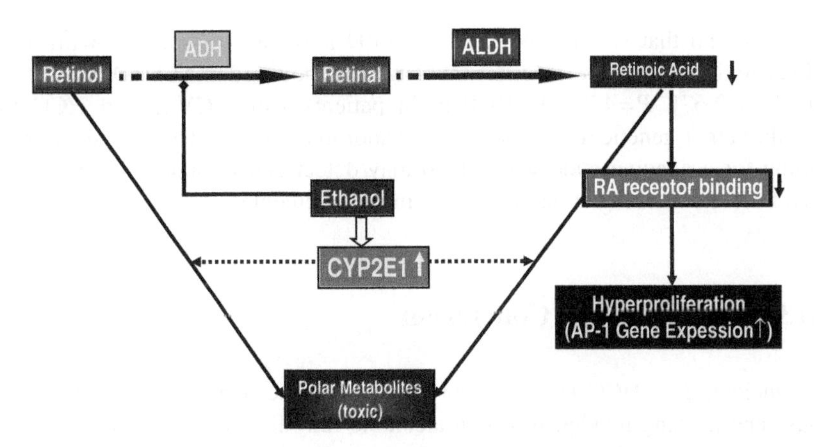

Fig. 7.5 Interaction between alcohol and retinoid turnover takes place at various steps and results in a depletion of retinoic acid (RA) which augment hyperproliferation. In addition, alcohol-induced CYP2E1 causes the formation of polar metabolites from RA contributing to hepatocellular damage

growth factors, oncogenes and tumour promoters. Supplementation of RA to animals not only results in a decrease of AP-1 gene expression but also in reduced hepatic proliferation.

In addition to an increased degradation of retinol and RA by CYP2E1, this catabolism leads to polar retinoid metabolites which are identified as 4-oxo- and 18-hydroxy RA some yet unidentified metabolites [89]. However, these metabolites cause liver cell damage by disrupting the mitochondrial membrane potential leading to the release of mitochondrial cytochrome C, caspase activation and finally apoptosis. This may explain why chronic alcohol consumption together with the administration of retinol or RA may aggravate alcoholic liver injury [90].

7.12 Host Genetics and Alcohol-Associated HCC

Only a relatively small proportion of 10–15 % of alcoholic cirrhotics develops HCC suggesting factors which render certain individuals more susceptible than others. In that regard, increasing evidence indicates a significant impact of host genetic factors on the risk of developing HCC in patients with cirrhosis, particularly regarding alcohol-associated HCC. While polymorphic variants of genes implicated in alcohol-related liver pathophysiology suggested as risk factors of ALD in earlier hypothesis-driven case-control studies were not robustly confirmed [71, 91], a sequence variation within the gene coding for patatin-like phospholipase encoding 3 (*PNPLA3*, rs738409) was found to modulate steatosis, necroinflammation and fibrosis in alcoholic liver disease [92]. Subsequently, the same variant was repeatedly suggested also as genetic risk factor for HCC in patients with alcoholic cirrhosis. A recent metanalysis of individual patient data from candidate gene association studies found that variant PNPLA3 rs738409 is strongly associated with overall HCC and found that this association was more pronounced in ALD (OR = 2.20 95 % CI [1.80–2.67], $P = 4.71 \times 10^{-15}$) than in patients with HCV-related HCC [93]. Whether other genetic host factors confer additional risk for HCC is yet unknown, calling for a genome wide scan which may detect genetic markers that could be used for better identification of subjects in HCC screening.

7.13 Summary and Conclusion

The incidence of HCC is rising worldwide. Chronic hepatitis B and C, alcohol abuse and a rising incidence of non-alcoholic fatty liver disease in many affluent countries are the major causes. While the causative role of alcohol in the evolution of liver cirrhosis is well established, our understanding of its importance as an etiologic factor in hepatocarcinogenesis has only recently been recognized. To date, a number of possible mechanisms are well investigated by which alcohol promotes the development of HCC. These include dietary or environmental carcinogens ingested along with alcoholic beverages, alcoholic cirrhosis as a priming condition and particularly the toxicity of acetaldehyde, increased lipid peroxidation due to reactive oxygen species, activation of procarcinogens via induction of cytochrome P450 2E1 and DNA lesions derived from oxidative stress by-products. Alterations of DNA methylation through interactions with one carbon metabolism can lead to aberrant methylation of tumour suppressor genes and oncogenes, and alcohol metabolism reduces hepatic RA levels and alleviates RA-mediated silencing on hyperproliferation. Lately, evidence for an important role of host genetics has accumulated with variation of PNPLA3 rs738409 as the first and robustly confirmed genetic modifier of HCC in alcoholic cirrhosis. These insights underscore the importance of alcohol as an important etiologic factor in hepatocarcinogenesis and potentially pave the way for preventive and therapeutic measures.

References

1. Nordenstedt H, White DL, El-Serag HB (2010) The changing pattern of epidemiology in hepatocellular carcinoma. Dig Liver Dis 42(Suppl 3):S206–S214
2. Fattovich G, Stroffolini T, Zagni I, Donato F (2004) Hepatocellular carcinoma in cirrhosis: incidence and risk factors. Gastroenterology 127(5 Suppl 1):S35–S50
3. McGlynn KA, London WT (2011) The global epidemiology of hepatocellular carcinoma: present and future. Clin Liver Dis 15:223–243
4. WHO. Mortality database. http://www.who.int/whosis/en
5. El-Serag HB (2012) Epidemiology of viral hepatitis and hepatocellular carcinoma. Gastroenterology 142:1264–1273
6. Stickel F, Schuppan D, Hahn EG, Seitz HK (2002) Cocarcinogenic effects of ethanol in hepatocarcinogenesis. Gut 51:132–139
7. Ascha MS, Hanouneh IA, Lopez R et al (2010) The incidence and risk factors of hepatocellular carcinoma in patients with nonalcoholic steatohepatitis. Hepatology 51:1972–1978

8. Seitz HK, Stickel F (2007) Molecular mechanisms of alcohol-mediated carcinogenesis. Nat Rev Cancer 7:599–612
9. Baan R, Straif K, Grosse Y et al (2007) Carcinogenicity of alcoholic beverages. Lancet Oncol 8:292–293
10. Rehm J, Klotsche J, Patra J (2007) Comparative quantification of alcohol exposure as risk factor for global burden of disease. Int J Methods Psychiatr Res 16:66–76
11. El-Serag HB, Davila JA, Petersen NJ, McGlynn KA (2003) The continuing increase in the incidence of hepatocellular carcinoma in the United States: an update. Ann Intern Med 139:817–823
12. Parkin DM (2002) Cancer incidence in five continents, vol VIII, IARC scientific publications. IARC, Lyon
13. El-Serag HB, Rudolph KL (2011) Hepatocellular carcinoma: epidemiology and molecular carcinogenesis. Gastroenterology 132:2557–2576
14. El-Serag HB, Mason AC (2000) Risk factors for the rising rates of primary liver cancer in the United States. Arch Intern Med 160:3227–3230
15. Nzeako UC, Goodman ZD, Ishak KG (1996) Hepatocellular carcinoma in cirrhotic and noncirrhotic livers. A clinico-histopathologic study of 804 North American patients. Am J Clin Pathol 105:65–75
16. Chiesa R, Donato F, Tagger A et al (2000) Etiology of hepatocellular carcinoma in Italian patients with and without cirrhosis. Cancer Epidemiol Biomarkers Prev 9:213–216
17. Grando-Lemaire V, Guettier C, Chevret S et al (1999) Hepatocellular carcinoma without cirrhosis in the West: epidemiological factors and histopathology of the non-tumorous liver. Groupe d'Etude et de Traitement du Carcinome Hepatocellulaire. J Hepatol 31:508–513
18. De Lima VM, Oliveira CP, Alves VA et al (2008) A rodent model of NASH with cirrhosis, oval cell proliferation and hepatocellular carcinoma. J Hepatol 49:1055–1061
19. Nakanuma Y, Ohta G (1985) Is Mallory body formation a preneoplastic change? A study of 181 cases of liver bearing hepatocellular carcinoma and 82 cases of cirrhosis. Cancer 55: 2400–2405
20. Smith P, Tee LBG, Yeoh GCT (1996) Appearance of oval cells in the liver of rats after long-term exposure to ethanol. Hepatology 23:145–154
21. Lin RS, Lee FY, Lee SD et al (1995) Endotoxemia in patients with chronic liver diseases: relationship to severity of liver diseases, presence of esophageal varices, and hyperdynamic circulation. J Hepatol 22:165–172
22. Yan AW, Fouts DE, Brandl J et al (2011) Enteric dysbiosis associated with a mouse model of alcoholic liver disease. Hepatology 53:96–105
23. Tilg H, Moschen AR, Kaneider NC (2011) Pathways of liver injury in alcoholic liver disease. J Hepatol 55(5):1159–1161
24. Yamada Y, Kirillova I, Peschon JJ et al (1997) Initiation of liver growth by tumor necrosis factor: deficient liver regeneration in mice lacking type I tumor necrosis factor receptor. Proc Natl Acad Sci U S A 94:1441–1446
25. Bharrhan S, Koul A, Chopra K, Rishi P (2011) Catechin suppresses an array of signalling molecules and modulates alcohol-induced endotoxin mediated liver injury in a rat model. PLoS One 6:e20635
26. Stärkel P, De Saeger C, Strain AJ et al (2010) NFkappaB, cytokines, TLR 3 and 7 expression in human end-stage HCV and alcoholic liver disease. Eur J Clin Invest 40:575–584
27. Qiao L, Zhang H, Yu J et al (2006) Constitutive activation of NF-kappaB in human hepatocellular carcinoma: evidence of a cytoprotective role. Hum Gene Ther 17:280–290
28. Siu L, Foont J, Wands JR (2009) Hepatitis C virus and alcohol. Semin Liver Dis 29:188–199
29. Kowdley KV (2004) Iron, hemochromatosis, and hepatocellular carcinoma. Gastroenterology 127:S79–S86
30. Marcellin P, Pequignot F, Delarocque-Astagneau E et al (2008) Mortality related to chronic hepatitis B and chronic hepatitis C in France: evidence for the role of HIV coinfection and alcohol consumption. J Hepatol 48:200–207
31. Welzel TM, Graubard BI, Quraishi S et al (2013) Population-attributable fractions of risk factors for hepatocellular carcinoma in the United States. Am J Gastroenterol 108:1314–1321

32. Donato F, Tagger A, Gelatti U et al (2002) Alcohol and hepatocellular carcinoma: the effect of lifetime intake and hepatitis virus infections in men and women. Am J Epidemiol 155:323–331
33. Bedogni G, Miglioli L, Masutti F et al (2008) Natural course of chronic HCV and HBV infection and role of alcohol in the general population: the Dionysos Study. Am J Gastroenterol 103:2248–2253
34. Chen CL, Yang HI, Yang WS et al (2008) Metabolic factors and risk of hepatocellular carcinoma by chronic hepatitis B/C infection: a follow-up study in Taiwan. Gastroenterology 135:111–121
35. Jee SH, Ohrr H, Sull JW, Samet JM (2004) Cigarette smoking, alcohol drinking, hepatitis B, and risk for hepatocellular carcinoma in Korea. J Natl Cancer Inst 96:1851–1855
36. Lambert MP, Paliwal A, Vaissière T et al (2011) Aberrant DNA methylation distinguishes hepatocellular carcinoma associated with HBV and HCV infection and alcohol intake. J Hepatol 54:705–715
37. Miyakawa H, Izumi N, Marumo F, Sato C (1996) Roles of alcohol, hepatitis virus infection, and gender in the development of hepatocellular carcinoma in patients with liver cirrhosis. Alcohol Clin Exp Res 20(1 Suppl):91A–94A
38. Peters MG, Terrault NA (2002) Alcohol use and hepatitis C. Hepatology 36(5 Suppl 1): S220–S225
39. Davila JA, Morgan RO, Shaib Y, McGlynn KA, El-Serag HB (2004) Hepatitis C infection and the increasing incidence of hepatocellular carcinoma: a population-based study. Gastroenterology 127:1372–1380
40. Bellentani S, Pozzato G, Saccoccio G et al (1999) Clinical course and risk factors of hepatitis C virus related liver disease in the general population: report from the Dionysos study. Gut 44:874–880
41. Hassan MM, Hwang LY, Hatten CJ et al (2002) Risk factors for hepatocellular carcinoma: synergism of alcohol with viral hepatitis and diabetes mellitus. Hepatology 36:1206–1213
42. Uetake S, Yamauchi M, Itoh S et al (2003) Analysis of risk factors for hepatocellular carcinoma in patients with HBs antigen- and anti-HCV antibody-negative alcoholic cirrhosis: clinical significance of prior hepatitis B virus infection. Alcohol Clin Exp Res 27(8 Suppl):47S–51S
43. Donato F, Gelatti U, Limina RM, Fattovich G (2006) Southern Europe as an example of interaction between various environmental factors: a systematic review of the epidemiologic evidence. Oncogene 25:3756–3770
44. Moriya K, Fujie H, Shintani Y et al (1998) Hepatitis C virus core protein induces hepatocellular carcinoma in transgenic mice. Nat Med 4:1065–1068
45. Koike K (2007) Hepatitis C, virus contributes to hepatocarcinogenesis by modulating metabolic and intracellular signaling pathways. J Gastroenterol Hepatol 22(Suppl 1):S108–S111
46. Hoek JB, Pastorino JG (2004) Cellular signaling mechanisms in alcohol-induced liver damage. Semin Liver Dis 24:257–272
47. Naveau S, Giraud V, Borotto E, Aubert A, Capron F, Chaput JC (1997) Excess weight is a risk factor for alcoholic liver disease. Hepatology 25:108–111
48. Raynard B, Balian A, Fallik D et al (2002) Risk factors of fibrosis in alcohol-induced liver disease. Hepatology 35:635–638
49. Ekstedt M, Franzén LE, Holmqvist M et al (2009) Alcohol consumption is associated with progression of hepatic fibrosis in non-alcoholic fatty liver disease. Scand J Gastroenterol 44:366–374
50. Dunn W, Sanyal AJ, Brunt EM et al (2012) Modest alcohol consumption is associated with decreased prevalence of steatohepatitis in patients with non-alcoholic fatty liver disease (NAFLD). J Hepatol 57:384–391
51. Moriya A, Iwasaki Y, Ohguchi S et al (2011) Alcohol consumption appears to protect against non-alcoholic fatty liver disease. Aliment Pharmacol Ther 33:378–388
52. Craddock VM, Henderson AR (1991) Potent inhibition of oesophageal metabolism of N-nitrosomethylbenzylamine, an oesophageal carcinogen, by higher alcohols present in alcoholic beverages. IARC Sci Publ 564–567
53. Kuper H, Tzonou A, Kaklamani E et al (2000) Tobacco smoking, alcohol consumption and their interaction in the causation of hepatocellular carcinoma. Int J Cancer 85:498–502

54. Seitz HK, Osswald B (1992) Effect of ethanol on procarcinogen activation. In: Watson R (ed) Alcohol and cancer. CRC, Boca Raton, pp 55–72
55. Aguilar F, Hussain SP, Cerutti P (1993) Aflatoxin B1 induces the transversion of G->T in codon 249 of the p53 tumor suppressor gene in human hepatocytes. Proc Natl Acad Sci U S A 90:8586–8590
56. Bulatao-Jayme J, Almero EM, Castro MC, Jardeleza MT, Salamat LA (1982) A case-control dietary study of primary liver cancer risk from aflatoxin exposure. Int J Epidemiol 11: 112–119
57. National Toxicology Program (2011) Acetaldehyde. Rep Carcinog 12:21–24
58. Helander A, Lindahl-Kiessling K (1991) Increased frequency of acetaldehyde-induced sister-chromatid exchanges in human lymphocytes treated with an aldehyde dehydrogenase inhibitor. Mutat Res 264:103–107
59. Maffei F, Fimognari C, Castelli E et al (2000) Increased cytogenetic damage detected by FISH analysis on micronuclei in peripheral lymphocytes from alcoholics. Mutagenesis 15:517–523
60. Maffei F, Forti GC, Castelli E, Stefanini GF, Mattioli S, Hrelia P (2002) Biomarkers to assess the genetic damage induced by alcohol abuse in human lymphocytes. Mutat Res 514:49–58
61. Matsuda T, Kawanishi M, Yagi T et al (1998) Specific tandem GG to TT base substitutions induced by acetaldehyde are due to intra-strand crosslinks between adjacent guanine bases. Nucleic Acids Res 26:1769–1774
62. Garro AJ, Espina N, Farinati F, Salvagnini M (1986) The effects of chronic ethanol consumption on carcinogen metabolism and on O6-methylguanine transferase-mediated repair of alkylated DNA. Alcohol Clin Exp Res 10:73S–77S
63. Theruvathu JA, Jaruga P, Nath RG, Dizdaroglu M, Brooks PJ (2005) Polyamines stimulate the formation of mutagenic 1, N2-propanodeoxyguanosine adducts from acetaldehyde. Nucleic Acids Res 33:3513–3520
64. Fang JL, Vaca CE (1995) Development of a 32P-postlabelling method for the analysis of adducts arising through the reaction of acetaldehyde with 2'-deoxyguanosine-3'-monophosphate and DNA. Carcinogenesis 16:2177–2185
65. Fang JL, Vaca CE (1997) Detection of DNA adducts of acetaldehyde in peripheral white blood cells of alcohol abusers. Carcinogenesis 18:627–632
66. Matsuda T, Matsumoto A, Uchida M et al (2007) Increased formation of hepatic N2-ethylidene-2′-deoxyguanosine DNA adducts in aldehyde dehydrogenase 2-knockout mice treated with ethanol. Carcinogenesis 28:2363–2366
67. Wang M, McIntee EJ, Cheng G, Shi Y, Villalta PW, Hecht SS (2000) Identification of DNA adducts of acetaldehyde. Chem Res Toxicol 13:1149–1157
68. Wang M, Yu N, Chen L, Villalta PW, Hochalter JB, Hecht SS (2006) Identification of an acetaldehyde adduct in human liver DNA and quantitation as N2-ethyldeoxyguanosine. Chem Res Toxicol 19:319–324
69. Stein S, Lao Y, Yang IY, Hecht SS, Moriya M (2006) Genotoxicity of acetaldehyde- and crotonaldehyde-induced 1, N2-propanodeoxyguanosine DNA adducts in human cells. Mutat Res 608:1–7
70. Yokoyama A, Muramatsu T, Ohmori T et al (1998) Alcohol-related cancers and aldehyde dehydrogenase-2 in Japanese alcoholics. Carcinogenesis 19:1383–1387
71. Homann N, Stickel F, Konig IR, Jacobs A, Junghanns K, Benesova M, Schuppan D, Himsel S, Zuber-Jerger I, Hellerbrand C, Ludwig D, Caselmann WH et al (2006) Alcohol dehydrogenase 1C*1 allele is a genetic marker for alcohol-associated cancer in heavy drinkers. Int J Cancer 118:1998–2002
72. Albano E (2002) Free radical mechanisms in immune reactions associated with alcoholic liver disease. Free Radic Biol Med 32:110–114
73. Garcia-Ruiz C, Colell A, Paris R, Fernandez-Checa JC (2000) Direct interaction of GD3 ganglioside with mitochondria generates reactive oxygen species followed by mitochondrial permeability transition, cytochrome c release, and caspase activation. FASEB J 14:847–858
74. Bautista AP (2002) Neutrophilic infiltration in alcoholic hepatitis. Alcohol 27:17–21
75. Chamulitrat W, Spitzer JJ (1996) Nitric oxide and liver injury in alcohol-fed rats after lipopolysaccharide administration. Alcohol Clin Exp Res 20:1065–1070

76. Oneta CM, Lieber CS, Li J, Ruttimann S, Schmid B, Lattmann J, Rosman AS, Seitz HK (2002) Dynamics of cytochrome P4502E1 activity in man: induction by ethanol and disappearance during withdrawal phase. J Hepatol 36:47–52
77. Gouillon Z, Lucas D, Li J, Hagbjork AL, French BA, Fu P, Fang C, Ingelman-Sundberg M, Donohue TM Jr, French SW (2000) Inhibition of ethanol-induced liver disease in the intragastric feeding rat model by chlormethiazole. Proc Soc Exp Biol Med 224:302–308
78. Bradford BU, Kono H, Isayama F et al (2005) Cytochrome P450 CYP2E1, but not nicotinamide adenine dinucleotide phosphate oxidase, is required for ethanol-induced oxidative DNA damage in rodent liver. Hepatology 41:336–344
79. Morgan K, French SW, Morgan TR (2002) Production of a cytochrome P450 2E1 transgenic mouse and initial evaluation of alcoholic liver damage. Hepatology 36:122–134
80. Frank A, Seitz HK, Bartsch H, Frank N, Nair J (2004) Immunohistochemical detection of 1, N6-ethenodeoxyadenosine in nuclei of human liver affected by diseases predisposing to hepato-carcinogenesis. Carcinogenesis 25:1027–1031
81. Fernando RC, Nair J, Barbin A, Miller JA, Bartsch H (1996) Detection of 1, N6-ethenodeoxyadenosine and 3, N4-ethenodeoxycytidine by immunoaffinity/32 P-postlabelling in liver and lung DNA of mice treated with ethyl carbamate (urethane) or its metabolites. Carcinogenesis 17:1711–1718
82. Wang Y, Millonig G, Nair J et al (2009) Ethanol-induced cytochrome P4502E1 causes carcinogenic etheno-DNA lesions in alcoholic liver disease. Hepatology 50:453–461
83. Nair J, Srivatanakul P, Haas C, Jedpiyawongse A, Khuhaprema T, Seitz HK, Bartsch H (2010) High urinary excretion of lipid peroxidation-derived DNA damage in patients with cancer-prone liver diseases. Mutat Res 683:23–28
84. Mandrekar P (2011) Epigenetic regulation in alcoholic liver disease. World J Gastroenterol 17:2456–2464
85. Stickel F, Herold C, Seitz HK, Schuppan D (2006) Alcohol and methyl transfer: Implication for alcohol related hepatocarcinogenesis. In: Ali S, Friedman SL, Mann DA (eds) Liver disease: biochemical mechanisms and new therapeutic insights. Science Publisher Enfield, Plymouth, pp 45–58
86. Leo MA, Lieber CS (1982) Hepatic vitamin A depletion in alcoholic liver injury. N Engl J Med 307:597–601
87. Wang XD, Liu C, Chung J et al (1998) Chronic alcohol intake reduces retinoic acid concentration and enhances AP-1 (c-Jun and c-Fos) expression in rat liver. Hepatology 28:744–750
88. Liu C, Russell RM, Seitz HK, Wang XD (2001) Ethanol enhances retinoic acid metabolism into polar metabolites in rat liver via induction of cytochrome P4502E1. Gastroenterology 120:179–189
89. Dan Z, Popov Y, Patsenker E, Preimel D, Liu C, Wang XD, Seitz HK, Schuppan D, Stickel F (2005) Hepatotoxicity of alcohol-induced polar retinol metabolites involves apoptosis via loss of mitochondrial membrane potential. FASEB J 19:845–847
90. Albanes D, Heinonen OP, Taylor PR et al (1996) Alpha-Tocopherol and beta-carotene supplements and lung cancer incidence in the alpha-tocopherol, beta-carotene cancer prevention study: effects of base-line characteristics and study compliance. J Natl Cancer Inst 88:1560–1570
91. Nahon P, Sutton A, Rufat P et al (2009) Myeloperoxidase and superoxide dismutase 2 polymorphisms comodulate the risk of hepatocellular carcinoma and death in alcoholic cirrhosis. Hepatology 50:1484–1493
92. Stickel F, Hampe J (2012) Genetic determinants of alcoholic liver disease. Gut 61:150–159
93. Trépo E, Nahon P, Bontempi G et al (2014) Association between the PNPLA3 (rs738409 C>G) variant and hepatocellular carcinoma: evidence from a meta-analysis of individual participant data. Hepatology 59(6):2170–2177. doi:10.1002/hep.26767

Chapter 8
TLR4-Dependent Tumor-Initiating Stem Cell-Like Cells (TICs) in Alcohol-Associated Hepatocellular Carcinogenesis

Keigo Machida, Douglas E. Feldman, and Hidekazu Tsukamoto

Abstract Alcohol abuse predisposes individuals to the development of hepatocellular carcinoma (HCC) and synergistically heightens the HCC risk in patients infected with hepatitis C virus (HCV). The mechanisms of this synergism have been elusive until our recent demonstration of the obligatory role of ectopically expressed TLR4 in liver tumorigenesis in alcohol-fed HCV *Ns5a* or *Core* transgenic mice. CD133+/CD49f+ tumor-initiating stem cell-like cells (TICs) isolated from these models are tumorigenic in a manner dependent on TLR4 and NANOG. TICs' tumor-initiating

K. Machida, Ph.D. (✉)
Southern California Research Center for ALPD and Cirrhosis, Keck School of Medicine of the University of Southern California, 1333 San Pablo Street, MMR-402, Los Angeles, CA 90089-9141, USA

Department of Molecular Microbiology and Immunology, University of Southern California, Los Angeles, CA, USA
e-mail: keigo.machida@med.usc.edu

D.E. Feldman, Ph.D.
Southern California Research Center for ALPD and Cirrhosis, Keck School of Medicine of the University of Southern California, 1333 San Pablo Street, MMR-402, Los Angeles, CA 90089-9141, USA

Department of Pathology, Keck School of Medicine of the University of Southern California, Los Angeles, CA, USA
e-mail: defeldma@usc.edu

H. Tsukamoto, D.V.M., Ph.D. (✉)
Southern California Research Center for ALPD and Cirrhosis, Keck School of Medicine of the University of Southern California, 1333 San Pablo Street, MMR-402, Los Angeles, CA 90089-9141, USA

Department of Pathology, Keck School of Medicine of the University of Southern California, Los Angeles, CA, USA

Department of Veterans Affairs Greater Los Angeles Healthcare System, Los Angeles, CA, USA
e-mail: htsukamo@usc.edu

V. Vasiliou et al. (eds.), *Biological Basis of Alcohol-Induced Cancer*, Advances in Experimental Medicine and Biology 815,
DOI 10.1007/978-3-319-09614-8_8

activity and chemoresistance are causally associated with inhibition of TGF-β tumor suppressor pathway due to NANOG-mediated expression of IGF2BP3 and YAP1. TLR4/NANOG activation causes p53 degradation via phosphorylation of the protective protein NUMB and its dissociation from p53 by the oncoprotein TBC1D15. Nutrient deprivation reduces overexpressed TBC1D15 in TICs via autophagy-mediated degradation, suggesting a possible role of this oncoprotein in linking metabolic reprogramming and self-renewal.

Keywords HCC • Cancer stem cells • NANOG • TGF-β • p53

8.1 Introduction

Hepatocellular carcinoma (HCC) is the most prevalent primary malignancy of the liver and the fifth most common cancer in men. HCC is diagnosed in over half a million patients globally every year and is the second leading cause of the cancer-related mortality. Cirrhosis is the single and most important risk factor for HCC, raising the risk by 40-fold, and about 70 % of HCC patients have underlying cirrhosis. With respect to HCC risk by etiology, viral hepatitis (HBV and HCV) is most common, followed by alcoholic liver disease (ALD). In particular, chronic infection with HCV represents a major global risk factor for HCC [1] as more than 170 million people are infected with HCV worldwide [1–3]. HCV produces proteins which are directly implicated in hepatocyte toxicity and transformation. For instance, the HCV core protein causes overproduction of reactive oxygen species which may cause mitochondrial or nuclear DNA damage [2, 4, 5]. The core protein also inhibits microsomal triglyceride transfer protein activity and VLDL secretion [6], which contributes the genesis of fatty liver. The core also induces insulin resistance in mice and cell lines, and this effect may be mediated by degradation of insulin receptor substrates (IRS) 1 and 2 via upregulation of SOCS3 [7] in a manner dependent on PA28γ 73, or via IRS serine phosphorylation [8]. These mechanisms may also be relevant to another etiological entity which promotes HCC risk: non-alcoholic fatty liver disease (NAFLD) [9] that is a liver phenotype of obesity-associated metabolic syndrome. HCV/HBV infection, ALD, and NAFLD share common pathophysiological events such as oxidant stress, organelle stress, and metabolic dysregulation which may contribute to their oncogenic activities. More importantly, an apparent synergism among HCV, alcohol, and obesity exists for the risk of HCC. Coexistence of alcohol abuse or obesity increases the HCV risk of developing HCC by additional eightfold, culminating to an overall 45–55-fold increase in the risk as compared to normal subjects [10, 11] (Fig. 8.1). As alcohol and obesity continue to dominate as leading life-style factors for disease burdens around the world [12], heightened HCC incidence, caused by synergistic interactions of these factors with hepatitis viruses, represents the most predictable and devastating global health issue.

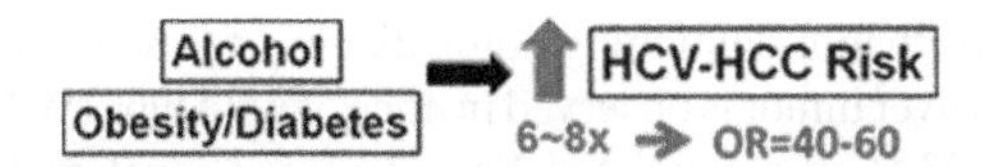

Fig. 8.1 Synergistically increased risk for HCV-mediated HCC by alcohol abuse or/and obesity. Comorbidity such as alcohol abuse and obesity synergistically heightens the risk of developing HCC by 6- to 8-fold in HCV-infected patients. As such HCV-infected, alcoholic or obese patients have 40–60 times higher HCC risk compared to healthy subjects

The most challenging aspect of HCC treatment is its refractoriness to chemotherapy. Among many potential mechanisms which underlie chemoresistance, the role of tumor-initiating stem cell-like cells (TICs) or the so-called cancer stem cells (CSCs) has received a spotlight. Stem cells have three major characteristics, self-renewal, asymmetric division (clonality), and plasticity. Forty percent of HCC are assumed to have clonality and to originate from progenitor/stem cells [13–16]. CD133+/CD49f+ cells in liver tumors correlate with tumorigenesity and the expression of "stemness" genes, such as Wnt/β-catenin, Notch, Hedgehog/SMO, and Oct3/4 [17–19]. Indeed, CD133+/CD49f+ HCC CSCs are chemoresistant [20] and survive during an initial therapy. Although an encouraging therapeutic response may be seen, survived CSCs eventually establish a clonal expansion and tumor recurrence. This chemoresistance may be caused by the plasticity of CSCs with dysregulated signaling and gene expression. Several oncogenic signaling pathways are activated in HCC or CSCs, including PI3K/AKT [21], signal transducer and activator of transcription 3 (STAT3) [22, 23], and hedgehog [24, 25] while defective tumor suppressor transforming growth factor-beta (TGF-β) pathway is also implicated [26, 27]. Another pivotal mechanism is asymmetric division of CSCs producing dormant daughter cells which are less sensitive to chemotherapeutic drugs.

8.2 Ectopic TLR4 Activation Underlies HCV-Alcohol Synergism

In our efforts to establish mouse models of HCV-alcohol synergism for HCC, we fed first alcohol to HCV *Ns5a* transgenic mice by using the mouse intragastric feeding (iG) model. This approach is prompted by the observation that TLR4 expression is induced in the liver by hepatocyte-specific NS5A expression [28]. As endotoxin-TLR4 pathway is established in the pathogenesis of ALD [29, 30], a synergism between alcohol-mediated endotoxemia and NS5A-induced TLR4 overexpression was predicted. Indeed, alcohol feeding to the *Ns5a* mice results in aggravated liver damage with severe hepatitis induced in some mice [31]. This pathology is dependent on TLR4 as it is abrogated in alcohol-fed *Ns5a:Tlr4−/−* mice. This synergism is also due to endotoxin as it is attenuated by the concomitant treatment of the mice with polymyxin B and neomycin and is conversely potentiated by intragastric administration of LPS. To extend this observation to an extended period of 12 months, a modified ethanol-containing liquid diet was fed to *Ns5a* and *Ns5a:Tlr4−/−* mice. Although no liver tumor is observed in none of wild type (WT) or *Ns5a:Tlr4−/−* mice fed alcohol, 23 % of alcohol-fed *Ns5a* Tg mice developed liver tumors [31]. This TLR4-dependent tumorigenic phenotype was subsequently reproduced in alcohol-fed HCV *Core* Tg mice [32].

8.3 Identification of TLR4/NANOG-Dependent TICs

Immunostaining of liver tumor sections from alcohol-fed *Ns5a* mice revealed cells double-positive for NANOG and CD133 or CD49f [31]. This prompted a FACS analysis of cells dissociated from liver tumors of these mice which detected a small yet significantly increased percentage of CD133+/CD49f+ cells as compared to WT mice (1.11 % vs. 0.05 %). Gene profiling analysis of sorted CD133+/CD49+ cells shows consistently upregulated stemness genes such as *Nanog, Oct4, Sox2* as compared to CD133−/CD49+ or CD133−/CD49f− cells.

This heightened expression of stemness genes and cell proliferation are largely reduced by TLR4 silencing with lentiviral short-hairpin RNA (shRNA) compared to control CD133−/CD49f+ cells [32]. The CD133+/CD49f+ cells form colonies in soft agar and spheroids in ultra-low-adhesion plates, demonstrating they have anchorage-independent growth and self-renewal ability [32]. Subcutaneous transplantation of CD133+/CD49f+ cells, but not CD133−/CD49f+ cells into immunocompromised (NOG) mice, following infection with a red-fluorescence (dsRed) lentiviral vector, results in tumor development; and this tumor growth, assessed by dsRed imaging, is inhibited by *Tlr4* or *Nanog* silencing by lentiviral shRNA transduced prior to transplantation [32] (Fig. 8.2), indicating that CD133+/CD49f+ cells are TLR4/NANOG-dependent TICs and that *Tlr4* is a putative proto-oncogene involved in the genesis/maintenance of TICs and liver tumor in HCV Tg models. Furthermore, these NOG mice derived tumor have tumor-initiation capacity since injection of serial transplantation of these tumors into another NOG mice generates tumors in NOG mice. Then one will raise a question where these TICs are generated. These TICs may have been generated from hepatoblasts in several etiologies since hepatoblastic HCC subtype with poor prognosis has a gene expression profile with markers of hepatic oval cells, suggesting that this subtype of HCC arises from LPCs [33–36]. Indeed, these HCC often recurs after chemotherapy presumably due to the presence of chemo-resistant TICs [37]. These aspects need further investigation.

8.4 *TLR4* as a Putative Proto-oncogene

The obligatory roles of endotoxin and TLR4 in HCC promotion are shown in various etiological settings including initiation in alcohol-HCV model and promotion [32] in the DEN/CCl_4 model [38]. Activated TLR4 pathway is responsible for

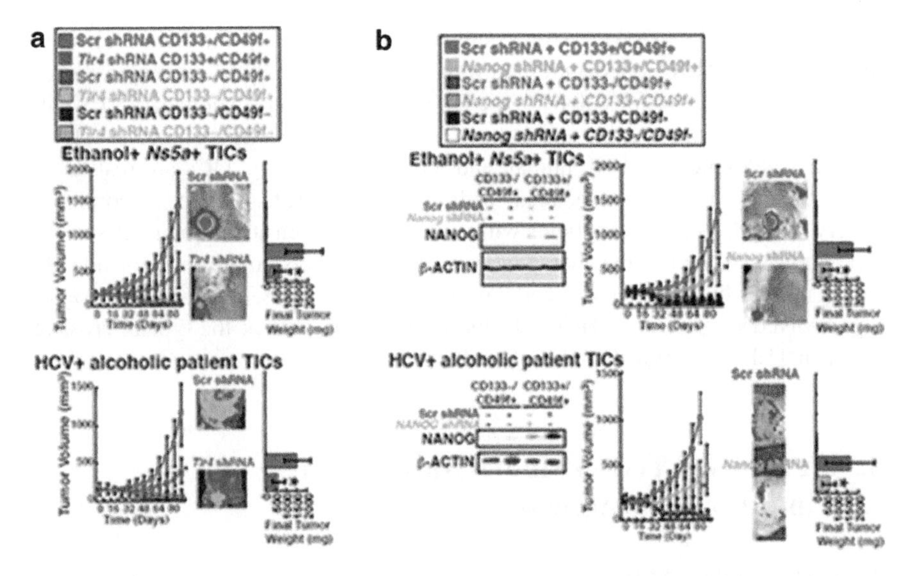

Fig. 8.2 TICs' tumorigenic activity is dependent on TLR4 and NANOG. CD133+/CD49f+ TICs isolated from liver tumors of alcohol-fed *Ns5a* Tg mice or alcoholic HCV patients were transplanted subcutaneously into NOG mice following lentiviral transduction of EGFP to allow a whole animal imaging for assessment of tumor growth for a period of 80 days. To test the dependence on TLR4 and NANOG, they were respectively knocked down by lentiviral expression of specific or control shRNA prior to transplantation. Tumor volume was calculated by three dimensional assessment of the tumor image, and final tumor weights were also compared at the end of the experiment. Note TLR4 (**a**) or NANOG (**b**) knockdown significantly attenuates tumor growth derived from mouse and patient TICs. Immunoblotting of cell lysates collected 10 days after the transplantation confirms expression of NANOG in the TICs and knockdown of this protein with specific shRNA. $^{*}p<0.05$ (Reproduced from the published figure of the reference [32] with permission)

promotion of HCCs from different animal models [38], which offer new therapeutic targets for HCC. As PAMP (pathogen-associated molecular pattern), including TLR4, mediates inflammatory responses to endotoxin and other ligands, inflammation is strongly associated with cancer in other cancers, including lung [39], colon [40], and skin carcinomas [41]. While TLR expression is very heightened in macrophages and lymphocytes, normal hepatocytes have less or nonfunctional TLR4, but ectopically expressed-TLR4 in epithelial cells is involved in oncogenesis as studies in other cancer models implicate liver has also similar oncogenic pathways since gut-derived endotoxin directly damages liver due to proximal anatomy of gut and liver through portal veins [31, 38, 42]. As we mentioned above, we have shown that long-term (12 months) feeding of alcohol diet induces liver tumors in HCV *Ns5a* Tg mice [43–46] and these incidences are reduced if the mice were crossbred with defective *Tlr4* mice [32]. Plasma LPS levels are elevated equally in both TLR4 sufficient and deficient mice fed these diets, indicating that ligand level is not changed even by disruption of its receptor *Tlr4* [31, 32]. Indeed, tumor tissues from *Tlr4+/+* models display accentuated expression of TLR4 and its activation as assessed by its downstream marker TRAF6-TAK1 complex formation [32]. Furthermore, activation of human TLR4 oncogenic pathway, especially NANOG overexpression, is also noted in HCC sections of alcoholic HCV as well as non-steatohepatitis (NASH) patients [32], supporting this activation of TLR4-NANOG axis is clinically relevant for the development of both human and mouse HCCs [31]. TICs are resistant to chemotherapy and are associated with metastatic HCC, which is commonly observed in HCV-infected patients with alcohol abuse. Sensitization of TICs to chemotherapy and identifying therapeutic molecular targets could be a considerable savings in morbidity, mortality, and cost. However, in HCCs, there are many signaling involved in genesis of HCCs. In the next section, one of the typical crosstalk between TLR4 and TGF-β pathways will be discussed.

8.5 TLR4 and TGF-β Mutual Antagonism in Liver Tumorigenesis

Our study identifies TLR4 signaling as a central mediator in synergistic liver tumor formation by HCV and alcohol via the genesis of TLR4/NANOG-dependent TICs. On the other hand, deficient TGF-β pathway caused by inactivation of at least one of the TGF-β signaling components is a well-known risk factor for HCC in man [26, 47] and a causal oncogenic mechanism in animal models [15, 48]. Thus, we wondered about the relationship between the two pathways. To investigate this question, we have used two complementary approaches. First, we looked for TIC-specific genes which may be involved in regulating TGF-β pathway by screening lentiviral cDNA library established from TIC vs. CD133−/CD49f+ control cells for transformation of the p53 deficient hepatoblast cell line PL4 [32]. This has identified *Yap1* and *Igf2bp3* as two TIC-associated genes which are under direct transcriptional control of NANOG and contribute to TICs' tumor-initiating activities both in vitro and in vivo [32]. Further, these two gene products are shown to inhibit the TGF-β tumor suppressor pathway at the two distinct levels but in an interactive manner. YAP1 associates with SMAD3 and SMAD7 to block nuclear translocation of p-Smad3. IGF2BP3, an mRNA binding protein, promotes IGF-II translation by binding to the 5′ UTR of *Igf-II* mRNA [49]. IGF-II activates AKT and subsequently mTOR, which suppresses SMAD3 activation [50]. Indeed, mTOR activation by IGF2BP3 inhibits phosphor-activation of SMAD3 as such Rapamycin increases p-SMAD3 in TICs or even abrogates suppressed p-SMAD3 level caused by a constitutively active AKT. This IGF2BP3-AKT-mTOR pathway also interfaces with the YAP1-SMAD3/SMAD7 pathway described above. Activated AKT phosphorylates Ser-127 of YAP1, and p-Ser127-YAP1 interacts most actively with p-SMAD3 for cytosolic SMAD3 retention. Thus AKT activated by IGF2BP3 facilitates dual actions of mTOR-mediated suppression of SMAD3 phosphor-activation and p-YAP1-mediated p-SMAD3 retention, resulting in effective blockade of TGF-β pathway. As expected, silencing of *Igf2bp3* and *Yap1* in TICs restores TGF-β pathway with increased nuclear p-SMAD3, reduces TICs' tumorigenic activity, and enhances the chemosensitivity of TICs in vivo [32].

We have also used a reverse approach to test the reciprocal TLR4-TGF-β antagonism by assessing TLR4 expression and activation in *Spnb2+/–* mice. In fact, it is well known in innate immunity that the lack of a functional TβRII [51, 52] or Smad3 [53] results in extensive inflammation due to increased TLR4 expression and LPS hyper-responsiveness [54]. We believe this reciprocal regulation of augmented TLR4 response due to deficient TGF-β signaling also plays a critical role in generation and oncogenic activity of Nanog+ CSCs. SPNB2 is the chaperone protein which recruits p-SMAD3/SMAD4 into the nucleus, and SPNB2 haploinsufficiency in *Spnb2+/–* mice reduces TGF-β signaling and causes spontaneous development of HCC [15]. TLR4 expression is induced in the liver of this genetic mouse model as compared with WT mice. Feeding *Spnb2+/–* mice with alcohol for 12 months results in conspicuous TLR4 activation as assessed by TAK1/TRAF6 interaction and doubles the liver tumor incidence as compared to *Spnb2+/–* mice fed with a control diet [32]. More importantly, this increment of the tumor incidence is completely abrogated in alcohol-fed *Spnb2+/-Tlr4–/–* compound mice, demonstrating reciprocally upregulated TLR4 in *Spnb2+/–* mice with reduced TGF-β signaling, is also responsible for alcohol-associated liver tumorigenesis in the model. We readily extend this concept to clinically more relevant cells, the Huh7 human HCC cell line. Knockdown of SPNB2 in these cells increases TLR4 expression and tumorigenic activity in NOG mice [32].

8.6 Anabolic Metabolism and TIC Self-Renewal

A critical event leading to deregulated TIC proliferation is inactivation of the p53 tumor suppressor [55, 56], which acts as a key barrier against pluripotency and stem cell proliferation. This function of p53 is carried out through direct repression of stemness-associated transcription factor (TF) network components [57]. Mutation or loss of p53 is found recurrently in diverse malignancies including HCC [58] and is associated with poor prognosis [59, 60]. Strikingly, genetic inactivation of p53 also leads to loss of cell polarity and aberrant execution of self-renewal [61–63]. The cell polarity determinant and tumor suppressor NUMB stably interacts directly with p53, protecting it from ubiquitin-mediated proteolysis catalyzed by the MDM2 E3 ubiquitin ligase [64]. In polarized epithelial cells and in untransformed progenitor cells, NUMB is distributed asymmetrically and segregates into the daughter cell that proceeds to differentiate. Cells deficient in p53 fail to correctly localize NUMB and lose this intrinsic polarity [65, 66], however little is understood about the composition and regulation of the Numb-p53 complex.

We examined biochemically the composition of the Numb-p53 complex in CD133+/CD49f+ TICs isolated from liver tumors of alcohol-fed HCV *Ns5aTg* mice, Fractionation of TIC lysates using sucrose density gradient centrifugation revealed that NUMB and p53 are the constituents of a high molecular mass (>700 kDa) complex, which is disintegrated upon NANOG-mediated activation of aPKCζ, a NUMB kinase [67]. Using affinity purification and tandem mass spectrometry, we identified the ATG8/LC3-binding protein TBC1D15 as a novel component of this high molecular mass complex. Enforced expression of TBC1D15 reduces steady state levels of p53 and this effect is blocked by a Nutlin-3 treatment, suggesting that TBC1D15 triggers the MDM2-dependent degradation of p53.

TBC1D15 is comprised of two distinct structural domains: a C-terminal GTPase activating protein (GAP) domain that inactivates the Rab7 GTPase, which mediates endosome/autophagosome-fusion to lysosomes [68, 69] and a functionally uncharacterized N-terminal domain. We expressed Flag-tagged variants of each domain individually with myc-tagged p53 (myc-p53), and found that the N-terminal domain (Flag-TBC1D15-N) recapitulated inhibition of myc-p53. Destabilization of myc-p53 corresponded closely with the extent of its displacement from NUMB. Sequence analysis of the N-terminal domain revealed a 50 amino acid region containing significant homology to the *Drosophila* protein CANOE, which regulates the localization of cell-fate determinants during asymmetric division and interacts genetically with *numb* [70, 71]. Coexpression of myc-p53 with GFP fusion proteins containing either the CANOE homology region (TBC-cno, aa 159–270) or the N-terminal region (TBC-N1, aa 2–158) revealed that GFP-TBC-N1 but not GFP-TBC-cno associated stably with NUMB, suggesting that a primary sequence or higher order structural motif within this region of TBC1D15 directly binds to NUMB and dissociates it from p53 to promote p53 degradation. However, the mutual requirements and relative contributions of TBC1D15 and aPKCζ-mediated NUMB phosphorylation for p53 dissociation and degradation are not yet fully understood and merit further investigation.

We also found that *Tbc1d15* is transcriptionally repressed by p53, revealing a mutually antagonistic regulation between these genes. In agreement with these findings, three human HCC cell lines express TBC1D15 at higher levels than primary hepatocytes. In particular, Hep3B cells with p53 deficiency and Huh7 cells with mutant p53 express substantially higher TBC1D15 than HepG2 cells which have wild type p53. Thus, p53 levels are inversely correlated with TBC1D15 expression in these cells. Similarly, TBC1D15 levels are increased in TICs compared to CD133– cells, and TBC1D15 is strongly expressed in tumors arising from TICs implanted subcutaneously into mice, as determined by immunohistochemical analysis of sectioned tumors.

Interestingly, the TCTP oncoprotein was also found in association with the Numb-p53 complex and shown to stimulate MDM2-mediated proteolysis of p53 [72]. These results, along with our recent data on TBC1D15, collectively suggest that the Numb-p53 complex may serve as a pivotal control platform that integrates multiple inputs to permit the rapid modulation of cellular p53 levels. As there appears to be no significant primary sequence homology between TCTP and TBC1D15, these proteins may dock with distinct subunits in the NUMB-p53 complex.

Cellular levels of TBC1D15 are diminished through starvation-induced autophagic degradation, triggered through acute nutrient deprivation, depletion of ATP or chemical inactivation of the mTOR kinase complex. Conversely, TBC1D15 accumulates when autophagic flux is blocked. These observations together suggest a scenario whereby accumulation of the TBC1D15 oncoprotein drives deregulated TIC proliferation and formation of liver tumors due to alcohol-induced suppression of autophagy. This proposal resonates with accumulating evidence that suppression of autophagy, including through targeted genetic ablation of core autophagic machinery components, promotes the accumulation of oncoproteins leading to tumor formation [73–75], underscoring the importance of autophagic degradation in the tonic suppression of cancer. These findings suggest that depletion of p53 levels through aPKCζ activation and TBC1D15 upregulation may in turn cause de-repression of *Tbc1d15* transcription, further accelerating p53 degradation and establishing a self-reinforcing feedback cycle. This cycle represents an attractive therapeutic target for inhibition of TICs in HCC, and developing a deeper understanding of the aPKCζ-NUMB-TBC1D15 regulatory axis in p53 degradation will further define optimal therapeutic targets.

More broadly, defining the machinery that controls the expansion of TICs will have important ramifications for cancer treatment. Conventional chemotherapy kills a large fraction of tumor cells, resulting in a transient reduction in tumor volume. However, it typically fails to eradicate TICs and may actually impose a strong selective pressure for TIC survival [76]. As a result, following chemotherapy tumors are often enriched with TICs resistant to subsequent treatments. To be effective in the long term, cancer therapies will need to include agents that target TIC survival and self-renewal. We propose that selective inhibition of the machinery that drives inappropriate, self-renewing, symmetrical divisions in TICs will lead to "sterilization" of the tumor and to a lasting, beneficial clinical response. The newly elucidated mechanistic framework for TIC proliferation described here represents an innovation that holds significant potential as a prospective therapeutic target.

8.7 Conclusions and Discussions

CD133+ TICs have previously been isolated from liver tumors of PTEN or MAT1A deficient mice [51, 52]. Using the same surface marker, we successfully isolated CD133+/CD49f+ TICs which activate a unique TLR4-NANOG pathway as an integral component for their self-renewal and tumorigenic activities. These TICs are also identified in HCC sections of alcoholic HCV patients by immunostaining [32] and isolated from such patients to validate induction of TLR4-dependent stemness genes and transformation [32]. These TICs respond to endotoxin to initiate tumorigenesis as shown in alcohol-fed HCV Tg mouse models, but TICs isolated from alcohol *Core* Tg mice and alcoholic HCV patients grow efficiently in vitro without addition of LPS but this growth is reduced by TLR4 knockdown, suggesting LPS-independent mechanisms of TLR4 activation in these cells which remain to be elucidated. Possibilities include non-LPS ligands such as HMGB1 released in inflammation activating TLR4 and protein–protein interactions leading to ligand-independent activation. Although we began and focused our studies on alcohol-HCV synergism, the oncogenic role of TLR4 has been extended to HCC of non-viral and non-alcohol etiology such as that in *Spnb+/–* mice and NAFLD patient [32]. A recent study demonstrates the critical role of endotoxin-activated TLR4 in promotion but not initiation of hepatocarcinogenesis induced by diethylnitrosamine and carbon tetrachloride [38]. We believe that the TLR4-dependent mechanisms of TIC generation actually contribute to or at least promote the initiation of HCC. A conceptual breakthrough of our findings is that the proinflammatory *TLR4* is now

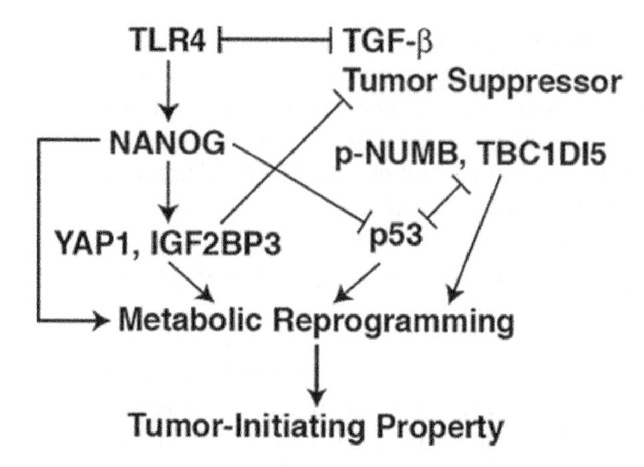

Fig. 8.3 TLR4-NANOG oncogenic pathway activated by alcohol, obesity, and HCV. TLR4 is ectopically activated in the liver by alcohol abuse, obesity, or/and HCV infection. This results in expression of the stem cell factor NANOG to mediate the genesis of TICs and tumorigenesis via antagonism of the TGF-β tumor suppressor pathway and degradation of p53 mediated by the oncoprotein TBC1D15. Upregulated NANOG and TBC1D15 and conversely suppressed p53 also contribute to anabolic metabolic reprogramming that is associated with the cancer stem cell fate regulation of TICs

considered as a putative proto-oncogene in hepatocarcinogenesis that links inflammation to carcinogenesis, the notion which has been entertained for the past 150 years. Based on this renewed concept, our studies have offered two novel insights into the molecular mechanisms of TLR4-mediated TICs' tumorigenic activity (see a schematic diagram shown in Fig. 8.3): NANOG-dependent upregulation of IGF2BP3 and YAP1 which in turn block the TGF-β tumor suppressor pathway; and NANOG-mediated p53 degradation by disengagement from the protective NUMB protein via TBC1D15 interaction. These studies are now exploring potential mechanistic connections to metabolic programming known to occur in cancer cells and TICs in promoting and maintaining "stem cell fate" via molecular, genetic, and epigenetic mechanisms.

Acknowledgments The authors' studies described in this review article have been supported by the NIH grants, 1R01AA018857, 5RC2AA019392, P50AA011999 (Animal Core, Morphology Core, and Pilot Project Program), R24AA012885 (Non-Parenchymal Liver Cell Core), and the Medical Research Service of the Department of Veterans Affairs.

References

1. Okuda K (2000) Hepatocellular carcinoma. J Hepatol 32:225–237
2. Okuda M, Li K, Beard MR, Showalter LA, Scholle F, Lemon SM, Weinman SA (2002) Mitochondrial injury, oxidative stress, and antioxidant gene expression are induced by hepatitis C virus core protein. Gastroenterology 122:366–375

3. Yao F, Terrault N (2001) Hepatitis C and hepatocellular carcinoma. Curr Treat Options Oncol 2:473–483
4. Korenaga M, Wang T, Li Y, Showalter LA, Chan T, Sun J, Weinman SA (2005) Hepatitis C virus core protein inhibits mitochondrial electron transport and increases reactive oxygen species (ROS) production. J Biol Chem 280:37481–37488
5. Moriya K, Nakagawa K, Santa T, Shintani Y, Fujie H, Miyoshi H, Tsutsumi T, Miyazawa T, Ishibashi K, Horie T, Imai K, Todoroki T, Kimura S, Koike K (2001) Oxidative stress in the absence of inflammation in a mouse model for hepatitis C virus-associated hepatocarcinogenesis. Cancer Res 61:4365–4370
6. Perlemuter G, Sabile A, Letteron P, Vona G, Topilco A, Chretien Y, Koike K, Pessayre D, Chapman J, Barba G, Brechot C (2002) Hepatitis C virus core protein inhibits microsomal triglyceride transfer protein activity and very low density lipoprotein secretion: a model of viral-related steatosis. FASEB J 16:185–194
7. Kawaguchi T, Yoshida T, Harada M, Hisamoto T, Nagao Y, Ide T, Taniguchi E, Kumemura H, Hanada S, Maeyama M, Baba S, Koga H, Kumashiro R, Ueno T, Ogata H, Yoshimura A, Sata M (2004) Hepatitis C virus down-regulates insulin receptor substrates 1 and 2 through up-regulation of suppressor of cytokine signaling 3. Am J Pathol 165:1499–1508
8. Banerjee S, Saito K, Ait-Goughoulte M, Meyer K, Ray RB, Ray R (2008) Hepatitis C virus core protein upregulates serine phosphorylation of insulin receptor substrate-1 and impairs the downstream akt/protein kinase B signaling pathway for insulin resistance. J Virol 82:2606–2612
9. Dyson J, Jaques B, Chattopadyhay D, Lochan R, Graham J, Das D, Aslam T, Patanwala I, Gaggar S, Cole M, Sumpter K, Stewart S, Rose J, Hudson M, Manas D, Reeves HL (2014) Hepatocellular cancer—the impact of obesity, type 2 diabetes and a multidisciplinary team. J Hepatol 60(1):110–117
10. Hassan MM, Hwang LY, Hatten CJ, Swaim M, Li D, Abbruzzese JL, Beasley P, Patt YZ (2002) Risk factors for hepatocellular carcinoma: synergism of alcohol with viral hepatitis and diabetes mellitus. Hepatology 36:1206–1213
11. Yuan JM, Govindarajan S, Arakawa K, Yu MC (2004) Synergism of alcohol, diabetes, and viral hepatitis on the risk of hepatocellular carcinoma in blacks and whites in the U.S. Cancer 101:1009–1017
12. Tsukamoto H (2007) Conceptual importance of identifying alcoholic liver disease as a lifestyle disease. J Gastroenterol 42:603–609
13. Alison MR (2005) Liver stem cells: implications for hepatocarcinogenesis. Stem Cell Rev 1:253–260
14. Roskams T (2006) Liver stem cells and their implication in hepatocellular and cholangiocarcinoma. Oncogene 25:3818–3822
15. Tang Y, Kitisin K, Jogunoori W, Li C, Deng CX, Mueller SC, Ressom HW, Rashid A, He AR, Mendelson JS, Jessup JM, Shetty K, Zasloff M, Mishra B, Reddy EP, Johnson L, Mishra L (2008) Progenitor/stem cells give rise to liver cancer due to aberrant TGF-beta and IL-6 signaling. Proc Natl Acad Sci U S A 105:2445–2450
16. Zender L, Spector MS, Xue W, Flemming P, Cordon-Cardo C, Silke J, Fan ST, Luk JM, Wigler M, Hannon GJ, Mu D, Lucito R, Powers S, Lowe SW (2006) Identification and validation of oncogenes in liver cancer using an integrative oncogenomic approach. Cell 125:1253–1267
17. Beachy PA, Karhadkar SS, Berman DM (2004) Tissue repair and stem cell renewal in carcinogenesis. Nature 432:324–331
18. Chambers I, Smith A (2004) Self-renewal of teratocarcinoma and embryonic stem cells. Oncogene 23:7150–7160
19. Valk-Lingbeek ME, Bruggeman SW, van Lohuizen M (2004) Stem cells and cancer; the polycomb connection. Cell 118:409–418
20. Rountree CB, Senadheera S, Mato JM, Crooks GM, Lu SC (2008) Expansion of liver cancer stem cells during aging in methionine adenosyltransferase 1A-deficient mice. Hepatology 47:1288–1297
21. Ma S, Lee TK, Zheng BJ, Chan KW, Guan XY (2008) CD133+ HCC cancer stem cells confer chemoresistance by preferential expression of the Akt/PKB survival pathway. Oncogene 27:1749–1758

22. Wurmbach E, Chen YB, Khitrov G, Zhang W, Roayaie S, Schwartz M, Fiel I, Thung S, Mazzaferro V, Bruix J, Bottinger E, Friedman S, Waxman S, Llovet JM (2007) Genome-wide molecular profiles of HCV-induced dysplasia and hepatocellular carcinoma. Hepatology 45:938–947
23. Yeoh GC, Ernst M, Rose-John S, Akhurst B, Payne C, Long S, Alexander W, Croker B, Grail D, Matthews VB (2007) Opposing roles of gp130-mediated STAT-3 and ERK-1/2 signaling in liver progenitor cell migration and proliferation. Hepatology 45:486–494
24. Sicklick JK, Li YX, Jayaraman A, Kannangai R, Qi Y, Vivekanandan P, Ludlow JW, Owzar K, Chen W, Torbenson MS, Diehl AM (2006) Dysregulation of the Hedgehog pathway in human hepatocarcinogenesis. Carcinogenesis 27:748–757
25. Sicklick JK, Li YX, Melhem A, Schmelzer E, Zdanowicz M, Huang J, Caballero M, Fair JH, Ludlow JW, McClelland RE, Reid LM, Diehl AM (2006) Hedgehog signaling maintains resident hepatic progenitors throughout life. Am J Physiol Gastrointest Liver Physiol 290:G859–G870
26. Kitisin K, Ganesan N, Tang Y, Jogunoori W, Volpe EA, Kim SS, Katuri V, Kallakury B, Pishvaian M, Albanese C, Mendelson J, Zasloff M, Rashid A, Fishbein T, Evans SR, Sidawy A, Reddy EP, Mishra B, Johnson LB, Shetty K, Mishra L (2007) Disruption of transforming growth factor-beta signaling through beta-spectrin ELF leads to hepatocellular cancer through cyclin D1 activation. Oncogene 26:7103–7110
27. Nguyen LN, Furuya MH, Wolfraim LA, Nguyen AP, Holdren MS, Campbell JS, Knight B, Yeoh GC, Fausto N, Parks WT (2007) Transforming growth factor-beta differentially regulates oval cell and hepatocyte proliferation. Hepatology 45:31–41
28. Petersen RK, Madsen L, Pedersen LM, Hallenborg P, Hagland H, Viste K, Doskeland SO, Kristiansen K (2008) Cyclic AMP (cAMP)-mediated stimulation of adipocyte differentiation requires the synergistic action of Epac- and cAMP-dependent protein kinase-dependent processes. Mol Cell Biol 28:3804–3816
29. Mathurin P, Deng QG, Keshavarzian A, Choudhary S, Holmes EW, Tsukamoto H (2000) Exacerbation of alcoholic liver injury by enteral endotoxin in rats. Hepatology 32:1008–1017
30. Uesugi T, Froh M, Arteel GE, Bradford BU, Thurman RG (2001) Toll-like receptor 4 is involved in the mechanism of early alcohol-induced liver injury in mice. Hepatology 34:101–108
31. Machida K, Tsukamoto H, Mkrtchyan H, Duan L, Dynnyk A, Liu HM, Asahina K, Govindarajan S, Ray R, Ou JH, Seki E, Deshaies R, Miyake K, Lai MM (2009) Toll-like receptor 4 mediates synergism between alcohol and HCV in hepatic oncogenesis involving stem cell marker Nanog. Proc Natl Acad Sci U S A 106:1548–1553
32. Chen CL, Tsukamoto H, Liu JC, Kashiwabara C, Feldman D, Sher L, Dooley S, French SW, Mishra L, Petrovic L, Jeong JH, Machida K (2013) Reciprocal regulation by TLR4 and TGF-beta in tumor-initiating stem-like cells. J Clin Invest 123:2832–2849
33. Andersen JB, Loi R, Perra A, Factor VM, Ledda-Columbano GM, Columbano A, Thorgeirsson SS (2010) Progenitor-derived hepatocellular carcinoma model in the rat. Hepatology 51:1401–1409
34. Cai X, Zhai J, Kaplan DE, Zhang Y, Zhou L, Chen X, Qian G, Zhao Q, Li Y, Gao L, Cong W, Zhu M, Yan Z, Shi L, Wu D, Wei L, Shen F, Wu M (2012) Background progenitor activation is associated with recurrence after hepatectomy of combined hepatocellular-cholangiocarcinoma. Hepatology 56:1804–1816
35. Lee JS, Heo J, Libbrecht L, Chu IS, Kaposi-Novak P, Calvisi DF, Mikaelyan A, Roberts LR, Demetris AJ, Sun Z, Nevens F, Roskams T, Thorgeirsson SS (2006) A novel prognostic subtype of human hepatocellular carcinoma derived from hepatic progenitor cells. Nat Med 12:410–416
36. Yamashita T, Ji J, Budhu A, Forgues M, Yang W, Wang HY, Jia H, Ye Q, Qin LX, Wauthier E, Reid LM, Minato H, Honda M, Kaneko S, Tang ZY, Wang XW (2009) EpCAM-positive hepatocellular carcinoma cells are tumor-initiating cells with stem/progenitor cell features. Gastroenterology 136:1012–1024
37. Reya T, Morrison SJ, Clarke MF, Weissman IL (2001) Stem cells, cancer, and cancer stem cells. Nature 414:105–111
38. Dapito DH, Mencin A, Gwak GY, Pradere JP, Jang MK, Mederacke I, Caviglia JM, Khiabanian H, Adeyemi A, Bataller R, Lefkowitch JH, Bower M, Friedman R, Sartor RB, Rabadan R, Schwabe RF (2012) Promotion of hepatocellular carcinoma by the intestinal microbiota and TLR4. Cancer Cell 21:504–516

39. Bauer AK, Dixon D, DeGraff LM, Cho HY, Walker CR, Malkinson AM, Kleeberger SR (2005) Toll-like receptor 4 in butylated hydroxytoluene-induced mouse pulmonary inflammation and tumorigenesis. J Natl Cancer Inst 97:1778–1781
40. Fukata M, Chen A, Vamadevan AS, Cohen J, Breglio K, Krishnareddy S, Hsu D, Xu R, Harpaz N, Dannenberg AJ, Subbaramaiah K, Cooper HS, Itzkowitz SH, Abreu MT (2007) Toll-like receptor-4 promotes the development of colitis-associated colorectal tumors. Gastroenterology 133:1869–1881
41. Mittal D, Saccheri F, Venereau E, Pusterla T, Bianchi ME, Rescigno M (2010) TLR4-mediated skin carcinogenesis is dependent on immune and radioresistant cells. EMBO J 29:2242–2252
42. Fukata M, Shang L, Santaolalla R, Sotolongo J, Pastorini C, Espana C, Ungaro R, Harpaz N, Cooper HS, Elson G, Kosco-Vilbois M, Zaias J, Perez MT, Mayer L, Vamadevan AS, Lira SA, Abreu MT (2011) Constitutive activation of epithelial TLR4 augments inflammatory responses to mucosal injury and drives colitis-associated tumorigenesis. Inflamm Bowel Dis 17:1464–1473
43. Kanda T, Steele R, Ray R, Ray RB (2009) Inhibition of intrahepatic gamma interferon production by hepatitis C virus nonstructural protein 5A in transgenic mice. J Virol 83:8463–8469
44. Majumder M, Ghosh AK, Steele R, Zhou XY, Phillips NJ, Ray R, Ray RB (2002) Hepatitis C virus NS5A protein impairs TNF-mediated hepatic apoptosis, but not by an anti-FAS antibody, in transgenic mice. Virology 294:94–105
45. Majumder M, Steele R, Ghosh AK, Zhou XY, Thornburg L, Ray R, Phillips NJ, Ray RB (2003) Expression of hepatitis C virus non-structural 5A protein in the liver of transgenic mice. FEBS Lett 555:528–532
46. Sarcar B, Ghosh AK, Steele R, Ray R, Ray RB (2004) Hepatitis C virus NS5A mediated STAT3 activation requires co-operation of Jak1 kinase. Virology 322:51–60
47. Park YN, Chae KJ, Oh BK, Choi J, Choi KS, Park C (2004) Expression of Smad7 in hepatocellular carcinoma and dysplastic nodules: resistance mechanism to transforming growth factor-beta. Hepatogastroenterology 51:396–400
48. Tang B, Bottinger EP, Jakowlew SB, Bagnall KM, Mariano J, Anver MR, Letterio JJ, Wakefield LM (1998) Transforming growth factor-beta1 is a new form of tumor suppressor with true haploid insufficiency. Nat Med 4:802–807
49. Liao B, Hu Y, Herrick DJ, Brewer G (2005) The RNA-binding protein IMP-3 is a translational activator of insulin-like growth factor II leader-3 mRNA during proliferation of human K562 leukemia cells. J Biol Chem 280:18517–18524
50. Song K, Wang H, Krebs TL, Danielpour D (2006) Novel roles of Akt and mTOR in suppressing TGF-beta/ALK5-mediated Smad3 activation. EMBO J 25:58–69
51. Lucas PJ, Kim SJ, Melby SJ, Gress RE (2000) Disruption of T cell homeostasis in mice expressing a T cell-specific dominant negative transforming growth factor beta II receptor. J Exp Med 191:1187–1196
52. Hahm KB, Im YH, Parks TW, Park SH, Markowitz S, Jung HY, Green J, Kim SJ (2001) Loss of transforming growth factor beta signalling in the intestine contributes to tissue injury in inflammatory bowel disease. Gut 49:190–198
53. Yang X, Letterio JJ, Lechleider RJ, Chen L, Hayman R, Gu H, Roberts AB, Deng C (1999) Targeted disruption of SMAD3 results in impaired mucosal immunity and diminished T cell responsiveness to TGF-beta. EMBO J 18:1280–1291
54. Cartney-Francis N, Jin W, Wahl SM (2004) Aberrant Toll receptor expression and endotoxin hypersensitivity in mice lacking a functional TGF-beta 1 signaling pathway. J Immunol 172:3814–3821
55. Aparicio S, Eaves CJ (2009) p53: a new kingpin in the stem cell arena. Cell 138:1060–1062
56. Bonizzi G, Cicalese A, Insinga A, Pelicci PG (2012) The emerging role of p53 in stem cells. Trends Mol Med 18:6–12
57. Li M, He Y, Dubois W, Wu X, Shi J, Huang J (2012) Distinct regulatory mechanisms and functions for p53-activated and p53-repressed DNA damage response genes in embryonic stem cells. Mol Cell 46:30–42
58. Bressac B, Galvin KM, Liang TJ, Isselbacher KJ, Wands JR, Ozturk M (1990) Abnormal structure and expression of p53 gene in human hepatocellular carcinoma. Proc Natl Acad Sci U S A 87:1973–1977

59. Gonzalez-Angulo AM, Sneige N, Buzdar AU, Valero V, Kau SW, Broglio K, Yamamura Y, Hortobagyi GN, Cristofanilli M (2004) p53 expression as a prognostic marker in inflammatory breast cancer. Clin Cancer Res 10:6215–6221
60. Resetkova E, Gonzalez-Angulo AM, Sneige N, Mcdonnell TJ, Buzdar AU, Kau SW, Yamamura Y, Reuben JM, Hortobagyi GN, Cristofanilli M (2004) Prognostic value of P53, MDM-2, and MUC-1 for patients with inflammatory breast carcinoma. Cancer 101:913–917
61. Cicalese A, Bonizzi G, Pasi CE, Faretta M, Ronzoni S, Giulini B, Brisken C, Minucci S, Di Fiore PP, Pelicci PG (2009) The tumor suppressor p53 regulates polarity of self-renewing divisions in mammary stem cells. Cell 138:1083–1095
62. Martin-Belmonte F, Perez-Moreno M (2012) Epithelial cell polarity, stem cells and cancer. Nat Rev Cancer 12:23–38
63. Zhao Z, Zuber J, Diaz-Flores E, Lintault L, Kogan SC, Shannon K, Lowe SW (2010) p53 loss promotes acute myeloid leukemia by enabling aberrant self-renewal. Genes Dev 24:1389–1402
64. Colaluca IN, Tosoni D, Nuciforo P, Senic-Matuglia F, Galimberti V, Viale G, Pece S, Di Fiore PP (2008) NUMB controls p53 tumour suppressor activity. Nature 451:76–80
65. Bric A, Miething C, Bialucha CU, Scuoppo C, Zender L, Krasnitz A, Xuan Z, Zuber J, Wigler M, Hicks J, McCombie RW, Hemann MT, Hannon GJ, Powers S, Lowe SW (2009) Functional identification of tumor-suppressor genes through an in vivo RNA interference screen in a mouse lymphoma model. Cancer Cell 16:324–335
66. March HN, Rust AG, Wright NA, Ten Hoeve J, de Ridder J, Eldridge M, van der Weyden L, Berns A, Gadiot J, Uren A, Kemp R, Arends MJ, Wessels LF, Winton DJ, Adams DJ (2011) Insertional mutagenesis identifies multiple networks of cooperating genes driving intestinal tumorigenesis. Nat Genet 43:1202–1209
67. Nishimura T, Kaibuchi K (2007) Numb controls integrin endocytosis for directional cell migration with aPKC and PAR-3. Dev Cell 13:15–28
68. Peralta ER, Martin BC, Edinger AL (2010) Differential effects of TBC1D15 and mammalian Vps39 on Rab7 activation state, lysosomal morphology, and growth factor dependence. J Biol Chem 285:16814–16821
69. Zhang XM, Walsh B, Mitchell CA, Rowe T (2005) TBC domain family, member 15 is a novel mammalian Rab GTPase-activating protein with substrate preference for Rab7. Biochem Biophys Res Commun 335:154–161
70. Speicher S, Fischer A, Knoblich J, Carmena A (2008) The PDZ protein Canoe regulates the asymmetric division of Drosophila neuroblasts and muscle progenitors. Curr Biol 18:831–837
71. Wee B, Johnston CA, Prehoda KE, Doe CQ (2011) Canoe binds RanGTP to promote Pins(TPR)/Mud-mediated spindle orientation. J Cell Biol 195:369–376
72. Amson R, Pece S, Lespagnol A, Vyas R, Mazzarol G, Tosoni D, Colaluca I, Viale G, Rodrigues-Ferreira S, Wynendaele J, Chaloin O, Hoebeke J, Marine JC, Di Fiore PP, Telerman A (2012) Reciprocal repression between P53 and TCTP. Nat Med 18:91–99
73. Mathew R, Karp CM, Beaudoin B, Vuong N, Chen G, Chen HY, Bray K, Reddy A, Bhanot G, Gelinas C, Dipaola RS, Karantza-Wadsworth V, White E (2009) Autophagy suppresses tumorigenesis through elimination of p62. Cell 137:1062–1075
74. Takamura A, Komatsu M, Hara T, Sakamoto A, Kishi C, Waguri S, Eishi Y, Hino O, Tanaka K, Mizushima N (2011) Autophagy-deficient mice develop multiple liver tumors. Genes Dev 25:795–800
75. Wei Y, Zou Z, Becker N, Anderson M, Sumpter R, Xiao G, Kinch L, Koduru P, Christudass CS, Veltri RW, Grishin NV, Peyton M, Minna J, Bhagat G, Levine B (2013) EGFR-mediated Beclin 1 phosphorylation in autophagy suppression, tumor progression, and tumor chemoresistance. Cell 154:1269–1284
76. Gupta PB, Onder TT, Jiang G, Tao K, Kuperwasser C, Weinberg RA, Lander ES (2009) Identification of selective inhibitors of cancer stem cells by high-throughput screening. Cell 138:645–659

Chapter 9
Synergistic Toxic Interactions Between CYP2E1, LPS/TNFα, and JNK/p38 MAP Kinase and Their Implications in Alcohol-Induced Liver Injury

Arthur I. Cederbaum, Yongke Lu, Xiaodong Wang, and Defeng Wu

Abstract The mechanisms by which alcohol causes cell injury are not clear. Many pathways have been suggested to play a role in how alcohol induces oxidative stress. Considerable attention has been given to alcohol-elevated production of lipopolysaccharide (LPS) and TNFα and to alcohol induction of CYP2E1. These two pathways are not exclusive of each other; however, associations and interactions between them, especially in vivo, have not been extensively evaluated. We have shown that increased oxidative stress from induction of CYP2E1 in vivo sensitizes hepatocytes to LPS and TNFα toxicity and that oxidative stress, activation of p38 and JNK MAP kinases, and mitochondrial dysfunction are downstream mediators of this CYP2E1-LPS/TNFα potentiated hepatotoxicity. This Review will summarize studies showing potentiated interactions between these two risk factors in promoting liver injury and the mechanisms involved including activation of the mitogen-activated kinase kinase kinase ASK-1 as a result of CYP2E1-derived reactive oxygen intermediates promoting dissociation of the inhibitory thioredoxin from ASK-1. This activation of ASK-1 is followed by activation of the mitogen-activated kinase kinases MKK3/MKK6 and MKK4/MMK7 and subsequently p38 and JNK MAP kinases. Synergistic toxicity occurs between CYP2E1 and the JNK1 but not the JNK2 isoform as JNK1 knockout mice are completely protected against CYP2E1 plus TNFα toxicity, elevated oxidative stress, and mitochondrial dysfunction. We hypothesize that similar interactions occur as a result of ethanol induction of CYP2E1 and TNFα.

Keywords Alcohol liver injury • CYP2E1-dependent toxicity • Lipopolysaccharide–CYP2E1 interactions • Oxidative stress and liver injury • CYP2E1-tumor necrosis factor alpha toxicity • JNK and p38 mitogen-activated kinases • Mitochondrial dysfunction

A.I. Cederbaum (✉) • Y. Lu • X. Wang • D. Wu
Department of Pharmacology and Systems Therapeutics, Mount Sinai School of Medicine, Box 1603, One Gustave L. Levy Place, New York, NY 10029, USA
e-mail: arthur.cederbaum@mssm.edu

V. Vasiliou et al. (eds.), *Biological Basis of Alcohol-Induced Cancer*, Advances in Experimental Medicine and Biology 815,
DOI 10.1007/978-3-319-09614-8_9

9.1 Introduction

The ability of acute and chronic ethanol treatment to increase production of reactive oxygen species (ROS) and enhance peroxidation of lipids, protein, and DNA has been demonstrated in a variety of systems [1]. The mechanism(s) by which alcohol causes cell injury are still not clear. Several mechanisms have been briefly summarized [2–4], and it is likely that many of them ultimately converge as they reflect a spectrum of the organism's response to the myriad of direct and indirect actions of alcohol. A major mechanism that is a focus of considerable research is the role of lipid peroxidation and oxidative stress in alcohol toxicity. Under certain conditions, such as acute or chronic alcohol exposure, production of reactive oxygen species (ROS) is enhanced and/or the level or activity of antioxidants is reduced. The resulting state—which is characterized by a disturbance in the balance between rates of ROS production, on one hand and rates of ROS removal and repair of damaged complex molecules, on the other—is called oxidative stress.

Many pathways have been suggested to play a key role in how ethanol induces "oxidative stress" [1–4]. Many of these pathways are not exclusive of one another and it is likely that several, indeed many, systems contribute to the ability of ethanol to induce a state of oxidative stress. Two major pathways include ethanol effects on the immune system with altered cytokine production due to ethanol-induced increase in bacterial-derived endotoxin with subsequent activation of Kupffer cells and ethanol induction of Cytochrome P4502E1, CYP2E1.

9.2 Kupffer Cells and Alcoholic Liver Disease

Kupffer cells are stimulated by chronic ethanol treatment to produce free radicals and cytokines, including TNFα, which plays a role in ALD [5, 6]. This stimulation is mediated by bacterial-derived endotoxin, and ALD is decreased when gram-negative bacteria are depleted from the gut by treatment with lactobacillus or antibiotics [7]. The TNFα receptor super-family consists of several members sharing a sequence homology, the death domain, located in the intracellular portion of the receptor. These "death" receptors, including Fas, TNF-R1, and TRAIL-R1/TRAIL-R2 are expressed in hepatocytes and when stimulated by their respective ligands, FasL, TNFα, or TRAIL, hepatocyte injury can occur [8]. Lipopolysaccharide (LPS) is a component of the outer wall of gram-negative bacteria that normally inhabit the gut. LPS penetrates the gut epithelium only in trace amounts; however, LPS absorption can be elevated under pathophysiological conditions such as alcoholic liver disease (ALD) [9]. LPS directly causes liver injury by mechanisms involving inflammatory cells such as Kupffer cells, and chemical mediators such as superoxide, nitric oxide, and tumor necrosis factor (TNFα) and other cytokines [10]. In addition, LPS potentiates liver damage induced by hepatotoxins including ethanol [11–14]. Ethanol alters gut microflora, the source of LPS, and ethanol increases the permeability of the gut, thus increasing the distribution of LPS from the gut into the portal circulation (endotoxemia). This causes activation of Kupffer cells, the resident macrophages in liver, resulting in release of chemical mediators including cytokines and ROS and subsequently, ALD [15]. Destruction of Kupffer cells with gadolinium chloride attenuated ALD [5] and anti-TNFα antibodies protect against ALD [6]. NADPH oxidase was identified as a key enzyme for generating ROS in Kupffer cells after ethanol treatment [16]. The role of TNFα in ALD was further validated by the findings that the ethanol-induced pathology was nearly blocked in TNFα receptor 1 knockout mice [17].

9.3 CYP2E1

CYP2E1 metabolizes a variety of small, hydrophobic substrates including solvents such as chloroform and carbon tetrachloride, aromatic hydrocarbons such as benzene and toluene, alcohols such as ethanol and pentanol, aldehydes such as acetaldehyde, halogenated anesthetics such as enflurane and halothane, nitrosamines such as *N,N*-dimethylnitrosamine, and drugs such as chlorzoxazone and acetaminophen [18–20]. CYP2E1 metabolizes and activates many toxicologically important compounds such as ethanol, carbon tetrachloride, acetaminophen, benzene, halothane, and many other halogenated substrates. Toxicity by the above compounds is enhanced after induction of CYP2E1 e.g., by ethanol treatment, and toxicity is reduced by inhibitors of CYP2E1 or in CYP2E1 knockout mice [21]. Molecular oxygen itself is likely to be a most important substrate for CYP2E1. CYP2E1, relative to several other P450 enzymes, displays high NADPH oxidase activity as it appears to be poorly coupled with NADPH-cytochrome P450 reductase [22, 23]. CYP2E1 was the most efficient P450 enzyme in the initiation of NADPH-dependent lipid peroxidation in reconstituted membranes among five different P450 forms investigated and anti-CYP2E1 IgG inhibited microsomal NADPH oxidase activity and microsomal lipid peroxidation dependent on P450 [22]. Microsomes isolated from rats fed ethanol chronically were about twofold to threefold more reactive in generating superoxide radical and H_2O_2, and in the presence of ferric complexes, in generating hydroxyl radical and undergoing lipid peroxidation than microsomes from pair-fed control rats [24, 25]. CYP2E1 levels were elevated about threefold to fivefold in liver microsomes after feeding rats the Lieber–DeCarli diet for 4 weeks and the enhanced effectiveness of microsomes isolated from the ethanol-fed rats was prevented by addition of chemical inhibitors of CYP2E1 and by polyclonal antibody raised against CYP2E1 purified from pyrazole-treated rats, confirming that the increased activity in these microsomes was due to CYP2E1.

CYP2E1 plays a major role in mechanisms by which alcohol promotes hepatocarcinogenesis. One prominent role for CYP2E1 is the oxidation of many low molecular weight cancer suspected agents such as benzene, styrene, halogenated hydrocarbons, nitriles, and nitrosamines [26]. Induction of CYP2E1 by alcohol results in enhanced activation of numerous procarcinogens to carcinogens [27]. Deletion of CYP2E1 or pharmacological inhibition of CYP2E1 by chlormethiazole reduces diethylnitroamine-induced hepatic tumors [28, 29]. Indeed, a recent editorial

[30] was entitled "Pharmacological Blockage of CYP2E1 and Alcohol-mediated Liver Cancer: is the Time Ready?" Further relevant to the role of CYP2E1 in alcohol-induced liver cancer are the studies showing that alcohol-induced DNA strand cleavage and etheno-adduct formation are dependent on CYP2E1 expression. For example, in HeLa cells overexpressing CYP2E1, ethanol produced oxidative stress and DNA strand breakage and these effects were blunted by a CYP2E1 inhibitor; no such effects were found with HeLa cells not expressing CYP2E [31]. In vivo administration of ethanol to male Wistar rats produced an increase in free radicals and DNA strand breaks, effects highly correlating with the induction of CYP2E1 [32]. Acute ethanol consumption increased oxidative DNA damage and production of 8-hydroxy-deoxyguanine especially in ALDH2 knockout mice and induction of CYP2E1 played a pivotal role in these effects [33]. We showed that incubation of plasmids with microsomes from chronic ethanol-fed rats in which CYP2E1 was induced results in DNA strand breaks which could be completely prevented by anti-CYP2E1 IgG [34]. Bradford et al. [35] showed an increase of oxidative DNA adducts and of mutagenic apurinic and aprimidinic DNA sites and induction of DNA repair enzymes in wild type mice and NADPH oxidase knockout mice chronically fed alcohol but not in CYP2E1 knockout mice. Seitz and collaborators [36] recently showed that etheno-DNA adducts strongly correlated with CYP2E1 expression in patients with alcohol liver disease. The number of nuclei in hepatocytes from alcohol-fed lean and obese mice stained positively for etheno-DNA adducts correlated with CYP2E1 expression and in vitro, etheno-DNA adducts were produced when ethanol was incubated with HepG2 cells expressing CYP2E1 but not with HepG2 cells not expressing CYP2E1. Taken as a whole, such results, and others, strongly support a role for ethanol induction of CYP2E1 in ethanol induction of DNA strand breakage and hepatocarcinogenesis.

Since CYP2E1 can generate ROS during its catalytic cycle, and its levels are elevated by chronic treatment with ethanol, CYP2E1 has been suggested as a major contributor to ethanol-induced oxidative stress, and to ethanol-induced liver injury [37–39]. Experimentally, a decrease in CYP2E1 induction was found to be associated with a reduction in alcohol-induced liver injury [40, 41]. A CYP2E1 transgenic mouse treated with ethanol displayed higher transaminase levels and histological features of liver injury compared with control mice [42]. Infection of HepG2 cells with an adenoviral vector expressing CYP2E1 adenovirus potentiated acetaminophen toxicity as compared to HepG2 cells infected with a LacZ expressing adenovirus [43]. Administration of the CYP2E1 adenovirus in vivo to mice elevated CYP2E1 levels and activity and produced significant liver injury compared to the LacZ-infected mice [44].

9.4 LPS/TNFα–CYP2E1 Interactions

Abnormal cytokine metabolism is a major feature of ALD. Rats chronically fed ethanol were more sensitive to the hepatotoxic effects of administration of LPS and had higher plasma levels of TNFα than control rats [45, 46]. Anti-TNFα antibody prevented alcohol liver injury in rats [6] and mice lacking the TNFR1 receptor did not develop alcohol liver injury [17]. These and other studies clearly implicate TNFα as a major risk factor for the development of alcoholic liver injury. One complication in this central role for TNFα is that hepatocytes are normally resistant to TNFα induced toxicity. This led to the hypothesis that besides elevating TNFα, alcohol somehow sensitizes or primes the liver to become susceptible to TNFα [47, 48]. Known factors which sensitize the liver to TNFα are inhibitors of mRNA or protein synthesis, which likely prevent the synthesis of protective factors, inhibition of NF-κB activation in hepatocytes to lower synthesis of such protective factors, depletion of GSH, especially mitochondrial GSH, lowering of *S*-adenosyl methionine coupled to elevation of *S*-adenosyl homocysteine or inhibition of the proteasome. Combined treatment with ethanol plus TNFα was more toxic to hepatocytes and HepG2 E47 cells which express high levels of CYP2E1 than control hepatocytes with lower levels of CYP2E1 or HepG2 C34 cells which do not express CYP2E1 [49]. RALA hepatocytes with increased expression of CYP2E1 were sensitized to TNFα mediated cell death [50]. These results suggest that increased oxidative stress from CYP2E1 may sensitize isolated hepatocytes to TNFα-induced toxicity. Since either LPS/TNFα or CYP2E1 is considered independent risk factors involved in ALD, but mutual relationships or interactions between them are unknown, we initiated studies to evaluate whether CYP2E1 contributes or potentiates LPS- or TNFα-mediated liver injury in vivo.

9.5 Pyrazole Potentiates LPS toxicity [51, 52]

Male, Sprague–Dawley rats (160–180 g) were injected intraperitonally with pyrazole (PY), 200 mg/kg body wt, once a day for 2 days to induce CYP2E1. After an overnight fast, either saline or LPS (Sigma, serotype 055: BS, 10 mg/kg body wt) was injected via the tail vein. Rats were killed 8–10 h after the LPS or saline injection and blood and liver tissue collected. Neither pyrazole nor LPS alone caused liver injury as reflected by transaminase (ALT, AST) levels or liver histopathology (Fig. 9.1). However, the combination of LPS plus pyrazole increased AST and ALT levels about four-fold over the levels in the pyrazole alone or LPS alone groups (Fig. 9.1A). LPS plus pyrazole treatment induced extensive necrosis of hepatocytes, mainly located both in periportal and pericentral zones of the liver, accompanied by strong infiltration of inflammatory cells (Fig. 9.1B). LPS alone treatment caused some apoptosis and activation of caspases 3 and 9, whereas pyrazole treatment alone had no effect. LPS plus pyrazole treatment was not any more effective than LPS alone in increasing apoptosis, unlike the increases in necrosis and inflammation.

To assess whether oxidative stress occurs after the various treatments, malondialdehyde (MDA) levels as a reflection of lipid peroxidation, were assayed. Whereas pyrazole alone or LPS alone did not elevate TBAR levels over those found with saline controls, the combination of LPS plus pyrazole increased MDA levels about 65 % ($p<0.05$ compared to the other three groups, Fig. 9.2A). Levels of 3 nitrotyrosine protein adducts as a marker for oxidized nitrated protein formation were

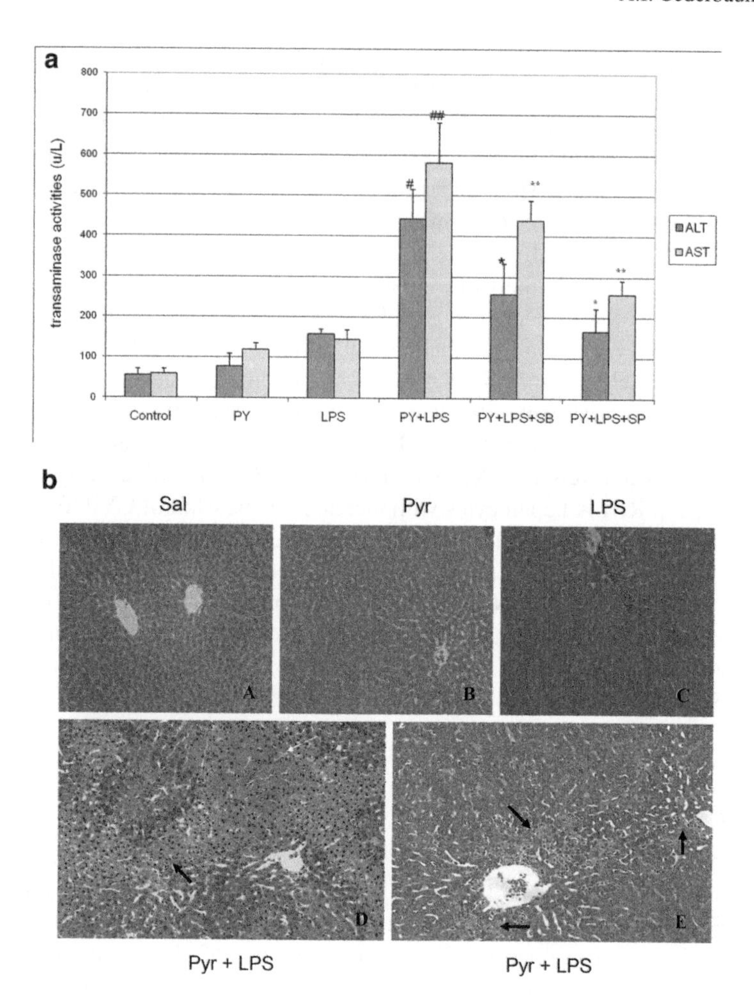

Fig. 9.1 Effect of pyrazole or LPS or LPS plus pyrazole on serum ALT or AST levels (**A**) and liver histopathology (**B**). Panels in (**B**) refer to (**A**) saline; (**B**) pyrazole-treated; (**C**) LPS-treated; (**D**) and (**E**) LPS plus pyrazole-treated. *Arrows* show necrotic foci with inflammatory cell infiltration. $^{**}p<0.01$ compared to all other groups. Either the JNK inhibitor SP600125 or the p38 MAPK inhibitor prevents the LPS plus pyrazole elevation of ALT and AST

determined by slot blot analysis. Low levels of 3-NT adducts were found in saline control livers. Treatment with either LPS alone or pyrazole alone slightly elevated 3-NT protein adduct levels; however, striking increases in protein carbonyls were found in the combined LPS plus pyrazole group (Fig. 9.2B). Thus oxidative/nitrosative stress was elevated in livers from the LPS plus pyrazole-treated mice.

CYP2E1 catalytic activity (oxidation of p-nitrophenol to p-nitrocatechol) was increased about twofold by either the pyrazole alone or the pyrazole plus LPS treatments. LPS alone slightly but not significantly decreased CYP2E1 activity.

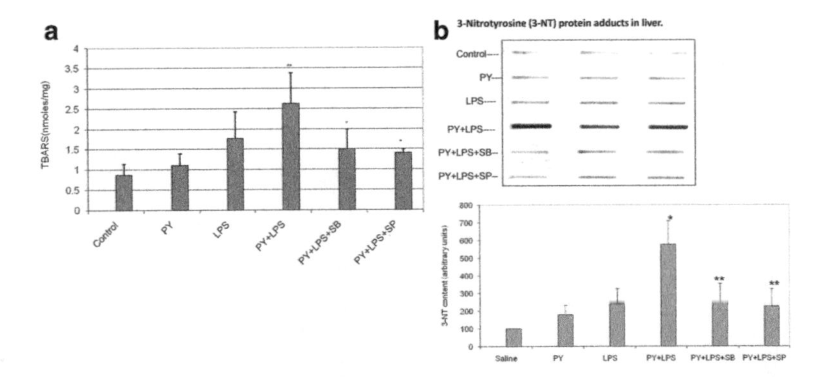

Fig. 9.2 Pyrazole plus LPS increases oxidative stress. Mice were treated with either saline or pyrazole alone or LPS alone or LPS plus pyrazole. (**A**) Lipid peroxidation was assayed as formation of thiobarbituric acid-reactive substances (TBARS). (**B**) Formation of three nitrotyrosine protein adducts was assayed by slot blot and results expressed as arbitrary units. The inhibitors of JNK (SP) and p38 MAPK (SB) lower the potentiated increase in oxidative stress produced by pyrazole plus LPS

Levels of CYP2E1 protein, measured by immunoblot analysis, showed similar trends, being increased about twofold by pyrazole or pyrazole plus LPS treatments. This enhanced liver injury is associated with elevated levels of CYP2E1 and increased oxidative/nitrosative stress generated by the combination of LPS plus CYP2E1.

9.6 Pyrazole Potentiates TNFα Toxicity [53, 54]

Since TNFα levels are elevated after LPS administration and TNFα plays an important role in the effects of LPS, we determined if pyrazole treatment to induce CYP2E1 potentiates TNFα toxicity as it did LPS toxicity. Basically, the same approaches described above for LPS were used, with injection of TNFα (50 μg/kg body wt) replacing the LPS treatment. Figure 9.3A shows that ALT and AST levels were low in the saline control mice and in the pyrazole-treated mice challenged with saline. Treatment of control mice with TNFα elevated transaminase levels by about 2–3-fold. Treatment of the pyrazole mice with TNFα elevated transaminase levels more than 3-fold over the TNFα-saline control treated mice (Fig. 9.3A). Liver sections were stained with H&E for morphological evaluation. The saline and TNFα treated mice showed normal liver morphology. Liver from pyrazole-treated mice showed some vacuolar degeneration. Liver from the TNFα plus pyrazole-treated mice showed several necrotic loci (arrows), typical pathology morphology changes including nuclear pyknosis, karyorrhexis, and karyolysis were observed (Fig. 9.3B). The treatment with pyrazole did not significantly alter the levels of thiobarbituric acid-reactive substrates (TBARS) in the total liver extract or the

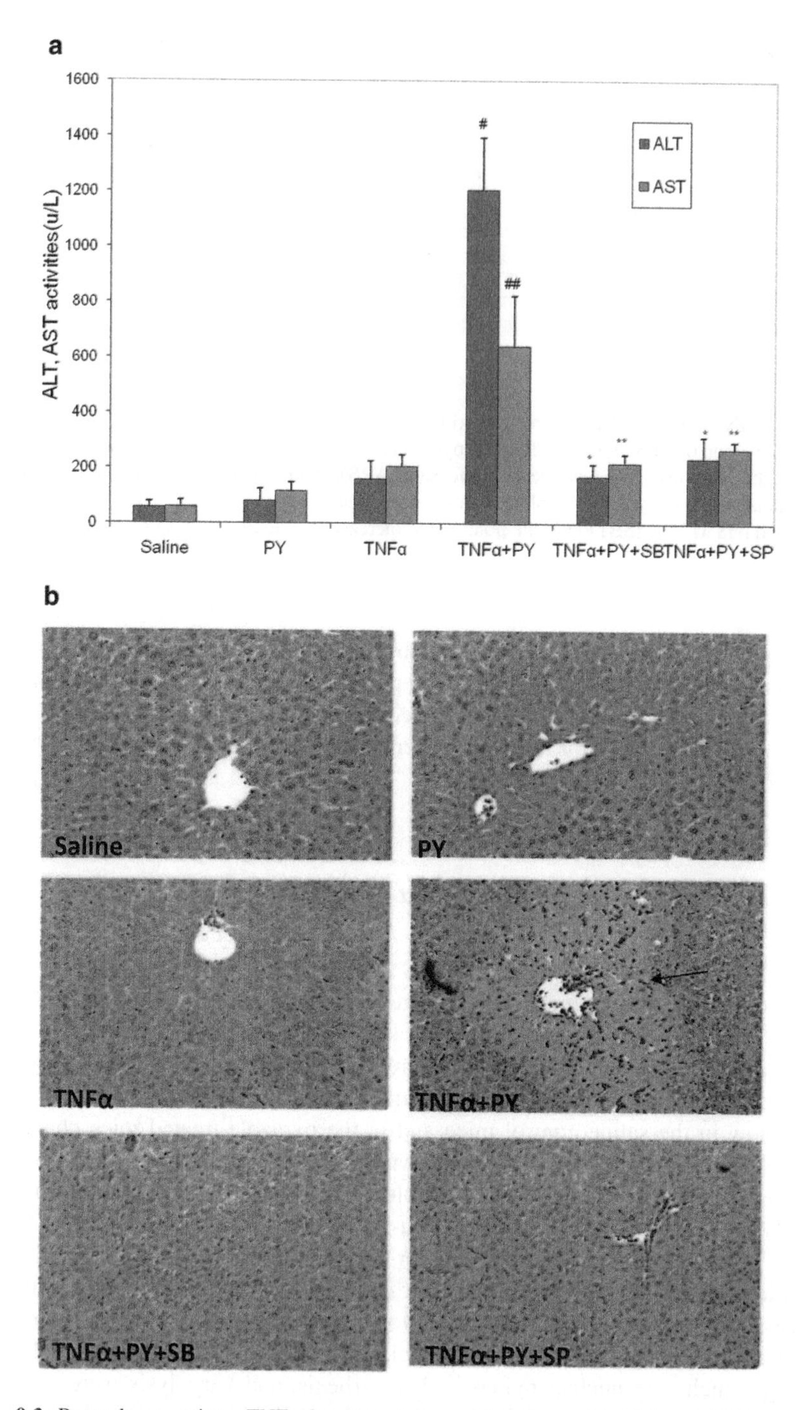

Fig. 9.3 Pyrazole potentiates TNF**α** hepatotoxicity and in mice and inhibitors of JNK or p38 MAPK block the increase in toxicity. Mice were treated with either saline or pyrazole alone or TNF**α** alone or pyrazole plus TNF**α** followed by assays for serum ALT/AST (**A**) or histopathology (**B**) (*arrows* show necrotic zones)

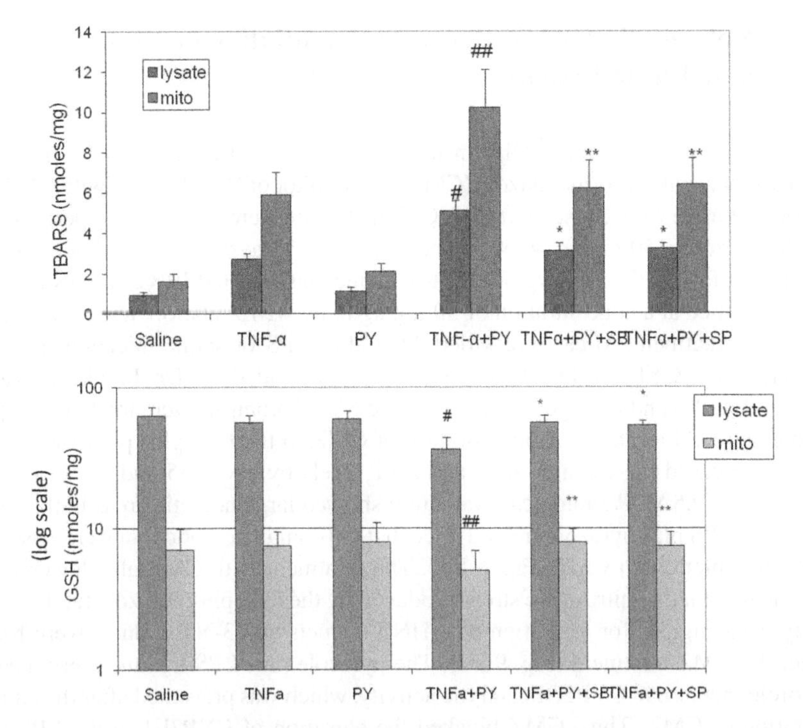

Fig. 9.4 Pyrazole potentiates TNFα-induced oxidative stress and inhibitors of JNK or p38 MAPK block the increase. Liver homogenates and mitochondria were isolated from mice treated as described in the legend to Fig. 9.3. Lipid peroxidation (**A**) and levels of GSH (**B**, note the log scale) were determined in the homogenates and the mitochondrial fraction

mitochondria (Fig. 9.4A). TNFα treatment of control mice elevated levels of TBARS about 2–3-fold. TBARS in the homogenates and the mitochondria were further elevated when TNFα was administered to the pyrazole-treated mice. Highest liver and mitochondrial TBARs levels were observed in the pyrazole plus TNFα treated mice. Liver GSH levels were similar in the saline, pyrazole-treated, and TNFα-treated mice but were decreased about 40 % in the liver extracts from the pyrazole plus TNFα treated mice (Fig. 9.4B). GSH levels were lowered 40 % in the liver mitochondria from the pyrazole plus TNFα-treated mice compared to the TNFα alone treated mice. These results suggest that the combined pyrazole plus TNFα treatment produces elevated oxidative stress in the liver compared to TNFα alone or pyrazole alone, and that mitochondrial oxidative stress may occur in livers of the pyrazole plus TNFα-treated mice.

As expected, CYP2E1 activity and the content of CYP2E1 were elevated 2–3-fold by pyrazole or by pyrazole plus TNFα treatment, over the saline or TNFα alone treated mice. However, induction of CYP2E1 alone by pyrazole is not sufficient to induce liver injury; rather, a second "hit" e.g., TNFα is required.

9.7 Role of CYP2E1 in Pyrazole Potentiation of LPS and TNFα Toxicity

To validate the role of CYP2E1 in the potentiation of LPS toxicity by pyrazole, experiments with chlormethiazole (CMZ) an inhibitor of CYP2E1 and with CYP2E1 knockout mice were carried out [51]. C57BL/6 mice were injected intraperitoneally with pyrazole, 150 mg/kg body wt once a day for 2 days or 0.9 % saline. After an overnight fast, LPS, 4 mg/kg body wt or saline was injected IP. CMZ was injected in some mice at a concentration of 50 mg/kg body wt 15 h before and 30 min after the LPS treatment. Mice were killed 24 h after LPS or saline injection. In other experiments, CYP2E1 knockout mice, kindly provided by Dr. Frank Gonzalez, NCI, NIH (20) and their genetic background SV129 controls were treated with pyrazole and LPS as above. Administration of CMZ to the LPS plus pyrazole-treated mice decreased the elevated ALT and AST levels by about 55 and 65 %, respectively (Fig. 9.5A). Pathological evaluation showed large necrotic areas in the livers from the LPS plus pyrazole-treated mice, but only small necrotic foci were observed after treatment with CMZ (Fig. 9.5B). The treatment with CMZ also lowered the elevated oxidative/nitrosative stress produced by the LPS plus pyrazole treatment as only weak signals for formation of 4-HNE adducts and 3-NT adducts were found after the CMZ treatment (Fig. 9.5C). The pyrazole plus LPS treatment produced a twofold increase in CYP2E1 catalytic activity, which was prevented after the administration of CMZ. Thus, CMZ blocked the elevation of CYP2E1 in the LPS plus pyrazole-treated mice, and this was associated with a decline in oxidative/nitrosative stress and blunting of liver injury.

CYP2E1 knockout or wild type control SV129 mice were treated with LPS plus pyrazole. As with C57Bl/6 mice, liver injury was observed in the wild type SV129 mice treated with LPS plus pyrazole, but not mice treated with LPS alone or pyrazole alone. Serum ALT and AST levels were about 50 % lower in LPS plus pyrazole-treated CYP2E1 knockout mice as compared to wild type mice (Fig. 9.6A). Pathological evaluation showed large necrotic areas and widespread necrotic foci in wild type mice whereas almost normal histology was found in the LPS plus pyrazole-treated CYP2E1 knockout mice (Fig.9.6B). Positive TUNEL staining was also significantly lower in the CYP2E1 null mice compared to wild type mice (Fig. 9.6C). Immunoblots confirmed the absence of CYP2E1 protein in the knockout mice while strong signals from CYP2E1 were detected in immunoblots of the wild type mice (Fig. 9.6D). Thus, in both rats and mice, the CYP2E1 inducer pyrazole potentiates LPS-induced liver injury. This potentiation is associated with elevated oxidative/nitrosative stress and is blocked by the CYP2E1 inhibitor CMZ, and blunted in CYP2E1 knockout mice. We hypothesize that CYP2E1-mediated oxidative stress may synergize with LPS-generated oxidative stress in this model to produce liver injury.

Similar results were found with the TNFα plus pyrazole potentiated toxicity model. Large increases in ALT and AST levels were found after TNFα administration to pyrazole-treated SV129 wild type mice. TNFα treatment of pyrazole-treated

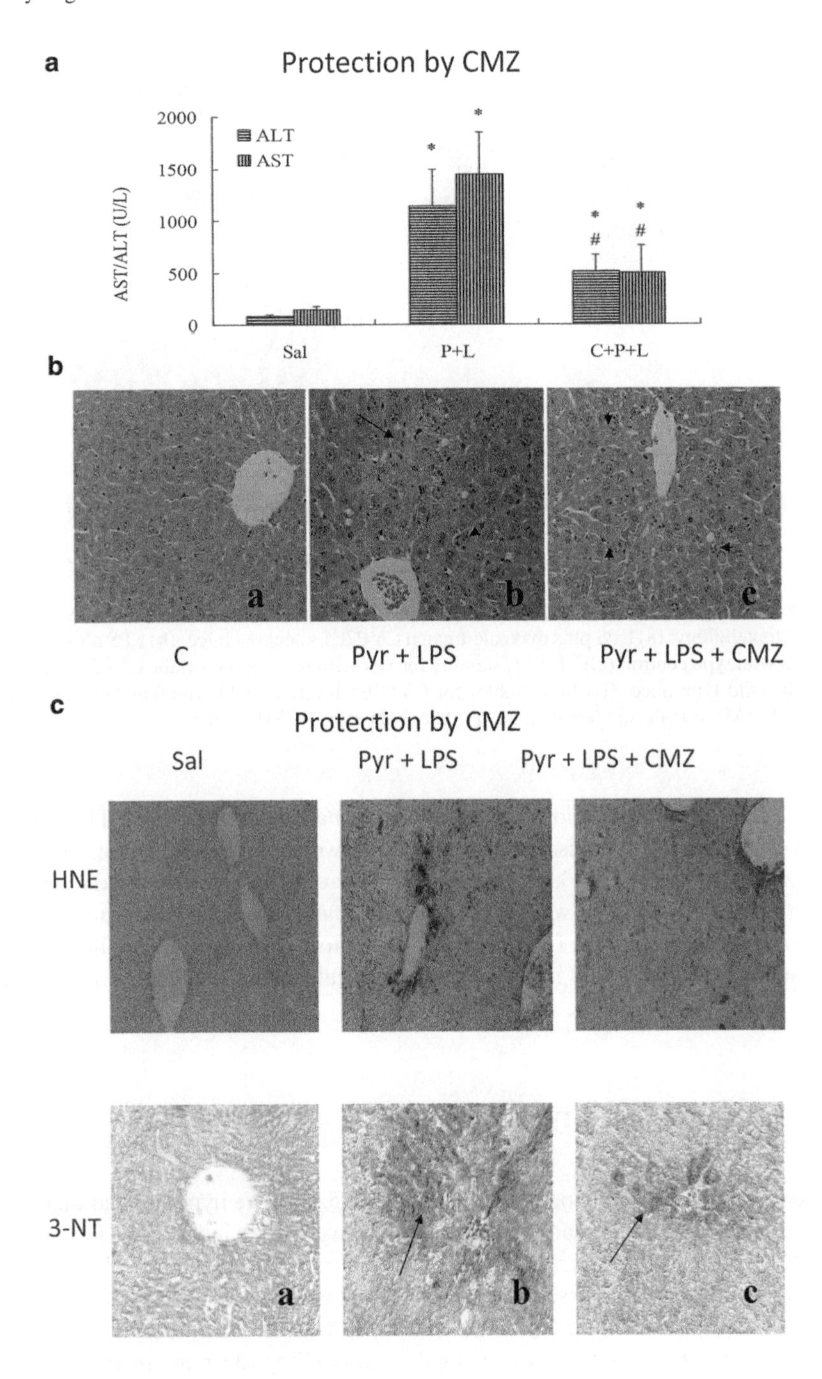

Fig. 9.5 The CYP2E1 inhibitor, chlormethiazole (CMZ) protects against LPS plus pyrazole toxicity and oxidative/nitrosative stress in mice. *Sal* Saline-treated; $P+L$ pyrazole plus LPS-treated, $C+P+L$ CMZ plus pyrazole plus LPS-treated. (**A**) ALT/AST levels, (**B**) histopathology; (**C**) 4-HNE adducts and 3 NT adducts

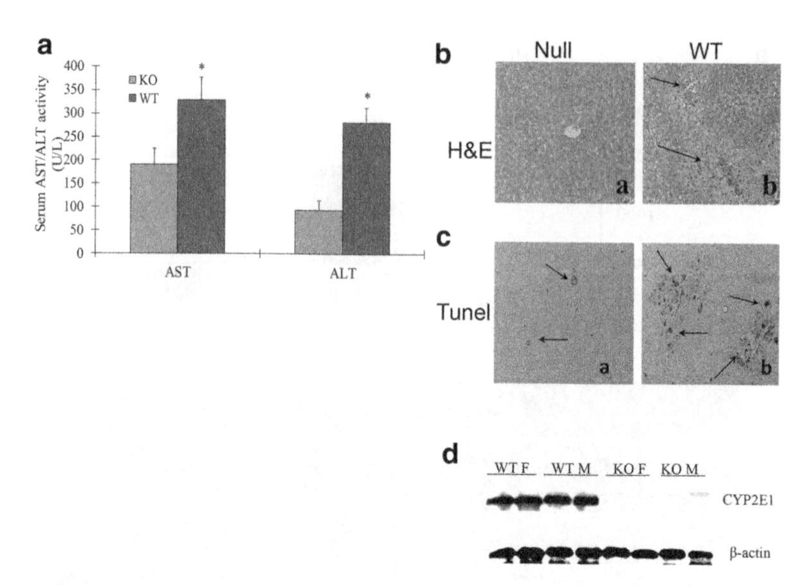

Fig. 9.6 LPS plus pyrazole toxicity is lowered in CYP2E1 knockout mice. (**A**) ALT/AST levels: (**B**) histopathology: (**a**) LPS plus pyrazole-treated CYP2E1 knockout mice; (**b**) LPS plus pyrazole-treated wild type control. (**C**) TUNEL staining for DNA fragmentation: (**a**) the CYP2E1 null mice and (**b**) wild type mice. (**D**) Immunoblot for CYP2E1 levels in wild type female or male mice (WTF, WTM) or male and female CYP2E1 knockout mice (KOM, KOF)

CYP2E1 knockout mice did not elevate transaminase levels. Similarly, TBAR levels in liver homogenates and isolated mitochondria were not elevated in the TNFα plus pyrazole-treated CYP2E1 knockout mice but were increased in the wild type mice. Normal liver pathology was observed after pyrazole plus TNFα treatment of CYP2E1 knockout mice. The failure of TNFα to induce liver injury in pyrazole-treated CYP2E1 knockout mice supports a critical role for CYP2E1 in the potentiated injury observed in the wild type mice.

9.8 Mitochondrial Dysfunction

Alcohol can cause mitochondrial dysfunction [55, 56]. We hypothesized that mitochondria are an eventual target for developing liver injury induced by TNFα when CYP2E1 is elevated by pyrazole. Initiation of a mitochondrial permeability transition was determined by assessing mitochondrial swelling in the absence and presence of 100 μM calcium. Succinate (10 mM) was the respiratory substrate. As shown in Fig. 9.7A, in the absence of calcium, swelling (decrease in absorbance at 540 nm) was low with all mitochondrial preparations although there was some basal swelling with the mitochondria from the pyrazole plus TNFα-treated mice. The addition of 100 μM calcium caused a low rate of swelling in the saline or TNFα

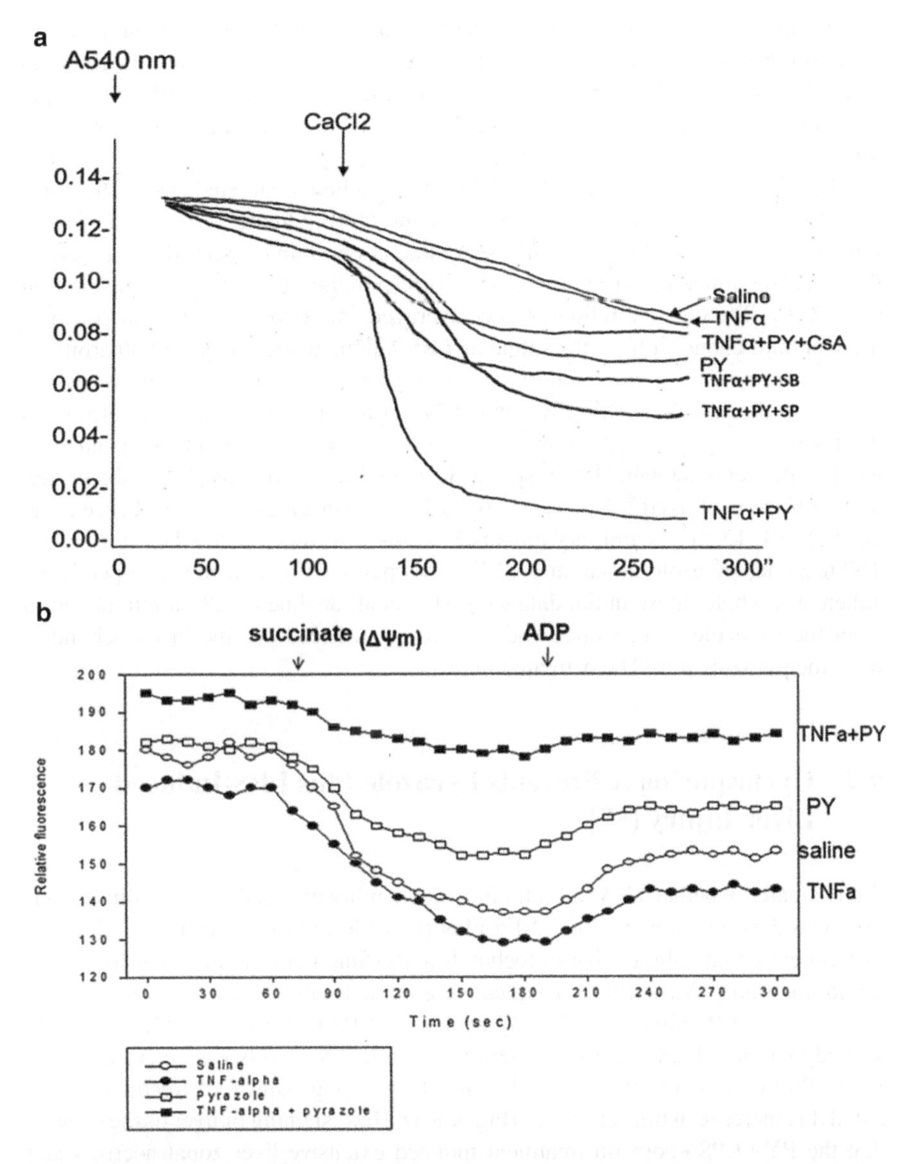

Fig. 9.7 Treatment with TNF**α** plus pyrazole causes mitochondrial injury as reflected by (**A**) increased mitochondrial swelling (decreased absorbance at 540 nm) or (**B**) decreased mitochondrial membrane potential as assayed by succinate-dependent decline in rhodamine 123 fluorescence. The increased swelling produced by TNF**α** plus pyrazole is prevented by cyclosporine (CsA), an inhibitor of the mitochondrial permeability transition as well as by the inhibitors of JNK and p38 MAPK

alone mitochondria; swelling was somewhat elevated in the pyrazole alone mitochondria. Swelling was very rapid without any lag phase with the mitochondria from the pyrazole plus TNFα-treated mice (Fig. 9.7A). Importantly, this rapid swelling was blocked by Cyclosporine A (CsA) (2 μM), a classic inhibitor of the mitochondrial permeability transition.

The electrochemical potential of the proton gradient generated across the mitochondrial membrane (ΔΨ) was assessed by monitoring fluorescence quenching of rhodamine 123. Addition of 10 mM succinate at one minute caused a decrease in fluorescence reflective of a high ΔΨ corresponding to state 4 of respiration (Fig. 9.7B). The decline in fluorescence averaged about 40 arbitrary units per minute with mitochondria from the saline or TNFα alone treated mice and 30 arbitrary units per minute with mitochondria from the pyrazole-treated mice. However, the decline in fluorescence was only about 14 arbitrary units with mitochondria from the TNFα plus pyrazole-treated mice. Addition of ADP at 3 min caused an enhancement of fluorescence which corresponds to state 3 respiration as part of the proton motive force is utilized to synthesize ATP. This enhancement of fluorescence averaged 15, 14, 12, and 4 arbitrary units per minute for mitochondria from the saline, TNFα alone, pyrazole alone, and TNFα plus pyrazole-treated mice, respectively. Taken as a whole, these initial data suggest a small decline in ΔΨ in mitochondria from the pyrazole-treated mice and a more pronounced decline in mitochondria from the pyrazole plus TNFα-treated mice.

9.9 Cyclosporine A Prevents Pyrazole Plus LPS-Induced Liver Injury [57]

We evaluated whether CsA, an inhibitor of the mitochondrial permeability transition, could protect against the TNFα plus pyrazole-induced liver injury. Such an experiment could validate that mitochondrial dysfunction is a key downstream target in this injury. Male C57BL/6 mice were treated with saline, pyrazole, LPS, or pyrazole plus LPS plus corn oil or pyrazole plus LPS plus 1 dose of CsA (100 mg/kg body wt, dissolved in corn oil). Serum ALT and AST levels were elevated in the PY + LPS + corn oil group compared to the other three groups. CsA treatment attenuated this increase in transaminases (Fig. 9.8A). H&E staining of liver tissue showed that the PY + LPS + corn oil treatment induced extensive liver zonal necrosis and that the CsA treatment prevented this (Fig. 9.8B). Mitochondrial swelling was increased in mitochondria isolated from the PY + LPS + corn oil treated mice compared to mitochondria from the saline + corn oil mice. The in vivo treatment with CsA prevented this increase in mitochondrial swelling, which likely explains the protection against LPS plus pyrazole-induced liver injury. The LPS plus pyrazole elevation of 4-HNE and 3-NT protein adducts were also decreased by CsA, suggesting that mitochondrial dysfunction plays an important role in the increase in oxidative/nitrosative stress.

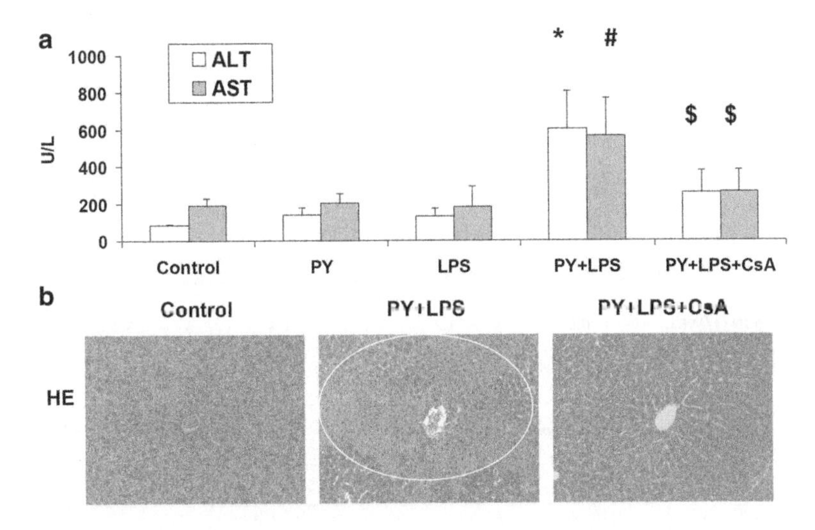

Fig. 9.8 In vivo administration of the mitochondrial permeability transition inhibitor, cyclosporine A (CsA) lowers LPS plus pyrazole-induced increase in transaminase levels (**A**) and liver injury as detected by H&E staining (**B**)

9.10 Activation of MAP Kinases

Mitogen-activated protein kinases (MAPKs) are serine-threonine kinases that mediate intracellular signaling associated with a variety of cellular activities including cell proliferation, differentiation, survival, death, and transformation. The mammalian MAPK family consists of extracellular signal-regulated kinase (ERK), p38 MAPK, and c-Jun NH_2-terminal kinase (JNK) [58]. The MAPK signaling cascade consists of three distinct members of the protein kinase family, including MAP kinase (MAPK), MAPK kinase (MAPKK), and MAPKK kinase (MAPKKK). MAPKKK phosphorylates and thereby activates MAPKK, and the activated form of MAPKK in turn phosphorylates and activates MAPK. Activated MAPK may translocate to the cell nucleus and regulate the activities of transcription factors and thereby control gene expression [59]. In either in vivo or in vitro models of ALD, an increase of gene expression and activation of the MAPK pathway was found [60–62].

JNK or p38 MAPK have been shown to play important roles in several models of liver injury, including CYP2E1-dependent toxicity [49, 50, 63–65]. We evaluated possible activation of MAP kinases in our pyrazole/LPS or pyrazole/TNFα hepatotoxicity models. LPS treatment alone did not cause significant JNK activation (Fig. 9.9A) or p38 MAPK activation (Fig. 9.9B) as reflected by the low p-JNK and pp38 MAPK levels relative to total JNK and p38 MAPK levels. Similar low ratios were found for the saline or the pyrazole alone treated mice (Fig. 9.9a, B). However, both JNK and p38 MAPK were activated in livers of the pyrazole plus LPS-treated mice. A similar activation of JNK and p38 MAPK was also observed after pyrazole plus TNFα but not in mice treated with TNFα or pyrazole alone [53]. ERK was

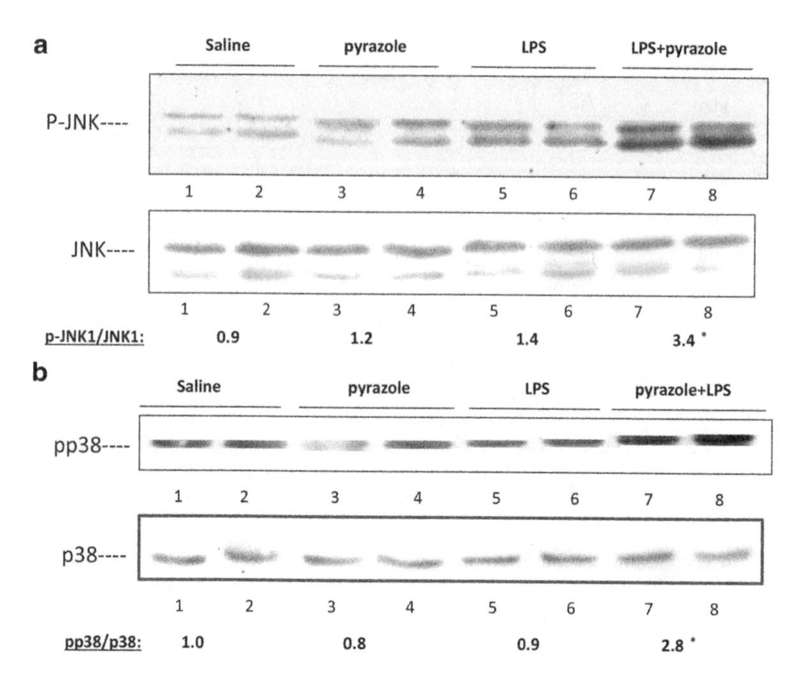

Fig. 9.9 LPS plus pyrazole treatment activates JNK (**A**) and p38 MAPK (**B**). The p-JNK/JNK and pp38 MAPK/p38 MAPK ratios are shown below the blots

not altered by TNFα alone or pyrazole plus TNFα treatment. To evaluate the significance of these changes in MAPK activation, the effect of SP600125, an inhibitor of JNK, and SB203580, an inhibitor of p38MAPK on the hepatotoxicity was determined. The LPS (Fig. 9.1) or TNFα (Fig. 9.3) plus pyrazole elevation of transaminases was blunted by administration of SP600125 (15 mg/kg) or SB203580 (15 mg/kg) (Figs. 9.1 and 9.3). The MAPK inhibitors also lowered the necrosis, the mitochondrial dysfunction (Fig. 9.7) and partially blocked the increased oxidative stress (Figs. 9.2 and 9.4) produced by the pyrazole plus LPS/TNFα treatment, but had no effect on CYP2E1 activity or protein levels. These results suggest the CYP2E1 elevation of LPS/TNFα liver injury and oxidative stress is MAPK dependent. The activation of JNK in the pyrazole plus TNFα group was blocked by SP600125 but not SB203580 whereas the activation of p38 MAPK was blocked by SB203580 but not SP600125, validating their specificity.

9.11 Activation of ASK-1 and Downstream MAP Kinase Kinases

The upstream mediators of JNK and p38 MAPK activation were not identified in these previous studies. Apoptosis signal-regulating kinase 1 (ASK-1) is a member of the MAP3K family which is responsive to stress-induced cell damage. Activation of ASK-1 can determine cell fate by regulation of both the MKK4/MKK7-JNK and the MKK3/MKK6-p38 MAPK signaling cascades [66]. ASK-1 is activated by oxidative stress, ER stress, and inflammatory cytokines such as TNFα [67]. In resting cells, ASK-1 forms an inactive complex with reduced thioredoxin (Trx). Under conditions of stress by TNFα or ROS, ASK-1 dissociates from Trx and becomes activated [68]. Oxidation of Trx by ROS causes dissociation of ASK-1 from the oxidized Trx which switches the inactive form of ASK-1 to the active kinase. The Trx-ASK complex is thought to be a redox sensor, which functions as a molecular switch turning the cellular redox state into a MAP kinase signaling pathway [69]. Activated ASK-1 then promotes activation (phosphorylation) of the downstream MAPKK, MKK4/MKK7 which can activate JNK, and MKK3/MKK6 which can activate p38 MAPK [70, 71]. We evaluated whether CYP2E1 plus TNFα induced ROS promote release of ASK-1 from the Trx-ASK1 complex and activate ASK-1 followed by the phosphorylation of MKK4/MKK7 and/or MKK3/MKK6 which subsequently regulate the phosphorylation of JNK and p38 MAPK and contribute to the liver injury [72].

Treatment with TNFα for 4–12 h or pretreatment with PY alone did not activate ASK-1 (Fig. 9.10a). TNFα plus PY treatment activated ASK-1 about threefold compared with the 0 h control at 4 h after TNFα treatment. Activation of ASK-1 decreased at 8 and 12 h. Immunoprecipitation of Trx1 showed that ASK-1 was bound to Trx-1 at 0 h but was released from the Trx-ASK1 complex at 4 h and remained free from binding to Trx1 at 8 and 12 h (Fig. 9.10b). No ASK-1 release from the Trx-ASK1 complex was found in TNFα alone treated mice. ASK-1 was not activated in PY plus TNFα treated CYP2E1−/− mice and no ASK-1 was released from the Trx-ASK1 complex in CYP2E1−/− mice. WT mice treated with pyrazole plus TNFα developed liver injury between 8 and 12 h after addition of the TNFα (Fig. 9.10C), Oxidative stress is a likely key factor to trigger signaling and liver injury in CYP2E1-mediated hepatotoxicity [73]. Thus, activation of ASK-1 by treatment with TNFα plus PY is associated with its release from the Trx-ASK1 complex, occurs prior to the liver injury, and requires CYP2E1.

MKK4/7 and MKK3/6 are the MAPKK which activate downstream JNK or p38 MAPK, respectively [71]. They are also targets for activation by ASK-1 [69–71]. Treatment of wild type mice with PY plus TNFα activated MKK4 at 4, 8, and 12 h compared with the TNFα alone groups [72]. No activation of MKK4 was found in TNFα or TNFα+PY treated CYP2E1−/− mice [72]. MKK7 was activated only at 12 h. MKK3 was activated as early as 4 h in the TNFα plus PY treated mice while MKK6 was activated at 8 h. JNK was activated in the TNFα+PY mice at 8 and 12 h and p38 MAPK was activated at 12 h when compared with TNFα alone. In CYP2E1−/− mice, neither MKK4/7, MKK3/6, JNK, nor p38 MAPK was activated. Thus, the time course experiments suggest MKK4 may be the MAPK responsible for activation of JNK, while either MKK3 or MKK6 may be the MAPKK responsible for the activation of p38 MAPK.

Our results implicate a role for ASK-1 in CYP2E1 potentiation of TNFα-induced liver injury. Future experiments with ASK-1 knockout mice [74] would be interesting to further validate the role of ASK-1 in the PY/TNFα model. In CYP2E1−/− mice, no MAPKK was activated at any observation time point. TNFα alone did not

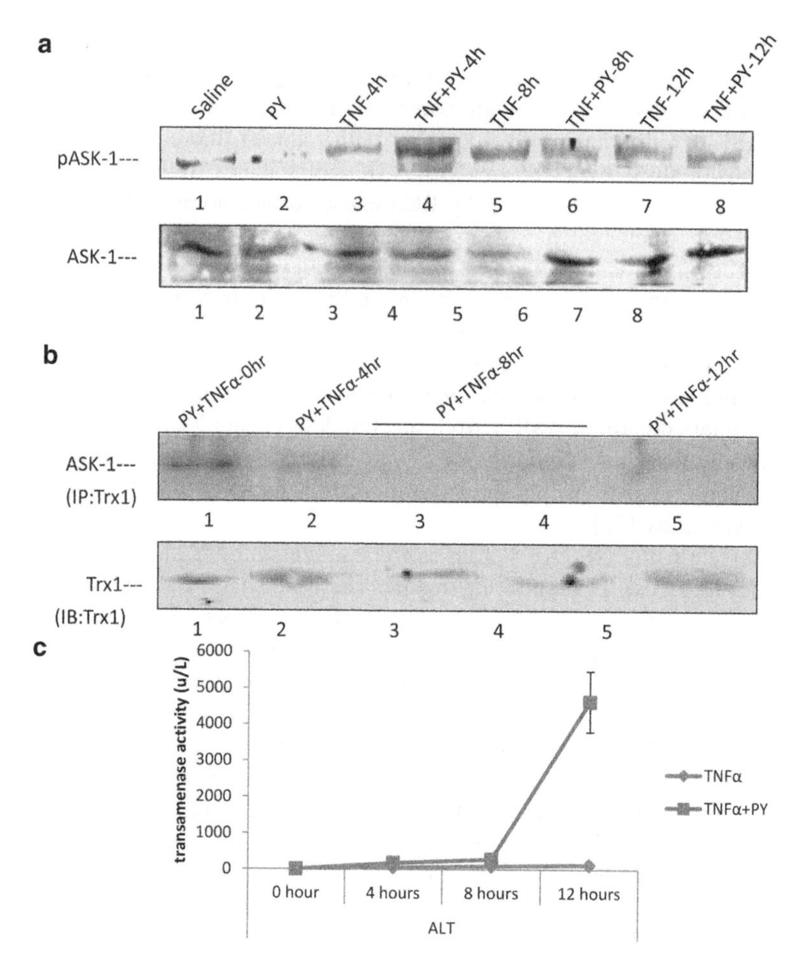

Fig. 9.10 Activation of the MAPKKK ASK-1 by pyrazole plus TNF**α** treatment. (**A**) Immunoblot showing activation (phosphorylation) of ASK-1 4 h after treatment with pyrazole plus TNF**α** but not by pyrazole alone or TNF**α** alone. (**B**) Dissociation of the inhibitory thioredoxin 1 (Trx1) from its complex with ASK-1 4 h after treatment with pyrazole plus TNF**α**. Note that Trx1 is bound to ASK-1 initially (0 h) but not 4–12 h after treatment with pyrazole plus TNF**α**. Liver homogenate-nates were immunoprecipitated with Trx1 antibody and then the immunoprecipitates were probed for ASK-1 by Western blot. (**C**) Time course for liver injury produced by pyrazole plus TNF **α**. Note injury occurs after 8 h, a time point when ASK-1 has already been activated

significantly activate the MAPKK in wild type or CYP2E1−/− mice. The activation of MKK4 and MKK3/6 (4–8 h) occurs prior to the onset of liver injury (8–12 h). We hypothesize that TNFα alone-or CYP2E1 alone-generated ROS stress is not sufficient to trigger the dissociation of ASK-1 from the Trx-ASK complex. The CYP2E1 sensitization of TNFα induced liver injury may occur through a synergistic effect with TNFα to produce an enhanced ROS stress consistent with the so call "Two Hit" hypothesis.

9.12 Effect of *N*-Acetylcysteine

We evaluated [54] the effect of *N*-acetylcysteine (NAC), a general antioxidant and a precursor of GSH, on the potentiation of TNFα toxicity by pyrazole as a proof of principle that oxidative stress plays an important role in the overall liver injury. C57BL/6 mice were treated with pyrazole for 2 days and then challenged with either saline or TNFα. Some mice in each group were also treated with 150 mg/kg NAC on the second day of treatment with pyrazole and on day 3 prior to the challenge with TNFα. The elevation in ALT and AST and the necrosis caused by the pyrazole plus TNFα treatment were lowered by NAC. The increase in TBARs produced by pyrazole plus TNFα and the decline in liver GSH were both prevented by NAC. Treatment with NAC had no effect on CYP2E1 protein levels or CYP2E1 catalytic activity. The activation of JNK or p38 MAPK by the pyrazole plus TNFα treatment was blocked by NAC. These results with NAC suggest that elevated oxidative stress is central to the activation of JNK and p38 MAPK and to the liver injury produced by treatment with pyrazole plus TNFα.

9.13 Hepatotoxicty by CYP2E1 Plus TNFα Occurs in JNK2 but not JNK1 Knockout Mice

The hepatotoxicity produced by pyrazole plus LPS/TNF-α was prevented by SP600125, an inhibitor of JNK (Figs. 9.1 and 9.3). JNK is encoded for by three genes, each of which is alternatively spliced to yield α and β forms of both a 54 and 46 kDa protein. In hepatocytes, only two of the genes, JNK1 and JNK2 are expressed [75]. Mice deficient in either JNK1 or JNK2 are viable but double knockouts are embryonic lethal suggesting some redundant functions [75]. JNK has been implicated in hepatic injury produced by TNF-α, ischemia-reperfusion, hepatitis virus, bile acids, alcohols and acetaminophen [64, 65, 76–78]. Recent studies have evaluated whether JNK1 or JNK2 play the more predominant role in potentiation of liver injury. In fibroblasts, JNK1 but not JNK2 appears to be essential for TNF-α-induced apoptosis [79]. However, liver injury produced by either LPS/D-galactosamine or TNF-α/D-galacosamine was the same in WT and JNK1 KO mice but lower in JNK2 KO mice [80]. JNK2 promoted the development of steatohepatitis in mice fed a methionine choline-deficient diet [77]. Singh et al. [81] reported that JNK1 KO mice fed a high fat diet did not gain weight or develop steatohepatitis as did the WT and JNK2 KO mice. JNK2 was found to be predominant in acetaminophen toxicity [65]. 6-Hydroxydopamine-induced apoptosis in PC12 cells was JNK2 but not JNK1 dependent [82]. It appears that depending on the toxin and cell type, either JNK1 or JNK2 or both play the major role in cell injury. We evaluated whether JNK1 or JNK2 or both play critical roles in the potentiation of TNF-α-induced hepatotoxicity and oxidative stress by pyrazole [83].

Male *jnk1*$^{-/-}$ (B6.129-*Mapk8*^{tm1Flv}/J), *jnk2*$^{-/-}$ (B6.129-*Mapk9*^{tm1Flv}/J), and wild type, C57BL/6 J mice (JNK1 KO, JNK2 KO and WT), weighing 24–26 g, at 8–10 weeks of age, were purchased from Jackson Laboratory (Bar Harbor, ME). Mice were divided into three groups and each strain received either pyrazole alone, TNF-α alone, or TNF-α following pyrazole pretreatment. At 24 h after administration of TNF-α, the mice were sacrificed. Serum ALT and percentage of ischemic liver injury were significantly higher in the WT and JNK2 KO groups treated with pyrazole plus TNF-α than that in the JNK1 KO group treated with pyrazole plus TNF-α or in the WT and JNK2 KO groups treated with pyrazole alone or TNF-α alone (Fig. 9.11A). Severe pathological changes were detected in the WT mice treated with pyrazole plus TNF-α but not in WT mice treated with pyrazole alone or TNF-α alone (Fig. 9.11B bottom panels). Severe pathological changes, similar to those found in WT mice, were observed in JNK2 KO mice treated with pyrazole plus TNF-α (Fig. 9.11B middle panels). Many hepatocytes appeared to display significant ischemic necrosis and inflammatory infiltration in the hepatic centrilobular zone. Only some mild lesions were observed in the JNK1 KO group treated with pyrazole plus TNF-α, mainly dilation and congestion in the sinusoid, swelling and focal necrosis of hepatocytes in the centrilobular area (Fig.9.11B top panels). Only mild degeneration of hepatocytes was found in JNK1 KO, JNK2 KO, or WT mice after pyrazole alone or TNF-α alone administration. TUNEL results showed that there were many hepatocytes with positive staining nuclei in the WT and JNK2 KO mice treated with pyrazole plus TNF-α compared to JNK1 KO group.

Western blot and immunohistochemistry showed that expression of CYP2E1 in situ was higher in all mice treated with pyrazole or pyrazole plus TNFα compared to mice treated with saline or with TNF-α alone. CYP2E1 catalytic activity was also higher in all pyrazole-treated mice, confirming the elevation of CYP2E1 by pyrazole in all three groups. Similarly, expression of CYP2E1 was higher in all mice treated with pyrazole plus TNF-α compared to TNF-α alone. 4-HNE immunohistochemical staining was significantly stronger in the WT and JNK2 KO groups treated with pyrazole plus TNF-α compared to the JNK1 KO group treated with pyrazole plus TNF-α. The elevation of 4-HNE adducts in the WT and JNK2 KO mice were mainly found in the central lobular zone of the liver, the zone showing necrotic injury and elevated CYP2E1. MDA was significantly higher while levels of GSH were significantly lower in homogenates of liver of the WT and JNK2 KO groups treated with pyrazole plus TNF-α compared to the JNK1 KO group treated with pyrazole plus TNF-α [83].

9.14 Conclusions

There are many mechanisms proposed as to how alcohol produces liver damage such as NAD^+/NADH redox state changes, reactive acetaldehyde production, protein adduct formation, damage to mitochondria, direct hydrophobic membrane effects of alcohol, hypoxia, alcohol–nutritional interactions, alcohol effects on the

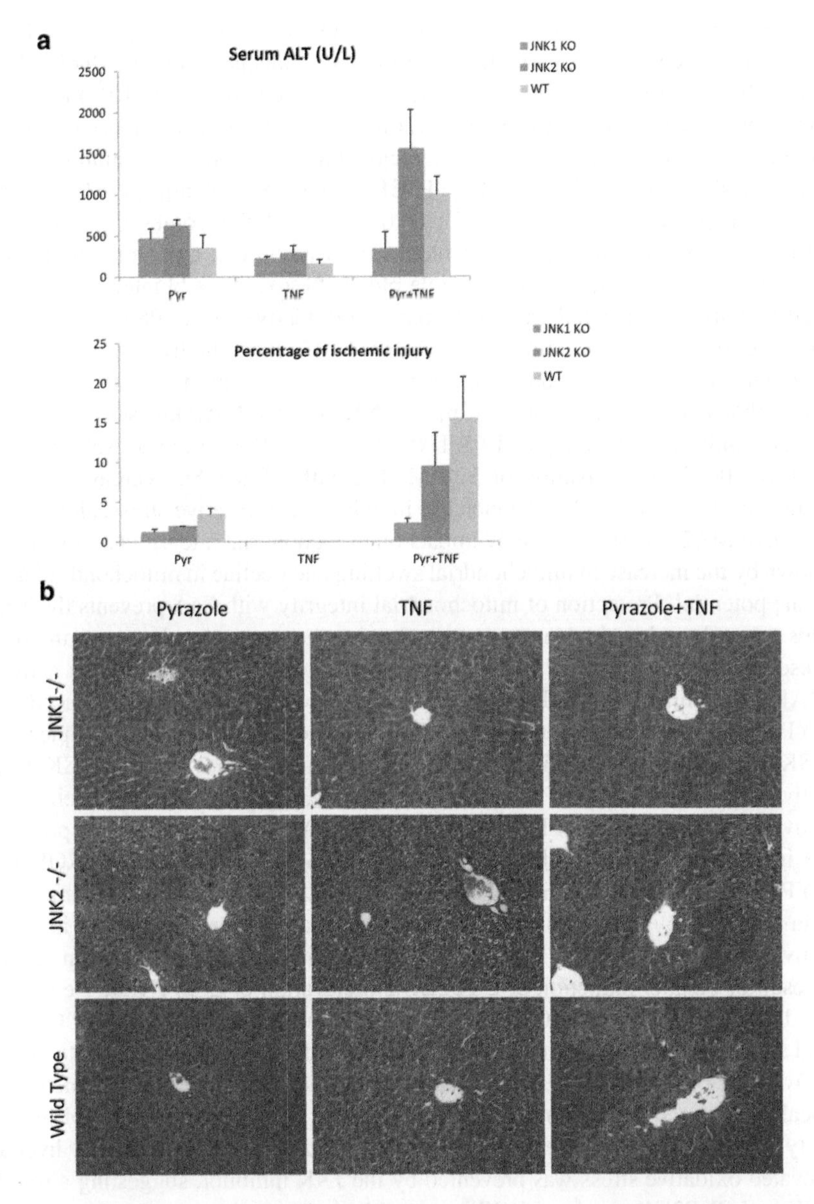

Fig. 9.11 Liver injury produced by pyrazole plus TNFα occurs in wild type mice and JNK2 knockout mice but not JNk1 knockout mice. (**A**) Serum ALT levels and percent ischemic liver cells. (**B**) histopathology after treatment of WT or JNK2 KO or JNK1 KO with pyrazole alone or TNFα alone or pyrazole plus TNFα. Note the ischemic necrosis and infiltration of inflammatory cells in the centrilobular zone of liver from the Pyrazole plus TNFα-treated WT and JNK2 KO mice (*right middle* and *bottom panels*) but normal liver in the JNK 1 KO mice (*right top panel*)

immune system, and others. These have been reviewed elsewhere [3, 84, 85]. This review has focused on two major contributors to mechanisms by which ethanol causes liver injury, induction of CYP2E1 and elevated endotoxin (LPS) levels followed by increased production of TNFα. Each of these has been extensively studied but there are few studies in which both factors have been evaluated simultaneously. We have shown that induction of CYP2E1 by pyrazole potentiates LPS- or TNF-induced hepatotoxicity. Evidence for a role for CYP2E1 comes from studies in which the CYP2E1 inhibitor CMZ blocks the liver injury, and from studies with CYP2E1 knockout mice where pyrazole plus LPS toxicity is blunted. The potentiated toxicity is associated with an increase in oxidative stress. Prevention of such increases e.g., treatment with the antioxidant NAC blunts the liver injury thus validating that the elevated oxidative stress plays a key role in producing the liver injury rather than occurs because of liver injury. JNK and P38 MAP kinases are activated by the combined pyrazole plus LPS/TNFα treatment. Preventing activation of JNK with SP600125 or activation of P38 MAPK with SB203580 decreases the liver injury. We hypothesize that the increase in oxidative/nitrosative stress, and the activation of MAP kinases ultimately impact on mitochondrial integrity and function as shown by the increase in mitochondrial swelling and decline in mitochondrial membrane potential. Protection of mitochondrial integrity with CsA prevents the TNFα plus pyrazole-induced hepatotoxicity and oxidative stress. A scheme summarizing these results is shown in Fig. 9.12. Upon binding to its receptor, TNFα activates MAPK signaling via complex 1 and caspases via complex II. ROS produced from CYP2E1 causes dissociation of the inhibitory thioredoxin from its complex with ASK-1, resulting in activation (phosphorylation) of ASK-1. Activated ASK-1 then activates downstream MAP kinase kinases MKK3/6 and MKK 4/7 which in turn activate JNK and p38 MAPK. Activation of the MAP kinases may also be promoted by inhibition of MAPK phosphatases such as MKP-1. Increases in ROS from CYP2E1 and TNFα and activated MAPK ultimately cause damage to the mitochondria followed by liver injury. We hypothesize that similar interactions involving activation of MAP kinases, oxidative stress, and mitochondrial dysfunction occur as a result of ethanol induction of CYP2E1 and elevation of LPS/TNFα. We recently used CYP2E1 knockout (KO) mice, and a JNK inhibitor to test the role of CYP2E1 and JNK in acute alcohol-induced steatosis [86]. In wild type mice, acute alcohol activated CYP2E1, produced fatty liver and increased oxidative stress, which reciprocally increased activation of the JNK signaling pathway. Acute alcohol-induced fatty liver and oxidative stress was blunted in CYP2E1 KO mice. The fatty liver and elevated oxidative stress was prevented by the JNK inhibitor, suggesting a role for JNK and CYP2E1-generated ROS in acute alcohol fatty liver.

JNK2 KO mice displayed hepatotoxicity similar to WT mice when treated with pyrazole plus TNF-α; however, JNK1 KO mice did not display this hepatotoxicity. A similar induction of CYP2E1 protein and activity by pyrazole was found with all three genotypes. While ROS play an important role in activating JNK, increases in JNK can further elevate ROS. Several parameters associated with oxidative/nitrosative stress were elevated in the WT and the JNK2 KO mice, but not the JNK1 KO mice treated with pyrazole plus TNF-α. It is likely that the elevated

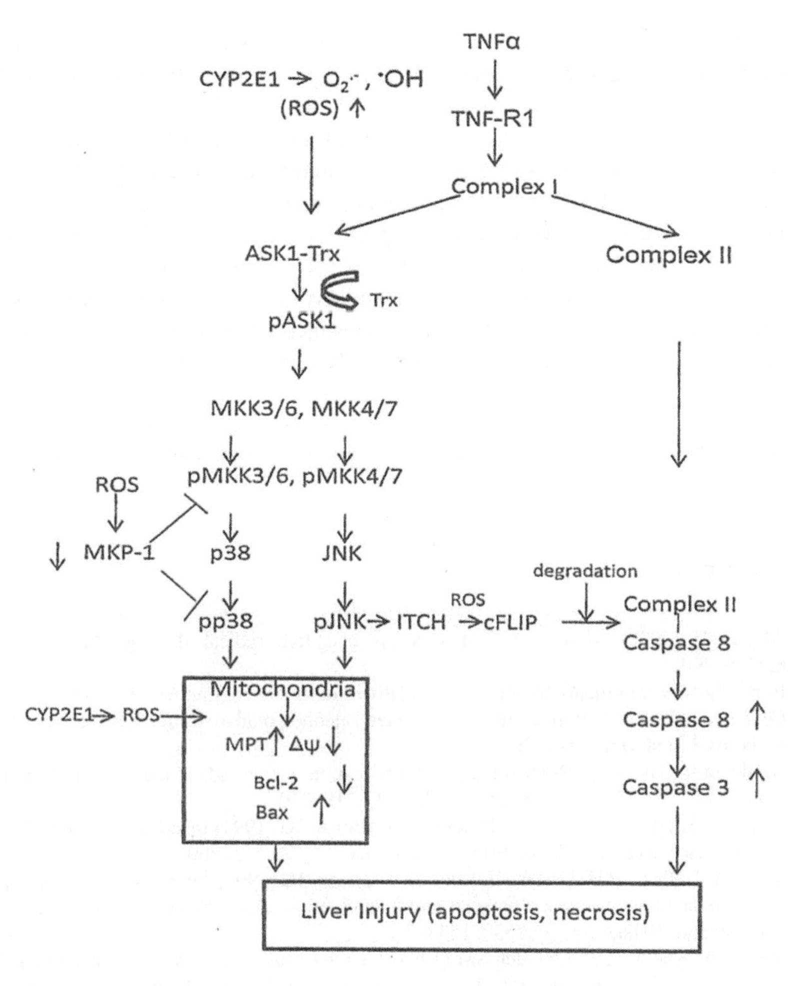

Fig. 9.12 Scheme for TNFα plus CYP2E1 activation of the MAP kinase kinase kinase ASK-1, MAP kinase kinases MKK3/6 or MKK4/7, and MAP kinases JNK or p38 and the role of CYP2E1-generated ROS in removal of inhibitory thioredoxin from its complex with ASK-1. The scheme is discussed in the text

oxidative/nitrosative stress is playing a central role in the hepatotoxicity found with the JNK2 KO mice, while the lack of an increase in oxidative/nitrosative stress in the JNK1 KO mice is preventive against hepatotoxicity. Our results show that JNK1 plays the major role in the pyrazole plus TNF-α-induced liver injury, elevated oxidative stress, and mitochondrial dysfunction. It is not clear what activity JNK1 may catalyze that is not being catalyzed by JNK2, or vice versa, to explain the predominant role of JNK1 or JNK2 in promoting hepatotoxicity of a particular toxin or condition. Proteomic and molecular modeling experiments may be helpful in identifying JNK1 target(s). Use of a liver specific JNK1 KO mouse model may also be helpful in determining whether differences between JNK1 and JNK2 in promoting the pyrazole plus TNF-α hepatotoxicity is specific to JNK actions in the liver, and not extrahepatic JNK actions. While elevated oxidative/nitrosative stress, decline in antioxidant protection, and mitochondrial dysfunction occur in the JNK2 but not JNK1 KO mice treated with pyrazole plus TNF-α, suggesting a role for JNK1 in interacting with and potentiating CYP2E1 toxicity, additional studies are necessary to identify specific target(s) for JNK1, but not JNK2, which play critical roles in the CYP2E1 plus TNF-α toxicity. Identifying JNK1 as the JNK isoform responsible for the liver injury, the elevated oxidative stress and the mitochondrial dysfunction, may have relevance in attempts to prevent or lower the hepatotoxicity by specifically inhibiting JNK1 without affecting JNK2.

Acknowledgement Studies from the author's lab were supported by USPHS grants RO1 AA 018790 and R21 AA 021362 from The National Institute on Alcohol Abuse and Alcoholism.

References

1. Albano E (2006) Alcohol, oxidative stress and free radical damage. Proc Nutr Soc 65:278–290
2. Bondy SC (1992) Ethanol toxicity and oxidative stress. Toxicol Lett 63:231–242
3. Cederbaum AI (2001) Introduction serial review: alcohol, oxidative stress and cell injury. Free Radic Biol Med 31:1524–1526
4. Nordman R, Riviere C, Rouach H (1992) Implication of free radical mechanisms in ethanol-induced cellular injury. Free Radic Biol Med 12:219–240
5. Adachi Y, Bradford BU, Gao W, Bojes HK, Thurman RG (1994) Inactivation of Kupffer cells prevents early alcohol-induced liver injury. Hepatology 20:453–460
6. Iimuro Y, Gallucci RM, Luster MI, Kono H, Thurman RG (1997) Antibodies to tumor necrosis factor alpha attenuate hepatic necrosis and inflammation caused by chronic exposure to ethanol in the rat. Hepatology 26:1530–1537
7. Nanji AA, Khettry U, Sadrzadeh SM (1994) Lactobacillus feeding reduces endotoxemia and severity of experimental alcoholic liver disease. Proc Soc Exp Biol Med 205:243–247
8. Yin XM, Ding WX (2003) Death receptor activation-induced hepatocyte apoptosis and liver injury. Curr Mol Med 3:491–508
9. Rao RK, Seth A, Sheth P (2004) Recent advances in alcoholic liver disease I. Role of intestinal permeability and endotoxemia in alcoholic liver disease. Am J Physiol Gastrointest Liver Physiol 286:G881–884
10. Su GL (2002) Lipopolysaccharides in liver injury: molecular mechanisms of Kupffer cell activation. Am J Physiol Gastrointest Liver Physiol 283:G256–G265
11. Deaciuc IV, Nikolova-Karakashian M, Fortunato F, Lee EY, Hill DB, McClain CJ (2000) Apoptosis and dysregulated ceramide metabolism in a murine model of alcohol-enhanced lipopolysaccharide hepatotoxicity. Alcohol Clin Exp Res 24:1557–1565
12. Hansen J, Cherwitz DL, Allen JI (1994) The role of tumor necrosis factor-alpha in acute endotoxin-induced hepatotoxicity in ethanol-fed rats. Hepatology 20:461–474
13. Koteish A, Yang S, Lin H, Huang X, Diehl AM (2002) Chronic ethanol exposure potentiates lipopolysaccharide liver injury despite inhibiting Jun N-terminal kinase and caspase 3 activation. J Biol Chem 277:13037–13044
14. Mathurin P, Deng QG, Keshavarzian A, Choudhary S, Holmes EW, Tsukamoto H (2000) Exacerbation of alcoholic liver injury by enteral endotoxin in rats. Hepatology 32:1008–1017

15. Kono H, Rusyn I, Uesugi T, Yamashina S, Connor HD, Dikalova A, Mason RP, Thurman RG (2001) Diphenyleneiodonium sulfate, an NADPH oxidase inhibitor, prevents early alcohol-induced liver injury in the rat. Am J Physiol Gastrointest Liver Physiol 280:G1005–G1012
16. Kono H, Rusyn I, Yin M, Gäbele E, Yamashina S, Dikalova A, Kadiiska MB, Connor HD, Mason RP, Segal BH, Bradford BU, Holland SM, Thurman RG (2000) NADPH oxidase-derived free radicals are key oxidants in alcohol-induced liver disease. J Clin Invest 106:867–872
17. Yin M, Wheeler MD, Kono H, Bradford BU, Galluci RM, Luster MI, Thurman RG (1999) Essential role of TNFα in alcohol-induced liver injury in mice. Gastroenterology 117:942–952
18. Lieber CS (1997) Cytochrome P4502E1; its physiological and pathological role. Physiol Rev 77:517–544
19. Koop DR (1992) Oxidative and reductive metabolism by cytochrome P4502E1. FASEB J 6:724–730
20. Song BJ, Cederbaum AI, Koop DR, Ingelman-Sundberg M, Nanji A (1996) Ethanol-inducible cytochrome P450 (CYP2E1): Biochemistry, molecular biology and clinical relevance. Alcohol Clin Exp Res 20(Suppl):138A–146A
21. Lee SST, Buters JTM, Pineau T, Fernandez-Salguero P, Gonzalez FJ (1996) Role of CYP2E1 in the hepatototoxicity of acetaminophen. J Biol Chem 271:12063–1206
22. Ekstrom G, Ingelman-Sundberg M (1989) Rat liver microsomal NADPH-supported oxidase activity and lipid peroxidation dependent on ethanol-inducible cytochrome P-450 (P-450IIE1). Biochem Pharmacol 38:1313–1319
23. Gorsky LD, Koop DR, Coon MJ (1984) On the stoichiometry of the oxidase and monooxygenase reactions catalyzed by liver microsomal cytochrome P450. J Biol Chem 259:6812–6817
24. Rashba-Step J, Turro NJ, Cederbaum AI (1993) Increased NADPH- and NADH-dependent production of superoxide and hydroxyl radical by microsomes after chronic ethanol treatment. Arch Biochem Biophys 300:401–408
25. Caro AA, Cederbaum AI (2004) Oxidative stress, toxicology, and pharmacology of CYP2E1. Annu Rev Pharmacol Toxicol 44:27–42
26. Guengerich FP, Kim DH, Iwasaki M (1991) Role of cytochrome P450 IIE1in the oxidation of many low molecular weight cancer suspects. Chem Res Toxicol 4:168–179
27. Seitz HK, Stickel F (2007) Molecular mechanisms of alcohol-mediated carcinogenesis. Nat Rev Cancer 7:599–612
28. Kang JS, Wanibuchi H, Morimura K, Gonzalez FJ, Fukushima S (2007) Role of CYP2E1 in diethylnitrosamine-induced hepatocarcinogenesis in vivo. Cancer Res 67:11141–11146
29. Ye Q, Lian F, Chavez PR, Chung J, Ling W, Qin H, Seitz HK, Wang SD (2012) Cytochrome P4502E1 inhibition prevents hepatic carcinogenesis induced by diethylitrosamine in alcohol-fed rats. Hepatobiliary Surg Nutr 1:5–18
30. Mueller S (2013) Pharmacological blockage of CYP2E1 and alcohol-mediated liver cancer: is the time ready? Chin J Cancer Res 25:269–271
31. Hodges NJ, Green RM, Chipman JK, Graham M (2007) Induction of DNA strand breaks and oxidative stress in HeLa cells by ethanol is dependent on CYP2E1 expression. Mutagenesis 22:189–194
32. Navasumrit P, Ward D, Dodd NJF, O'Connor JO (2000) Ethanol-induced free radicals and hepatic DNA strand breaks are prevented in vivo by antioxidants: effects of acute and chronic ethanol exposure. Carcinogenesis 21:93–99
33. Kim YD, Eom SY, Ogawa M, Oyama T, Isse T, Kang JW, Zhang YW, Kawamoto T, Kim H (2007) Ethanol-induced oxidative DNA damage and CYP2E1 expression in liver tissue of ALDH2 knockout mice. J Occup Health 49:363–369
34. Kukielka E, Cederbaum AI (1992) The effect of ethanol consumption on NADH- and NADPH-dependent generation of reactive oxygen intermediates by isolated rat liver nuclei. Alcohol Alcohol 27:233–239

35. Bradford BU, Kono H, Isayama F, Kosyk O, Wheeler MD, Akiyama TE, Bleye L, Krausz K, Gonzalez FJ, Koop DR, Rusyn I (2005) Cytochrome P450 CYP2E1, but not nicotinamide adenine dinucleotide phosphate oxidase, is required for ethanol-induced oxidative DNA damage in rodent liver. Hepatology 41:336–344
36. Wang Y, Millonig G, Nair J, Patsenker E, Stickel F, Mueller S, Bartsch H, Seitz HK (2009) Ethaol-induced cytochrome P4502E1 causes carcinogenic ethano-DNA lesions in alcoholic liver disease. Hepatology 50:453–461
37. Morimoto M, Zern MA, Hagbjork AL, Ingelman-Sundberg M, French SW (1994) Fish oil, alcohol and liver pathology: role of cytochrome P450 2E1. Proc Soc Exp Biol Med 207:197–205
38. Nanji AA, Zhao S, Sadrzadeh SMH, Dannenberg AJ, Tahan SR, Waxman DJ (1994) Markedly enhanced cytochrome P4502E1 induction and lipid peroxidation is associated with severe liver injury in fish oil-ethanol-fed rats. Alcohol Clin Exp Res 18:1280–1285
39. French SW, Wong K, Jui L, Albano E, Hagbjork AL, Ingelman-Sundberg M (1993) Effect of ethanol on cytochrome P450 2E1 (CYP2E1), lipid peroxidation, and serum protein adduct formation in relation to liver pathology pathogenesis. Exp Mol Pathol 58:61–75
40. Morimoto M, Hagbjork AL, Wan YJ, Fu PC, Clot P, Albano E, Ingelman-Sundberg M, French SW (1995) Modulation of experimental alcohol-induced liver disease by cytochrome P450 2E1 inhibitors. Hepatology 21:1610–1617
41. Gouillon Z, Lucas D, Li J, Hagbjork AL, French BA, Fu P, Fang C, Ingelman-Sundberg M, Donohue TM Jr, French SW (2000) Inhibition of ethanol-induced liver disease in the intragastric feeding rat model by chlormethiazole. Proc Soc Exp Biol Med 224:302–308
42. Morgan K, French SW, Morgan TR (2002) Production of a cytochrome P450 2E1 transgenic mouse and initial evaluation of alcoholic liver damage. Hepatology 36:122–134
43. Bai JX, Cederbaum AI (2004) Adenovirus-mediated overexpression of CYP2E1 increases sensitivity of HepG2 cells to acetaminophen induced cytotoxicity. Mol Cell Biochem 262:165–176
44. Bai JX, Cederbaum AI (2006) Adenovirus-mediated expression of CYP2E1 produces liver toxicity in mice. Toxicol Sci 91:365–371
45. Kamimura S, Tsukamoto H (1995) Cytokine gene expression by Kupffer cells in experimental alcoholic liver disease. Hepatology 22:1304–1309
46. Honchel R, Ray MB, Marsano L, Cohen D, Lee E, Shedlofsky S, McClain CJ (1992) Tumor necrosis factor in alcohol enhanced endotoxin liver injury. Alcohol Clin Exp Res 16:665–669
47. Purohit V, Brenner DA (2006) Mechanisms of alcohol-induced hepatic fibrosis: a summary of the Ron Thurman Symposium. Hepatology 43:872–878
48. Tsukamoto H (2001) How is the liver primed or sensitized for alcoholic liver disease? Alcohol Clin Exp Res 25:171S–181S
49. Pastorino JG, Hoek JB (2000) Ethanol potentiates tumor necrosis factor-alpha cytotoxicity in hepatoma cells and primary rat hepatocytes by promoting induction of the mitochondrial permeability transition. Hepatology 31:1141–1152
50. Liu H, Jones BE, Bradham C, Czaja MJ (2002) Increased cytochrome P-450 2E1 expression sensitizes hepatocytes to c-Jun-mediated cell death from TNF-alpha. Am J Physiol Gastrointest Liver Physiol 282:G257–G266
51. Lu Y, Cederbaum AI (2006) Enhancement by pyrazole of lipopolysaccharide-induced liver injury in mice: role of cytochrome P450 2E1 and 2A5. Hepatology 44:263–274
52. Lu Y, Wang X, Cederbaum AI (2005) Lipopolysaccharide-induced liver injury in rats treated with the CYP2E1 inducer pyrazole. Am J Physiol Gastrointest Liver Physiol 289:G308–G319
53. Wu D, Cederbaum AI (2008) Cytochrome P4502E1 sensitizes to tumor necrosis factor alpha-induced liver injury through activation of mitogen-activated protein kinases in mice. Hepatology 47:1005–1017
54. Wu D, Xu C, Cederbaum A (2009) Role of nitric oxide and nuclear factor-kappaB in the CYP2E1 potentiation of tumor necrosis factor alpha hepatotoxicity in mice. Free Radic Biol Med 46:480–491

55. Bailey SM (2003) A review of the role of reactive oxygen and nitrogen species in alcohol-induced mitochondrial dysfunction. Free Radic Res 37:585–596
56. Hoek JB, Cahill A, Pastorino JG (2002) Alcohol and mitochondria: a dysfunctional relationship. Gastroenterology 122:2049–2063
57. Zhuge J, Cederbaum AI (2009) Inhibition of the mitochondrial permeability transition by cyclosporin A prevents pyrazole plus lipopolysaccharide-induced liver injury in mice. Free Radic Biol Med 46:406–413
58. McCubrey JA, Franklin RA (2006) Reactive oxygen intermediates and signaling through kinase pathways. Antioxid Redox Signal 8:1745–1748
59. Ichijo H, Nishida E, Irie K, ten Dijke P, Saitoh M, Moriguchi T, Takagi M, Matsumoto K, Miyazono K, Gotoh Y (1997) Induction of apoptosis by ASK1, a mammalian MAPKKK that activates SAPK/JNK and p38 signaling pathways. Science 275:90–94
60. Bardag-Gorce F, French BA, Dedes J, Li J, French SW (2006) Gene expression patterns of the liver in response to alcohol: in vivo and in vitro models compared. Exp Mol Pathol 80:241–251
61. Li J, Bardag-Gorce FJ, Oliva J, Dedes J, French BA, French SW (2009) Gene expression modifications in the liver caused by binge drinking and S-adenosylmethionine feeding. The role of epigenetic changes. Genes Nutr 5:169–179
62. Aroor AR, James TT, Jackson DE, Shukla SD (2010) Differential changes in MAP kinases, histone modifications, and liver injury in rats acutely treated with ethanol. Alcohol Clin Exp Res 34:1543–1551
63. Pastorino JG, Shulga N, Hoek JB (2003) TNF-alpha-induced cell death in ethanol-exposed cells depends on p38 MAPK signaling but is independent of Bid and caspase-8. Am J Physiol Gastrointest Liver Physiol 285:G503–G516
64. Czaja MJ (2003) The future of GI and liver research: editorial perspectives. III. JNK/AP-1 regulation of hepatocyte death. Am J Physiol Gastrointest Liver Physiol 284:G875–G879
65. Gunawan BK, Liu ZX, Han D, Hanawa N, Gaarde WA, Kaplowitz N (2006) c-Jun N-terminal kinase plays a major role in murine acetaminophen hepatotoxicity. Gastroenterology 131:165–178
66. Matsuzawa A, Ichijo H (2008) Redox control of cell fate by MAP kinase: physiological roles of ASK-1MAP kinase pathway in stress signaling. Biochim Biophys Acta 1780:1325–1336
67. Kyraiakis JM, Avruch J (2001) Mammalian mitogen-activated protein kinase signal transduction pathway activated by stress and inflammation. Physiol Rev 81:807–869
68. Fujino G, Noguchi T, Matsuzawa A, Yamauchi S, Saitoh M, Takeda K, Ichijo H (2007) Thioredoxin and TRAF family proteins regulate reactive oxygen species-dependent activation of ASK1 through reciprocal modulation of the N-terminal homophilic interaction of ASK1. Mol Cell Biol 27:8152–8163
69. Liu H, Nishitoh H, Ichijo H, Kyriakis M (2000) Activation of apoptosis signal-regulating kinase 1 (ASK1) by tumor necrosis factor receptor-associated factor 2 requires prior dissociation of the ASK1 inhibitor thioredoxin. Mol Cell Biol 20:2198–2208
70. Wang X, Destrument A, Tournier C (2007) Physiological roles of MKK4 and MKK7: insights from animal models. Biochim Biophys Acta 1773:1349–1357
71. Matsukawa J, Matsuzawa A, Takeda K, Ichijo H (2004) The ASK-1-MAP kinase cascades in mammalian stress response. J Biochem 136:261–265
72. Wu D, Cederbaum AI (2010) Activation of ASK-1 and downstream MAP kinases in cytochrome P4502E1 potentiated tumor necrosis factor alpha liver injury. Free Radic Biol Med 49:348–60
73. Wu D, Cederbaum AI (2009) Oxidative stress and alcoholic liver disease. Semin Liver Dis 29:141–154
74. Nakagawa H, Maeda S, Hikiba Y, Ohmae T, Shibata W, Yanai KS, Ogura K, Noguchi T, Karin M, Ichijo H, Omata M (2008) Deletion of apoptosis signal-regulating kinase 1 attenuates acetaminophen-induced liver injury by inhibiting c-Jun N—Terminal kinase activation. Gastroenterology 135:1311–132

75. Davis RJ (2000) Signal transduction by the JNK group of MAP kinases. Cell 103:239–252
76. Schwabe RF, Uchinami H, Qian T, Bennett BL, Lemasters JJ, Brenner DA (2004) Differential requirement for c-Jun NH_2-terminal kinase in TNFα-and Fas-mediated apoptosis in hepatocytes. FASEB J 18:720–722
77. Liu H, Lo CR, Czaja MJ (2002) NF-kappaB inhibition sensitizes hepatocytes to TNF-induced apoptosis through a stained activation of JNK and c-Jun. Hepatology 35:772–778
78. Schattenberg JM, Singh R, Wang Y, Lefkowitch JH, Rigoli RM, Scherer PE, Czaja M (2006) JNK1 but not JNK2 promotes the development of steatohepatitis in mice. Hepatology 43:163–176
79. Liu J, Minemoto Y, Lin A (2004) C-Jun N-terminal kinase1 (JNK1) but not JNK2 is essential for TNFα-induced c-Jun kinase activation and apoptosis. Mol Cell Biol 24:10844–10856
80. Wang Y, Singh R, Lefkowitch JH, Rigoli RM, Scherer PE, Czaja MJ (2006) Tumor necrosis factor induced liver injury results from JNK2-dependent activation of caspase-8 and the mitochondrial death pathway. J Biol Chem 281:15258–15267
81. Singh R, Wang Y, Xiang Y, Tanaka KE, Gaarde WA, Czaja MJ (2009) Differential effects of JNK1 and JNK2 inhibition on murine steatohepatitis and insulin resistance. Hepatology 49:87–96
82. Eminel S, Klettner K, Roemer L, Herdegan T, Waetzig V (2004) JNK2 translocates to the mitochondria and mediates cytochrome release in PC12 cells in response to 6-hydroxydopamine. J Biol Chem 279:55385–55392
83. Wang X, Wu D, Yang L, Cederbaum AI (2011) Hepatotoxicity mediated by cytochrome P4502E plus TNFα occurs in cJun-N terminal kinase 2–/– but not in cJun-N terminal 1–/– mice. Hepatology 54:1753–1766
84. Dey A, Cederbaum AI (2006) Alcohol and oxidative liver injury. Hepatology 43:S63–S74
85. Cederbaum AI (2009) Nrf2 and antioxidant defense against CYP2E1 toxicity. Expert Opin Drug Metab Toxicol 5:1–22
86. Yang L, Wu D, Wang X, Cederbaum AI (2012) Cytochrome P4502E1, oxidative stress. JNK and autophagy in acute-alcohol-induced fatty liver. Free Radic Biol Med 53:1170–1180

Chapter 10
Understanding the Tumor Suppressor PTEN in Chronic Alcoholism and Hepatocellular Carcinoma

Colin T. Shearn and Dennis R. Petersen

Abstract The tumor suppressor phosphatase and tensin homolog deleted on chromosome 10 (PTEN) is a phosphatidylinositol (PtdIns) phosphatase that regulates Akt activation via PtdIns 3 kinase. Changes in PTEN expression and/or activity have been identified in a variety of chronic hepatocellular disorders including obesity, NAFLD, NASH, and alcoholism. In cancer biology, PTEN is frequently mutated or deleted in a wide variety of tumors. Mutations, decreased promoter activity, and decreased expression in PTEN are frequently identified in patients with hepatocellular carcinoma. While the majority of research on PTEN concerns obesity and NASH, PTEN clearly has a role in hepatic insulin sensitivity and in the development of steatosis during chronic alcoholism. Yet, in chronic alcoholics and HCC, very little is known concerning PTEN mutation/deletion or low PTEN expression. This review is focused on an overview of the current knowledge on molecular mechanisms of dysregulation of PTEN expression/activity in the liver and their relationship to development of ethanol-induced hepatocellular damage and cancer.

Keywords PTEN • Chronic ethanol • Fatty acid synthesis • Posttranslational modification • Oxidative stress • Hepatic steatosis

10.1 ALD and Cancer

Alcoholic liver disease (ALD) affects millions of patients worldwide every year and ranks among the leaders in morbidity and mortality. In ALD, disease progression is initially characterized by an increase in steatosis followed by progressive hepatocellular damage as evidenced by steatohepatitis, fibrosis, and ultimately cirrhosis [1].

C.T. Shearn, Ph.D. (✉) • D.R. Petersen
Department of Pharmaceutical Sciences, University of Colorado Denver Anchutz Medical Campus, 12850 East Montview Blvd Box C238, Building V20 Room 2460B, Aurora, CO 80045, USA
e-mail: Colin.Shearn@ucdenver.edu

V. Vasiliou et al. (eds.), *Biological Basis of Alcohol-Induced Cancer*, Advances in Experimental Medicine and Biology 815,
DOI 10.1007/978-3-319-09614-8_10

Based on current data, ALD accounts for approximately ~3.8 % of all global deaths with 4.6 % global disability-adjusted life-years attributable to alcohol [2]. In the United States, in 2011 there were 22,073 deaths related to alcohol, with 15,000 specifically to ALD [3, 4] with direct costs due to ALD of approximately 29 billion dollars (US Department of Health and Human Services [5]).The relationship between chronic alcohol abuse and hepatocellular cancer (HCC) is currently under intense study. Patients who report regular alcohol use are at an increased risk for the development of HCC [6]. Clearly, however, HCC in alcoholics occurs after formation of cirrhosis. In patients that develop alcoholic cirrhosis, estimated mortality risks are approximately 30 % (1 year) and 60 % (5 years). When HCC is diagnosed it frequently occurs in association with concurrent hepatitis C infection [7]. In active alcoholics, HCC is relatively rare with only 3–10 % of patients with alcoholic cirrhosis developing HCC [8]. It is hypothesized that this may be due to the death of patients with ALD due to alcoholic cirrhosis before they develop cancer [9]. This is supported by the fact that current statistics indicate alcoholic patients who die from HCC are typically significantly older than alcoholic patients who die from liver failure [10]. Upon autopsy, however, has been estimated that as many as half of alcoholics will have early evidence of cancer.

10.2 The PTEN Tumor Suppressor

The tumor suppressor phosphatase and tensin homolog deleted on chromosome 10 (PTEN) is a 47,167 Da protein that frequently is found mutated or deleted in many late stage tumors and is considered the second most prevalent genetic mutation found in all cancer. In humans, three primary disorders have been identified containing germline PTEN mutations: Cowden disease (G129E), Lhermitte–Duclos syndrome, and Bannayan–Zonana syndrome [11]. All three of these diseases show birth defects and an increased prevalence of multiple benign tumors including breast and thyroid malignancies. When examining human hepatocellular carcinoma, the PTEN gene is mutated in 5 % and PTEN expression is downregulated in almost half of HCC cases [12–15]. In addition, the expression of PTEN is downregulated as the stage of cancer progresses [16]. In patients with HCC, increased Akt expression and activation (as evidenced by phosphoSer^{473}Akt) was identified in early stage 1 and stage 2 HCC. This phenotype did persist in more advanced stages [17]. Surprisingly, there has only been one documented case of a patient with Cowden Disease developing NASH that progresses into hepatocellular carcinoma [18].

PTEN suppresses growth via its enzymatic activity. PTEN is a PtdIns 3-phosphatase, as it will convert PtdIns (3,4,5) P_3 to PtdIns (4,5) P_2 altering the association of a variety of lipid-binding proteins. Foremost among these proteins is the protein kinases Akt1, 2, and 3. Akt kinases regulate cell growth, proliferation as well as metabolic functions such as gluconeogenesis and lipogenesis [19, 20]. In the liver, only two isoforms of Akt (Akt1 and Akt2) are expressed [21]. Examining overall expression, Akt1 accounts for 15 % of Akt activity and Akt2 85 %.

Using murine deletion models, Akt2 regulates insulin control over glucose metabolism as well as de novo lipogenesis. Recently concurrent deletion of Akt2 in Lep(ob/ob) mice resulted in insulin resistance without hepatic triglyceride accumulation [21]. Furthermore, feeding of a high-fat diet resulted in reduction of hepatic triglycerides but there was no change in lipogenesis nor was there evidence of an increase in β-oxidation. Not surprisingly, Akt2 expression is increased in HCC and correlates with poorer prognosis [22]. In the same study, changes in Akt1 were not evident suggesting that Akt2 may play a more important role in HCC.

Overexpression of PTEN results in decreased Akt phosphorylation, cell growth, and proliferation in cell lines [23]. Cell lines containing catalytically inactive forms of PTEN exhibit elevated Akt activity leading to increased cellular survival and/or proliferation [24–26]. Other proteins that are regulated by PTEN include PDK1, p70S6 kinase and in T-cells, the Tek homology kinase ITK [27, 28]. These proteins are all downstream effectors of PtdIns 3-kinase signaling and are either activated by proteins that can interact with phosphatidylinositides or interact with phosphatidylinositides themselves. Thus, PTEN is a negative regulator of PtdIns 3-kinase signaling. In the murine models, hepatospecific deletion of PTEN (PTEN$^{f/f}$) leads to insulin hypersensitivity, severe steatosis, increased hepatic and serum triglycerides, and progression into steatohepatitis in mice [29–31]. Over time, these mice developed hepatocellular carcinoma, but did not display insulin sensitivity and hyperlipidemia that are more characteristic of NASH. Therefore, PTEN$^{f/f}$ mice appear to be a good model for specific types of NASH that do not possess insulin resistance or obesity. This is in part due to the fact that PTEN is a negative regulator of the pro-growth, pro-survival kinase Akt. Activation of Akt regulates the activation of pro-steatotic genes such as SREBP1, Forkhead (FoxO1), and PPARγ transcription factors. In the PTEN$^{f/f}$ model, Akt activation was increased and cellular processes such as *de novo* lipogenesis were constitutively activated [32]. Not surprisingly, when crossed with Akt2$^{f/f}$ mice, steatosis and the development of HCC were significantly decreased [33].

10.3 PTEN and ALD

In alcoholism, the PtdIns 3 kinase/PTEN/Akt pathway has been studied for over a decade primarily in rat models of chronic ethanol feeding. The data accumulated from these studies, however, are somewhat conflicting. In an early study using male Sprague Dawley rats with 35 % ETOH for 2 weeks, PtdIns 3-kinase activity was increased in both the acute (3.2-fold) and chronic (2.8-fold) ETOH groups [34]. This is also evidenced by increased phosphotyrosine on IRS-1 and IRS-2. The authors concluded that insulin resistance occurring in chronic alcoholism is downstream of PtdIns 3-kinase in both acute and chronic ethanol models. Yet, this study only spanned 2 weeks and data regarding hepatocellular damage was not significant. In another study using female Long Evans rats, an opposing result occurred [35]. Chronic ethanol consumption for 8 weeks with a ramp to 35.4 %, resulted in decreased PtdIns 3-kinase activity, an increase in PTEN expression and activity corresponding to a decrease in pSer^{473}Akt, Akt activity and insulin resistance [35]. Concurrently, PTEN phosphorylation was reduced contributing to an increase in PTEN activity. In a follow-up study, it was determined that this result was more pronounced in Long Evans rats compared to Sprague Dawley and nonexistent in Fischer rats [36]. In the latter study, increased PTEN expression corresponded to an increase in p53 expression, hepatocellular death, and decreased insulin signaling. Yet, in non-ethanol models, it has been demonstrated that in order to preserve the half-life of 3-phosphoinositides, PtdIns 3-kinase is a positive regulator of PTEN and a decrease in p85 expression results in decreased PTEN activity [37]. This suggests that in the Yeon et al. and Derdak et al. studies, PtdIns 3-kinase and PTEN would have opposing effects and cellular concentrations of PtdIns (3,4,5P)$_3$ would be severely suppressed. Given that Akt is activated by membrane association of its PH domain to PtdIns (3,4,5) P$_3$, these data are not reflected in total Akt activity following ethanol consumption which is decreased by approximately 25 %. Global Akt activity also did not correspond to GSK3β activity which was also decreased. A decrease in GSK3β activity is reflective of an increase in phosphorylation on Ser^9 by Akt [19]. It is interesting to note that although PTEN levels increased more than twofold, overall activity only increased by 25 %. One possibility is that PTEN activity is partially inhibited by posttranslational modification of its active site via reactive electrophiles resulting in differences in activity and expression. This study is partially in agreement with a recent study in male Sprague Dawley rats fed 36 % ETOH for 4 weeks [38]. Activation of Akt (pSer^{473}) was determined to be decreased and levels of Fox01 decreased [38]. Alcohol also increased hepatic mRNA expression of FoxO1 and p53 which was not reflected in corresponding protein levels. In addition, acetylation of the tumor suppressor p53 was increased in the nucleus and there was evidence of increased DNA damage in ethanol-treated animals. The tumor suppressor p53 is involved in a positive feedback loop with PTEN [37]. The increase in PTEN is thought to arrest cell cycle or to promote cell death. Therefore although not shown, PTEN concentrations and activity should be suppressed in this system.

The discrepancy for these opposing results might be explained by two recent companion papers [39, 40]. Using male Sprague Dawley rats with intragastric addition of ethanol (13 g/kg/day), chronic ethanol resulted in decreased pSer^{9}GSK3β and decreased pThr^{308}Akt but increased pSer^{473}Akt. Ethanol feeding resulted in decreased hepatic Akt activity and reduced nuclear SREBP-1. In this model there were no differences in either PTEN or PTEN phosphorylation indicating an alternative mechanism of regulation of Akt via ethanol. Examining other PtdIns-regulated proteins, there was significant recruitment of PDK1, PtdIns 3-kinase p110α, p85 and Rictor to the plasma membrane in ETOH animals. Chronic ETOH consumption inhibited recruitment of Akt to the plasma membrane and increased cytosolic Akt. The adaptor protein TRB3 has been demonstrated to associate with Akt in the cytosol [41]. This association inhibits Akt phosphorylation on Thr^{308} via PDK1. TRB3 protein and mRNA was increased in ETOH fed rats. Further experiments demonstrated that TRB3 associates with Akt2 via the Akt2-PH domain preventing plasma membrane association. This manuscript was followed by an additional set of experiments that examined the effects of either high-dose (13 g/kg/dL) or a low-dose ETOH (4 g/kg/dL) in Sprague Dawley rats. With respect to PTEN, they found no differences. At low concentrations of ethanol, Akt phosphorylation at Thr^{308}/Ser^{473} was increased whereas at high doses phosphorylation was decreased on Thr^{308}. At high concentrations of ethanol, nSREBP1 and $pSer^{9}GSK3\beta$ were decreased indicative of decreased Akt activity. Further experiments also demonstrated at low doses, expression of the PtdIns 3-kinase P55 subunit is decreased by ethanol. Thus, at least in rats, Akt signaling is dysregulated by ethanol at higher concentrations but at low concentrations, Akt is activated. This provides a plausible explanation to the Yeon et al. and the Lieber et al. studies. The change in Akt activation due to changing blood ethanol concentrations may have resulted in the discrepancies in Akt activation. The ramifications of Akt activation at the lower concentrations are not currently known. It can be hypothesized that given the role of Akt2 in the activation of de novo lipogenesis, early Akt2 activation at low ethanol concentrations may contribute to the early formation of steatosis following ethanol consumption.

It should be noted that all of these studies were performed in rats. We have examined the effects of ethanol consumption on PTEN signaling in C57BL6/J mice [42]. Using a modified Lieber–DeCarli high-fat diet and chronic ethanol for 9 weeks (ramping to 31.5 % ETOH) we determined that PTEN activity was decreased and phosphorylation increased by ethanol. This corresponded to an increase in both total Akt activity and in Akt2 activity. No differences were identified in Akt1 expression and activity. The blood ethanol concentrations also were similar to those found in the biphasic rat study (250 mg/dL). Thus, in mice, blood ethanol concentrations may need to be high than 250 mg/dL to increase PTEN activity and decrease Akt activity. This suggests that in mice, mechanisms of PTEN regulation by ethanol may be different. One difference is that in the biphasic study, ethanol was administered via intragastric cannulation where as in our study ethanol was administered *ad libitum*. These cellular processes might not be as straightforward as they appear. PTEN expression also can be downregulated following the addition of unsaturated free fatty acids [43]. Downregulation was more significant in steatotic hepatocytes compared to surrounding non-steatotic hepatocytes. In our study dietary fat was largely derived from corn oil and therefore contained primarily polyunsaturated fatty acids.

10.4 Posttranslational Modifications of PTEN

A primary mechanism for PTEN regulation is via posttranslational modifications. PTEN has been demonstrated to be susceptible to a wide variety of posttranslational modifications (Fig. 10.1). These include glutathionylation, S-nitrosylation, acetylation, phosphorylation, ubiquitinylation, and nitrosylation as well as undergo electrophilic modification by oxidative species [44–49]. Not shown in Fig. 10.1 are additional phosphorylation sites: Ser^{229}, Thr^{232}, Thr^{319}, Thr^{321}, Tyr^{336}, Thr^{366}, Ser^{370},

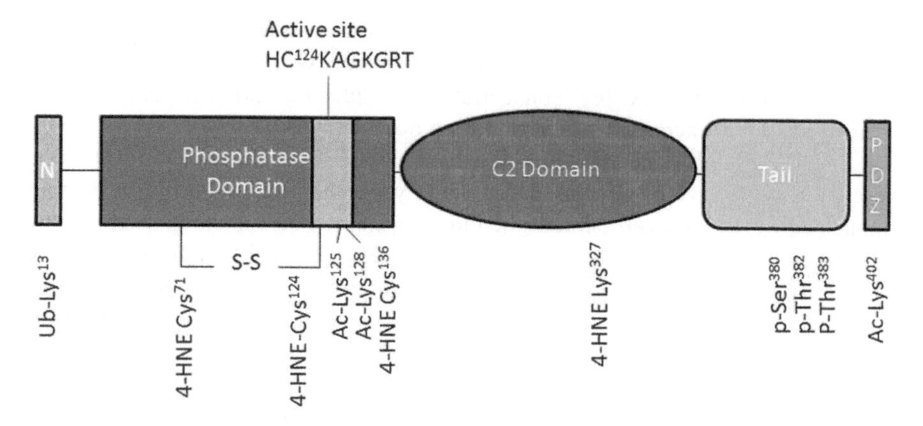

Fig. 10.1 Schematic model showing the overall domain structure and posttranslational modifications of PTEN. PTEN is a 403 amino acid protein that contains five domains. On the N-terminal side is a PtdIns (4,5) P_2-binding domain that can be monoubiquitinated. This is followed by the catalytic domain that contains the signature CX_5RT PTP phosphatase motif that is the active site. Following the catalytic domain is a topology II C2 domain that interacts with negatively charged phospholipids. On the C-terminus is a Pro-, Glu-, Ser-, Thr-rich domain (PEST) and a PDZ-binding motif

and Ser^{385}. Of these modifications, only carbonylation and phosphorylation of Ser^{380} and $Thr^{382/3}$ have been examined in ethanol models.

The active site of PTEN contains a reactive cysteine that is required for catalysis (**HC**$^{\mathbf{124}}$KAGKGRT) [50]. The pKa of Cys^{124} is 4.7–5.4 and is deprotonated at physiological pH making it a strong nucleophile. This is due to the presence of Lys residues ($Lys^{125,\ 128}$) within the active site pocket. This makes Cys^{124} susceptible to oxidation by hydrogen peroxide via the formation of a disulfide with Cys^{71} resulting in inhibition of enzymatic activity [46]. These cysteine disulfides may be subsequently reduced via the thioredoxin system and the peroxiredoxin system [46, 51, 52]. In addition Cys^{124} is thought to be also susceptible to glutathionylation and carbonylation [42, 49, 53].

Another source of electrophiles in cells is reactive aldehydes that are formed from lipid peroxidation [54, 55]. In alcoholism, a focus of research has been the identification of the effects of lipid peroxidation in chronic alcoholics as well as in murine models [56]. The lipid aldehyde, 4-hydroxynonenal (4-HNE) irreversibly modifies PTEN both in its active site as well as within its C2 domain [42]. Studies have shown that inhibition of PTEN activity by direct 4-HNE modification occurs in numerous cell types (HepG2, MCF7, HEK-293) as well as in a murine chronic ethanol model [42, 53, 57, 58]. One deficit of these studies was that the site of lipid aldehyde modification of PTEN was not directly identified in vivo. Current technology was not sufficient to reliably identify the site of PTEN carbonylation from livers isolated from ethanol fed mice.

PTEN is reversibly modified by the electrophile S-nitrosoglutathione in pulmonary epithelial cells. This modification decreased PTEN activity, increased

AKT activity, and increased HIF1α protein accumulation [59]. In the liver, using a NASH model, a high-fat diet resulted in increased glutathionylation of PTEN and decreased activity [44]. We have demonstrated increased glutathionylation occurs following chronic ethanol consumption. Utilizing immunohistochemistry, pericentral increases in protein-SSG were identified in mice chronically fed with ethanol for 6-weeks [60]. Based on these data, it is not unreasonable to predict increased glutathionylation of PTEN in alcoholism.

PTEN negatively regulated by acetylation. Acetylation of Lys^{125}, Lys^{128} by the histone acetyltransferase PCAF results in decreased PTEN activity and increased Akt activation following growth factor stimulation [61]. Acetylation of Lys^{402} in the PTEN C-terminal PDZ-binding domain by CREB-binding protein results in enhanced binding to PDZ containing proteins [62]. Acetylation is significantly altered by ethanol via regulation of the deacetylases Sirtuin 1 and Sirtuin 3 [63–66]. Following chronic ethanol consumption, acetylation of cytosolic proteins as well as mitochondrial proteins is increased and Sirtuin 1/3 activities are decreased [38, 63, 67]. Based on these data, acetylation of PTEN may be altered during chronic ethanol consumption.

PTEN has been demonstrated to localize to the nucleus. Nuclear localization is increased following both monoubiquitination of Lys^{13} and Lys^{289} as well as under conditions of increased oxidative stress. Once in the nucleus, PTEN reduces p53-dependent tumor progression [68]. In a murine high-fat diet model of NASH, PTEN monoubiquitination was not increased [44]. In rat ethanol models, PTEN has been demonstrated to be downstream of p53 and increased p53 nuclear localization correlated with increased PTEN expression [36]. The authors concluded that these factors contributed to insulin resistance. Monoubiquitination of PTEN was not examined in this system.

Phosphorylation plays an integral role in regulating PTEN activity. Although PTEN has been demonstrated to be phosphorylated on numerous residues within its c-terminal tail as well as its C2 domain, of particular importance to this review is phosphorylation on Ser^{380}, Thr^{382}, and Thr^{383}. Substituting these residues for alanine results in increased membrane association as well as PtdIns 3-phosphatase activity demonstrating an inhibitory role for these residues [69].

10.5 Conclusions

In summary as presented in Fig. 10.2, from the presented data it is clear that the role that PTEN plays in ALD as well as ALD-induced HCC has not been fully elucidated. What is known is that there appears to be concentration-dependent ETOH effects on PTEN inhibition/Akt activation. Given the preponderance of data is in rat models, additional experiments are necessary in alternative rodent models to validate findings in the rat. It is also not known if current data translates into humans. In other models, PTEN has been demonstrated to localize to not only the cytosol but

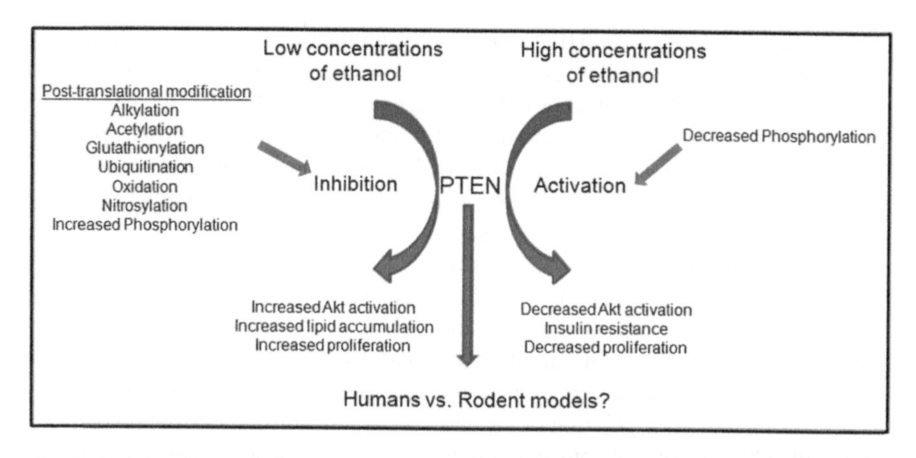

Fig. 10.2 Schematic summary of the effects of ethanol on PTEN and future directions. PTEN is clearly an important regulator of hepatic steatosis as well as insulin resistance. In chronic alcoholism, the effects of ethanol have not been fully elucidated. It is apparent that different concentrations of ethanol induce differential responses in PTEN activation. At low ETOH concentrations, PTEN is inhibited resulting in increased Akt signaling and increased lipid accumulation. At high ETOH concentrations, PTEN is activated contributing to decreased Akt signaling and insulin resistance. Future research will need to be focused on elucidating the ability of ethanol to affect PTEN spatiotemporally in the liver and to determine ethanol-dependent changes in posttranslational modification of PTEN

also to the mitochondria and the nucleus. What are the effects of chronic ethanol consumption on hepatic PTEN subcellular localization? Does monoubiquitination play a role? Given the fact that ethanol induces a phase III response in the liver, are there hepatozonal-specific effects on PTEN signaling as well? Given changes in PtdIns 3 kinase signaling over time during ethanol consumption, more studies need to be performed using a spaciotemporal model to fully elucidate the contribution of the PTEN/Akt pathway to ethanol-induced steatosis and hepatocellular damage.

As mentioned, we have demonstrated an increase in overall protein glutathionylation in our chronic ethanol models [60]. An unexplored question is whether PTEN is glutathionylated following chronic ethanol consumption and what are the ramifications? In addition, does the acetylation status of PTEN change following ethanol consumption? Posttranslational modifications of PTEN have not been examined in patients with HCC. Furthermore, if PTEN is downregulated following ethanol consumption at low concentrations of ethanol, mild or moderate consumption may promote HCC. In alcoholics who consume high amounts of alcohol, via its ability to downregulate Akt, PTEN activation may actually have a "beneficial" inhibitory effect towards development of HCC.

Grants and Funding This research was funded by the following grants from the National Institutes of Health; 5F32AA018613-03 CTS.

References

1. De Minicis S, Brenner DA (2008) Oxidative stress in alcoholic liver disease: role of NADPH oxidase complex. J Gastroenterol Hepatol 23(Suppl 1):S98–S103
2. Rehm J, Mathers C, Popova S, Thavorncharoensap M, Teerawattananon Y, Patra J (2009) Global burden of disease and injury and economic cost attributable to alcohol use and alcohol-use disorders. Lancet 373(9682):2223–2233
3. Beier JI, McClain CJ (2010) Mechanisms and cell signaling in alcoholic liver disease. Biol Chem 391(11):1249–1264
4. Kochanek KD, Murphy SL, Anderson RN, Scott C (2004) Deaths: final data for 2002. Natl Vital Stat Rep 53(5):1–115
5. Mohapatra S, Patra J, Popova S, Duhig A, Rehm J (2009) Social cost of heavy drinking and alcohol dependence in high-income countries. Int J Public Health 55(3):149–157
6. Ascha MS, Hanouneh IA, Lopez R, Tamimi TA, Feldstein AF, Zein NN (2010) The incidence and risk factors of hepatocellular carcinoma in patients with nonalcoholic steatohepatitis. Hepatology 51(6):1972–1978
7. Koike K (2013) The oncogenic role of hepatitis C virus. Recent Results Cancer Res 193:97–111
8. Mathurin P (2012) EASL clinical practical guidelines: management of alcoholic liver disease. J Hepatol 57(2):399–420
9. Liew CT (1990) The clinicopathological spectrum of alcoholic liver disease—an autopsy survey of 441 cases. Changgeng Yi Xue Za Zhi 13(2):72–85
10. Davis GL, Dempster J, Meler JD, Orr DW, Walberg MW, Brown B, Berger BD, O'Connor JK, Goldstein RM (2008) Hepatocellular carcinoma: management of an increasingly common problem. Proc (Bayl Univ Med Cent) 21(3):266–280
11. Di Cristofano A, Pandolfi PP (2000) The multiple roles of PTEN in tumor suppression. Cell 100:387–390
12. Cotler SJ, Hay N, Xie H, Chen ML, Xu PZ, Layden TJ, Guzman G (2008) Immunohistochemical expression of components of the Akt-mTORC1 pathway is associated with hepatocellular carcinoma in patients with chronic liver disease. Dig Dis Sci 53(3):844–849
13. Sze KM, Wong KL, Chu GK, Lee JM, Yau TO, Ng IO (2011) Loss of phosphatase and tensin homolog enhances cell invasion and migration through AKT/Sp-1 transcription factor/matrix metalloproteinase 2 activation in hepatocellular carcinoma and has clinicopathologic significance. Hepatology 53(5):1558–1569
14. Wang L, Wang WL, Zhang Y, Guo SP, Zhang J, Li QL (2007) Epigenetic and genetic alterations of PTEN in hepatocellular carcinoma. Hepatol Res 37(5):389–396
15. Whittaker S, Marais R, Zhu AX (2010) The role of signaling pathways in the development and treatment of hepatocellular carcinoma. Oncogene 29(36):4989–5005
16. Wu SK, Wang BJ, Yang Y, Feng XH, Zhao XP, Yang DL (2007) Expression of PTEN, PPM1A and P-Smad2 in hepatocellular carcinomas and adjacent liver tissues. World J Gastroenterol 13(34):4554–4559
17. Boyault S, Rickman DS, de Reynies A, Balabaud C, Rebouissou S, Jeannot E, Herault A, Saric J, Belghiti J, Franco D, Bioulac-Sage P, Laurent-Puig P, Zucman-Rossi J (2007) Transcriptome classification of HCC is related to gene alterations and to new therapeutic targets. Hepatology 45(1):42–52
18. Sugihara T, Mandai M, Koda M, Matono T, Nagahara T, Ueki M, Murawaki Y (2011) Cowden syndrome complicated with hepatocellular carcinoma possibly originating from non-alcoholic steatohepatitis (NASH). Hepatol Res 41(2):189–193
19. Krycer JR, Sharpe LJ, Luu W, Brown AJ (2010) The Akt-SREBP nexus: cell signaling meets lipid metabolism. Trends Endocrinol Metab 21(5):268–276
20. Marte BM, Downward J (1997) PKB/Akt: connecting phosphoinositide 3-kinase to cell survival and beyond. Trends Biochem Sci 22(9):355–358
21. Leavens KF, Easton RM, Shulman GI, Previs SF, Birnbaum MJ (2009) Akt2 is required for hepatic lipid accumulation in models of insulin resistance. Cell Metab 10(5):405–418

22. Xu X, Sakon M, Nagano H, Hiraoka N, Yamamoto H, Hayashi N, Dono K, Nakamori S, Umeshita K, Ito Y, Matsuura N, Monden M (2004) Akt2 expression correlates with prognosis of human hepatocellular carcinoma. Oncol Rep 11(1):25–32
23. Cantley LC, Neel BG (1999) New insights into tumor suppression: PTEN suppresses tumor formation by restraining the phosphoinositide 3-kinase/AKT pathway. Proc Natl Acad Sci U S A 96(8):4240–4245
24. Choi Y, Zhang J, Murga C, Yu H, Koller E, Monia BP, Gutkind JS, Li W (2002) PTEN, but not SHIP and SHIP2, suppresses the PI3K/Akt pathway and induces growth inhibition and apoptosis of myeloma cells. Oncogene 21(34):5289–5300
25. Stambolic V, Suzuki A, de la Pompa JL, Brothers GM, Mirtsos C, Sasaki T, Ruland J, Penninger JM, Siderovski DP, Mak TW (1998) Negative regulation of PKB/Akt-dependent cell survival by the tumor suppressor PTEN. Cell 95(1):29–39
26. Stiles B, Gilman V, Khanzenzon N, Lesche R, Li A, Qiao R, Liu X, Wu H (2002) Essential role of AKT-1/protein kinase B alpha in PTEN-controlled tumorigenesis. Mol Cell Biol 22(11):3842–3851
27. Nakashima N, Sharma PM, Imamura T, Bookstein R, Olefsky JM (2000) The tumor suppressor PTEN negatively regulates insulin signaling in 3T3-L1 adipocytes. J Biol Chem 275(17):12889–12895
28. Shan X, Czar MJ, Bunnell SC, Liu P, Liu Y, Schwartzberg PL, Wange RL (2000) Deficiency of PTEN in Jurkat T cells causes constitutive localization of Itk to the plasma membrane and hyperresponsiveness to CD3 stimulation. Mol Cell Biol 20(18):6945–6957
29. Horie Y, Suzuki A, Kataoka E, Sasaki T, Hamada K, Sasaki J, Mizuno K, Hasegawa G, Kishimoto H, Iizuka M, Naito M, Enomoto K, Watanabe S, Mak TW, Nakano T (2004) Hepatocyte-specific PTEN deficiency results in steatohepatitis and hepatocellular carcinomas. J Clin Invest 113(12):1774–1783
30. Stiles B, Wang Y, Stahl A, Bassilian S, Lee WP, Kim YJ, Sherwin R, Devaskar S, Lesche R, Magnuson MA, Wu H (2004) Liver-specific deletion of negative regulator PTEN results in fatty liver and insulin hypersensitivity [corrected]. Proc Natl Acad Sci U S A 101(7):2082–2087
31. Watanabe S, Horie Y, Suzuki A (2005) Hepatocyte-specific PTEN-deficient mice as a novel model for nonalcoholic steatohepatitis and hepatocellular carcinoma. Hepatol Res 33(2):161–166
32. Sato W, Horie Y, Kataoka E, Ohshima S, Dohmen T, Iizuka M, Sasaki J, Sasaki T, Hamada K, Kishimoto H, Suzuki A, Watanabe S (2006) Hepatic gene expression in hepatocyte-specific PTEN deficient mice showing steatohepatitis without ethanol challenge. Hepatol Res 34(4):256–265
33. He L, Hou X, Kanel G, Zeng N, Galicia V, Wang Y, Yang J, Wu H, Birnbaum MJ, Stiles BL (2010) The critical role of AKT2 in hepatic steatosis induced by PTEN loss. Am J Pathol 176(5):2302–2308
34. Onishi Y, Honda M, Ogihara T, Sakoda H, Anai M, Fujishiro M, Ono H, Shojima N, Fukushima Y, Inukai K, Katagiri H, Kikuchi M, Oka Y, Asano T (2003) Ethanol feeding induces insulin resistance with enhanced PI 3-kinase activation. Biochem Biophys Res Commun 303(3): 788–794
35. Yeon JE, Califano S, Xu J, Wands JR, De La Monte SM (2003) Potential role of PTEN phosphatase in ethanol-impaired survival signaling in the liver. Hepatology 38(3):703–714
36. Derdak Z, Lang CH, Villegas KA, Tong M, Mark NM, de la Monte SM, Wands JR (2011) Activation of p53 enhances apoptosis and insulin resistance in a rat model of alcoholic liver disease. J Hepatol 54(1):164–172
37. Taniguchi CM, Tran TT, Kondo T, Luo J, Ueki K, Cantley LC, Kahn CR (2006) Phosphoinositide 3-kinase regulatory subunit p85alpha suppresses insulin action via positive regulation of PTEN. Proc Natl Acad Sci U S A 103(32):12093–12097
38. Lieber CS, Leo MA, Wang X, Decarli LM (2008) Alcohol alters hepatic FoxO1, p53, and mitochondrial SIRT5 deacetylation function. Biochem Biophys Res Commun 373(2):246–252
39. He L, Marecki JC, Serrero G, Simmen FA, Ronis MJ, Badger TM (2007) Dose-dependent effects of alcohol on insulin signaling: partial explanation for biphasic alcohol impact on human health. Mol Endocrinol 21(10):2541–2550

40. He L, Simmen FA, Mehendale HM, Ronis MJ, Badger TM (2006) Chronic ethanol intake impairs insulin signaling in rats by disrupting Akt association with the cell membrane. Role of TRB3 in inhibition of Akt/protein kinase B activation. J Biol Chem 281(16):11126–11134
41. Du K, Herzig S, Kulkarni RN, Montminy M (2003) TRB3: a tribbles homolog that inhibits Akt/PKB activation by insulin in liver. Science 300(5625):1574–1577
42. Shearn CT, Smathers RL, Backos DS, Reigan P, Orlicky DJ, Petersen DR (2013) Increased carbonylation of the lipid phosphatase PTEN contributes to Akt2 activation in a murine model of early alcohol-induced steatosis. Free Radic Biol Med 65C:680–692
43. Vinciguerra M, Veyrat-Durebex C, Moukil MA, Rubbia-Brandt L, Rohner-Jeanrenaud F, Foti M (2008) PTEN down-regulation by unsaturated fatty acids triggers hepatic steatosis via an NF-kappaBp65/mTOR dependent mechanism. Gastroenterology 134(1):268–280
44. Alisi A, Bruscalupi G, Pastore A, Petrini S, Panera N, Massimi M, Tozzi G, Leoni S, Piemonte F, Nobili V (2011) Redox homeostasis and posttranslational modifications/activity of phosphatase and tensin homolog in hepatocytes from rats with diet-induced hepatosteatosis. J Nutr Biochem 23(2):169–178
45. Kwon J, Lee SR, Yang KS, Ahn Y, Kim YJ, Stadtman ER, Rhee SG (2004) Reversible oxidation and inactivation of the tumor suppressor PTEN in cells stimulated with peptide growth factors. Proc Natl Acad Sci U S A 101(47):16419–16424
46. Lee SR, Yang KS, Kwon J, Lee C, Jeong W, Rhee SG (2002) Reversible inactivation of the tumor suppressor PTEN by H_2O_2. J Biol Chem 277(23):20336–20342
47. Shi Y, Paluch BE, Wang X, Jiang X (2012) PTEN at a glance. J Cell Sci 125(Pt 20): 4687–4692
48. Song MS, Salmena L, Pandolfi PP (2012) The functions and regulation of the PTEN tumour suppressor. Nat Rev Mol Cell Biol 13(5):283–296
49. Yu CX, Li S, Whorton AR (2005) Redox regulation of PTEN by S-nitrosothiols. Mol Pharmacol 68(3):847–854
50. Xiao Y, Yeong Chit Chia J, Gajewski JE, Sio Seng Lio D, Mulhern TD, Zhu HJ, Nandurkar H, Cheng HC (2007) PTEN catalysis of phospholipid dephosphorylation reaction follows a two-step mechanism in which the conserved aspartate-92 does not function as the general acid–mechanistic analysis of a familial Cowden disease-associated PTEN mutation. Cell Signal 19(7):1434–1445
51. Cao J, Schulte J, Knight A, Leslie NR, Zagozdzon A, Bronson R, Manevich Y, Beeson C, Neumann CA (2009) Prdx1 inhibits tumorigenesis via regulating PTEN/AKT activity. EMBO J 28(10):1505–1517
52. Hui ST, Andres AM, Miller AK, Spann NJ, Potter DW, Post NM, Chen AZ, Sachithanantham S, Jung DY, Kim JK, Davis RA (2008) Txnip balances metabolic and growth signaling via PTEN disulfide reduction. Proc Natl Acad Sci U S A 105(10):3921–3926
53. Covey TM, Edes K, Coombs GS, Virshup DM, Fitzpatrick FA (2010) Alkylation of the tumor suppressor PTEN activates Akt and beta-catenin signaling: a mechanism linking inflammation and oxidative stress with cancer. PLoS One 5(10):e13545
54. Esterbauer H, Schaur RJ, Zollner H (1991) Chemistry and biochemistry of 4-hydroxynonenal, malonaldehyde and related aldehydes. Free Radic Biol Med 11(1):81–128
55. Schaur RJ (2003) Basic aspects of the biochemical reactivity of 4-hydroxynonenal. Mol Aspects Med 24(4–5):149–159
56. Smathers RL, Galligan JJ, Stewart BJ, Petersen DR (2011) Overview of lipid peroxidation products and hepatic protein modification in alcoholic liver disease. Chem Biol Interact 192(1–2):107–112
57. Shearn CT, Fritz KS, Reigan P, Petersen DR (2011) Modification of Akt2 by 4-hydroxynonenal inhibits insulin-dependent Akt signaling in HepG2 cells. Biochemistry 50(19):3984–3996
58. Shearn CT, Smathers RL, Stewart BJ, Fritz KS, Galligan JJ, Hail N Jr, Petersen DR (2011) Phosphatase and tensin homolog deleted on chromosome 10 (PTEN) inhibition by 4-hydroxynonenal leads to increased Akt activation in hepatocytes. Mol Pharmacol 79(6): 941–952

59. Carver DJ, Gaston B, Deronde K, Palmer LA (2007) Akt-mediated activation of HIF-1 in pulmonary vascular endothelial cells by S-nitrosoglutathione. Am J Respir Cell Mol Biol 37(3):255–263
60. Galligan JJ, Smathers RL, Shearn CT, Fritz KS, Backos DS, Jiang H, Franklin CC, Orlicky DJ, Maclean KN, Petersen DR (2012) Oxidative stress and the ER stress response in a murine model for early-stage alcoholic liver disease. J Toxicol 2012:207594
61. Okumura K, Mendoza M, Bachoo RM, DePinho RA, Cavenee WK, Furnari FB (2006) PCAF modulates PTEN activity. J Biol Chem 281(36):26562–26568
62. Ikenoue T, Inoki K, Zhao B, Guan KL (2008) PTEN acetylation modulates its interaction with PDZ domain. Cancer Res 68(17):6908–6912
63. Fritz KS, Galligan JJ, Hirschey MD, Verdin E, Petersen DR (2012) Mitochondrial acetylome analysis in a mouse model of alcohol-induced liver injury utilizing SIRT3 knockout mice. J Proteome Res 11(3):1633–1643
64. Hirschey MD, Shimazu T, Jing E, Grueter CA, Collins AM, Aouizerat B, Stancakova A, Goetzman E, Lam MM, Schwer B, Stevens RD, Muehlbauer MJ, Kakar S, Bass NM, Kuusisto J, Laakso M, Alt FW, Newgard CB, Farese RV Jr, Kahn CR, Verdin E (2011) SIRT3 deficiency and mitochondrial protein hyperacetylation accelerate the development of the metabolic syndrome. Mol Cell 44(2):177–190
65. Lan F, Cacicedo JM, Ruderman N, Ido Y (2008) SIRT1 modulation of the acetylation status, cytosolic localization, and activity of LKB1. Possible role in AMP-activated protein kinase activation. J Biol Chem 283(41):27628–27635
66. Yao XH, Nyomba BL (2008) Hepatic insulin resistance induced by prenatal alcohol exposure is associated with reduced PTEN and TRB3 acetylation in adult rat offspring. Am J Physiol Regul Integr Comp Physiol 294(6):R1797–R1806
67. Shen Z, Liang X, Rogers CQ, Rideout D, You M (2010) Involvement of adiponectin-SIRT1-AMPK signaling in the protective action of rosiglitazone against alcoholic fatty liver in mice. Am J Physiol Gastrointest Liver Physiol 298(3):G364–374
68. Chang CJ, Mulholland DJ, Valamehr B, Mosessian S, Sellers WR, Wu H (2008) PTEN nuclear localization is regulated by oxidative stress and mediates p53-dependent tumor suppression. Mol Cell Biol 28(10):3281–3289
69. Vazquez F, Grossman SR, Takahashi Y, Rokas MV, Nakamura N, Sellers WR (2001) Phosphorylation of the PTEN tail acts as an inhibitory switch by preventing its recruitment into a protein complex. J Biol Chem 276(52):48627–48630

Chapter 11
Alcohol Consumption, Wnt/β-Catenin Signaling, and Hepatocarcinogenesis

K.E. Mercer, L. Hennings, and M.J.J. Ronis

Abstract Alcohol is a well-established risk factor for hepatocellular carcinoma, and the mechanisms by which alcohol liver cancer is complex. It has been suggested that ethanol (EtOH) metabolism may enhance tumor progression by increasing hepatocyte proliferation. To test this hypothesis, ethanol (EtOH) feeding of male mice began 7 weeks post-injection of the chemical carcinogen diethylnitrosamine (DEN), and continued for 16 weeks, with a final EtOH concentration of 28 % of total calories. As expected, EtOH increased the total number of cancerous foci and liver tumors identified in situ fixed livers from the EtOH+DEN group compared to corresponding pair-fed (PF)+DEN and chow+DEN control groups. In the EtOH+DEN group, tumor multiplicity corresponded to a 3- to 4-fold increase in proliferation and immunohistochemical staining of β-catenin in non-tumorigenic hepatocytes when compared to the PF+DEN and chow+DEN groups, $p<0.05$. Analysis of EtOH-treated livers from a previously published rat model of chronic liver disease revealed increases in hepatocyte proliferation accompanied by a hepatic depletion of retinol and retinoic acid stores ($p<0.05$), nuclear accumulation of β-catenin ($p<0.05$), increased cytosolic expression p-GSK3β ($p<0.05$), significant upregulation of soluble Wnts, Wnt2, and Wnt7a, and increased expression of several β-catenin targets involved in tumor promotion and progression, cyclin D1, c-myc, WISP1, and MMP7 ($p<0.05$). These data suggest that chronic EtOH consumption activates the Wnt/β-catenin signaling pathway, which increases hepatocyte proliferation thus promoting tumorigenesis following an initiating insult in the liver.

Keywords DEN • Hepatocarcinogenesis • Alcohol • Wnt signaling • β-Catenin • Proliferation

---

K.E. Mercer (✉) • M.J.J. Ronis
Arkansas Children's Nutrition Center, University of Arkansas for Medical Sciences, Little Rock, AR, USA

Department of Pediatrics, University of Arkansas for Medical Sciences, Little Rock, AR, USA
e-mail: kmercer@uams.edu

L. Hennings
Department of Pathology, University of Arkansas for Medical Sciences, Little Rock, AR, USA

V. Vasiliou et al. (eds.), *Biological Basis of Alcohol-Induced Cancer*, Advances in Experimental Medicine and Biology 815,
DOI 10.1007/978-3-319-09614-8_11

11.1 Introduction

Cancer progression in an alcoholic liver occurs through the complex interaction of initiating and promoting mechanisms [1, 2]. Experimentally, ethanol (EtOH) feeding induces the hepatic alcohol metabolizing enzyme, cytochrome P450 (CYP) CYP2E1. Increased CYP2E1 activity produces reactive oxygen species and lipid peroxidation products, as well as activates environmental pro-carcinogens, i.e. dietary nitrosamines and polycyclic hydrocarbons, all of which damage DNA, increase mutagenicity, and initiate hepatocarcinogenesis in rodent models. EtOH metabolism by CYP2E1 and by alcohol dehydrogenase produces the reactive metabolite acetaldehyde, and reduces DNA methylation as a result of disruption of one-carbon metabolism, which may also contribute to tumor initiation [3]. In addition to these initiating mechanisms, chronic EtOH feeding also increases hepatocyte proliferation in animal models of alcoholic liver disease and HCC [4–9]. Several signaling pathways have been implicated in this process, one of which is decreased in retinoic acid receptor (RAR) signaling resulting from vitamin A depletion in alcoholic livers [6]. Downregulation of RAR signaling in EtOH-treated mice has been reported to increase expression of Wnt signaling targets, cyclin D1 and c-Jun. [6] Interestingly, a study using transgenic mice expressing a liver-specific, dominant-negative form of RAR, the loss of RAR signaling also resulted in subsequent upregulation TCF-4/β-catenin-regulated targets such as cyclin D1, increased hepatocyte proliferation, and increased tumor incidence [10]. Clinically, a significant proportion of hepatocellular carcinomas (HCCs) including those from alcoholics have been shown to be β-catenin positive [11]. In these tumors, aberrant activation occurs through mutations found either in the β-catenin gene, primarily in the phosphorylation site for GSK3β, or in Axin1/2, a scaffolding protein necessary for targeting β-catenin for degradation [12, 13]. However, studies using transgenic mice expressing mutant forms of β-catenin have shown that activation through mutations is not sufficient to initiate tumorigenesis, and suggest instead that Wnt/β-catenin signaling participates in tumor promotion [12].

These data support a hypothesis linking EtOH-mediated loss of retinoic acids and RAR signaling, activation of Wnt/β-catenin pathways, and tumor promotion. In the present study we chronically fed a Lieber–DeCarli EtOH liquid diet to a diethylnitrosamine (DEN)-initiated HCC mouse model for 4 months to test if long-term EtOH ingestion increases tumor multiplicity through increased Wnt/β-catenin signaling. At the same time, we utilized liver tissue from a previously published rat model of long-term, chronic alcoholic liver disease [8] to mechanistically validate the hypothesis that chronic EtOH ingestion alone increases hepatocyte proliferation through increased Wnt/β-catenin signaling, resulting in increased expression of known progression markers of HCC.

11.2 Materials and Methods

11.2.1 In Vivo Model of EtOH Promotion of DEN-Induced Hepatocarcinogenesis

C57Bl/6 male mice ($n=51$) received an i.p. injection 10 mg/kg DEN in saline on postnatal day (PND) 14. The mice were weaned on PND28 and maintained on rodent chow until PND53 at which the mice were randomly assigned to three weight-matched diet groups: a standard chow diet ($n=17$, chow+DEN), an EtOH-containing Lieber–DeCarli liquid diet ($n=15$, EtOH+DEN), and a corresponding pair-fed (PF) control Lieber–DeCarli diet ($n=18$, PF+DEN). PF and PF+DEN mice were fed the Lieber–DeCarli control diet (Dyets#710027), which were isocalorically matched to their corresponding EtOH group based on the diet consumptions of the previous day. At sacrifice, mice livers were perfused with 10 % neutral buffered formalin for in situ fixation [14]. Liver samples acquired from a previously published rat model of alcoholic liver disease were also analyzed in this study [8]. Blood alcohol concentrations were analyzed using an Analox analyzer as previously reported [15].

11.2.2 Pathological Evaluation

For each mouse, formalin-fixed lobes were separated, embedded into paraffin, sectioned at 4 μm, stained with H&E, and examined under a light microscope and scored by a veterinary pathologist. Within each lobe, lesions were counted at 40× magnification. Foci and tumors were defined as follows: basophilic focus, a non-compressive lesion less than the width of 4 hepatic lobules in which hepatocytes are smaller and stain more basophilic than normal; eosinophilic focus, a non-compressive lesion less than the width of four hepatic lobules in which hepatocytes are slightly larger than normal with more acidophilic cytoplasm; adenoma, a compressive lesion of any size without evidence of invasion or other criteria of malignancy; and HCC, a compressive and invasive lesion with criteria of malignancy [16].

11.2.3 Immunohistochemistry

β-Catenin expression was assessed in paraffin-embedded mouse liver sections by immunohistochemistry as using a monoclonal β-catenin antibody detecting the active, dephosphorylated (Ser37 or Thr41) form (Anti-Active-β-Catenin, clone 8E7, EMD Millipore, Billerica, MA) as previously described [14]. Quantification of β-catenin staining was achieved by color deconvolution using Aperio Technologies

Spectrum analysis algorithm package and ImageScope analysis software. Hepatocyte proliferation was also measured by immunohistochemical analysis of proliferating cell nuclear antigen (PCNA) staining as described previously [8]

11.2.4 Retinoid Extraction and LC/MS/MS Analysis

Retinoids were extracted from control and EtOH-treated rat livers as previously described [14]. Retinoid extracts were separated by HPLC using an Agilent 1100 HPLC system (Agilent Technologies, Santa Clara, CA) using Phenomenex Synergi 4 μm Max-RP 80A column (4 μm, 3 × 150 mm) as previously described [6]. Retinol and retinoic acid were identified using a 4000 Q TRAP mass spectrometer (Applied Biosystems) coupled with the HPLC. The mass spectrometer was controlled using Analyst version 1.5.1 software and was operated in a MRM mode.

11.2.5 Protein Isolation and Western Blotting

Nuclear and cytosolic protein fractions were isolated from TEN control and EtOH-treated rat livers [8] using NE-PER Nuclear and Cytoplasmic Extraction reagent kit (Thermo Fisher Scientific) as per manufacturer's instructions. Proteins (30 μg) were separated by SDS-polyacrylamide gel electrophoresis using standard methods. Blotted cytosolic proteins were incubated with either an anti-active β-catenin antibody used in immunohistochemistry, or with a polyclonal antibody recognizing the phosphorylated form of GSK3β, (Phospho-GSK-3α/β (Ser21/9), Cell Signaling Technology, Beverly, MA). Total nuclear β-catenin expression was determined using standard procedures and the anti-active β-catenin antibody previously described [14]. Protein bands were quantified using a densitometer and band densities were corrected for total protein loaded by staining with 0.1 % amido black.

11.2.6 Gene Expression

RNA was isolated from livers from a previously published rat model of alcoholic liver disease [8]. Samples from the control and EtOH groups were combined separately into $n=3$ pools/treatment, reverse transcribed using iScript cDNA synthesis kit (Bio-Rad Laboratories, Hercules, CA), and then analyzed for gene expression using a WNT signaling pathway RT^2 profiler PCR array (#PARN 043Z, Sabiosciences, Qiagen, Valencia, CA) following manufacturer's instructions. Array gene expression was confirmed by real time RT-PCR of individual cDNA samples from each group ($n=9$) using SYBR green and an ABI 7500 sequence detection system (Applied Biosystems, Foster City, CA).

11.2.7 Data and Statistical Analysis

Data presented as means ± SE comparisons between two groups were accomplished using either Student's *t*-test or Mann–Whitney *U* rank-sum test. Number of lesions was compared across groups using negative binomial regression which generalizes Poisson regression to account for over-dispersion of the count data as previously described [14]. Continuous outcomes, such as PCNA and β-catenin were compared across groups using nonparametric Kruskal–Wallis rank test, and significant findings were followed by Bonferroni adjusted Wilcoxon/Mann–Whitney *U* rank-sum test for post hoc comparisons. Statistical analysis was performed using the SigmaPlot software package 11.0 (Systat Software, Inc., San Jose, CA) and Stata statistical software 13.1 (Stata Corporation, College Station, TX). Statistical significance was set at $P<0.05$.

11.3 Results and Discussion

It is clear that alcohol consumption can result in the initiation and progression of HCC. In a recent case-control study involving US HCC patients, heavy alcohol consumption (≥80 g daily) was a primary contributing factor for one-third of the reported cases [17]. It was also reported that chronic drinking combined with other risk factors, particularly HCV/HBV infections or diabetes mellitus, increased risk an additional twofold. Although mechanisms underlying alcohol-induced initiation have been well characterized, alcohol-related signaling pathways involved in tumor promotion and progression are poorly described. In this study, we developed a mouse model of tumor promotion using DEN (i.p. 10 mg/kg at PND13) as the initiating agent, followed by chronic EtOH feeding for 16 weeks. To eliminate any possible EtOH-associated initiating effects, primarily through increased CYP2E1 expression and activity in response to EtOH feeding, liquid diets were started 7 weeks post DEN injection. EtOH was added to the Lieber–DeCarli liquid diet by slowly substituting carbohydrate calories for EtOH calories in a stepwise manner until 28 % total calories were reached, which constitutes a final EtOH concentration of 5.0 % (v/v) respectively, and was maintained until sacrifice. On average, the EtOH + DEN group received 21 g/kg/day of EtOH, which resulted in a blood alcohol concentration of 75 ± 29 mg/dL (range 13–393) at sacrifice. Histological analysis was performed using hematoxylin and eosin (H&E)-stained liver sections for each mouse receiving the chow + DEN, PF + DEN, and EtOH + DEN diets. Overall, EtOH-feeding significantly increased the total number of cancer foci, which encompassed basophilic and eosinophilic foci and adenomas, present in the EtOH + DEN group compared to the PF + DEN and chow + DEN group (Table 11.1). The number of carcinomas identified the PF + DEN and EtOH + DEN groups were small and insufficient for statistical analysis. In the EtOH + DEN-treated mice, we also observed a fourfold increase in hepatocyte proliferation in the non-tumor hepatic

Table 11.1 Pathological assessment of tumor foci in Chow + DEN, PF + DEN and EtOH + DEN mice

Group	Tumor foci		Adenoma	
	% Incidence	Multiplicity	% Incidence	Multiplicity
Chow + DEN (n = 17)	16/17 = 94	100/17 = 5.8[a]	1/17 = 6	2/17 = 0.1[a]
PF + DEN (n = 18)	18/18 = 100*	209/18 = 11.6[b]	10/18 = 55*	15/18 = 0.8[b]
EtOH + DEN (n = 15)	16/15 = 100*	266/15 = 17.7[c]	9/15 = 60*	33/15 = 2.2[c]

% Incidence = number of tumor-bearing animals/$n \times 100$ (n = total number of animals per group), significance was determined by Fisher's exact test, *$p < 0.05$ vs. Chow+DEN; Multiplicity = total number of lesions/n (n = total number of animals per group), significance was determined by One-way ANOVA followed by Student Newman-Keuls post hoc analysis, $p < 0.05$ a<b<c. [14]

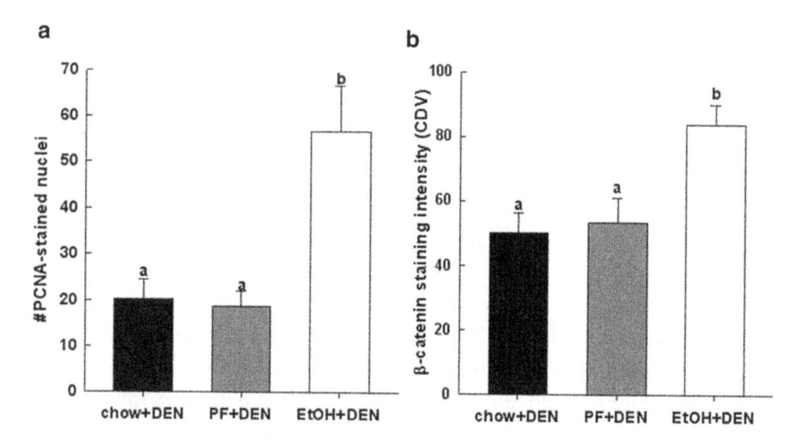

Fig. 11.1 Quantification of immunohistochemical staining of (**a**) PCNA and (**b**) un-phosphorylated (Ser41/33) β-catenin in paraffin-embedded liver sections from chow + DEN (n = 17), PF + DEN (n = 18) and EtOH + DEN (n = 15) diet groups. Data is expressed as means ± SE. Statistical significance was determined by using Bonferroni-adjusted Wilcoxon/Mann–Whitney U rank-sum for post hoc analysis. Groups with different *letters* are significantly different from each other ($p < 0.05$) [14]

tissue as measured by PCNA immunohistochemistry when compared to the PF + DEN, and chow + DEN-treated mice, $p < 0.05$ (Fig. 11.1a). These results are consistent with Yip-Schneider et al. who reported an increase in tumor multiplicity in male alcohol-preferring rats receiving EtOH in their drinking water compared to pair-matched rats on water alone [9]. Brandon-Warner et al., also observed increased tumor burden in male DEN-treated mice receiving EtOH in the drinking water, and reported an association between tumor burden and hepatic PCNA and cyclinD1 expression in the EtOH-drinking DEN-treated mice, suggesting an EtOH-related promotional effect [5]. However, unlike these published models, we did not observe a significant increase in tumor incidence in the EtOH + DEN group compared to all other groups (Table 11.1). In contrast to EtOH feeding, the PF + DEN diet, composed of 47 % carbohydrates, 35 % fat and 18 % protein, increased both adenoma

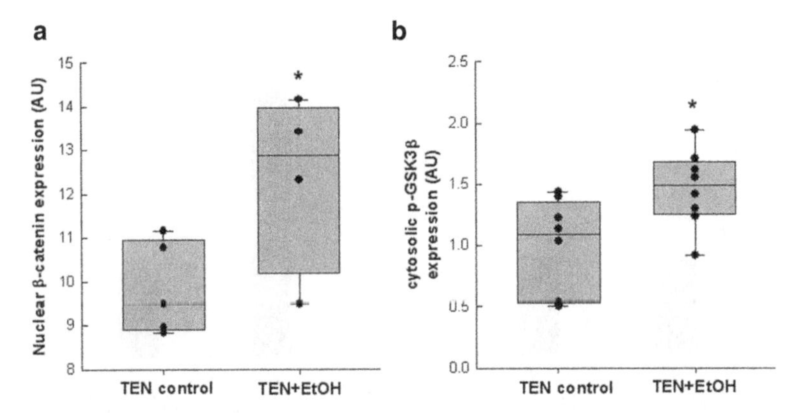

Fig. 11.2 Chronic alcohol consumption increased (a) nuclear expression of β-catenin which corresponded to (b) increased cytosolic expression of phosphorylated (Ser 21/9) GSK3β in the TEN+EtOH-treated rats compared to TEN controls. Statistical analysis between the two groups was performed by Student's *t*-test, $*p<0.05$) [14]

incidence and multiplicity when compared to the DEN+chow fed mice (Table 11.1). These results are not surprising given that disease progression of alcoholic and non-alcoholic steatohepatitis share many signaling pathways [18, 19]. Recent epidemiological studies have also shown that obesity-related diseases like metabolic syndrome, contribute to HCC risk in patient populations [17, 20]. However, tumor-promoting mechanisms between EtOH exposure and high-fat feeding may be different. In this study, tumor incidence corresponded to increased TNFα signaling associated with high-fat feeding [14]. In Fig. 11.1a, we observed no significant differences in PCNA staining between the chow+DEN and PF+DEN controls, suggesting that high-fat feeding does not increase proliferation as observed in the EtOH+DEN group.

Immunohistochemical staining of β-catenin was significantly increased in the non-tumor hepatic tissue of the EtOH+DEN-treated mice in comparison to both the chow+DEN and PF+DEN controls (Fig. 11.1b). Localization of β-catenin expression was different between diet groups. Previously we have reported that DEN alone increased membrane expression of β-catenin in the chow-fed mice, and similar to the chow+DEN group, β-catenin expression remains localized in the membrane of the PF+DEN-treated group [14]. Unlike the other groups, in the EtOH+DEN group we observed β-catenin staining in the membrane, cytosol, and in some areas, nuclear β-catenin accumulation, which is a staining pattern similar to what we observed in tumor tissue [14]. Equally as important, in a separate rodent model of alcohol liver disease [8], prolonged feeding of EtOH alone (TEN+EtOH) also increased nuclear β-catenin expression in rat hepatocytes compared to TEN controls (Fig. 11.2a). For the most part, aberrant β-catenin activity in liver tumors corresponds to mutations in the *β-catenin* gene [13]. Interestingly, the use of Phenobarbital (Pb) as a promoting agent in the DEN-induced HCC mouse model produces a subset of β-catenin positive tumors which contains mutations in exon

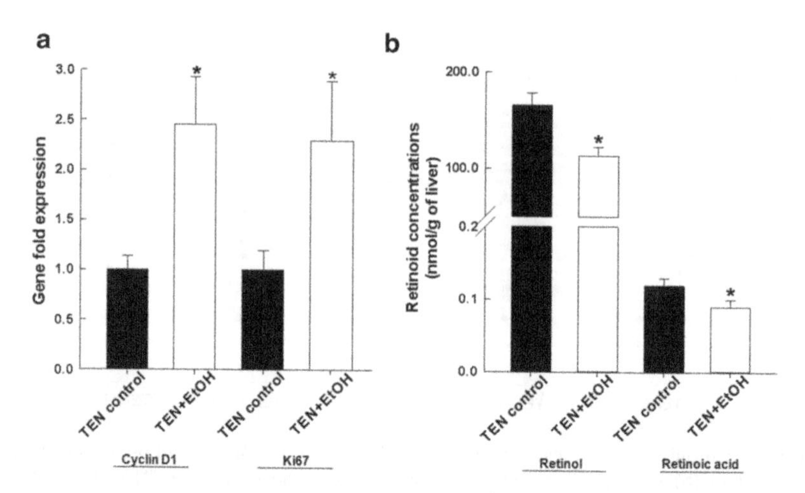

Fig. 11.3 Biochemical analysis of hepatic retinoid concentrations, (**a**) retinol and (**b**) retinoic acid in TEN + EtOH-treated rats, and compared to TEN controls. Tissue extraction and quantification of retinol and retinoic acid was performed as previously described. Statistical analysis between the two groups was performed by Student's *t*-test, $*p<0.05$ [14]

3 of *β-catenin* [21, 22] Further analysis of this DEN/Pb model has shown that the chromosomal instability which produces these mutations occurs after 32 weeks of Pb treatment [22]. In our DEN/EtOH model, EtOH feeding occurs over 16 weeks, which suggests the possibility that early exposure to EtOH may activate β-catenin through a different mechanism. In addition to increased β-catenin accumulation in the TEN + EtOH group, we also observed increased cytosolic expression of the phosphorylated (Ser 21/9) GSK3β, $p<0.05$, when compared to the TEN controls (Fig. 11.2b). This phosphorylated form is unable to complex with β-catenin, APC, axin1, thus preventing the targeting β-catenin for degradation [13]. As expected in the TEN + EtOH rats, nuclear β-catenin expression also correlated with a twofold increase in mRNA expression of proliferative markers, Ki67 and cyclin D1, (Fig. 11.3a) compared to TEN controls, $p<0.05$, and increased PCNA staining previously reported [8]. In addition, chronic EtOH feeding significantly reduced hepatic retinol and retinoic acid concentrations by 31 % and 24 % respectively, compared to the TEN control group (Fig. 11.3b). Loss of hepatic retinoid storage is a well-established event in the progression of alcoholic liver disease and HCC development [23, 24]. In rodents, Chung et al. reported that EtOH-mediated loss of hepatic retinoids increased proliferation through upregulation of cyclin D1 [6]. Our current study provides a link between retinoid depletion, upregulation of Wnt signaling, increased hepatocyte proliferation, and tumor promotion in response to EtOH feeding. It also identifies other potential targets that may participate in the promoting effects observed in our DEN-treated mice receiving EtOH, which include mediators of the Wnt signaling pathway, soluble Wnt2 and Wnt7a, transcription factors associated with proliferative signaling pathways, c-fos and c-myc and known β-catenin targets like MMP7 and WISP1 (Table 11.2).

Table 11.2 Relative expression of Wnt signaling components and β-catenin targets following chronic EtOH consumption in rats using the TEN feeding system

Gene	TEN control (fold change)	TEN + EtOH (fold change)
Wnt2	1.0 (0.12)[a]	2.6 (0.53)[b]
Wnt7a	1.0 (0.28)[a]	2.8 (0.49)[b]
c-Jun	1.0 (0.75)[a]	0.7 (0.13)[a]
c-fos	1.0 (0.19)[a]	2.0 (0.46)[b]
Tcf3	1.0 (0.86)[a]	1.3 (0.08)[b]
c-myc	1.0 (0.14)[a]	1.5 (0.17)[b]
WISP-1	1.0 (0.38)[a]	3.3 (1.20)[b]
MMP7	1.0 (0.23)[a]	2.6 (0.67)[b]

Gene expression was normalized to GAPDH; fold expression, relative to TEN controls, and reported as mean ± SE. Significance between groups was determined by Student's T-test, $p<0.05$, $a<b$ [14].

In conclusion, we have developed a new DEN-induced hepatocarcinogenesis mouse model designed to identify the EtOH-specific mechanisms involved in tumor promotion and progression. In response to EtOH feeding, DEN-treated mice have increased tumor burden compared to DEN-treated chow-fed and PF controls, which corresponded to increased hepatocyte proliferation and increased β-catenin expression and nuclear localization in non-tumor hepatic tissue. In a separate study, EtOH feeding alone reduced hepatic retinoid concentrations, increased hepatocyte proliferation, and nuclear expression of β-catenin, and increased expression of established markers of HCC progression. Clinical administration of synthetic retinoids in primary liver cancer patients has been shown to prevent tumor occurrence, and may also have a chemopreventive effect through restoration of RXRα signaling [24]. We believe future studies using this novel animal model will provide valuable information on the molecular mechanisms whereby EtOH acts as a tumor promoter, and also potentially reveal additional targets in the hepatic Wnt/β-catenin signaling system for cancer prevention studies.

References

1. Morgan TR, Mandayam S, Jamal MM (2004) Alcohol and hepatocellular carcinoma. Gastroenterology 127:S87–S96
2. Poschl G, Seitz HK (2004) Alcohol and cancer. Alcohol Alcohol 39:155–165
3. Seitz HK, Stickel F (2007) Molecular mechanisms of alcohol-mediated carcinogenesis. Nat Rev Cancer 7:599–612
4. Baumgardner JN, Shankar K, Korourian S, Badger TM, Ronis MJ (2007) Undernutrition enhances alcohol-induced hepatocyte proliferation in the liver of rats fed via total enteral nutrition. Am J Physiol Gastrointest Liver Physiol 293:G355–G364

5. Brandon-Warner E, Walling TL, Schrum LW, McKillop IH (2012) Chronic ethanol feeding accelerates hepatocellular carcinoma progression in a sex-dependent manner in a mouse model of hepatocarcinogenesis. Alcohol Clin Exp Res 36:641–653
6. Chung J, Liu C, Smith DE, Seitz HK, Russell RM, Wang XD (2001) Restoration of retinoic acid concentration suppresses ethanol-enhanced c-Jun expression and hepatocyte proliferation in rat liver. Carcinogenesis 22:1213–1219
7. Isayama F, Froh M, Yin M, Conzelmann LO, Milton RJ, McKim SE, Wheeler MD (2004) TNF alpha-induced Ras activation due to ethanol promotes hepatocyte proliferation independently of liver injury in the mouse. Hepatology 39:721–731
8. Ronis MJ, Hennings L, Stewart B, Basnakian AG, Apostolov EO, Albano E, Badger TM, Petersen DR (2011) Effects of long-term ethanol administration in a rat total enteral nutrition model of alcoholic liver disease. Am J Physiol Gastrointest Liver Physiol 300:G109–G119
9. Yip-Schneider MT, Doyle CJ, McKillop IH, Wentz SC, Brandon-Warner E, Matos JM, Sandrasegaran K, Saxena R, Hennig ME, Wu H, Waters JA, Klein PJ, Froehlich JC, Schmidt CM (2011) Alcohol induces liver neoplasia in a novel alcohol-preferring rat model. Alcohol Clin Exp Res 35:2216–2225
10. Yanagitani A, Yamada S, Yasui S, Shimomura T, Murai R, Murawaki Y, Hashiguchi K, Kanbe T, Saeki T, Ichiba M, Tanabe Y, Yoshida Y, Morino S, Kurimasa A, Usuda N, Yamazaki H, Kunisada T, Ito H, Murawaki Y, Shiota G (2004) Retinoic acid receptor alpha dominant negative form causes steatohepatitis and liver tumors in transgenic mice. Hepatology 40:366–375
11. Edamoto Y, Hara A, Biernat W, Terracciano L, Cathomas G, Riehle HM, Matsuda M, Fujii H, Scoazec JY, Ohgaki H (2003) Alterations of RB1, p53 and Wnt pathways in hepatocellular carcinomas associated with hepatitis C, hepatitis B and alcoholic liver cirrhosis. Int J Cancer 106:334–341
12. Nejak-Bowen KN, Monga SP (2011) Beta-catenin signaling, liver regeneration and hepatocellular cancer: sorting the good from the bad. Semin Cancer Biol 21:44–58
13. Villanueva A, Newell P, Chiang DY, Friedman SL, Llovet JM (2007) Genomics and signaling pathways in hepatocellular carcinoma. Semin Liver Dis 27:55–76
14. Mercer, K.E., Hennings, L., Sharma, N., Lai, K., Cleves, M.A., Wynne, R.A., Badger, T.M., and Ronis, M.J.J. (2014) Alcohol consumption promotes diethylnitrosamine-induced hepatocarcinogenesis in male mice through the activation of the Wnt/β-catenin signaling pathway. Cancer Prev Res 7:675–685
15. Mercer KE, Wynne RA, Lazarenko OP, Lumpkin CK, Hogue WR, Suva LJ, Chen JR, Mason AZ, Badger TM, Ronis MJ (2012) Vitamin D supplementation protects against bone loss associated with chronic alcohol administration in female mice. J Pharmacol Exp Ther 343:401–412
16. Cardiff RD, Anver MR, Boivin GP, Bosenberg MW, Maronpot RR, Molinolo AA, Nikitin AY, Rehg JE, Thomas GV, Russell RG, Ward JM (2006) Precancer in mice: animal models used to understand, prevent, and treat human precancers. Toxicol Pathol 34:699–707
17. Hassan MM, Hwang LY, Hatten CJ, Swaim M, Li D, Abbruzzese JL, Beasley P, Patt YZ (2002) Risk factors for hepatocellular carcinoma: synergism of alcohol with viral hepatitis and diabetes mellitus. Hepatology 36:1206–1213
18. Abdelmegeed MA, Banerjee A, Yoo SH, Jang S, Gonzalez FJ, Song BJ (2012) Critical role of cytochrome P450 2E1 (CYP2E1) in the development of high fat-induced non-alcoholic steatohepatitis. J Hepatol 57:860–866
19. Farrell GC, Larter CZ (2006) Nonalcoholic fatty liver disease: from steatosis to cirrhosis. Hepatology 43:S99–S112
20. Welzel TM, Graubard BI, Zeuzem S, El Serag HB, Davila JA, McGlynn KA (2011) Metabolic syndrome increases the risk of primary liver cancer in the United States: a study in the SEER-Medicare database. Hepatology 54:463–471
21. Aleksic K, Lackner C, Geigl JB, Schwarz M, Auer M, Ulz P, Fischer M, Trajanoski Z, Otte M, Speicher MR (2011) Evolution of genomic instability in diethylnitrosamine-induced hepatocarcinogenesis in mice. Hepatology 53:895–904

22. Stahl S, Ittrich C, Marx-Stoelting P, Kohle C, Altug-Teber O, Riess O, Bonin M, Jobst J, Kaiser S, Buchmann A, Schwarz M (2005) Genotype-phenotype relationships in hepatocellular tumors from mice and man. Hepatology 42:353–361
23. Leo, M.A, Lieber CS (1982) Hepatic vitamin A depletion in alcoholic liver injury. N. Engl. J. Med. 307, 597-601.
24. Shimizu M, Shirakami Y, Imai K, Takai K, Moriwaki H (2012) Acyclic retinoid in chemoprevention of hepatocellular carcinoma: targeting phosphorylated retinoid X receptor-alpha for prevention of liver carcinogenesis. J Carcinog 11:11

Chapter 12
Alcohol and HCV: Implications for Liver Cancer

Gyongyi Szabo, Banishree Saha, and Terence N. Bukong

Abstract Liver cancers are one of the deadliest known malignancies which are increasingly becoming a major public health problem in both developed and developing countries. Overwhelming evidence suggests a strong role of infection with hepatitis B and C virus (HBV and HCV), alcohol abuse, as well as metabolic diseases such as obesity and diabetes either individually or synergistically to cause or exacerbate the development of liver cancers. Although numerous etiologic mechanisms for liver cancer development have been advanced and well characterized, the lack of definite curative treatments means that gaps in knowledge still exist in identifying key molecular mechanisms and pathways in the pathophysiology of liver cancers. Given the limited success with current therapies and preventive strategies against liver cancer, there is an urgent need to identify new therapeutic options for patients. Targeting HCV and or alcohol-induced signal transduction, or virus–host protein interactions may offer novel therapies for liver cancer. This review summarizes current knowledge on the mechanistic development of liver cancer associated with HCV infection and alcohol abuse as well as highlights potential novel therapeutic strategies.

Keywords Hepatitis C virus • Alcohol • Cancer • Therapy • Inflammation • Molecular signaling • Immunity • Liver

12.1 Introduction

Alcohol use, hepatitis B (HBV), and hepatitis C (HCV) infection are the most common etiologies for chronic liver disease, liver fibrosis, and hepatocellular cancer (HCC) [1–4]. HCV infection affects over 170 million worldwide and leads to

---

G. Szabo, M.D., Ph.D. (✉) • B. Saha • T.N. Bukong
Department of Medicine, University of Massachusetts Medical School,
364 Plantation Street, Worcester, MA 01605, USA
e-mail: Gyongyi.Szabo@umassmed.edu; Banishree.Saha@umassmed.edu;
Terence.Bukong@umassmed.edu

V. Vasiliou et al. (eds.), *Biological Basis of Alcohol-Induced Cancer*,
Advances in Experimental Medicine and Biology 815,
DOI 10.1007/978-3-319-09614-8_12

chronic infection in 50–80 % of infected individuals [5, 6]. Excessive alcohol use is common in all parts of the world and according to WHO, there are about 140 million people who are chronic alcoholics and many of them develop alcoholic liver disease (ALD), alcoholic hepatitis, liver fibrosis, and HCC [7–9]. While HCV and alcohol independently increase the risk of liver cancer, they have a synergistic effect and liver cancer odds ratio increases to 47.8 from 8.6 by having concomitant alcohol abuse in HBV- or HCV-infected patients [10, 11]. There are multiple overlaps between alcohol use and HCV infection. About 30 % of alcoholics are infected with HCV [12] and about 70 % of people with HCV infection have heavy alcohol use history [13]. The clinical progression of ALD is accelerated in individuals with HCV infection and alcohol use is an independent risk factor for HCV progression [14, 15]. The mechanisms of the synergistic effects of alcohol and HCV on liver damage, fibrosis, and HCC are not fully understood. It has been found that even ongoing moderate alcohol consumption increases inflammatory activity in the HCV-infected liver and interferon (IFN)-based therapies are ineffective in individuals with active alcohol use [16, 17]. There are several common targets of alcohol and HCV in the pathophysiology of HCC such as immune surveillance, host factors, cell proliferation, and regeneration [18]. It is well known that chronic alcohol consumption increase gut bacteria flora growth as well as increase guts permeability that allows the translocation of gut bacteria to the liver compromising the immune barrier functions of the gut [19–21]. The transfer of bacterial products leads to the activation the hepatic immune cells driving inflammatory cytokine production [21–24]. Additionally, increased alcohol in the liver leads to oxidative stress induced by alcohol and its metabolites [22–24]. In association with HCV infection, which can independently induce liver injury and inflammation, alcohol use creates a dangerous mix which can additionally induce and sustain liver injury with an increased risk of liver cancer development [11, 25].

In this communication, we will review some of the recent advances in the understanding of the combined effects of alcohol and HCV on the liver as it relates to innate immunity, HCV replication, and HCC (Fig. 12.1).

12.2 Alcohol, HCV and Innate Immunity

The immune response is a common target of both alcohol and the HCV. HCV is a typical tissue-tropic virus that replicates in hepatocytes and triggers innate and adaptive immune responses [26, 27]. Recognition of HCV by innate immune cells such as monocytes and macrophages leads to inflammatory cytokine induction and type I IFN production [28–30]. Recognition of HCV by dendritic cells and their antigen presentation to T cells is pivotal in a robust anti-HCV immune response that leads to viral clearance. In effective resolution of acute HCV, DCs induce a strong CD4 Th1-type T cell activation that is HCV specific to lead to viral elimination. In contrast, during chronic HCV infection, DC functions are impaired and T cell activation is predominantly Th2 type and not antigen specific [31–33].

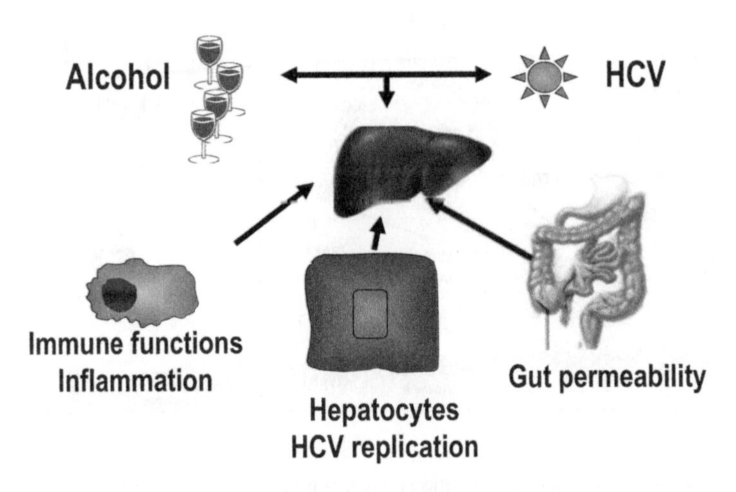

Fig. 12.1 Alcohol and HCV in liver disease: common sites of action

Importantly, alcohol also impairs immune responses. Acute alcohol binge results in attenuation of inflammatory cytokine induction and attenuation of Type I IFN induction [34–36]. Chronic alcohol equally impairs immunity and increases non-specific inflammatory responses and decreases virus-induced Type I IFN production [35, 37].

These immune abnormalities in the local environment of alcohol exposed, HCV-infected liver can then contribute to a detrimental imbalance of innate immune responses that otherwise are critical in HCV elimination. Moreover, the chronic inflammatory signals and inflammatory cell presence and activation may promote HCC development.

12.2.1 Antiviral Immunity in HCV Infection and Alcohol Use

HCV is a single stranded RNA virus that is recognized by pattern recognition receptors (PRRs) on the host cells, that induce Type I IFNs and inflammatory mediators [29, 35]. In the HCV, several vital components represent "dangerous signals" for the host immune system (Fig. 12.2). The ssRNA of HCV is recognized by TLR8 while during HCV replication, the dsRNA is sensed by the host TLR3, RIG-I, and perhaps Mda-5. All of these PRRs induce type I IFNs, particularly IFNß in hepatocytes. In immune cells, HCV-induced activation of these PRRs can also trigger IFNα and IFNß (type I IFNs), IFN-gamma (γ) (type II), and IFN-lambdas (Type III IFNs) [38–40]. HCV, however, has several ways to undermine these host immune

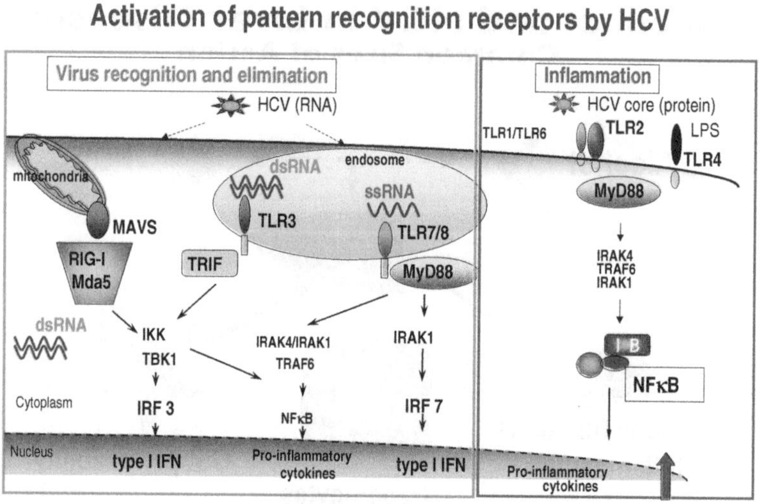

Activation of pattern recognition receptors by HCV

Fig. 12.2 Activation of pattern recognition receptors by HCV

alert mechanisms. For example, the HCV NS3-4a serine protease cleaves the host adapter molecules that are critical in TLR3 and RIG-I activation, respectively [41, 42]. Current therapies with protease inhibitors target this activity and now are in clinical practice for treatment of HCV [43–45].

Antiviral immunity is compromised by both acute and chronic alcohol use [46–49]. In a binge drinking model, blood mononuclear cells of human volunteers showed impaired type I IFN production in response to stimulation with viral danger signals via TLR8 or with bacterial danger molecule (LPS) that is recognized by TLR4 [35]. More important, similar defects were seen after chronic alcohol treatment of immune cells suggesting that alcohol, whether acute or chronic, impairs Type I IFN induction [35, 50, 51]. The combined negative effects of alcohol and HCV were also seen on IFN induction in hepatoma cells [52].

12.2.2 Mechanisms of Inflammation

HCV components, such as the core and NS3 proteins can be recognized by TLR2 and induce downstream activation of pro-inflammatory cascades in immune cells [53–56]. This has direct clinical relevance because in the HCV-infected liver there is increased infiltration and activity of innate immune cells including tissue macrophages and Kupffer cells that produce inflammatory mediators [57–60]. Even in peripheral blood monocytes that can serve as precursors of liver macrophages, there is increased in vivo activation and hyper-inflammatory response in patients with chronic HCV infection [60, 61]. It has been shown that the inflammatory cell and cytokine milieu contribute to triggering and promoting liver cirrhosis and cancerous transformation in hepatocytes [62, 63].

Chronic alcohol abuse is another trigger of inflammation. Activation of the inflammatory cascade, increased liver, and circulating pro-inflammatory cytokine levels are the characteristics of ALD and alcoholic steatohepatitis [64–66]. Several key elements have been identified in the pathomechanisms of ALD including LPS/TLR4-mediated activation of inflammation, upregulation of the inflammatory cytokine cascade (TNFα, IL-1ß, MCP-1, IL-6), that are linked to an in vivo Kupffer cell activation and sensitization to LPS. There is evidence for increased circulating levels of LPS, a gram negative bacterial innate immune danger signal in ALD in humans as well as in animal models [34, 67, 68]. Interestingly, serum levels of LPS are also increased in patients with treatment-naïve chronic HCV infection [60]. In vitro experiments demonstrated that chronic alcohol exposure augments TLR-induced TNFα production in human monocytes [35].

12.3 Cancer Surveillance in HCV Infection and Alcohol Use

12.3.1 NK Cells

Host immunity serves as a "controller" to prevent the development of cancer by recognizing and eliminating cells with cancerous malformation [69, 70]. However, alcohol is a major inhibitor of many of the key functions of the innate immune system that are pivotal in this process. For example, alcohol interferes with function of natural killer cells (NK cells) [71–73]. Deregulated activation of NK cells through a reduction of systemic β-endorphins production can also occur with alcohol consumption leading to increased hepatocyte damage [74]. Alternatively, recent reports have suggested a critical inhibitory role of alcohol on NK cells to effectively carry out immune surveillance. For example, alcohol consumption can suppress the effective function of NK cells in the liver by decreasing the expression of TRAIL, IFN-γ, and NKG2D—the activating NK cell receptor [75, 76]. Additionally, alcohol use blocks NK cell release from the bone marrow and significantly induces splenic NK cell apoptosis [77] which can compromise viral clearance and exacerbate disease progression during HCV infection. Finally, alcohol has also been shown to increase serum corticosterone levels which can impair the function of NK cells [78] and enhance HCV disease progression. Given the important role of NK cell in viral clearance, patients with a genetic predisposition for lower NK cell function usually progress to chronic HCV infection [79].

During HCV infection of hepatocytes, NK cells are rapidly activated providing an important role in the resolution of early infection [80, 81]. However, HCV infection can also interfere with NK cell activation and inhibit NK cell IFNγ production [82, 83]. Impaired activity of NK cells has been proposed as a mechanism contributing to HCV persistence. HCV chronic infection severely affects NK phenotype and function. It exhibits a polarized phenotype with increased cytotoxicity [84, 85].

In addition to their antiviral function, NK cells can suppress the development of fibrosis by directly killing activated myofibroblasts (MFB) and IFN-α production which induces cell cycle arrest and apoptosis of MFB [86, 87]. Chronic ethanol feeding diminished the inhibitory effect of NK/IFN-γ in liver fibrosis, which has been suggested to be an important mechanism contributing to alcohol acceleration of liver fibrosis in patients with chronic HCV infection [75, 87, 88].

12.3.2 Antigen Presenting Cells in HCV Infection

In addition to NK cells, antigen presenting cells are also important in recognition of injured and abnormal cells. However, both alcohol and HCV infection alone and together attenuate the capacity of myeloid dendritic cells to fulfill their antigen presenting and T cell stimulatory function [89–92]. Some of the mechanistic aspects of alcohol- and HCV-induced impairment of DC phenotype and function are similar and additive. For example, both alcohol and HCV infection results in impaired capacity of monocyte-derived DCs to reach full maturation after an external stimulation [91, 93]. This involves decreased expression of co-stimulatory molecules, decreased IL-12 and increased IL-10 production [91, 93]. In HCV infection, there is also overexpression of the PD1 ligand that is an inhibitory cell surface molecule for T cells [94–97]. Other types of the dendritic cell population are also negatively impacted in HCV infection. Plasmacytoid DCs, that are the major producers of IFNα, have reduced numbers in the periphery and most important, decreased capacity to produce IFNα [30, 98]. The mechanisms for this involve some of the inflammatory mediators produced by monocytes such as IL-10 and TNFα [30, 99, 100]. Most recently, IFN-lambda (IL-18 and IL-29) production was found in a unique DC population, the M2 DCs [40]. In HCV infection, however, this population has impaired production of IFN-lambda [40]. The biological and clinical consequence of this finding is under active investigation, however, one of the effects of IFN-lambda is to trigger and amplify production of other interferons [40]. Thus, decreased IFN-lambda in chronic HCV infection can have broad and magnified effects.

12.4 Alcohol and HCV Replication, Role of Micro-RNAs

12.4.1 Oxidative Stress

Despite the substantial epidemiologic data on HCV, alcohol and liver disease, the molecular mechanisms modulating chronic liver disease development and progress to fibrosis, cirrhosis, and even HCC are not fully known. Recent in vitro and mice studies expressing HCV proteins have been instrumental in deciphering some of the mechanisms underlying the synergism of alcohol abuse and HCV infection in liver disease [101–106]. The expression of the HCV core protein in mice caused significant production of reactive oxygen species (ROS) which appears to be responsible for mitochondrial DNA damage during HCV infection [105–108]. These observations clearly demonstrated a mechanistic insight by which the HCV core protein through synergistic significant oxidative stress induction can exacerbate liver injury in alcohol-fed core transgenic mice. Additionally, chronic alcohol use and HCV infection independently increase TLR2 and TLR4 expression in hepatocytes, Kupffer cells, and peripheral monocytes in both in vitro and in vivo studies [25, 35, 109–111]. Increased synergistic TLR expression during HCV infection and alcohol will mechanistically enhance TLR-mediated signal activations resulting in liver disease progression ultimately resulting in HCC [25, 110]. Mechanistic evidence has been demonstrated by chronic ethanol feeding in mice expressing the HCV NS5A protein in a hepatocyte-specific manner were mice fed with alcohol were prone to liver tumor development in a TLR4 dependent manner [10]. The tumorigenic effect of alcohol in a TLR4 dependent manner in hepatic HCV NS5A transgenic mice was also associated with the upregulation of the NANOG usually associated with HCC [112]. NANOG is transcription factor essential for self-renewal of stem cells [113, 114] and is associated with tumor malignancy and metastasis [115].

Chronic alcohol and HCV infection independently or in association also induce numerous biochemical and metabolic changes in hepatocytes that could directly or indirect initiate or potentiate HCC. Increased production of ROS and deposition of iron and their downstream effect have been advanced to at least in part account for the synergism of alcohol and HCV in mediating liver disease including HCC. The generation of ROS has been associated with both ALD and HCC. While there may be differences in their generation their downstream effects are quite similar. HCV core [116] and NS5A [117] have been shown to induce the generation of ROS via mitochondrial damage and calcium release. ROS also play an important role in alcohol-induced liver injury and in hepatocarcinogenesis [118]. In the liver, alcohol is metabolized by alcohol dehydrogenase and by cytochrome P450 2E1 (CYP2E1) to acetaldehyde—a carcinogen in mouse studies. Chronic alcohol use increases alcohol dehydrogenase and CYP2E1 which produces ROS in the presence or absence of alcohol. We and others have demonstrated the importance of CYP2E1 in enhancing HCV replication during alcohol exposure in CYP2E1-expressing hepatoma cells [119, 120]. Human studies on genetic polymorphisms alcohol dehydrogenase and CYP2E1 which metabolize alcohol in the development of HCC have provided conflicting findings. A Japanese study demonstrated an association between alcohol dehydrogenase and CYP2E1 gene polymorphism with the risk of HCC in humans while earlier a Korean study found no such association [121–123].

12.4.2 Modulation of HCV Replication by Alcohol Via HSP90

Alcohol abuse has been shown by numerous studies to exacerbate disease outcome during HCV infection. Recent studies have increasingly identified Heat-shock protein 90 (HSP90) as an important mediator of HCV disease progression associated with alcohol abuse. HSP90 is an evolutionary conserved chaperone protein that plays a key role with other co-chaperone proteins in assisting newly synthesized proteins to fold properly. They also assist in stabilizing proteins under stress and play a vital role in modulating protein degradation. HSP90 is normally induced during cellular stress conditions and numerous studies have demonstrated the diverse functions of HSP90 including those related to viral-pathogenesis, transcription regulation, neo-vascularization, and cancer development/metastasis [124–132]. It is estimated that over 16–20 % of cancers worldwide are caused by viruses [133, 134] including HCV. The molecular mechanisms by which HCV infection induces cancer development has been linked the expression viral oncogenes as well as the capacity for virus to induces sustained inflammation that promotes neoplastic transformation of infected cells [135]. While some oncogenic viruses have vaccines to prevent human infections, there is no vaccine against HCV. In this regard, it is therefore imperative to better understand the molecular mechanism by which HCV alone and in association with exacerbating factors like alcohol can promote cancer development. We and others have demonstrated that HSP90 increases during ALD, HCV infection, and even HCC development [136–141]. Mandrekar et al. recently demonstrated that alcohol-induced hepatic stress modulates alcohol liver disease in an HSP90 dependent manner [136, 138]. Additionally, we and others have demonstrated an important role of HSP90 modulating HCV replication with and without alcohol exposure [137, 142–146]. During the HCV life cycle, the HCV RNA is translated to form a single viral polyprotein. This HCV viral polyprotein is cleaved by cellular and viral proteins to form its nine functional viral proteins. HSP90 has been shown to play a critical role in the cleavage and activation of HCV NS3 and NS2 proteins [147]. The mechanism by which HSP90 induces such cleavage is yet to be identified but reports have speculated that indirect interaction between HSP90 and viral proteins might protect them from proteolysis [142, 147]. HSP90 also interacts with HCV NS5A protein and enhances viral replication in association with other host proteins [144, 148] which can promote cancer development.

In addition to its role in modulating HCV replication and viral protein stability, HSP90 has been shown to enhance the proliferative potential of malignant cells and even enable neoplastic cells to escape cell death ultimately enhancing neoplastic development [149, 150]. HSP90 also plays an important role in maintaining the integrity of NF-κB and Akt which are two main cell survival pathways that attenuate the anticancer efficacy of some cancer therapeutic drugs [151]. Inhibition of HSP90 has been shown to improve the efficacy of anticancer agents [152]. Given the important role played by HSP90 in ALD, HCV, and even HCC development, our findings and those by other groups show a great promise for HSP90 inhibitors as novel pharmacological agents for ALD, HCV, and anticancer therapy [136–138, 143, 153, 154]. Treatments against cancer targeting HSP90 are of prime importance and likely to be more successful since HSP90 is of great importance in maintaining the integrity, conformation, stability, and functional properties of important oncogenic proteins.

12.4.3 Role of Micro-RNA 122

Previous studies indicated that alcohol use can increase serum HCV levels [155, 156]; however, it remained unclear whether this was related to increased HCV replication. In a recent study, we demonstrated that alcohol increases HCV replication in vitro and identified a critical role of micro-RNA-122 (miRNA-122) in the process [119, 137] (Fig. 12.3), however a meta-analysis study showed that alcohol had no effect on HCV replication [157]. MiRNAs are non-coding RNAs that are increasingly being recognized as major players in the regulation of most physiologic and pathological processes. In the liver, miRNA-122 has the highest abundance in hepatocytes compared to other cell types [158, 159]. Furthermore, miRNA-122 is a host factor that has been shown to modulate the HCV replication machinery by binding to the HCV 5′-UTR [160–162]. Recently we discovered that alcohol increases miRNA-122 levels in hepatoma cells and through this mechanism, increases HCV replication [119]. Additionally, miRNA-122 regulation of HCV has been shown to be enhanced by the RISC-complex molecules Argonaute 2 (Ago2) [158] and GW182 which is increased during alcohol exposure in cultured human hepatic cell line Huh7.5 [137]. While RISC-complex proteins have been shown to enhance HCV replication in association with miRNA-122, a recent report indicates that Ago2 might be dispensable when miRNA-122 is overexpressed [163]. Interestingly miRNA-122 expression is decreased in the liver in HCC [164, 165]. In a mouse model of ALD, we found that chronic alcohol feeding in mice results in decreased miRNA-122 levels in the liver but increased levels in the serum [166]. Because serum miRNAs are highly stable, they have been proposed as potential biomarkers. Indeed, increased levels of miRNA-122 appear to correlate with the extent of liver damage and serum ALT levels in different mouse models of liver disease [166–170].

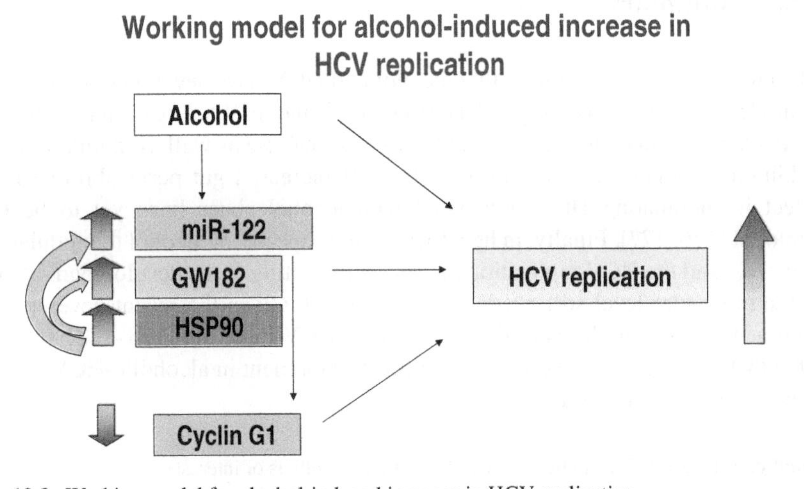

Fig. 12.3 Working model for alcohol-induced increase in HCV replication

Table 12.1 Characteristic micro-RNAs in HCC

Liver reduced levels of:	Serum reduced levels of:
miR-24a	miR-21
miR-26a	miR-16
miR-15a/b	miR-199a
miR-150	miR-122
miR-195	miR-22
miR-122	
miR-20 family	
miR-124	
let-7 family members	

MiRNA-122 also regulates genes involved in lipid metabolism and in cell cycle. Recent findings have described the phenotype of miRNA-122 loss in hepatic lipid modulation [171, 172]. These findings demonstrated that miRNA-122 loss was associated with modulation of multiple hepatic pathways involved in fat metabolism, tumor suppression, inflammation, and even hepatic fibrosis. While loss of miRNA-122 in these studies induces steatohepatitis, it is unknown whether exogenous miRNA-122 might ameliorate fatty liver disease induced by diet or virus infection. Additionally, the tumor-suppressive function of miRNA-122 illustrated by both groups suggests that the restoration of miRNA-122 expression may be of benefit against HCC and even enhance response to therapy as recently proposed [173]. While this might prove beneficial for some HCC patients, such applications might need to carefully assess findings from current trails using miRNA-122 inhibitors to treat HCV infection [174, 175]. In addition, several other miRNAs are also dysregulated in HCC that may represent biomarkers or therapeutic targets (Table 12.1).

12.5 Summary

The current literature suggests that alcohol and HCV have several common targets in modeling liver disease (Fig. 12.4). Both affect and disable key immune functions particularly in innate immunity in antiviral host defense as well as in inflammation. Additional common element on the effect of increased gut permeability and the effect in modulating HCV associated with alcohol abuse have yet to be fully explored [176, 177]. Finally, in hepatocytes, the capacity of alcohol to modulate the cell cycle and the HCV replication process contributing to the development of HCC at the molecular level still needs to be addressed. Given the recent developments with new and potent drugs in the treatment of HCV infection [178–182], it is possible that the negative trend in liver cancer development in alcoholic-HCV patients may be significantly reduced.

Conflict of Interest The authors declare there are no conflicts of interest.

Grant Acknowledgment 5R37AA014372-10

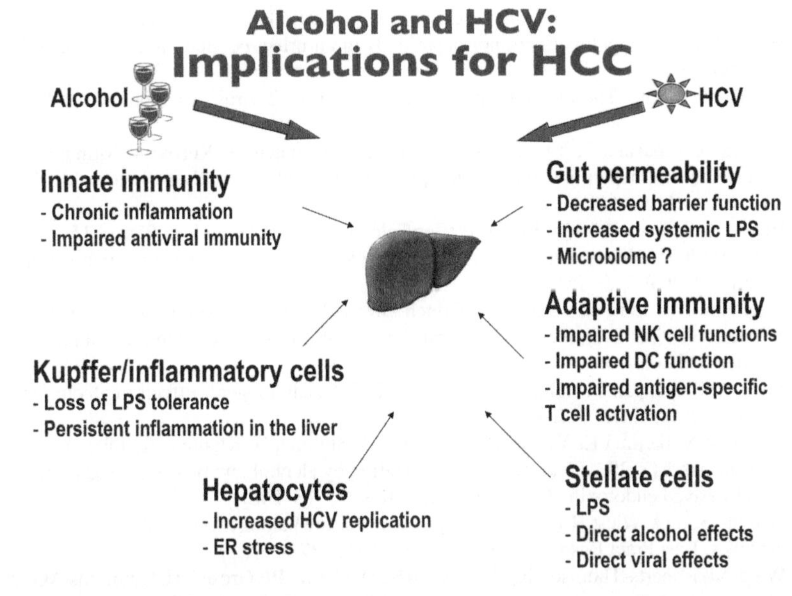

Fig. 12.4 Alcohol and HCV: implications for HCC

References

1. Davis GL, Dempster J, Meler JD, Orr DW, Walberg MW, Brown B, Berger BD, O'connor JK, Goldstein RM (2008) Hepatocellular carcinoma: management of an increasingly common problem. Proc (Bayl Univ Med Cent) 21:266–280
2. Morgan TR, Mandayam S, Jamal MM (2004) Alcohol and hepatocellular carcinoma. Gastroenterology 127:S87–S96
3. Sanyal AJ, Yoon SK, Lencioni R (2010) The etiology of hepatocellular carcinoma and consequences for treatment. Oncologist 15(Suppl 4):14–22
4. Tornesello ML, Buonaguro L, Tatangelo F, Botti G, Izzo F, Buonaguro FM (2013) Mutations in TP53, CTNNB1 and PIK3CA genes in hepatocellular carcinoma associated with hepatitis B and hepatitis C virus infections. Genomics 102:74–83
5. Levrero M (2006) Viral hepatitis and liver cancer: the case of hepatitis C. Oncogene 25:3834–3847
6. Bartenschlager R, Penin F, Lohmann V, Andre P (2011) Assembly of infectious hepatitis C virus particles. Trends Microbiol 19:95–103
7. O'Shea RS, Dasarathy S, McCullough AJ, Practice Guideline Committee of the American Association for the Study of Liver Diseases, Practice Parameters Committee of the American College of Gastroenterology (2010) Alcoholic liver disease. Hepatology 51:307–328
8. Frazier TH, Stocker AM, Kershner NA, Marsano LS, McClain CJ (2011) Treatment of alcoholic liver disease. Therap Adv Gastroenterol 4:63–81
9. Szabo G, Lippai D (2012) Molecular hepatic carcinogenesis: impact of inflammation. Dig Dis 30:243–248
10. Machida K (2010) TLRs, alcohol, HCV, and tumorigenesis. Gastroenterol Res Pract 2010:518674
11. Koike K, Tsutsumi T, Miyoshi H, Shinzawa S, Shintani Y, Fujie H, Yotsuyanagi H, Moriya K (2008) Molecular basis for the synergy between alcohol and hepatitis C virus in hepatocarcinogenesis. J Gastroenterol Hepatol 23(Suppl 1):S87–S91

12. Rosman AS, Waraich A, Galvin K, Casiano J, Paronetto F, Lieber CS (1996) Alcoholism is associated with hepatitis C but not hepatitis B in an urban population. Am J Gastroenterol 91:498–505
13. Schiff ER (1999) The alcoholic patient with hepatitis C virus infection. Am J Med 107: 95S–99S
14. Tanaka T, Yabusako T, Yamashita T, Kondo K, Nishiguchi S, Kuroki T, Monna T (2000) Contribution of hepatitis C virus to the progression of alcoholic liver disease. Alcohol Clin Exp Res 24:112S–116S
15. Regev A, Jeffers LJ (1999) Hepatitis C and alcohol. Alcohol Clin Exp Res 23:1543–1551
16. Day CP (2001) Heavy drinking greatly increases the risk of cirrhosis in patients with HCV hepatitis. Gut 49:750–751
17. Neumann AU, Lam NP, Dahari H, Gretch DR, Wiley TE, Layden TJ, Perelson AS (1998) Hepatitis C viral dynamics in vivo and the antiviral efficacy of interferon-alpha therapy. Science 282:103–107
18. Stickel F, Schuppan D, Hahn EG, Seitz HK (2002) Cocarcinogenic effects of alcohol in hepatocarcinogenesis. Gut 51:132–139
19. Enomoto N, Ikejima K, Yamashina S, Hirose M, Shimizu H, Kitamura T, Takei Y, Sato And N, Thurman RG (2001) Kupffer cell sensitization by alcohol involves increased permeability to gut-derived endotoxin. Alcohol Clin Exp Res 25:51S–54S
20. Garcia-Tsao G, Wiest R (2004) Gut microflora in the pathogenesis of the complications of cirrhosis. Best Pract Res Clin Gastroenterol 18:353–372
21. Wigg AJ, Roberts-Thomson IC, Dymock RB, McCarthy PJ, Grose RH, Cummins AG (2001) The role of small intestinal bacterial overgrowth, intestinal permeability, endotoxaemia, and tumour necrosis factor alpha in the pathogenesis of non-alcoholic steatohepatitis. Gut 48: 206–211
22. Purohit V, Bode JC, Bode C, Brenner DA, Choudhry MA, Hamilton F, Kang YJ, Keshavarzian A, Rao R, Sartor RB, Swanson C, Turner JR (2008) Alcohol, intestinal bacterial growth, intestinal permeability to endotoxin, and medical consequences: summary of a symposium. Alcohol 42:349–361
23. Szabo G, Bala S (2010) Alcoholic liver disease and the gut-liver axis. World J Gastroenterol 16:1321–1329
24. Petrasek J, Mandrekar P, Szabo G (2010) Toll-like receptors in the pathogenesis of alcoholic liver disease. Gastroenterol Res Pract 2010:pii: 710381
25. Szabo G, Aloman C, Polyak SJ, Weinman SA, Wands J, Zakhari S (2006) Hepatitis C infection and alcohol use: a dangerous mix for the liver and antiviral immunity. Alcohol Clin Exp Res 30:709–719
26. Protzer U, Maini MK, Knolle PA (2012) Living in the liver: hepatic infections. Nat Rev Immunol 12:201–213
27. Horner SM, Gale M Jr (2013) Regulation of hepatic innate immunity by hepatitis C virus. Nat Med 19:879–888
28. Heydtmann M (2009) Macrophages in hepatitis B and hepatitis C virus infections. J Virol 83:2796–2802
29. Liu BS, Janssen HL, Boonstra A (2012) Type I and III interferons enhance IL-10R expression on human monocytes and macrophages, resulting in IL-10-mediated suppression of TLR-induced IL-12. Eur J Immunol 42:2431–2440
30. Dolganiuc A, Chang S, Kodys K, Mandrekar P, Bakis G, Cormier M, Szabo G (2006) Hepatitis C virus (HCV) core protein-induced, monocyte-mediated mechanisms of reduced IFN-alpha and plasmacytoid dendritic cell loss in chronic HCV infection. J Immunol 177:6758–6768
31. Fan Z, Huang XL, Kalinski P, Young S, Rinaldo CR Jr (2007) Dendritic cell function during chronic hepatitis C virus and human immunodeficiency virus type 1 infection. Clin Vaccine Immunol 14:1127–1137
32. MacDonald AJ, Semper AE, Libri NA, Rosenberg WM (2007) Monocyte-derived dendritic cell function in chronic hepatitis C is impaired at physiological numbers of dendritic cells. Clin Exp Immunol 148:494–500

33. Kanto T, Inoue M, Miyatake H, Sato A, Sakakibara M, Yakushijin T, Oki C, Itose I, Hiramatsu N, Takehara T, Kasahara A, Hayashi N (2004) Reduced numbers and impaired ability of myeloid and plasmacytoid dendritic cells to polarize T helper cells in chronic hepatitis C virus infection. J Infect Dis 190:1919–1926
34. Bala S, Tang A, Catalano D, Petrasek J, Taha O, Kodys K, Szabo G (2012) Induction of Bcl-3 by acute binge alcohol results in toll-like receptor 4/LPS tolerance. J Leukoc Biol 92:611–620
35. Pang M, Bala S, Kodys K, Catalano D, Szabo G (2011) Inhibition of TLR8- and TLR4-induced Type I IFN induction by alcohol is different from its effects on inflammatory cytokine production in monocytes. BMC Immunol 12:55
36. Norkina O, Dolganiuc A, Catalano D, Kodys K, Mandrekar P, Syed A, Efros M, Szabo G (2008) Acute alcohol intake induces SOCS1 and SOCS3 and inhibits cytokine-induced STAT1 and STAT3 signaling in human monocytes. Alcohol Clin Exp Res 32:1565–1573
37. Ye L, Wang S, Wang X, Zhou Y, Li J, Persidsky Y, Ho W (2010) Alcohol impairs interferon signaling and enhances full cycle hepatitis C virus JFH-1 infection of human hepatocytes. Drug Alcohol Depend 112:107–116
38. Stone AE, Giugliano S, Schnell G, Cheng L, Leahy KF, Golden-Mason L, Gale M Jr, Rosen HR (2013) Hepatitis C virus pathogen associated molecular pattern (PAMP) triggers production of lambda-interferons by human plasmacytoid dendritic cells. PLoS Pathog 9: e1003316
39. Liu HM, Gale M (2010) Hepatitis C virus evasion from RIG-I-dependent hepatic innate immunity. Gastroenterol Res Pract 2010:548390
40. Zhang S, Kodys K, Li K, Szabo G (2013) Human type 2 myeloid dendritic cells produce interferon-lambda and amplify interferon-alpha in response to hepatitis C virus infection. Gastroenterology 144(414–425):e7
41. Nitta S, Sakamoto N, Nakagawa M, Kakinuma S, Mishima K, Kusano-Kitazume A, Kiyohashi K, Murakawa M, Nishimura-Sakurai Y, Azuma S, Tasaka-Fujita M, Asahina Y, Yoneyama M, Fujita T, Watanabe M (2013) Hepatitis C virus NS4B protein targets STING and abrogates RIG-I-mediated type I interferon-dependent innate immunity. Hepatology 57:46–58
42. Li K, Foy E, Ferreon JC, Nakamura M, Ferreon AC, Ikeda M, Ray SC, Gale M Jr, Lemon SM (2005) Immune evasion by hepatitis C virus NS3/4A protease-mediated cleavage of the Toll-like receptor 3 adaptor protein TRIF. Proc Natl Acad Sci U S A 102:2992–2997
43. Pawlotsky JM (2013) Treatment of chronic hepatitis C: current and future. Curr Top Microbiol Immunol 369:321–342
44. Shah N, Pierce T, Kowdley KV (2013) Review of direct-acting antiviral agents for the treatment of chronic hepatitis C. Expert Opin Investig Drugs 22:1107–1121
45. Chou R, Hartung D, Rahman B, Wasson N, Cottrell EB, Fu R (2013) Comparative effectiveness of antiviral treatment for hepatitis C virus infection in adults: a systematic review. Ann Intern Med 158:114–123
46. Singal AK, Anand BS (2007) Mechanisms of synergy between alcohol and hepatitis C virus. J Clin Gastroenterol 41:761–772
47. Siu L, Foont J, Wands JR (2009) Hepatitis C virus and alcohol. Semin Liver Dis 29:188–199
48. Shuper PA, Neuman M, Kanteres F, Baliunas D, Joharchi N, Rehm J (2010) Causal considerations on alcohol and HIV/AIDS—a systematic review. Alcohol Alcohol 45:159–166
49. Molina PE, Happel KI, Zhang P, Kolls JK, Nelson S (2010) Focus on: alcohol and the immune system. Alcohol Res Health 33:97–108
50. Jerrells TR, Pavlik JA, Devasure J, Vidlak D, Costello A, Strachota JM, Wyatt TA (2007) Association of chronic alcohol consumption and increased susceptibility to and pathogenic effects of pulmonary infection with respiratory syncytial virus in mice. Alcohol 41:357–369
51. Le Strat Y, Grant BF, Ramoz N, Gorwood P (2010) A new definition of early age at onset in alcohol dependence. Drug Alcohol Depend 108:43–48
52. Plumlee CR, Lazaro CA, Fausto N, Polyak SJ (2005) Effect of ethanol on innate antiviral pathways and HCV replication in human liver cells. Virol J 2:89

53. Imran M, Waheed Y, Manzoor S, Bilal M, Ashraf W, Ali M, Ashraf M (2012) Interaction of Hepatitis C virus proteins with pattern recognition receptors. Virol J 9:126
54. Chung H, Watanabe T, Kudo M, Chiba T (2010) Hepatitis C virus core protein induces homotolerance and cross-tolerance to Toll-like receptor ligands by activation of Toll-like receptor 2. J Infect Dis 202:853–861
55. Chang S, Dolganiuc A, Szabo G (2007) Toll-like receptors 1 and 6 are involved in TLR2-mediated macrophage activation by hepatitis C virus core and NS3 proteins. J Leukoc Biol 82:479–487
56. Dolganiuc A, Oak S, Kodys K, Golenbock DT, Finberg RW, Kurt-Jones E, Szabo G (2004) Hepatitis C core and nonstructural 3 proteins trigger toll-like receptor 2-mediated pathways and inflammatory activation. Gastroenterology 127:1513–1524
57. Shrivastava S, Mukherjee A, Ray R, Ray RB (2013) Hepatitis C virus induces interleukin-1beta (IL-1beta)/IL-18 in circulatory and resident liver macrophages. J Virol 87:12284–12290
58. Negash AA, Ramos HJ, Crochet N, Lau DT, Doehle B, Papic N, Delker DA, Jo J, Bertoletti A, Hagedorn CH, Gale M Jr (2013) IL-1beta production through the NLRP3 inflammasome by hepatic macrophages links hepatitis C virus infection with liver inflammation and disease. PLoS Pathog 9:e1003330
59. Hosomura N, Kono H, Tsuchiya M, Ishii K, Ogiku M, Matsuda M, Fujii H (2011) HCV-related proteins activate Kupffer cells isolated from human liver tissues. Dig Dis Sci 56:1057–1064
60. Dolganiuc A, Norkina O, Kodys K, Catalano D, Bakis G, Marshall C, Mandrekar P, Szabo G (2007) Viral and host factors induce macrophage activation and loss of toll-like receptor tolerance in chronic HCV infection. Gastroenterology 133:1627–1636
61. Bala S, Tilahun Y, Taha O, Alao H, Kodys K, Catalano D, Szabo G (2012) Increased microRNA-155 expression in the serum and peripheral monocytes in chronic HCV infection. J Transl Med 10:151
62. Coulouarn C, Corlu A, Glaise D, Guenon I, Thorgeirsson SS, Clement B (2012) Hepatocyte-stellate cell cross-talk in the liver engenders a permissive inflammatory microenvironment that drives progression in hepatocellular carcinoma. Cancer Res 72:2533–2542
63. Brownell J, Polyak SJ (2013) Molecular pathways: hepatitis C virus, CXCL10, and the inflammatory road to liver cancer. Clin Cancer Res 19:1347–1352
64. Wang HJ, Zakhari S, Jung MK (2010) Alcohol, inflammation, and gut-liver-brain interactions in tissue damage and disease development. World J Gastroenterol 16:1304–1313
65. Szabo G (2010) The 40th anniversary of the National Institute on Alcoholism and Alcohol Abuse: the impact on liver disease. Hepatology 52:10–12
66. Szabo G, Mandrekar P (2010) Focus on: alcohol and the liver. Alcohol Res Health 33:87–96
67. Inokuchi S, Tsukamoto H, Park E, Liu ZX, Brenner DA, Seki E (2011) Toll-like receptor 4 mediates alcohol-induced steatohepatitis through bone marrow-derived and endogenous liver cells in mice. Alcohol Clin Exp Res 35:1509–1518
68. Schafer C, Parlesak A, Schutt C, Bode JC, Bode C (2002) Concentrations of lipopolysaccharide-binding protein, bactericidal/permeability-increasing protein, soluble CD14 and plasma lipids in relation to endotoxaemia in patients with alcoholic liver disease. Alcohol Alcohol 37:81–86
69. Blair GE, Cook GP (2008) Cancer and the immune system: an overview. Oncogene 27:5868
70. Whiteside TL (2006) Immune suppression in cancer: effects on immune cells, mechanisms and future therapeutic intervention. Semin Cancer Biol 16:3–15
71. Zhang T, Guo CJ, Douglas SD, Metzger DS, O'brien CP, Li Y, Wang YJ, Wang X, Ho WZ (2005) Alcohol suppresses IL-2-induced CC chemokine production by natural killer cells. Alcohol Clin Exp Res 29:1559–1567
72. Ben-Eliyahu S, Page GG, Yirmiya R, Taylor AN (1996) Acute alcohol intoxication suppresses natural killer cell activity and promotes tumor metastasis. Nat Med 2:457–460

73. Zhang H, Meadows GG (2008) Chronic alcohol consumption perturbs the balance between thymus-derived and bone marrow-derived natural killer cells in the spleen. J Leukoc Biol 83:41–47
74. Boyadjieva NI, Chaturvedi K, Poplawski MM, Sarkar DK (2004) Opioid antagonist naltrexone disrupts feedback interaction between mu and delta opioid receptors in splenocytes to prevent alcohol inhibition of NK cell function. J Immunol 173:42–49
75. Jeong WI, Park O, Gao B (2008) Abrogation of the antifibrotic effects of natural killer cells/interferon-gamma contributes to alcohol acceleration of liver fibrosis. Gastroenterology 134:248–258
76. Jeong WI, Gao B (2008) Innate immunity and alcoholic liver fibrosis. J Gastroenterol Hepatol 23(Suppl 1):S112–S118
77. Zhang H, Meadows GG (2009) Exogenous IL-15 in combination with IL-15R alpha rescues natural killer cells from apoptosis induced by chronic alcohol consumption. Alcohol Clin Exp Res 33:419–427
78. Ljunggren HG, Karre K (1990) In search of the 'missing self': MHC molecules and NK cell recognition. Immunol Today 11:237–244
79. Harrison RJ, Ettorre A, Little AM, Khakoo SI (2010) Association of NKG2A with treatment for chronic hepatitis C virus infection. Clin Exp Immunol 161:306–314
80. Ahlenstiel G, Edlich B, Hogdal LJ, Rotman Y, Noureddin M, Feld JJ, Holz LE, Titerence RH, Liang TJ, Rehermann B (2011) Early changes in natural killer cell function indicate virologic response to interferon therapy for hepatitis C. Gastroenterology 141:1231–1239e1-2
81. Amadei B, Urbani S, Cazaly A, Fisicaro P, Zerbini A, Ahmed P, Missale G, Ferrari C, Khakoo SI (2010) Activation of natural killer cells during acute infection with hepatitis C virus. Gastroenterology 138:1536–1545
82. Zhang S, Saha B, Kodys K, Szabo G (2013) IFN-gamma production by human natural killer cells in response to HCV-infected hepatoma cells is dependent on accessory cells. J Hepatol 59:442–449
83. Dessouki O, Kamiya Y, Nagahama H, Tanaka M, Suzu S, Sasaki Y, Okada S (2010) Chronic hepatitis C viral infection reduces NK cell frequency and suppresses cytokine secretion: Reversion by anti-viral treatment. Biochem Biophys Res Commun 393:331–337
84. Edlich B, Ahlenstiel G, Zabaleta Azpiroz A, Stoltzfus J, Noureddin M, Serti E, Feld JJ, Liang TJ, Rotman Y, Rehermann B (2012) Early changes in interferon signaling define natural killer cell response and refractoriness to interferon-based therapy of hepatitis C patients. Hepatology 55:39–48
85. Gonzalez VD, Falconer K, Michaelsson J, Moll M, Reichard O, Alaeus A, Sandberg JK (2008) Expansion of CD56- NK cells in chronic HCV/HIV-1 co-infection: reversion by antiviral treatment with pegylated IFNalpha and ribavirin. Clin Immunol 128:46–56
86. Gao B, Radaeva S, Jeong WI (2007) Activation of natural killer cells inhibits liver fibrosis: a novel strategy to treat liver fibrosis. Expert Rev Gastroenterol Hepatol 1:173–180
87. Gao B, Radaeva S, Park O (2009) Liver natural killer and natural killer T cells: immunobiology and emerging roles in liver diseases. J Leukoc Biol 86:513–528
88. Arteel GE (2008) Silencing a killer among us: ethanol impairs immune surveillance of activated stellate cells by natural killer cells. Gastroenterology 134:351–353
89. Szabo G, Dolganiuc A (2005) Subversion of plasmacytoid and myeloid dendritic cell functions in chronic HCV infection. Immunobiology 210:237–247
90. Szabo G, Dolganiuc A, Mandrekar P, White B (2004) Inhibition of antigen-presenting cell functions by alcohol: implications for hepatitis C virus infection. Alcohol 33:241–249
91. Dolganiuc A, Kodys K, Kopasz A, Marshall C, Mandrekar P, Szabo G (2003) Additive inhibition of dendritic cell allostimulatory capacity by alcohol and hepatitis C is not restored by DC maturation and involves abnormal IL-10 and IL-2 induction. Alcohol Clin Exp Res 27:1023–1031
92. Dolganiuc A, Kodys K, Kopasz A, Marshall C, Do T, Romics L Jr, Mandrekar P, Zapp M, Szabo G (2003) Hepatitis C virus core and nonstructural protein 3 proteins induce pro- and

anti-inflammatory cytokines and inhibit dendritic cell differentiation. J Immunol 170: 5615–5624

93. Mandrekar P, Catalano D, Dolganiuc A, Kodys K, Szabo G (2004) Inhibition of myeloid dendritic cell accessory cell function and induction of T cell anergy by alcohol correlates with decreased IL-12 production. J Immunol 173:3398–3407
94. Golden-Mason L, Palmer B, Klarquist J, Mengshol JA, Castelblanco N, Rosen HR (2007) Upregulation of PD-1 expression on circulating and intrahepatic hepatitis C virus-specific CD8+ T cells associated with reversible immune dysfunction. J Virol 81:9249–9258
95. Saha B, Choudhary MC, Sarin SK (2013) Expression of inhibitory markers is increased on effector memory T cells during hepatitis C virus/HIV coinfection as compared to hepatitis C virus or HIV monoinfection. AIDS 27:2191–2200
96. Urbani S, Amadei B, Tola D, Pedrazzi G, Sacchelli L, Cavallo MC, Orlandini A, Missale G, Ferrari C (2008) Restoration of HCV-specific T cell functions by PD-1/PD-L1 blockade in HCV infection: effect of viremia levels and antiviral treatment. J Hepatol 48:548–558
97. McMahan RH, Golden-Mason L, Nishimura MI, McMahon BJ, Kemper M, Allen TM, Gretch DR, Rosen HR (2010) Tim-3 expression on PD-1+ HCV-specific human CTLs is associated with viral persistence, and its blockade restores hepatocyte-directed in vitro cytotoxicity. J Clin Invest 120:4546–4557
98. Gondois-Rey F, Dental C, Halfon P, Baumert TF, Olive D, Hirsch I (2009) Hepatitis C virus is a weak inducer of interferon alpha in plasmacytoid dendritic cells in comparison with influenza and human herpesvirus type-1. PLoS One 4:e4319
99. Liu BS, Groothuismink ZM, Janssen HL, Boonstra A (2011) Role for IL-10 in inducing functional impairment of monocytes upon TLR4 ligation in patients with chronic HCV infections. J Leukoc Biol 89:981–988
100. Wegert M, La Monica N, Tripodi M, Adler G, Dikopoulos N (2009) Impaired interferon type I signalling in the liver modulates the hepatic acute phase response in hepatitis C virus transgenic mice. J Hepatol 51:271–278
101. Korenaga M, Wang T, Li Y, Showalter LA, Chan T, Sun J, Weinman SA (2005) Hepatitis C virus core protein inhibits mitochondrial electron transport and increases reactive oxygen species (ROS) production. J Biol Chem 280:37481–37488
102. Korenaga M, Okuda M, Otani K, Wang T, Li Y, Weinman SA (2005) Mitochondrial dysfunction in hepatitis C. J Clin Gastroenterol 39:S162–S166
103. Mas VR, Fassnacht R, Archer KJ, Maluf D (2010) Molecular mechanisms involved in the interaction effects of alcohol and hepatitis C virus in liver cirrhosis. Mol Med 16:287–297
104. Moriya K, Fujie H, Shintani Y, Yotsuyanagi H, Tsutsumi T, Ishibashi K, Matsuura Y, Kimura S, Miyamura T, Koike K (1998) The core protein of hepatitis C virus induces hepatocellular carcinoma in transgenic mice. Nat Med 4:1065–1067
105. Moriya K, Nakagawa K, Santa T, Shintani Y, Fujie H, Miyoshi H, Tsutsumi T, Miyazawa T, Ishibashi K, Horie T, Imai K, Todoroki T, Kimura S, Koike K (2001) Oxidative stress in the absence of inflammation in a mouse model for hepatitis C virus-associated hepatocarcinogenesis. Cancer Res 61:4365–4370
106. Okuda M, Li K, Beard MR, Showalter LA, Scholle F, Lemon SM, Weinman SA (2002) Mitochondrial injury, oxidative stress, and antioxidant gene expression are induced by hepatitis C virus core protein. Gastroenterology 122:366–375
107. Dionisio N, Garcia-Mediavilla MV, Sanchez-Campos S, Majano PL, Benedicto I, Rosado JA, Salido GM, Gonzalez-Gallego J (2009) Hepatitis C virus NS5A and core proteins induce oxidative stress-mediated calcium signalling alterations in hepatocytes. J Hepatol 50:872–882
108. Otani K, Korenaga M, Beard MR, Li K, Qian T, Showalter LA, Singh AK, Wang T, Weinman SA (2005) Hepatitis C virus core protein, cytochrome P450 2E1, and alcohol produce combined mitochondrial injury and cytotoxicity in hepatoma cells. Gastroenterology 128:96–107
109. Dolganiuc A, Bakis G, Kodys K, Mandrekar P, Szabo G (2006) Acute ethanol treatment modulates Toll-like receptor-4 association with lipid rafts. Alcohol Clin Exp Res 30:76–85

110. Testro AG, Gow PJ, Angus PW, Wongseelashote S, Skinner N, Markovska V, Visvanathan K (2010) Effects of antibiotics on expression and function of Toll-like receptors 2 and 4 on mononuclear cells in patients with advanced cirrhosis. J Hepatol 52:199–205
111. Szabo G, Velayudham A, Romics L Jr, Mandrekar P (2005) Modulation of non-alcoholic steatohepatitis by pattern recognition receptors in mice: the role of toll-like receptors 2 and 4. Alcohol Clin Exp Res 29:140S–145S
112. Machida K, Tsukamoto H, Mkrtchyan H, Duan L, Dynnyk A, Liu HM, Asahina K, Govindarajan S, Ray R, Ou JH, Seki E, Deshaies R, Miyake K, Lai MM (2009) Toll-like receptor 4 mediates synergism between alcohol and HCV in hepatic oncogenesis involving stem cell marker Nanog. Proc Natl Acad Sci U S A 106:1548–1553
113. Yu J, Vodyanik MA, Smuga-Otto K, Antosiewicz-Bourget J, Frane JL, Tian S, Nie J, Jonsdottir GA, Ruotti V, Stewart R, Slukvin II, Thomson JA (2007) Induced pluripotent stem cell lines derived from human somatic cells. Science 318:1917–1920
114. Shan J, Shen J, Liu L, Xia F, Xu C, Duan G, Xu Y, Ma Q, Yang Z, Zhang Q, Ma L, Liu J, Xu S, Yan X, Bie P, Cui Y, Bian XW, Qian C (2012) Nanog regulates self-renewal of cancer stem cells through the insulin-like growth factor pathway in human hepatocellular carcinoma. Hepatology 56:1004–1014
115. Sun C, Sun L, Jiang K, Gao DM, Kang XN, Wang C, Zhang S, Huang S, Qin X, Li Y, Liu YK (2013) NANOG promotes liver cancer cell invasion by inducing epithelial-mesenchymal transition through NODAL/SMAD3 signaling pathway. Int J Biochem Cell Biol 45:1099–1108
116. Hara Y, Hino K, Okuda M, Furutani T, Hidaka I, Yamaguchi Y, Korenaga M, Li K, Weinman SA, Lemon SM, Okita K (2006) Hepatitis C virus core protein inhibits deoxycholic acid-mediated apoptosis despite generating mitochondrial reactive oxygen species. J Gastroenterol 41:257–268
117. Gong G, Waris G, Tanveer R, Siddiqui A (2001) Human hepatitis C virus NS5A protein alters intracellular calcium levels, induces oxidative stress, and activates STAT-3 and NF-kappa B. Proc Natl Acad Sci U S A 98:9599–9604
118. Brault C, Levy PL, Bartosch B (2013) Hepatitis C virus-induced mitochondrial dysfunctions. Viruses 5:954–980
119. Hou W, Bukong TN, Kodys K, Szabo G (2013) Alcohol facilitates HCV RNA replication via up-regulation of miR-122 expression and inhibition of cyclin G1 in human hepatoma cells. Alcohol Clin Exp Res 37:599–608
120. Seronello S, Ito C, Wakita T, Choi J (2010) Ethanol enhances hepatitis C virus replication through lipid metabolism and elevated NADH/NAD+. J Biol Chem 285:845–854
121. Lee HS, Yoon JH, Kamimura S, Iwata K, Watanabe H, Kim CY (1997) Lack of association of cytochrome P450 2E1 genetic polymorphisms with the risk of human hepatocellular carcinoma. Int J Cancer 71:737–740
122. Hirose Y, Naito Z, Kato S, Onda M, Sugisaki Y (2002) Immunohistochemical study of CYP2E1 in hepatocellular carcinoma carcinogenesis: examination with newly prepared anti-human CYP2E1 antibody. J Nippon Med Sch 69:243–251
123. Liu C, Wang H, Pan C, Shen J, Liang Y (2012) CYP2E1 PstI/RsaI polymorphism and interaction with alcohol consumption in hepatocellular carcinoma susceptibility: evidence from 1,661 cases and 2,317 controls. Tumour Biol 33:979–984
124. Solit DB, Chiosis G (2008) Development and application of Hsp90 inhibitors. Drug Discov Today 13:38–43
125. Trepel J, Mollapour M, Giaccone G, Neckers L (2010) Targeting the dynamic HSP90 complex in cancer. Nat Rev Cancer 10:537–549
126. Schlesinger MJ (1990) Heat shock proteins. J Biol Chem 265:12111–12114
127. Zhao R, Houry WA (2005) Hsp90: a chaperone for protein folding and gene regulation. Biochem Cell Biol 83:703–710
128. Bohonowych JE, Gopal U, Isaacs JS (2010) Hsp90 as a gatekeeper of tumor angiogenesis: clinical promise and potential pitfalls. J Oncol 2010:412985
129. Freeman BC, Yamamoto KR (2002) Disassembly of transcriptional regulatory complexes by molecular chaperones. Science 296:2232–2235

130. Tariq M, Nussbaumer U, Chen Y, Beisel C, Paro R (2009) Trithorax requires Hsp90 for maintenance of active chromatin at sites of gene expression. Proc Natl Acad Sci U S A 106:1157–1162
131. Bagatell R, Whitesell L (2004) Altered Hsp90 function in cancer: a unique therapeutic opportunity. Mol Cancer Ther 3:1021–1030
132. Eustace BK, Sakurai T, Stewart JK, Yimlamai D, Unger C, Zehetmeier C, Lain B, Torella C, Henning SW, Beste G, Scroggins BT, Neckers L, Ilag LL, Jay DG (2004) Functional proteomic screens reveal an essential extracellular role for hsp90 alpha in cancer cell invasiveness. Nat Cell Biol 6:507–514
133. Parkin DM (2006) The global health burden of infection-associated cancers in the year 2002. Int J Cancer 118:3030–3044
134. De Martel C, Ferlay J, Franceschi S, Vignat J, Bray F, Forman D, Plummer M (2012) Global burden of cancers attributable to infections in 2008: a review and synthetic analysis. Lancet Oncol 13:607–615
135. McGivern DR, Lemon SM (2011) Virus-specific mechanisms of carcinogenesis in hepatitis C virus associated liver cancer. Oncogene 30:1969–1983
136. Mandrekar P, Catalano D, Jeliazkova V, Kodys K (2008) Alcohol exposure regulates heat shock transcription factor binding and heat shock proteins 70 and 90 in monocytes and macrophages: implication for TNF-alpha regulation. J Leukoc Biol 84:1335–1345
137. Bukong TN, Hou W, Kodys K, Szabo G (2013) Ethanol facilitates hepatitis C virus replication via up-regulation of GW182 and heat shock protein 90 in human hepatoma cells. Hepatology 57:70–80
138. Ambade A, Catalano D, Lim A, Mandrekar P (2012) Inhibition of heat shock protein (molecular weight 90 kDa) attenuates proinflammatory cytokines and prevents lipopolysaccharide-induced liver injury in mice. Hepatology 55:1585–1595
139. Sun Y, Zang Z, Xu X, Zhang Z, Zhong L, Zan W, Zhao Y, Sun L (2010) Differential proteomics identification of HSP90 as potential serum biomarker in hepatocellular carcinoma by two-dimensional electrophoresis and mass spectrometry. Int J Mol Sci 11:1423–1433
140. Breinig M, Caldas-Lopes E, Goeppert B, Malz M, Rieker R, Bergmann F, Schirmacher P, Mayer M, Chiosis G, Kern MA (2009) Targeting heat shock protein 90 with non-quinone inhibitors: a novel chemotherapeutic approach in human hepatocellular carcinoma. Hepatology 50:102–112
141. Pascale RM, Simile MM, Calvisi DF, Frau M, Muroni MR, Seddaiu MA, Daino L, Muntoni MD, De Miglio MR, Thorgeirsson SS, Feo F (2005) Role of HSP90, CDC37, and CRM1 as modulators of P16(INK4A) activity in rat liver carcinogenesis and human liver cancer. Hepatology 42:1310–1319
142. Taguwa S, Kambara H, Omori H, Tani H, Abe T, Mori Y, Suzuki T, Yoshimori T, Moriishi K, Matsuura Y (2009) Cochaperone activity of human butyrate-induced transcript 1 facilitates hepatitis C virus replication through an Hsp90-dependent pathway. J Virol 83:10427–10436
143. Nakagawa S, Umehara T, Matsuda C, Kuge S, Sudoh M, Kohara M (2007) Hsp90 inhibitors suppress HCV replication in replicon cells and humanized liver mice. Biochem Biophys Res Commun 353:882–888
144. Okamoto T, Nishimura Y, Ichimura T, Suzuki K, Miyamura T, Suzuki T, Moriishi K, Matsuura Y (2006) Hepatitis C virus RNA replication is regulated by FKBP8 and Hsp90. EMBO J 25:5015–5025
145. Ujino S, Nishitsuji H, Sugiyama R, Suzuki H, Hishiki T, Sugiyama K, Shimotohno K, Takaku H (2012) The interaction between human initiation factor eIF3 subunit c and heat-shock protein 90: a necessary factor for translation mediated by the hepatitis C virus internal ribosome entry site. Virus Res 163:390–395
146. Ujino S, Yamaguchi S, Shimotohno K, Takaku H (2009) Heat-shock protein 90 is essential for stabilization of the hepatitis C virus nonstructural protein NS3. J Biol Chem 284:6841–6846
147. Varmus HE (1984) The molecular genetics of cellular oncogenes. Annu Rev Genet 18: 553–612

148. Okamoto T, Omori H, Kaname Y, Abe T, Nishimura Y, Suzuki T, Miyamura T, Yoshimori T, Moriishi K, Matsuura Y (2008) A single-amino-acid mutation in hepatitis C virus NS5A disrupting FKBP8 interaction impairs viral replication. J Virol 82:3480–3489
149. Takayama S, Reed JC, Homma S (2003) Heat-shock proteins as regulators of apoptosis. Oncogene 22:9041–9047
150. Lanneau D, Brunet M, Frisan E, Solary E, Fontenay M, Garrido C (2008) Heat shock proteins: essential proteins for apoptosis regulation. J Cell Mol Med 12:743–761
151. Bai L, Xu S, Chen W, Li Z, Wang X, Tang H, Lin Y (2011) Blocking Nf-kappaB and Akt by Hsp90 inhibition sensitizes Smac mimetic compound 3-induced extrinsic apoptosis pathway and results in synergistic cancer cell death. Apoptosis 16:45–54
152. Leng AM, Liu T, Yang J, Cui JF, Li XH, Zhu YN, Xiong T, Zhang G, Chen Y (2012) The apoptotic effect and associated signalling of HSP90 inhibitor 17-DMAG in hepatocellular carcinoma cells. Cell Biol Int 36:893–899
153. Lang SA, Moser C, Fichnter-Feigl S, Schachtschneider P, Hellerbrand C, Schmitz V, Schlitt HJ, Geissler EK, Stoeltzing O (2009) Targeting heat-shock protein 90 improves efficacy of rapamycin in a model of hepatocellular carcinoma in mice. Hepatology 49:523–532
154. Lachenmayer A, Toffanin S, Cabellos L, Alsinet C, Hoshida Y, Villanueva A, Minguez B, Tsai HW, Ward SC, Thung S, Friedman SL, Llovet JM (2012) Combination therapy for hepatocellular carcinoma: additive preclinical efficacy of the HDAC inhibitor panobinostat with sorafenib. J Hepatol 56:1343–1350
155. Pessione F, Degos F, Marcellin P, Duchatelle V, Njapoum C, Martinot-Peignoux M, Degott C, Valla D, Erlinger S, Rueff B (1998) Effect of alcohol consumption on serum hepatitis C virus RNA and histological lesions in chronic hepatitis C. Hepatology 27:1717–1722
156. Safdar K, Schiff ER (2004) Alcohol and hepatitis C. Semin Liver Dis 24:305–315
157. Anand BS, Thornby J (2005) Alcohol has no effect on hepatitis C virus replication: a meta-analysis. Gut 54:1468–1472
158. Chang J, Nicolas E, Marks D, Sander C, Lerro A, Buendia MA, Xu C, Mason WS, Moloshok T, Bort R, Zaret KS, Taylor JM (2004) miR-122, a mammalian liver-specific microRNA, is processed from hcr mRNA and may downregulate the high affinity cationic amino acid transporter CAT-1. RNA Biol 1:106–113
159. Lagos-Quintana M, Rauhut R, Yalcin A, Meyer J, Lendeckel W, Tuschl T (2002) Identification of tissue-specific microRNAs from mouse. Curr Biol 12:735–739
160. Jopling CL (2008) Regulation of hepatitis C virus by microRNA-122. Biochem Soc Trans 36:1220–1223
161. Jopling CL, Schutz S, Sarnow P (2008) Position-dependent function for a tandem microRNA miR-122-binding site located in the hepatitis C virus RNA genome. Cell Host Microbe 4:77–85
162. Jopling CL, Yi M, Lancaster AM, Lemon SM, Sarnow P (2005) Modulation of hepatitis C virus RNA abundance by a liver-specific microRNA. Science 309:1577–1581
163. Machlin ES, Sarnow P, Sagan SM (2011) Masking the 5′ terminal nucleotides of the hepatitis C virus genome by an unconventional microRNA-target RNA complex. Proc Natl Acad Sci U S A 108:3193–3198
164. Coulouarn C, Factor VM, Andersen JB, Durkin ME, Thorgeirsson SS (2009) Loss of miR-122 expression in liver cancer correlates with suppression of the hepatic phenotype and gain of metastatic properties. Oncogene 28:3526–3536
165. Kutay H, Bai S, Datta J, Motiwala T, Pogribny I, Frankel W, Jacob ST, Ghoshal K (2006) Downregulation of miR-122 in the rodent and human hepatocellular carcinomas. J Cell Biochem 99:671–678
166. Bala S, Szabo G (2012) MicroRNA signature in alcoholic liver disease. Int J Hepatol 2012:498232
167. Wang K, Zhang S, Marzolf B, Troisch P, Brightman A, Hu Z, Hood LE, Galas DJ (2009) Circulating microRNAs, potential biomarkers for drug-induced liver injury. Proc Natl Acad Sci U S A 106:4402–4407

168. Laterza OF, Scott MG, Garrett-Engele PW, Korenblat KM, Lockwood CM (2013) Circulating miR-122 as a potential biomarker of liver disease. Biomark Med 7:205–210
169. Van Der Meer AJ, Farid WR, Sonneveld MJ, De Ruiter PE, Boonstra A, Van Vuuren AJ, Verheij J, Hansen BE, De Knegt RJ, Van Der Laan LJ, Janssen HL (2013) Sensitive detection of hepatocellular injury in chronic hepatitis C patients with circulating hepatocyte-derived microRNA-122. J Viral Hepat 20:158–166
170. Szabo G, Bala S (2013) MicroRNAs in liver disease. Nat Rev Gastroenterol Hepatol 10:542–552
171. Tsai WC, Hsu SD, Hsu CS, Lai TC, Chen SJ, Shen R, Huang Y, Chen HC, Lee CH, Tsai TF, Hsu MT, Wu JC, Huang HD, Shiao MS, Hsiao M, Tsou AP (2012) MicroRNA-122 plays a critical role in liver homeostasis and hepatocarcinogenesis. J Clin Invest 122:2884–2897
172. Hsu SH, Wang B, Kota J, Yu J, Costinean S, Kutay H, Yu L, Bai S, La Perle K, Chivukula RR, Mao H, Wei M, Clark KR, Mendell JR, Caligiuri MA, Jacob ST, Mendell JT, Ghoshal K (2012) Essential metabolic, anti-inflammatory, and anti-tumorigenic functions of miR-122 in liver. J Clin Invest 122:2871–2883
173. Bai S, Nasser MW, Wang B, Hsu SH, Datta J, Kutay H, Yadav A, Nuovo G, Kumar P, Ghoshal K (2009) MicroRNA-122 inhibits tumorigenic properties of hepatocellular carcinoma cells and sensitizes these cells to sorafenib. J Biol Chem 284:32015–32027
174. Lanford RE, Hildebrandt-Eriksen ES, Petri A, Persson R, Lindow M, Munk ME, Kauppinen S, Orum H (2010) Therapeutic silencing of microRNA-122 in primates with chronic hepatitis C virus infection. Science 327:198–201
175. Janssen HL, Reesink HW, Lawitz EJ, Zeuzem S, Rodriguez-Torres M, Patel K, Van Der Meer AJ, Patick AK, Chen A, Zhou Y, Persson R, King BD, Kauppinen S, Levin AA, Hodges MR (2013) Treatment of HCV infection by targeting microRNA. N Engl J Med 368:1685–1694
176. Sandler NG, Koh C, Roque A, Eccleston JL, Siegel RB, Demino M, Kleiner DE, Deeks SG, Liang TJ, Heller T, Douek DC (2011) Host response to translocated microbial products predicts outcomes of patients with HBV or HCV infection. Gastroenterology 141: 1220–1230e1-3
177. Caradonna L, Mastronardi ML, Magrone T, Cozzolongo R, Cuppone R, Manghisi OG, Caccavo D, Pellegrino NM, Amoroso A, Jirillo E, Amati L (2002) Biological and clinical significance of endotoxemia in the course of hepatitis C virus infection. Curr Pharm Des 8:995–1005
178. Welsch C, Jesudian A, Zeuzem S, Jacobson I (2012) New direct-acting antiviral agents for the treatment of hepatitis C virus infection and perspectives. Gut 61(Suppl 1):i36–i46
179. Sarrazin C, Hezode C, Zeuzem S, Pawlotsky JM (2012) Antiviral strategies in hepatitis C virus infection. J Hepatol 56(Suppl 1):S88–S100
180. Klibanov OM, Williams SH, Smith LS, Olin JL, Vickery SB (2011) Telaprevir: a novel NS3/4 protease inhibitor for the treatment of hepatitis C. Pharmacotherapy 31:951–974
181. Klibanov OM, Vickery SB, Olin JL, Smith LS, Williams SH (2012) Boceprevir: a novel NS3/4 protease inhibitor for the treatment of hepatitis C. Pharmacotherapy 32:173–190
182. Schinazi R, Halfon P, Marcellin P, Asselah T (2014) HCV direct-acting antiviral agents: the best interferon-free combinations. Liver Int 34(Suppl 1):69–78

Chapter 13
Application of Mass Spectrometry-Based Metabolomics in Identification of Early Noninvasive Biomarkers of Alcohol-Induced Liver Disease Using Mouse Model

Soumen K. Manna, Matthew D. Thompson, and Frank J. Gonzalez

Abstract A rapid, non-invasive urine test for early stage alcohol-induced liver disease (ALD) would permit risk stratification and treatment of high-risk individuals before ALD leads to irreversible liver damage and death. Urinary metabolomic studies were carried out to identify ALD-associated metabolic biomarkers using *Ppara*-null mouse model that is susceptible to ALD development on chronic alcohol consumption. Two successive studies were conducted to evaluate the applicability of mass spectrometry-based metabolomics in identification of ALD-specific signatures and to examine the robustness of these biomarkers against genetic background. Principal components analysis of ultraperformance liquid chromatography coupled with electrospray ionization quadrupole time-of-flight mass spectrometry (UPLC-ESI-QTOFMS)-generated urinary metabolic fingerprints showed that alcohol-treated wild-type and *Ppara*-null mice could be distinguished from control animals. It also showed that a combined endogenous biomarker panel helps to identify subjects with ALD as well as those at risk of developing ALD even without any information on alcohol intake or genetics. Quantitative analysis showed that increased excretion of indole-3-lactic acid and phenyllactic acid was a genetic background-independent signature exclusively associated with ALD pathogenesis in *Ppara*-null mice that showed liver pathologies similar to those observed in early stages of human ALD. These findings demonstrated that mass spectrometry-based metabolomic analysis could help in the identification of ALD-specific signatures, and that metabolites such as indole-3-lactic acid and phenyllactic acid, may serve as robust noninvasive biomarkers for early stages of ALD.

Keywords Alcohol-induced liver disease • PPARα • *Ppara*-null mouse • Steatosis • Metabolomics • UPLC-ESI-QTOFMS • Multivariate data analysis • Biomarker • Genetic background • Indole-3-lactic acid • Phenyllactic acid

S.K. Manna • M.D. Thompson • F.J. Gonzalez (✉)
Laboratory of Metabolism, Center for Cancer Research, National Cancer Institute, Building 37, Room 3106, Bethesda, MD 20892, USA
e-mail: soumenmanna@gmail.com; matthew.thompson2@nih.gov; gonzalef@mail.nih.gov

V. Vasiliou et al. (eds.), *Biological Basis of Alcohol-Induced Cancer*, Advances in Experimental Medicine and Biology 815,
DOI 10.1007/978-3-319-09614-8_13

Abbreviations

ALD	Alcohol-induced liver disease
ALT	Alanine aminotransferase
ANOVA	Analysis of variance
AST	Aspartate aminotransferase
ESI+	Electrospray ionization in positive mode
ESI–	Electrospray ionization in negative mode
MRM	Multiple reaction monitoring
NAD^+	Oxidized nicotinamide adenine dinucleotide
NADH	Reduced nicotinamide adenine dinucleotide
OPLS	Orthogonal projection to latent structures
PCA	Principal components analysis
Ppara-null	Peroxisome proliferator-activated receptor alpha knock-out mouse model
PPARα	Peroxisome proliferator-activated receptor alpha
UPLC-ESI-QTOF-MS	Ultraperformance liquid chromatography coupled with electrospray ionization quadrupole time-of-flight mass spectrometry

13.1 Introduction

13.1.1 Alcohol and Alcohol-Induced Liver Disease

Alcohol consumption is the third most common cause of lifestyle-associated mortality in the United States 2003 [1]. Alcohol consumption is also an emerging problem in developing countries [2]. The 2011 World Health Organization (WHO) status report [3] stated that "almost 4 % of all deaths worldwide are attributed to alcohol, greater than deaths caused by HIV/AIDS, violence or tuberculosis." Additionally, epidemiological studies have shown significant variation exists in susceptibility to alcohol use and alcohol-dependent health conditions depending on an individual's genetic background [2, 4–11]. Genetic polymorphisms related to alcohol metabolism affect incidence of alcoholism and physiological response [8, 10, 12], as well the development alcohol-induced liver disease (ALD) and associated outcomes [4, 5, 9, 13, 14]. Even in developed countries such as United States, more than half of alcoholism-related deaths are attributable to ALD [1]. Thus ALD poses a significant challenge to public health all over the world.

ALD pathogenesis is characterized by three stages; steatosis, alcoholic hepatitis, and fibrosis/cirrhosis [2]. Approximately 90 % of alcoholics develop fatty liver (steatosis) that resolves when alcohol consumption is discontinued [2]. However, continued excessive drinking with concomitant steatosis increases the risk of developing cirrhosis by 37 % [15], an irreversible stage of ALD [16]. Overall 5-year survival rates for patients with cirrhosis are as low as 35 %. Liver cirrhosis is also associated with increased risk of development of liver cancer [17, 18]. Although, at earlier stages of ALD (steatosis), liver damage is reversible and patients can recover completely [2, 11, 19, 20], it is largely asymptomatic and, thus, evades diagnosis to proceed to irreversible liver damage. Detection of ALD at this stage is, therefore, key to improve quality of life, maximize therapeutic benefit, and reduce mortality and healthcare burden.

13.1.2 The Role of PPARα in ALD

Since the first observable change in ALD pathogenesis is the deposition of free fatty acids in the liver [21], many scientific studies have focused on understanding pathways involved in fatty acid metabolism. The nuclear receptor peroxisome proliferator-activated receptor alpha (PPARα) [22] is a key regulator of the genes involved in lipid metabolism [23, 24], particularly catabolism of fatty acids in the liver. Expression of PPARα and its target genes are attenuated on chronic alcohol consumption [25]. Consistent with these observations, chronic alcohol treatment of the peroxisome proliferator-activated receptor alpha knockout (*Ppara*-null) mice was shown to result in the development of liver pathologies very similar to the early stages of the human ALD whereas wild-type animals remained protected [26].

13.1.3 Diagnosis of ALD

Currently, ALD diagnosis is based on biochemical assays including enzymatic activities of alanine aminotransferase (ALT), aspartate aminotransferase (AST), and gamma-glutamyl transpeptidase (GGT), along with patient history and other clinical symptoms [20, 27]. Serum-based enzymatic activity assays are non-specific with respect to etiology [11, 19]. This leaves liver biopsy as the only confirmatory tool for diagnosis [11, 19, 28]. However, in the absence of detailed life-style associated information, especially acknowledgement of alcohol consumption, often biopsy alone cannot be used to readily distinguish ALD from other liver disorders [29, 30]. The invasiveness of biopsies, and its associated complications [31] also precludes it as a routine screening and diagnostic tool, particularly, given the fact that ALD is largely asymptomatic initial stages [20]. Therefore, an early, noninvasive, high-throughput, ALD-specific biomarker is highly warranted.

13.1.4 Scope of Metabolomics

Metabolomics is an emerging field in chemical biology that seeks to identify and quantify changes in distribution of all endogenous and exogenous biochemicals (metabolites) in the biological matrix of interest. Since the production of a metabolite is dependent on interaction among of biological molecules (i.e. DNA, RNA, and proteins), the collection of metabolites (e.g. the metabolome), is essentially a reflection of physiological state of an organism at systems level. Therefore, in principle, every physiological state is expected to have a characteristic biochemical fingerprint represented by the metabolome and differences between these signatures can be used to predict a pathology. The latent signatures can also be used to elucidate the changes in biochemical landscape during pathogenesis. Metabolomics has yielded promising findings in recent studies of complex systems including pharmacometabolomics, radiation biodosimetry, and cancer biology [32–35]. As such, the application of metabolomics to elucidate biochemical changes associated with ALD represents a powerful approach to identify early biomarkers of the disease that could reveal novel aspects of underlying biology.

13.2 Methodological Overview for Urinary Metabolomics

13.2.1 Animal Model

Since, wild-type mice remain protected whereas *Ppara*-null mice develop ALD on chronic alcohol consumption; they together represent an excellent model for delineating ALD-specific changes. The studies discussed herein combine the power of metabolomics with the well-characterized *Ppara*-null mouse model to search for ALD-specific changes in urinary metabolome. Age-matched male wild-type and *Ppara*-null mice were fed control or an alcohol-containing liquid diet. Urine samples collected from these mice were analyzed using ultraperformance liquid chromatography coupled with electrospray ionization quadrupole time-of-flight mass spectrometry (UPLC-ESI-QTOFMS) to identify metabolomic changes associated with the development of ALD and to differentiate them from those related to the metabolomic changes due to of alcohol consumption [36–38]. A summary of the workflow is shown in Fig. 13.1.

13.2.2 Step 1: Preparation of Urine Samples for UPLC-ESI-QTOFMS Analysis

Urine was diluted 1:2 (v/v) with 50 % aqueous acetonitrile containing internal standards (50 μM 4-nitrobenzoic acid and 1 μM debrisoquine) in a Sirroco™ protein precipitation plate (Waters Corp.) and briefly vortexed. The deproteinated extracts were collected into 96-well collection plates under vacuum, and a 5 μL aliquot was injected into a Waters UPLC-ESI-QTOFMS system.

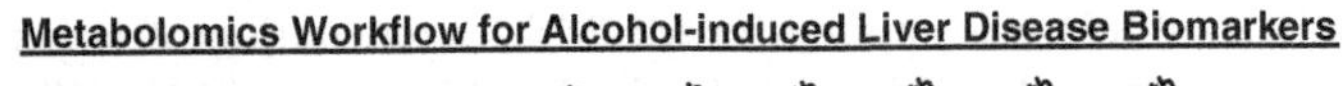

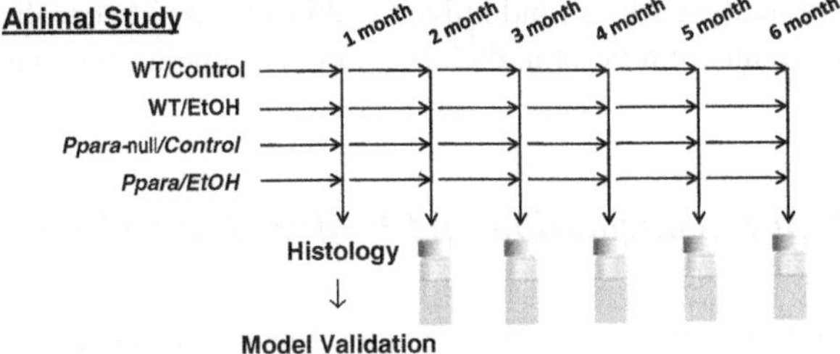

Metabolomic Analysis

Step 1: Preparation of Urine Samples for UPLC-ESI-QTOFMS Analysis

Step 2: UPLC-ESI-QTOFMS Analysis of Samples

Step 3: Data Deconvolution and Feature Extraction (MarkerLynx)

Step 4: Multivariate Data Analysis (SIMCA-P+)

Step 5: Metabolic Pathway Analysis (MassTrix)

Step 6: Identification of Urinary Biomarkers (UPLC-ESI-MS/MS and Chemical Modification)

Step 7: Quantitation of Urinary Biomarkers (UPLC-ESI-TQMS and MRM)

Fig. 13.1 Step-wise workflow for urinary metabolomic analysis to identify biomarkers of alcohol-induced liver disease (ALD). Wild-type and *Ppara*-null B6 and 129S mice were placed in the 4 % EtOH or control group. After 1 month of alcohol treatment, histological and biochemical analysis was performed to confirm ALD onset in *Ppara*-null mice. From 2 to 6 months, 24 h urine samples were collected monthly and subjected to UPLC-ESI-QTOFMS analysis, multivariate and pathway analyses, and identification and quantitation of urinary metabolites

13.2.3 Step 2: UPLC-ESI-QTOFMS Analysis of Urine Samples

An Acquity UPLC BEH C18 column (1.7 μm, 2.1 × 50 mm, Waters Corp.) was used for chromatographic separation of metabolites before introduction into electrospray. The mobile phase comprised of a mixture of 0.1 % aqueous formic acid (A) and acetonitrile containing 0.1 % formic acid (B). A 0.5 mL/min flow rate was maintained over a 10-min run with a gradient elution: 2 % B for 0.5 min, 2–20 % B in 4 min, 20–95 % B in 8 min, 95–99 % B in 8.1 min, holding at 99% B up to 9.0 min, bringing back to 2 % at 9.1 min and holding at 2 % till end. Column temperature was maintained at 40 °C throughout sample runs. The QTOF Premier mass spectrometer was operated in electrospray ionization positive (ESI+) and negative (ESI−) mode. Capillary voltage and cone voltage were maintained at 3 kV and 20 V, respectively. Source and desolvation temperatures were set at 120 °C and 350 °C, respectively. Nitrogen was used as both cone gas (50 L/h) and desolvation gas (600 L/h), and argon was used as collision gas. Sulfadimethoxine was used as the lock mass (m/z 311.0814$^+$) for accurate mass calibration in real time. Collision energy ranging from 10 to 40 eV was applied for MS/MS fragmentation of target ions. All urine samples were analyzed in a randomized fashion to avoid complications due to artifacts related to injection order and changes in instrument efficiency.

13.2.4 Step 3: Data Deconvolution and Feature Extraction

Ion chromatogram and mass spectral data were acquired using MassLynx software (Waters Corp.) in centroid format. Chromatograms were inspected for consistency of sample injection, reproducibility of retention time, and mass accuracy using internal standards and quality control samples. Data was binned, features extracted and area under the peak was calculated through integration using MarkerLynx software (Waters Corp.)

13.2.5 Step 4: Multivariate Data Analysis

Individual ion intensities were normalized with respect to the total ion count (TIC) in order to generate a data matrix consisting of the retention time, m/z value, and the normalized peak area. The data matrix was analyzed by SIMCA-P+12 software (Umetrics, Kennelon, NJ). Unsupervised segregation of control and alcohol-treated metabolomes was checked by principal components analysis (PCA) using Pareto-scaled data [39]. The supervised orthogonal projection to latent structures (OPLS) model was used to identify ions that contributed significantly to group discrimination. OPLS analysis concentrated group discrimination into the first component with remaining unrelated variation contained in subsequent components. The magnitude of the parameter p(corr)[1] obtained from the loadings S-plot generated by OPLS analysis correlates with the group discriminating power of a variable. A list of ions was then generated from the loadings S-plot showing considerable group discriminating power ($-0.8 > p(\text{corr})[1]$ or $p(\text{corr})[1] > 0.8$) (statistically significant ($P < 0.05$) difference in relative abundance between control and alcohol-treated animals). The p(corr)[1] values represent the interclass difference and $w(1)$ values indicate the relative abundance of the ions. Ions that contribute highly to the interclass differences were selected for further identification and quantitation as candidate biomarkers.

13.2.6 Step 5: Metabolic Pathway Analysis

MassTRIX (http://metabolomics.helmholtz-muenchen.de/masstrix/) is a web-based tool designed to assign ions of interest from a metabolomics experiment to annotated pathways [40]. It can be used to find metabolic pathways even without any systematic identification [41]. This was used to identify metabolic pathways affected by alcohol treatment. The masses of the ions that are significantly elevated (p(corr)[1] > 0.8) or depleted (p(corr)[1] < −0.8) upon alcohol treatment were used to identify metabolic pathways of interest using the KEGG (http://www.genome.jp/kegg/) database (including HMDB, Lipidmaps, and updated KEGG). A mass error of <5 ppm in the respective ionization modes and the possibility of formation of Na^+-adducts in the electrosprayer (ESI+ mode) was also taken into account.

13.2.7 Step 6: Identification of Urinary Biomarkers

Elemental compositions were derived using the Seven Golden Rules [42] considering a mass error <5 ppm. Possible candidates were also searched using metabolomic databases [43, 44]. Finally, authentic standards were used to confirm the identities of these ions by comparison of retention time (UPLC) and fragmentation pattern (ESI-MS/MS). Sulfatase (Sigma-Aldrich) treatment followed by retention time and fragmentation comparison of deconjugated metabolites with authentic standards, were used to confirm sulfate conjugates. Urine samples and standards were incubated with 40 U/mL of the enzyme solution in 200 mM sodium acetate buffer (pH 5.0) overnight at 37 °C. The enzyme and other particulates were precipitated with 50 % aqueous acetonitrile, and the supernatant was analyzed by UPLC-ESI-QTOFMS/MS. 4-Nitrocatechol sulfate was used as a positive control for the sulfatase activity. Deconjugation was also carried out using acid hydrolysis by heating the urine samples with 6 M HCl at 100 °C for 1 h under refluxing conditions.

13.2.8 Step 7: Quantitation of Urinary Metabolites

An Acquity® UPLC system coupled with a XEVO™ triple-quadrupole tandem mass spectrometer (Waters Corp.) was used to quantitate urinary metabolites by multiple reaction monitoring (MRM). Standard compounds were mixed together to optimize the condition for separation and detection of metabolites from a complex mixture such as urine. Standard calibration plots for quantitation were generated using authentic standards. Deproteinated urine samples containing 0.5 μM debrisoquine (internal standard) were analyzed in the same fashion as that of authentic compounds. The mobile phase was comprised of 0.1 % aqueous formic acid (A) and acetonitrile containing 0.1 % formic acid (B). The gradient elution was performed over 6 min at a flow rate of 0.3 mL using: 1–99 % B in 4 min, holding at 99%B up to 5.0 min, bringing back to 1 % at 5.5 min and holding at 1 % till end. The area under the peak for each metabolite was divided by that for the internal standard to calculate response and a serial dilution was performed to generate a standard calibration plot of response vs. concentration. Serially diluted urine samples containing 0.5 μM debrisoquine were analyzed in the same way as the authentic standards.

The quantitative abundances were calculated from the response using the linear range of detection of the calibration plot. All analyses were performed using TargetLynx software (Waters Corp.). One-way ANOVA with Bonferroni's correction for multiple comparisons was performed using GraphPad Prism 4 software (San Diego, CA) with a two-sided $P < 0.05$ considered statistically significant.

According to their fragmentation pattern, the following MRM transitions were monitored for the respective compounds: indole-3-lactic acid (206→118; ESI+), indole-3-pyruvic acid (204→130; ESI+), tryptophan (205→118; ESI+), 2-hydroxyphenylacetic acid (151→107; ESI−), 4-hydroxyphenylacetic acid (151→107; ESI−), adipic acid (147→101; ESI+), pimelic acid (159→97; ESI−), debrisoquine (176→134; ESI+), phenylalanine (166→120; ESI+), phenyllactic acid (165→103; ESI−), suberic acid (173→111; ESI−), *N*-hexanoylglycine (174→76; ESI+), xanthurenic acid (206→160, ESI−), *N*-acetylglycine (116→74, ESI−), taurine (124→80, ESI−), and creatinine (114→86; ESI+). All concentrations were normalized with respect to creatinine to account for any change in glomerular filtration rates.

13.2.9 Effect of Genetic Background on Metabolomic Signatures

Genetic background is well-known to influence outcome of alcoholism including alcohol-induced liver disease [4, 5, 13, 14]. Since, metabolome reflects the phenotype; robustness of metabolomic biomarkers against genetic background needs to be investigated. C57BL/6 (B6) and 129/Sv has earlier shown to differ considerabl with respect to physiological functions [45] as well as the biochemical response and outcome of xenobiotic insults [46, 47]. Thus these two strains of mice were used to characterize the influence of genetic background on overall metabolome and ALD biomarkers.

13.3 Animal Study Design

- *Study 1*: Identification of ALD-associated metabolic signatures in *Ppara*-null Mice.
  Male (6- to-8-week-old, $N=4$/group) wild-type and *Ppara*-null on 129/Sv background were fed a 4 % ethanol-containing liquid diet ad libitum (Lieber-DeCarli Diet, Dyets, Inc.). Control animals ($N=4$/group) were fed an isocaloric diet supplemented with maltose dextran ad libitum (Dyets, Inc.).
- *Study 2*: Identification of genetic background-independent ALD biomarkers.
  The design in Study 1 was replicated but with the addition of two genetic backgrounds: wild-type and *Ppara*-null mice (6- to 8-week-old male, $N=4$/group) on

B6 (C57BL/6 N-Ppara<tm1Gonz>/*N*) as well as their counterparts on a 129/Sv (129S4/SvJae-Ppara<tm1Gonz>/*N*) background.

In both studies, a subset of mice were euthanized after 1 month on the alcohol diet, serum was collected, and portions of the liver were harvested for histology to confirm that *Ppara*-null mice were developing steatosis. Livers were formalin-fixed, paraffin-embedded, sectioned, and stained with hematoxylin and eosin. Serum AST and ALT activities were measured using VetSpec™ kits (Catachem, Inc.). Liver and serum triglycerides were estimated using a colorimetric assay kit from Wako. At 2 months, after mice were accustomed well to the liquid diets, they were transferred to a urinary metabolomics protocol where urine samples were collected monthly using Nalgene metabolic cages (Tecniplast USA, Inc.). Urines were collected over 24 h and stored at −80 °C in glass vials until analyzed. All mice were acclimated to the metabolic cages by placing them in the metabolic cages before the actual sample collection.

13.4 Results and Discussion

13.4.1 PCA Analysis of Metabolomic Data

In agreement with an earlier report [26], only *Ppara*-null mice on alcohol treatment showed lipid accumulation after 1 month (Fig. 13.2a) indicating ALD onset [48, 49]. Mass spectrometry-based metabolomic analysis revealed that alcohol exposed *Ppara*-null mice had a distinct urinary metabolic profile compared to those on control diet even at the earliest time point, i.e., after 2 months of alcohol treatment (Fig. 13.2b). After 6 months of alcohol treatment, when *Ppara*-null mice exclusively developed alcoholic steatosis, the urinary metabolomic data showed distinct segregation of control and alcohol-treated mice as well as wild-type and *Ppara*-null animals on the scores-scatter plot for unsupervised principal components analysis (Fig. 13.2c). These data indicated that each of these four groups represents a distinct metabolic signature. The separation of these mice along first principal component was according to their alcohol exposure. Interestingly, the separation along second principal component, which was influenced by the genotype, was more prominent in *Ppara*-null mice compared to wild-type. This indicated that in agreement with the liver pathology, the *Ppara*-null metabolome is also more susceptible to chronic alcohol consumption. Subsequently, supervised orthogonal projection to latent structures (OPLS) analysis was performed. As the loadings S-plots showed (Fig. 13.2d, e) there were a number of ions that showed similar trends of elevation (such as P1, P1a, and P8) or depletion (such as P4, P5, and P5a) on alcohol treatment in both wild-type and *Ppara*-null mice. However, few ions were found to be exclusively elevated (such as P2) in the urine of alcohol-treated *Ppara*-null mice (Fig. 13.2e) that developed ALD. These ions might represent ALD-specific metabolic derangements.

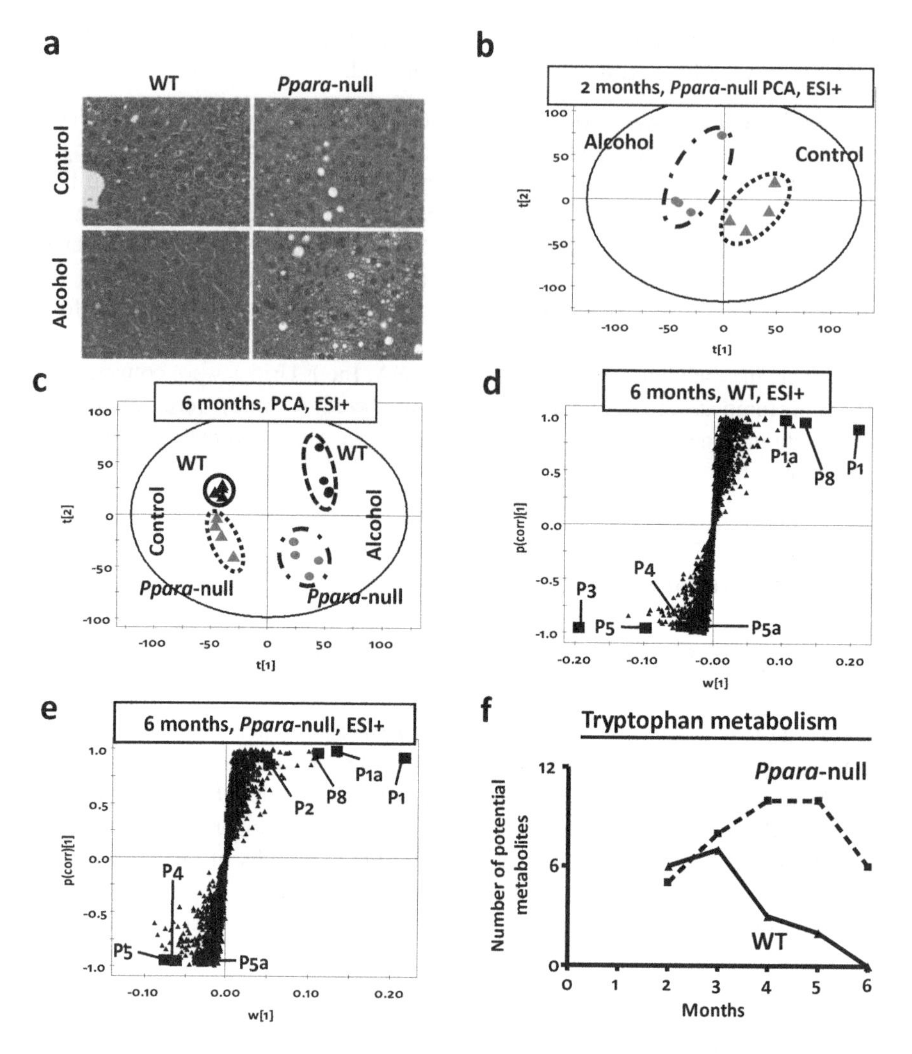

Fig. 13.2 (**a**) Liver histology (HE stain) of wild-type (WT, *left panel*) and peroxisome proliferator-activated receptor alpha knockout (*Ppara*-null, *right panel*) mice after a duration of 1 month on control (*upper panel*) or 4 % alcohol-containing liquid diet (*lower panel*). Histology shows increased fat deposition in *Ppara*-null animals on the 4 % alcohol containing liquid diet. (**b**) Scores scatter plot from principal components analysis (PCA) showing unsupervised segregation of the urinary metabolome (ESI+ mode) from control and alcohol-treated *Ppara*-null mice at 2 months. (**c**) PCA scores scatter plot showing a larger difference between wild-type and *Ppara*-null metabolomic data as a result of chronic alcohol treatment (over 6 months). The *triangles* and *dots* indicate mice on control and alcoholic diet, respectively, with *black* and *red color* representing wild-type and *Ppara*-null mice, respectively. (**d**) Loading S-plots from the supervised orthogonal projection to latent structures (OPLS) analysis of ESI+ mode metabolic signatures (at 6 months) for candidate markers of chronic alcohol exposure in wild-type and (**e**) *Ppara*-null mice. Each *triangle* represents an ion characterized by unique mass and retention time. Representative candidates have been highlighted (*solid box*) in the plots. A differential response was characterized by biomarkers that were exclusive to wild-type (P3) or *Ppara*-null (P2) mice. (**f**) MassTRIX analysis of putative metabolites related to tryptophan metabolism detected in ESI+ mode show variation over time during alcohol treatment. The *solid* and *dotted lines* represent wild-type and *Ppara*-null animals, respectively

13.4.2 Metabolic Pathway Analysis

To further identify possible metabolic pathways affected by alcohol treatment and ALD pathogenesis, ions that significantly contributed to the separation of alcohol-treated and control animals were analyzed using MassTRIX. Following alcohol exposure, metabolites potentially originating from tryptophan metabolism were found to be significantly significantly elevated (Fig. 13.2f). However, in wild-type animals, the number of such metabolites gradually decreased over time, while in *Ppara*-null animals, the corresponding number of metabolites increased. Thus, the MassTRIX analysis indicated that alcohol consumption impacted tryptophan metabolism more in the *Ppara*-null mice as compared to wild-type mice.

13.4.3 Identification and Quantitation of Metabolites

Identities of a number of these ions were subsequently confirmed using authentic standard and their concentrations were measured. Tables 13.1 and 13.2 show metabolites deranged in the urine of wild-type and *Ppara*-null mice on alcohol treatment. Both wild-type and *Ppara*-null animals showed an elevation of ethanol metabolites such as ethyl sulfate and ethyl-β-D-glucuronide, albeit to a different extent. In addition, metabolites such as 2-hydroxyphenylacetic acid, 4-hydroxyphenylacetic acid, 4-hydroxyphenylacetic acid sulfate and xanthurenic acid were elevated, whereas adipic acid and pimelic acid were depleted in the urine of alcohol-treated mice. Similar to alcohol metabolites, many of these endogenous metabolites also showed significant difference in their excretion in wild-type and *Ppara*-null animals. However, it was interesting to note that indole-3-lactic acid was exclusively elevated in the urine of alcohol-treated *Ppara*-null mice (Fig. 13.2a, b).

For biomarker discovery, reproducibility of measurements is a very important issue. Nuclear magnetic resonance (NMR)-based metabolic profiling has an advantage of being very reproducible as well as for giving direct structural information about the metabolite. However, it is interesting to note that concentrations of indole-3-lactic acid in these urine samples were in the low micromolar range. The sensitivity of the analytical method used for measuring changes in the metabolic profile becomes crucial to detect changes in the excretion of such metabolites. NMR typically fails to capture changes in abundance of metabolites at these concentration levels whereas mass spectrometry, as evident from these results, is sensitive enough to measure concentrations down to nanomolar and even picomolar ranges. Thus in spite of inferior reproducibility compared to NMR, mass spectrometry has a distinct advantage in increasing sensitivity and capturing miniscule changes in excretion of larger number of metabolites present in such low concentrations. Mass spectrometry can also increase the chance of identification novel metabolites that may be low in abundance but specific to the pathology. On the other hand, NMR typically measures only few hundreds of known and relatively abundant metabolites. Thus, mass spectrometry is often superior as a platform, particularly, for discovery of metabolic biomarkers.

Table 13.1 Metabolic signature of chronic alcohol exposure in the wild-type mice

Identity	Putative origin	Trend in B6	Trend in 129S
Ethyl sulfate	Alcohol metabolism	–	↑
Ethyl-β-D-glucuronide	Alcohol metabolism	↑	↑
N-Acetylglycine	Alcohol metabolism	↑	↑
4-Hydroxyphenylacetic acid	Phenylalanine metabolism and gut flora	–	↑
4-Hydroxyphenylacetic acid sulfate	Phenylalanine metabolism and gut flora	–	↑
2-Hydroxyphenylacetic acid	Phenylalanine metabolism and gut flora	–	↓
Xanthurenic acid	Tryptophan metabolism and gut flora	↑	–
Adipic acid	Fatty acid ω-oxidation	–	↓
Pimelic acid	Fatty acid ω-oxidation	–	↓
Taurine	Cysteine metabolism	↓	↑
N-hexanoylglycine	Fatty acid β-oxidation and gut flora	↑	–

Table 13.2 Metabolic signature of chronic alcohol exposure in the *Ppara*-null mice

Identity	Putative origin	Trend in B6	Trend in 129S
Ethyl sulfate	Alcohol metabolism	–	↑
Ethyl-β-D-glucuronide	Alcohol metabolism	↑	↑
N-Acetylglycine	Alcohol metabolism	↑	↑
4-Hydroxyphenylacetic acid	Phenylalanine metabolism and gut flora	–	↑
4-Hydroxyphenylacetic acid sulfate	Phenylalanine metabolism and gut flora	–	↑
2-Hydroxyphenylacetic acid	Phenylalanine metabolism and gut flora	–	↓
Xanthurenic acid	Tryptophan metabolism and gut flora	↑	–
Adipic acid	Fatty acid ω-oxidation	–	↓
Pimelic acid	Fatty acid ω-oxidation	–	↓
Taurine	Cysteine metabolism	↓	↑
Indole-3-lactic acid	Tryptophan metabolism	↑	↑
Phenyllactic acid	Phenylalanine metabolism	↑	↑

13.4.4 Potential Use of Metabolic Signature in Detection of Alcohol Intake and ALD Susceptibility

Diagnosis of ALD is often complicated due to lack or fidelity of information on alcohol intake. The principal components analysis including all endogenous and alcohol metabolites showed clear clustering of mice according to genotype and alcohol exposure as early as after 2 months of alcohol treatment (Fig. 13.3c). At 3 months, these clusters separated into different quadrants with first principal

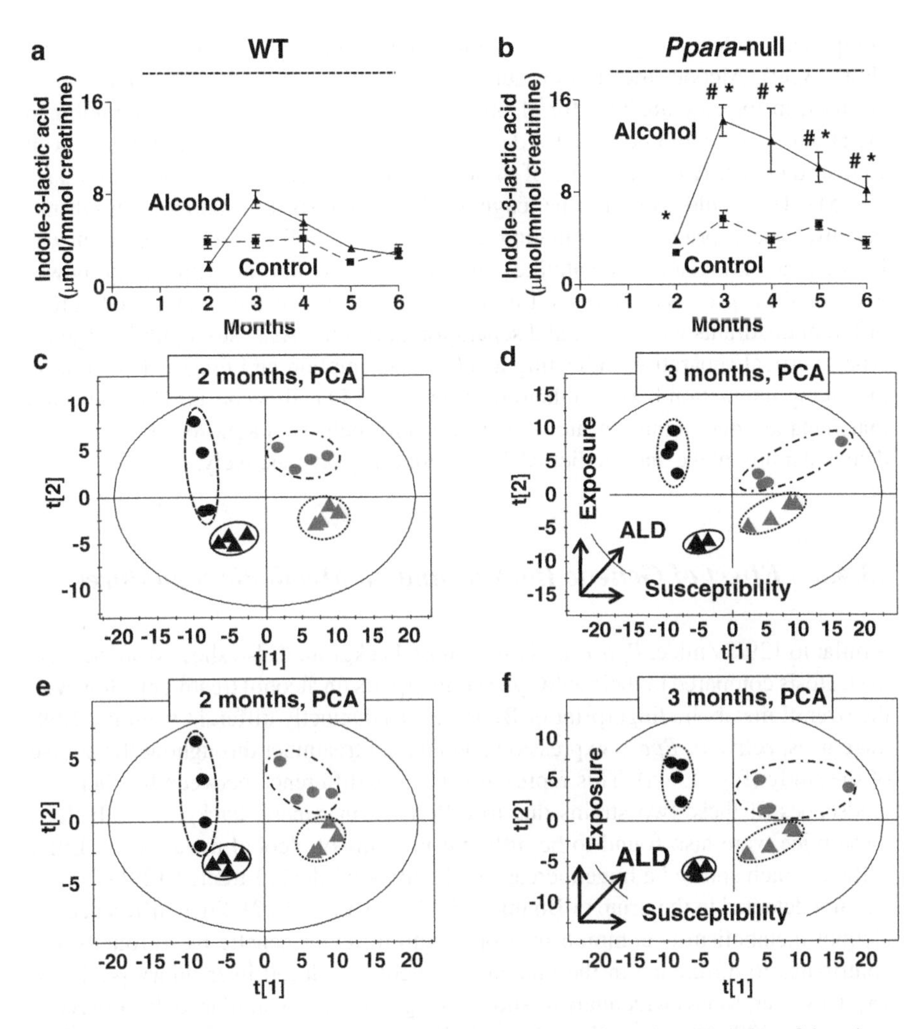

Fig. 13.3 Discriminatory power of non-invasive ALD urinary biomarkers. Variation of the urinary excretion of indole-3-lactic acid in wild-type mice (**a**) and *Ppara*-null mice (**b**) that develop alcohol-induced liver disease (ALD). The *dashed* and *solid lines* represent the variation in the concentration of urinary indole-3-lactic acid from control and alcohol-treated mice, respectively (One-way ANOVA with Bonferroni's correction for multiple comparisons, significance at $P<0.05$; #, significantly different from control *Ppara*-null mice; *, significantly different from the alcohol-treated wild-type mice). PCA scores scatter plot for the variation in the urinary excretion of the endogenous metabolites (indole-3-lactic acid, 4-hydroxyphenylacetic acid, 4-hydroxyphenylacetic acid sulfate, 2-hydroxyphenylacetic acid, adipic acid, and pimelic acid) as well as alcohol metabolites (ethyl sulfate, ethyl-β-D-glucuronide) at 2 months (**c**) and at 3 months (**d**) after beginning of alcohol treatment. The *triangles* and *dots* indicate mice on control and alcoholic diet, respectively, with *black* and *red color* representing wild-type and *Ppara*-null mice, respectively. Horizontal separation in the scatter plots correlates with *Ppara*-null expression that determines the susceptibility towards ALD with the *horizontal arrow* indicating a decrease in PPARα expression and an increase in ALD susceptibility. Vertical separation in the scatter plots correlates with alcohol exposure (in the direction of the *vertical arrow*). (**e**) The scores scatter plot for the PCA of endogenous urinary metabolites shows their collective discriminatory power to identify phenotypes at 2 months and (**f**) at 3 months of alcohol treatment. The *triangles* and *dots* indicate mice on control and alcoholic diet, respectively, with *black* and *red color* representing wild-type and *Ppara*-null mice, respectively

component reflecting genotype and the second representing alcohol exposure (Fig. 13.3d). Alcohol metabolites such as ethyl sulfate and ethyl-β-D-glucuronide are used in forensic analysis for alcohol consumption [50, 51]. Thus, a combined metabolic panel could be used for detection of recent alcohol consumption as well as liver damage. However, none of these metabolites are detectable beyond 3 days [50, 51]. This could present a challenge in ALD diagnosis if the patient stops drinking just 3 days prior to examination and denies alcoholism. Interestingly, it was found that endogenous metabolites also showed similar discriminatory power between these groups of animals (Fig. 13.3e, f). At 3 months, mice clustered in four different quadrants with horizontal separation indicating ALD susceptibility (genotype), vertical separation indicating alcohol consumption and diagonal separation indicating interaction between them resulting in ALD pathogenesis. This indicated that metabolomic signature alone may not only help to diagnose ALD but also detect alcohol intake and predict ALD susceptibility prospectively.

13.4.5 Effect of Genetic Background on Metabolic Signatures

Similar to 129/Sv mice, *Ppara*-null mice on B6 background also showed an increase in steatosis compared to their wild-type counterparts on alcohol treatment. However, the overall metabolic fingerprint of B6 mice was distinctly different from the 129S mice irrespective of *Ppara* expression and ethanol treatment throughout the course of the study (Fig. 13.4a). This represents intrinsic difference between biochemical landscapes of these two strains due to difference in genetic background. In fact, these mice were also found to be different in terms of alcohol metabolism. Ethyl sulfate, which showed a huge increase in the urine of alcohol-treated 129/Sv mice, was not detected in the urine of B6 mice (Tables 13.1 and 13.2). Such differences in alcohol metabolism is common in people with different genetic backgrounds and contributes to difference in the outcome of alcohol-induced liver injury [4, 5, 13, 14]. However, MassTRIX analysis showed (Fig. 13.4a) that similar to that observed in case of 129S mice, a number of metabolites potentially belonging to tryptophan metabolism were elevated on alcohol treatment in B6 mice and the number of elevated metabolites in *Ppara*-null was also higher than that in the wild-type mice. In addition, *Ppara*-null mice also showed a progressive increase in number of potential metabolites belonging to phenylalanine metabolism on alcohol treatment whereas wild-type mice showed a decrease. All metabolites were identified and quantitated using authentic standards. The results showed an elevation in urinary excretion of indole-3-lactic acid exclusively in alcohol-treated *Ppara*-null mice (Fig. 13.5a, b) on B6 background similar to that observed in 129/Sv mice. This was accompanied with an elevation in the urinary excretion of phenyllactic acid exclusively in alcohol-treated *Ppara*-null mice (Fig. 13.5c, d). Phenyllactic acid was also measured and found to be elevated in the urine of alcohol-treated *Ppara*-null mice on 129/Sv background (Fig. 13.5e, f).

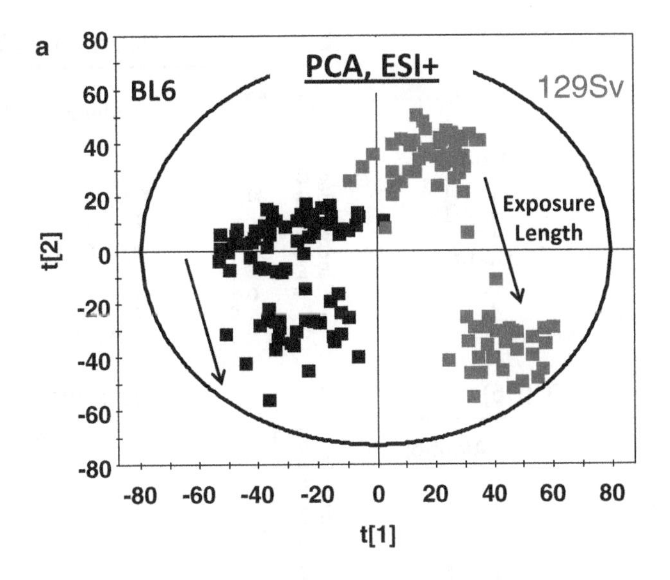

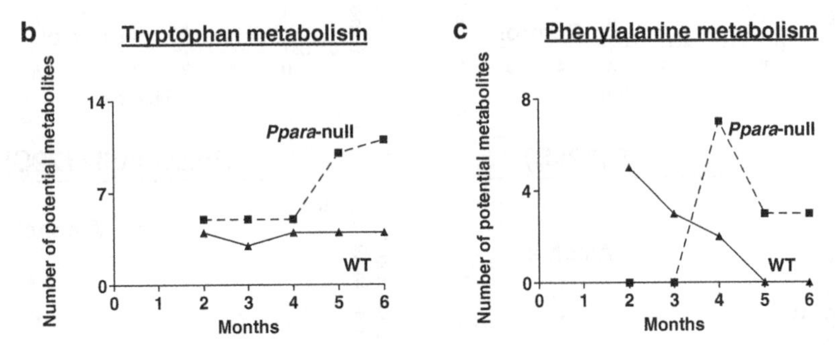

Fig. 13.4 (**a**) PCA scores scatter plot showing mice on B6 (*black*) and 129/Sc (*red*) backgrounds possess distinct metabotypes associated throughout the study duration, irrespective of genotype and alcohol exposure. The *solid arrows* indicate the shift along second principal component over time due to alcohol exposure. MassTRIX analysis of putative metabolites related to (**b**) tryptophan and (**c**) phenylalanine metabolism detected in ESI+ mode show variation over time during alcohol treatment. The *solid* and *dotted lines* represent wild-type and *Ppara*-null animals, respectively

13.4.6 The Biochemical Origin of ALD Biomarkers

These results showed a significant difference between B6 and 129/Sv animals with respect to alcohol metabolism, fatty acid metabolism, amino acid metabolism and gut flora metabolism as shown in Fig. 13.6. In spite such widespread difference due to genetic background, two α-hydroxy acid metabolites, namely, indole-3-lactic acid and phenyllactic acid were exclusively elevated in the urine of alcohol-treated *Ppara*-null mice of both backgrounds. This indicates plausible mechanistic

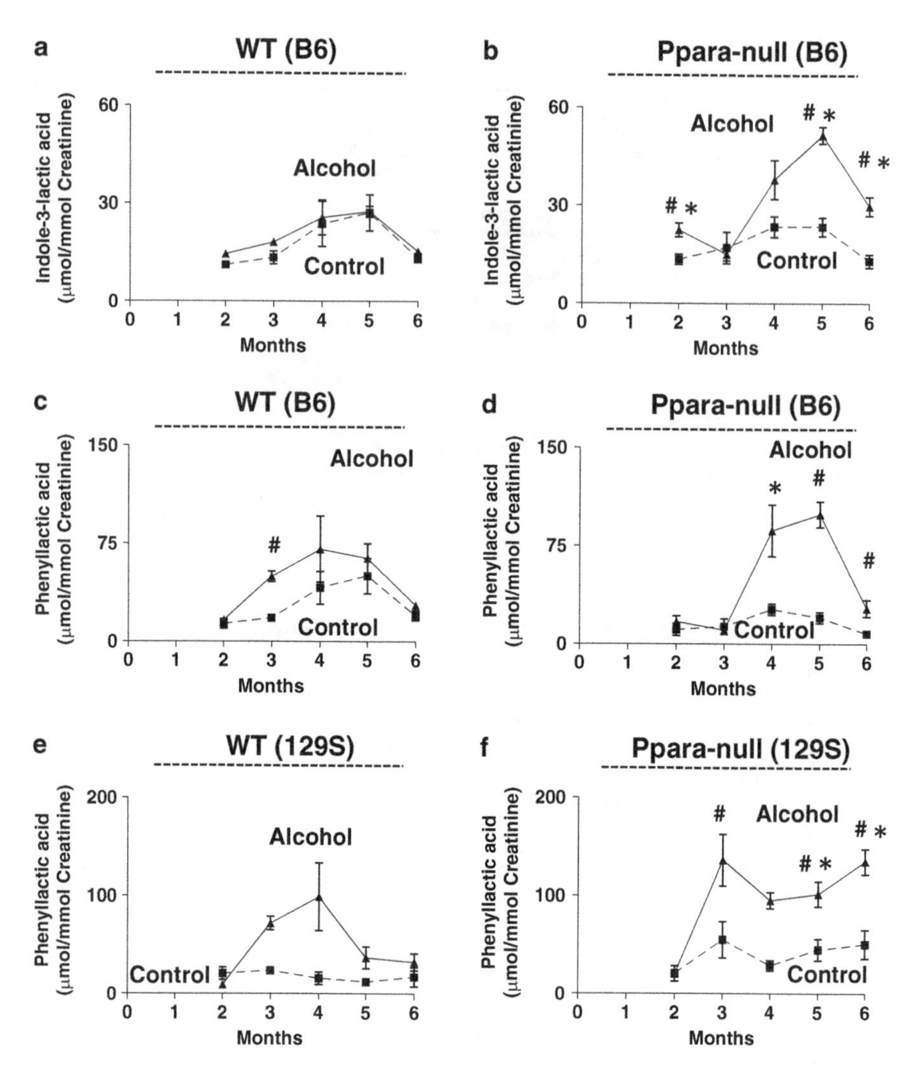

Fig. 13.5 Genetic background-independent increases in phenylalanine and tryptophan metabolites in an ALD mouse model. Urinary excretion of indole-3-lactic acid in (**a**) wild-type and (**b**) *Ppara-null* mice on the B6 background. Urinary excretion of phenyllactic acid in (**c**) wild-type and (**d**) *Ppara*-null mice on the B6 background. Urinary excretion of phenyllactic acid in (**e**) wild-type and (**f**) *Ppara*-null mice on the 129S background. Dashed and solid lines represent control and alcoholic diet-treated mice, respectively. (One-way ANOVA with Bonferroni's correction for multiple comparisons, significance at $P<0.05$ with ‡, significantly different from *Ppara*-null mice of same treatment group; #, significantly different from control mice of same genotype; *, significantly different from alcohol-treated wild-type mice)

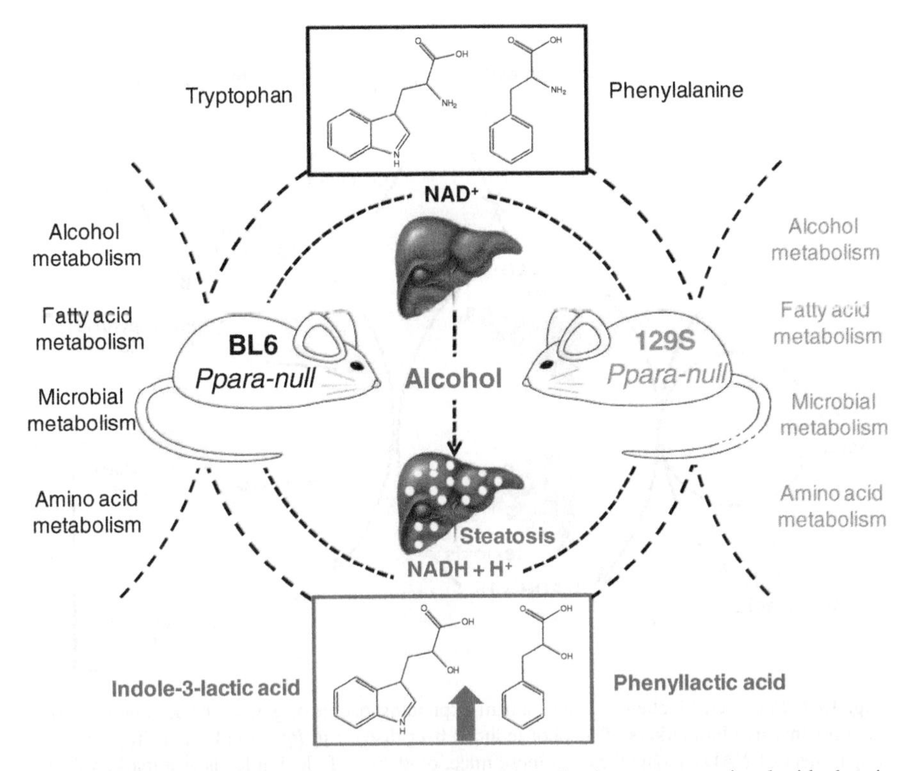

Fig. 13.6 Effect of genetic background on urinary metabolic signatures associated with chronic alcohol treatment of *Ppara*-null mice. These mice differed with respect to derangement of urinary excretion of metabolites related to alcohol metabolism, fatty acid metabolism, amino acid metabolism and gut flora were affected. However, elevation of indole-3-lactic acid and phenyllactic acid was exclusively associated with ALD pathogenesis irrespective of their genetic background

association of these metabolites with molecular events associated with ALD pathogenesis in *Ppara*-null mice. PPARα, which is a master regulator of genes involved in fatty acid β-oxidation, activates the tryptophan-quinolinic acid-NAD^+ pathway by down-regulating α-amino-β-carboxymuconate-ε-semialdehyde decarboxylase [52]. This results in the attenuation of NAD^+ production in *Ppara*-null. Since NAD^+ is a cofactor for fatty acid oxidation by both β- and ω-oxidation pathways, the reduced NAD+ biosynthesis makes *Ppara*-null mice more susceptible to fat deposition in the liver compared to their wild-type counterparts.

Alcohol is oxidized stepwise by alcohol dehydrogenase (EC 1.1.1.2) into acetaldehyde in the liver and acetaldehyde to acetic acid by aldehyde dehydrogenase (EC 1.2.1.3) (Fig. 13.7). Acetic acid can enter the TCA cycle or be a substrate for fatty acid synthesis [53]. However, both reactions consume NAD^+ and produce NADH. Therefore, chronic alcohol consumption further decreases the ratio of NAD^+/NADH [54] in *Ppara*-null mice, shifting cellular redox balance more towards reduced state to impair fatty acid catabolism and results in fat deposition in the liver (Fig. 13.7).

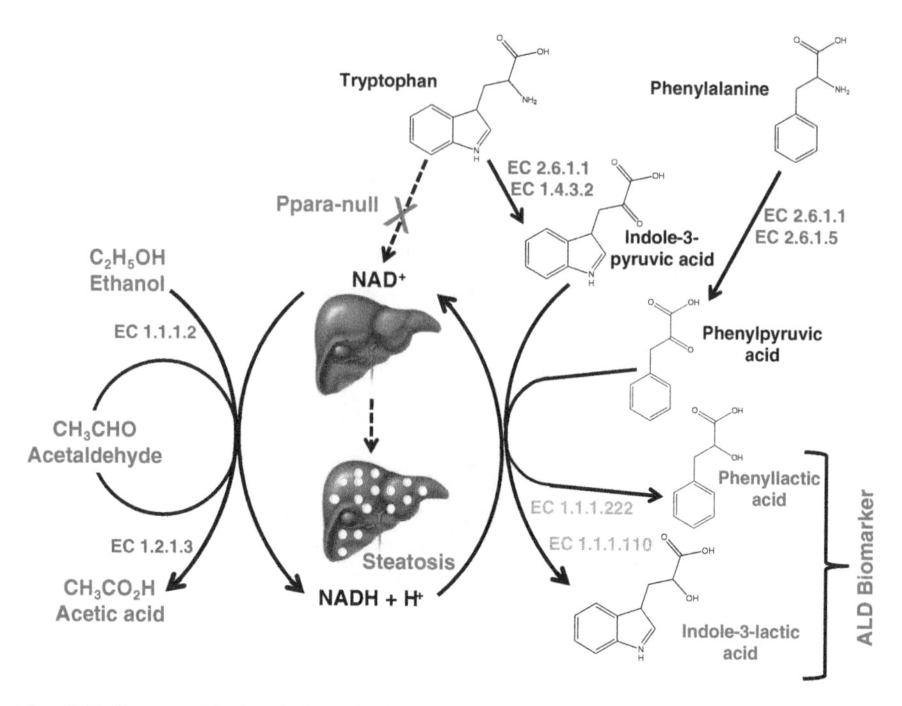

Fig. 13.7 Proposed biochemical mechanism explaining origin of genetic background-independent noninvasive biomarkers of alcohol-induced liver disease in *Ppara*-null mice. Together with impairment of NAD+ biosynthesis in these mice, oxidation of alcohol leads to a marked shift in redox balance occurs with an increased NADH/NAD+ ratio resulting in impairment of fatty acid oxidation and steatosis in alcohol-treated *Ppara*-null mice. Concurrent increase in aspartate aminotransferase activity (EC 2.6.1.1) due to liver injury might enhance the deamination of phenylalanine and tryptophan to produce α-keto acids: phenylpyruvic acid and indole-3-pyruvic acid. The elevated NADH/NAD+ ratio could drive the reduction α-keto acid intermediates to the corresponding α-hydroxy acids: phenyllactic acid and indole-3-lactic acid. Enzyme numbers mentioned in *blue* and *light gray* indicate enzymes annotated and unannotated in mammals, respectively

Tryptophan is an essential amino acid that is metabolized in the liver. Tryptophan is typically deaminated by L-amino acid oxidase (EC 1.4.3.2) to indole-3-pyruvic acid, an α-keto acid intermediate. It was shown that some microbial AST (EC 2.6.1.1) can also catalyze this reaction, albeit with lower efficiency [55, 56]. Interestingly, aspartate aminotransferase (AST) level also increases during liver injury. Thus it may also contribute to increase in production of to indole-3-pyruvic acid. In the presence of elevated NADH due to alcohol consumption, indole-3-pyruvic acid may be readily reduced to corresponding α-hydroxy acid, i.e., indole-3-lactic acid (Fig. 13.7). Enzymes responsible for this interconversion have been reported in microbes (EC 1.1.1.120, indolelactate dehydrogenase; and EC 1.1.1.222 (R)-4-hydroxyphenyllactate dehydrogenase) [57]. Tryptophan conversion to indole-3-lactate has been shown in protozoa [58].

Another α-hydroxy acid, phenyllactic acid, was also found to be elevated in the urine of *Ppara*-null mice following alcohol-treatment. Phenyllactic acid is a product of reduction of the deaminated phenylalanine. Apart from tyrosine aminotransferase (EC 2.6.1.5), aspartate aminotransferase can also catalyze the deamination of phenylalanine to phenylpyruvic acid [56, 59]. However, the shift in the redox balance in the alcohol-treated *Ppara*-null mice may drive the reduction of this intermediate to phenyllactate, possibly by the action of (R)-4-hydroxyphenyllactate dehydrogenase (EC 1.1.1.222) [60] as depicted in Fig. 13.7. This enzyme is also not known in mammals.

Taken together, the elevation of these α-hydroxy acids in the alcohol-treated *Ppara-null* mice strongly suggests common enzymatic pathways linking alcohol-induced liver injury and shift in redox balance. It shows that the increase in metabolic biomarkers is essentially driven by same biochemical events that are associated with alcohol-induced liver damage and steatosis. However, the proposed biochemical pathways highlight lacunae in our understanding of metabolic pathways, particularly, under pathological conditions.

13.5 Summary and Future Directions

This study showed that mass spectrometry-based metabolic fingerprinting can be a powerful tool in identification of noninvasive signature for detection of ALD at early stage. The simultaneous use of wild-type and *Ppara*-null animals helped to distinguish signatures of alcohol exposure from those associated with ALD pathogenesis. It showed that an endogenous metabolomic signature may be helpful in prediction of ALD susceptibility as well as ALD diagnosis even in absence of reliable information on alcohol consumption. Validation of metabolomic signatures in different genetic backgrounds helped to identify robust biomarkers for ALD. These noninvasive biomarkers appeared to simultaneously reflect change in redox balance and ongoing liver injury due to chronic alcohol consumption. In conclusion, this study demonstrated that metabolic signatures can be helpful in early noninvasive screening and diagnosis of ALD. However, these results are yet to be validated in human samples. In addition, to the effect of human genetic background, careful analysis on the effect of food habit, life-style as well as orthogonality of metabolic biomarkers to other liver disease/disorders remains to be undertaken.

Acknowledgements This work was supported by the National Cancer Institute Intramural Research Program, the National Institute of Environmental Health Sciences grant (U01ES016013).

References

1. Hoyert DL, Heron MP, Murphy SL, Kung HC (2006) Deaths: final data for 2003. Natl Vital Stat Rep 54(13):1–120
2. Mandayam S, Jamal MM, Morgan TR (2004) Epidemiology of alcoholic liver disease. Semin Liver Dis 24(3):217–232
3. WHO (2011) Global status report on alcohol and health. World Health Organization, Geneva

4. Shibuya A, Yoshida A (1988) Genotypes of alcohol-metabolizing enzymes in Japanese with alcohol liver diseases: a strong association of the usual Caucasian-type aldehyde dehydrogenase gene (ALDH1(2)) with the disease. Am J Hum Genet 43(5):744–748
5. Pirmohamed M, Kitteringham NR, Quest LJ et al (1995) Genetic polymorphism of cytochrome P4502E1 and risk of alcoholic liver disease in Caucasians. Pharmacogenetics 5(6):351–357
6. Tanaka F, Shiratori Y, Yokosuka O, Imazeki F, Tsukada Y, Omata M (1996) High incidence of ADH2*1/ALDH2*1 genes among Japanese alcohol dependents and patients with alcoholic liver disease. Hepatology 23(2):234–239
7. Zintzaras E, Stefanidis I, Santos M, Vidal F (2006) Do alcohol-metabolizing enzyme gene polymorphisms increase the risk of alcoholism and alcoholic liver disease? Hepatology 43(2):352–361
8. Sherva R, Rice JP, Neuman RJ, Rochberg N, Saccone NL, Bierut LJ (2009) Associations and interactions between SNPs in the alcohol metabolizing genes and alcoholism phenotypes in European Americans. Alcohol Clin Exp Res 33(5):848–857
9. Auguet T, Vidal F, Broch M et al (2010) Polymorphisms in the interleukin-10 gene promoter and the risk of alcoholism and alcoholic liver disease in Caucasian Spaniard men. Alcohol 44(3):211–216
10. Linneberg A, Gonzalez-Quintela A, Vidal C et al (2010) Genetic determinants of both ethanol and acetaldehyde metabolism influence alcohol hypersensitivity and drinking behaviour among Scandinavians. Clin Exp Allergy 40(1):123–130
11. O'Shea RS, Dasarathy S, McCullough AJ (2010) Alcoholic liver disease. Am J Gastroenterol 105(1):14–32; quiz 33
12. Chen AC, Manz N, Tang Y et al (2010) Single-nucleotide polymorphisms in corticotropin releasing hormone receptor 1 gene (CRHR1) are associated with quantitative trait of event-related potential and alcohol dependence. Alcohol Clin Exp Res 34(6):988–996
13. Grove J, Brown AS, Daly AK, Bassendine MF, James OF, Day CP (1998) The RsaI polymorphism of CYP2E1 and susceptibility to alcoholic liver disease in Caucasians: effect on age of presentation and dependence on alcohol dehydrogenase genotype. Pharmacogenetics 8(4):335–342
14. Wong NA, Rae F, Bathgate A, Smith CA, Harrison DJ (2000) Polymorphisms of the gene for microsomal epoxide hydrolase and susceptibility to alcoholic liver disease and hepatocellular carcinoma in a Caucasian population. Toxicol Lett 115(1):17–22
15. Teli MR, Day CP, Burt AD, Bennett MK, James OF (1995) Determinants of progression to cirrhosis or fibrosis in pure alcoholic fatty liver. Lancet 346(8981):987–990
16. MacSween RN, Burt AD (1986) Histologic spectrum of alcoholic liver disease. Semin Liver Dis 6(3):221–232
17. Schutte K, Bornschein J, Malfertheiner P (2009) Hepatocellular carcinoma—epidemiological trends and risk factors. Dig Dis 27(2):80–92
18. Sherman M (2010) Hepatocellular carcinoma: New and emerging risks. Dig Liver Dis 42(Suppl 3):S215–S222
19. Levitsky J, Mailliard ME (2004) Diagnosis and therapy of alcoholic liver disease. Semin Liver Dis 24(3):233–247
20. Menon KV, Gores GJ, Shah VH (2001) Pathogenesis, diagnosis, and treatment of alcoholic liver disease. Mayo Clin Proc 76(10):1021–1029
21. Crabb DW, Liangpunsakul S (2006) Alcohol and lipid metabolism. J Gastroenterol Hepatol 21(Suppl 3):S56–S60
22. Lee SS, Pineau T, Drago J et al (1995) Targeted disruption of the alpha isoform of the peroxisome proliferator-activated receptor gene in mice results in abolishment of the pleiotropic effects of peroxisome proliferators. Mol Cell Biol 15(6):3012–3022
23. Martin PG, Guillou H, Lasserre F et al (2007) Novel aspects of PPARalpha-mediated regulation of lipid and xenobiotic metabolism revealed through a nutrigenomic study. Hepatology 45(3):767–777

24. Rakhshandehroo M, Sanderson LM, Matilainen M et al (2007) Comprehensive analysis of PPARalpha-dependent regulation of hepatic lipid metabolism by expression profiling. PPAR Res 2007:26839
25. Sozio M, Crabb DW (2008) Alcohol and lipid metabolism. Am J Physiol Endocrinol Metab 295(1):E10–E16
26. Nakajima T, Kamijo Y, Tanaka N et al (2004) Peroxisome proliferator-activated receptor alpha protects against alcohol-induced liver damage. Hepatology 40(4):972–980
27. Mancinelli R, Ceccanti M (2009) Biomarkers in alcohol misuse: their role in the prevention and detection of thiamine deficiency. Alcohol Alcohol 44(2):177–182
28. Sharpe PC (2001) Biochemical detection and monitoring of alcohol abuse and abstinence. Ann Clin Biochem 38(Pt 6):652–664
29. Saadeh S (2007) Nonalcoholic Fatty liver disease and obesity. Nutr Clin Pract 22(1):1–10
30. Calvaruso V, Craxi A (2009) Implication of normal liver enzymes in liver disease. J Viral Hepat 16(8):529–536
31. Cadranel JF, Rufat P, Degos F (2000) Practices of liver biopsy in France: results of a prospective nationwide survey. For the Group of Epidemiology of the French Association for the Study of the Liver (AFEF). Hepatology 32(3):477–481
32. Tyburski JB, Patterson AD, Krausz KW et al (2008) Radiation metabolomics. 1. Identification of minimally invasive urine biomarkers for gamma-radiation exposure in mice. Radiat Res 170(1):1–14
33. Patterson AD, Lanz C, Gonzalez FJ, Idle JR (2009) The role of mass spectrometry-based metabolomics in medical countermeasures against radiation. Mass Spectrom Rev 29(3): 503–521
34. Sreekumar A, Poisson LM, Rajendiran TM et al (2009) Metabolomic profiles delineate potential role for sarcosine in prostate cancer progression. Nature 457(7231):910–914
35. MacIntyre DA, Jimenez B, Lewintre EJ et al (2010) Serum metabolome analysis by 1H-NMR reveals differences between chronic lymphocytic leukaemia molecular subgroups. Leukemia 24(4):788–797
36. Loftus N, Barnes A, Ashton S et al (2011) Metabonomic investigation of liver profiles of non-polar metabolites obtained from alcohol-dosed rats and mice using high mass accuracy MSn analysis. J Proteome Res 10(2):705–713
37. Bradford BU, O'Connell TM, Han J et al (2008) Metabolomic profiling of a modified alcohol liquid diet model for liver injury in the mouse uncovers new markers of disease. Toxicol Appl Pharmacol 232(2):236–243
38. Fernando H, Kondraganti S, Bhopale KK et al (2010) (1)H and (3)(1)P NMR lipidome of ethanol-induced fatty liver. Alcohol Clin Exp Res 34(11):1937–1947
39. Pearson K (1901) On lines and planes of closest fit to systems of points in space. Philos Mag 2(6):559–572
40. Suhre K, Schmitt-Kopplin P (2008) MassTRIX: mass translator into pathways. Nucleic Acids Res 36(Web Server Issue):W481–W484
41. Jansson J, Willing B, Lucio M et al (2009) Metabolomics reveals metabolic biomarkers of Crohn's disease. PLoS One 4(7):e6386
42. Kind T, Fiehn O (2007) Seven Golden Rules for heuristic filtering of molecular formulas obtained by accurate mass spectrometry. BMC Bioinformatics 8:105
43. Cui Q, Lewis IA, Hegeman AD et al (2008) Metabolite identification via the Madison Metabolomics Consortium Database. Nat Biotechnol 26(2):162–164
44. Smith CA, O'Maille G, Want EJ et al (2005) METLIN: a metabolite mass spectral database. Ther Drug Monit 27(6):747–751
45. Nguyen PV, Abel T, Kandel ER, Bourtchouladze R (2000) Strain-dependent differences in LTP and hippocampus-dependent memory in inbred mice. Learn Mem 7(3):170–179
46. Syn WK, Yang L, Chiang DJ et al (2009) Genetic differences in oxidative stress and inflammatory responses to diet-induced obesity do not alter liver fibrosis in mice. Liver Int 29(8): 1262–1272

47. Liu J, Corton C, Dix DJ, Liu Y, Waalkes MP, Klaassen CD (2001) Genetic background but not metallothionein phenotype dictates sensitivity to cadmium-induced testicular injury in mice. Toxicol Appl Pharmacol 176(1):1–9
48. Manna SK, Patterson AD, Yang Q et al (2010) Identification of noninvasive biomarkers for alcohol-induced liver disease using urinary metabolomics and the Ppara-null mouse. J Proteome Res 9(8):4176–4188
49. Manna SK, Patterson AD, Yang Q et al (2011) UPLC–MS-based urine metabolomics reveals indole-3-lactic acid and phenyllactic acid as conserved biomarkers for alcohol-induced liver disease in the Ppara-null mouse model. J Proteome Res 10(9):4120–4133
50. Helander A, Bottcher M, Fehr C, Dahmen N, Beck O (2009) Detection times for urinary ethyl glucuronide and ethyl sulfate in heavy drinkers during alcohol detoxification. Alcohol Alcohol 44(1):55–61
51. Hoiseth G, Bernard JP, Stephanson N et al (2008) Comparison between the urinary alcohol markers EtG, EtS, and GTOL/5-HIAA in a controlled drinking experiment. Alcohol Alcohol 43(2):187–191
52. Shin M, Kim I, Inoue Y, Kimura S, Gonzalez FJ (2006) Regulation of mouse hepatic alpha-amino-beta-carboxymuconate-epsilon-semialdehyde decarboxylase, a key enzyme in the tryptophan-nicotinamide adenine dinucleotide pathway, by hepatocyte nuclear factor 4alpha and peroxisome proliferator-activated receptor alpha. Mol Pharmacol 70(4):1281–1290
53. Lumeng L, Crabb DW (2001) Alcoholic liver disease. Curr Opin Gastroenterol 17(3): 211–220
54. Kalant H, Khanna JM, Loth J (1970) Effect of chronic intake of ethanol on pyridine nucleotide levels in rat liver and kidney. Can J Physiol Pharmacol 48(8):542–549
55. Recasens M, Benezra R, Basset P, Mandel P (1980) Cysteine sulfinate aminotransferase and aspartate aminotransferase isoenzymes of rat brain. Purification, characterization, and further evidence for identity. Biochemistry 19(20):4583–4589
56. Yagi T, Kagamiyama H, Motosugi K, Nozaki M, Soda K (1979) Crystallization and properties of aspartate aminotransferase from Escherichia coli B. FEBS Lett 100(1):81–84
57. Jean M, DeMoss RD (1968) Indolelactate dehydrogenase from Clostridium sporogenes. Can J Microbiol 14(4):429–435
58. Leelayoova S, Marbury D, Rainey PM, Mackenzie NE, Hall JE (1992) In vitro tryptophan catabolism by Leishmania donovani donovani promastigotes. J Protozool 39(2):350–358
59. Owen TG, Hochachka PW (1974) Purification and properties of dolphin muscle aspartate and alanine transaminases and their possible roles in the energy metabolism of diving mammals. Biochem J 143(3):541–553
60. Bode R, Lippoldt A, Birnbaum D (1986) Purification and properties of D-aromatic lactate dehydrogenase, an enzyme involved in the catabolism of the aromatic amino acids of Candida maltosa. Biochem Physiol Pflanz 181:189–198

Chapter 14
Alcohol Metabolism by Oral Streptococci and Interaction with Human Papillomavirus Leads to Malignant Transformation of Oral Keratinocytes

Lin Tao, Sylvia I. Pavlova, Stephen R. Gasparovich, Ling Jin, and Joel Schwartz

Abstract Poor oral hygiene, ethanol consumption, and human papillomavirus (HPV) are associated with oral and esophageal cancers. However, the mechanism is not fully known. This study examines alcohol metabolism in *Streptococcus* and its interaction with HPV-16 in the malignant transformation of oral keratinocytes. The acetaldehyde-producing strain *Streptococcus gordonii* V2016 was analyzed for *adh* genes and activities of alcohol and aldehyde dehydrogenases. *Streptococcus* attachment to immortalized HPV-16 infected human oral keratinocytes, HOK (HPV/HOK-16B), human oral buccal keratinocytes, and foreskin keratinocytes was studied. Acetaldehyde, malondialdehyde, DNA damage, and abnormal proliferation among keratinocytes were also quantified. We found that *S. gordonii* V2016 expressed three primary alcohol dehydrogenases, AdhA, AdhB, and AdhE, which all oxidize ethanol to acetaldehyde, but their preferred substrates were 1-propanol, 1-butanol, and ethanol, respectively. *S. gordonii* V2016 did not show a detectable aldehyde dehydrogenase. AdhE is the major alcohol dehydrogenase in *S. gordonii*. Acetaldehyde and malondialdehyde production from permissible *Streptococcus* species significantly increased the bacterial attachment to keratinocytes, which was associated with an enhanced expression of furin to facilitate HPV infection and several malignant phenotypes including acetaldehyde adduct formation, abnormal proliferation, and enhanced migration through integrin-coated basement membrane

L. Tao (✉) • S.I. Pavlova • L. Jin
Department of Oral Biology, College of Dentistry, University of Illinois at Chicago, Chicago, IL 60612, USA
e-mail: ltao@uic.edu

S.R. Gasparovich
81st Dental Squadron, Keesler Air Force Base, Biloxi, MS, USA

J. Schwartz
Department of Oral Medicine and Diagnostic Sciences, College of Dentistry, University of Illinois at Chicago, Chicago, IL 60612, USA

V. Vasiliou et al. (eds.), *Biological Basis of Alcohol-Induced Cancer*, Advances in Experimental Medicine and Biology 815,
DOI 10.1007/978-3-319-09614-8_14

by HPV-infected oral keratinocytes. Therefore, expression of multiple alcohol dehydrogenases with no functional aldehyde dehydrogenase contributes to excessive production of acetaldehyde from ethanol by oral streptococci. Oral *Streptococcus* species and HPV may cooperate to transform oral keratinocytes after ethanol exposure. These results suggest a significant clinical interaction, but further validation is warranted.

Keywords Alcohol • Ethanol • Acetaldehyde • Dehydrogenase • ADH • ALDH • Cancer • Carcinogenesis • Keratinocytes • Human papillomavirus • HPV • *Streptococcus*

14.1 Ethanol, Bacteria, Human Papillomavirus, and Oral Cancer

Several types of cancers, such as the cancers of the head and neck, liver, colorectal, and female breast, have been linked to chronic ethanol consumption [1]. The strongest association of ethanol and increased cancer risk is seen for the upper aerodigestive tract including the oral cavity, throat, voice box, and esophagus. Recently, poor oral hygiene and tooth loss have been associated with an increased risk of oral and esophageal cancers [2–4], suggesting a role of oral microorganisms in carcinogenesis. Recent epidemiologic studies show an increased incidence of human papillomavirus (HPV)-related cancers, such as oropharyngeal cancers, in North America and Northern Europe [5, 6]. Although the exact mechanism by which ethanol consumption, bacterial and/or viral infections cause cancer is not fully known, local production of carcinogenic agents by microorganisms is suspected [7]. Additionally, we observed that ethanol metabolism by certain oral *Streptococcus* strains promoted the function and activity of furin, a serine proprotein convertase in mammalian cells, which is required for the entry of HPV subtype 16 (oncogenic mucotropic form) into human oral keratinocytes (HOK) [8] and has a variety of other functions related to malignant transformation [9].

Ethanol itself is not carcinogenic, but it can be metabolized to carcinogenic acetaldehyde in the oral cavity by oral microorganisms [10, 11]. In the oral cavity, *Streptococcus* species are dominant bacteria [12]. It has been reported that many species of oral streptococci can produce acetaldehyde from ethanol [7, 13]. Also, alcohol dehydrogenase (ADH) activities have been observed in oral *Streptococcus* species [14]. However, little is known about genes encoding these ethanol metabolic enzymes in these bacteria and whether ethanol metabolism by bacteria play a role in facilitating HPV infection in the oral and esophageal carcinogenesis.

The metabolite of ethanol is acetaldehyde, which is a carcinogen in animal models [15] and causes chromosomal damage, including sister-chromatid exchanges and chromosomal aberrations [16]. It reacts with 2′-deoxyguanosine to form *N2-ethyl-2′-deoxyguanosine* (N^2-EtdG) to form DNA adduct in animal models of ethanol exposure and in white blood cells of human alcoholics [17]. Additionally, acetaldehyde inhibits DNA repair enzymes [18]. Recently, acetaldehyde associated with alcoholic beverages has been named as a Class I carcinogen for humans by the International Agency for Research on Cancer of WHO [19].

ADH catalyzes the conversion of ethanol to acetaldehyde, which can be further converted to acetic acid by aldehyde dehydrogenase (ALDH). Therefore, if a bacterium has both active ADH and ALDH, it can metabolize ethanol fully to the harmless acetic acid. However, if a bacterium has active ADH without ALDH, excessive acetaldehyde can be produced from ethanol. Acetaldehyde can readily diffuse through eukaryote and prokaryote membranes, and increase the long-term risk for DNA damage to epithelial cells with formation of bulky adducts and mutations. In both bacterial species and HOK, ethanol can be metabolized by ADH to produce acetaldehyde and then to acetyl-coenzyme A or acetate by ALDH [20]. Human genes encoding ADH and ALDH, as well as other associated enzymes, show different activities in ethanol metabolism, release of acetaldehyde, differentiation of oral epithelium, and risk in developing cancer depending on individual gene polymorphism [21].

In addition to production of acetaldehyde from ethanol, the oral fungus *Candida albicans* can also produce acetaldehyde from glucose through the pyruvate-bypass pathway [22, 23], but most oral bacteria, including *Streptococcus*, do not have pyruvate decarboxylase, the enzyme required for this pathway. Therefore, excessive production of acetaldehyde from ethanol by oral bacteria [3, 13] and from ethanol and/or glucose by oral fungi may contribute to an increased risk of oral-esophageal cancer. However, unless in immune compromised patients, the oral-esophageal cavity of a healthy adult human often has a low number of fungi but a high number of bacteria.

The human body carries about 10^{14} bacteria in the oral-digestive tract [24, 25]. The metabolic activities performed by these bacteria resemble those of an organ [26]. Like host cells, many bacteria in the oral cavity and gut can produce acetaldehyde from ingested ethanol [7, 11, 13, 14, 27–29]. Even though active ALDH is present in the human host tissues, nearly all acetaldehyde accumulated in the saliva is of microbial origin due to abundance of oral bacteria [14, 29]. Therefore, acetaldehyde production by oral bacteria from ethanol is significant to the etiology of oral-esophageal cancers.

14.2 Ethanol Metabolism by Oral Streptococci

14.2.1 Selection of a Representative Strain

Streptococci readily colonize mucosal tissues in the nasopharynx and the respiratory and gastrointestinal tracts [12]. The oral cavity contains teeth, which provide non-shedding surfaces. This allows bacteria to adhere to the surface of teeth to form dental plaque, or biofilm, which contains high density of bacteria, mainly streptococci. Therefore, we focused on oral streptococci to study the microbial ethanol metabolism [30].

Two groups of oral streptococcal strains were analyzed. The first group included 14 laboratory strains: *Streptococcus sanguinis* ATCC 10556, S7, Blackburn, 1239b, 133-79, V2020, V2053, V2054, and V2650 (SK36), *Streptococcus gordonii* V685, 488, CHI, V288 and V2016. The second group included 38 clinical strains isolated from the saliva of 12 healthy volunteers. Their species were identified by 16S rRNA gene sequence to be *Streptococcus gordonii, S. mitis, S. oralis, S. salivarius,* and *S. sanguinis*. The clinical study was approved by Institutional Review Board of the University of Illinois at Chicago.

Bacterial metabolism of ethanol involves two steps. The first step is to convert ethanol to acetaldehyde by ADH. The second step is to convert acetaldehyde to acetic acid by ALDH. If a bacterium has both high ADH and ALDH, it can convert ethanol to acetic acid, which reduces pH and can be detected by color change in the purple broth [30]. However, if a bacterium has only ADH but no ALDH, it can only produce acetaldehyde from ethanol without further converting it to acetic acid.

The growth of streptococcal strains and detection of acetaldehyde production from ethanol by oral *Streptococcus* were described previously [30]. A total of 52 oral *Streptococcus* strains were analyzed for their capacity of acetic acid and acetaldehyde production from ethanol. Only 17 strains of *S. mitis, S. oralis* and *S. salivarius* produced acetic acid from ethanol, while 19 strains of all five of these *Streptococcus* species (*S. gordonii, S. mitis, S. oralis, S. salivarirus* and *S. sanguinis*) produced acetaldehyde. Among these *Streptococcus* strains some produced only acetic acid without detectable acetaldehyde, while others produced only acetaldehyde without detectable acetic acids, and still others produced both or neither products from ethanol. Among all the strains tested, *S. gordonii* V2016, *S. oralis* 108, and *S. mitis* 110-5 showed the most abundant production of acetaldehyde. However, *S. oralis* 108 and *S. mitis* 110-5 also showed production of acetic acid from ethanol, but *S. gordonii* V2016 showed only acetaldehyde production from ethanol. Therefore, V2016 was selected for further study of its enzymes involved in acetaldehyde production from ethanol.

14.2.2 Construction of **adh** *Mutants in* **S. gordonii** *V2016*

By in silico analysis of the *S. gordonii* genome [31], we have identified three genes: *adhA* (SGO_0565; 1,023 bp), *adhB* (SGO_1774; 1,038 bp), and *adhE* (*acdH,* SGO_0113; 2,652 bp) that encode putative ADHs. These three genes were subject to PCR-ligation mutagenesis with different antibiotic resistance markers to achieve allelic exchange [32]. Standard recombinant DNA techniques were employed as described [33]. Multiple pairs of oligonucleotides (Integrated DNA Technologies) used in this study and the *S. gordonii* V2016 *adh* deletion mutants were constructed and confirmed as previously described [30].

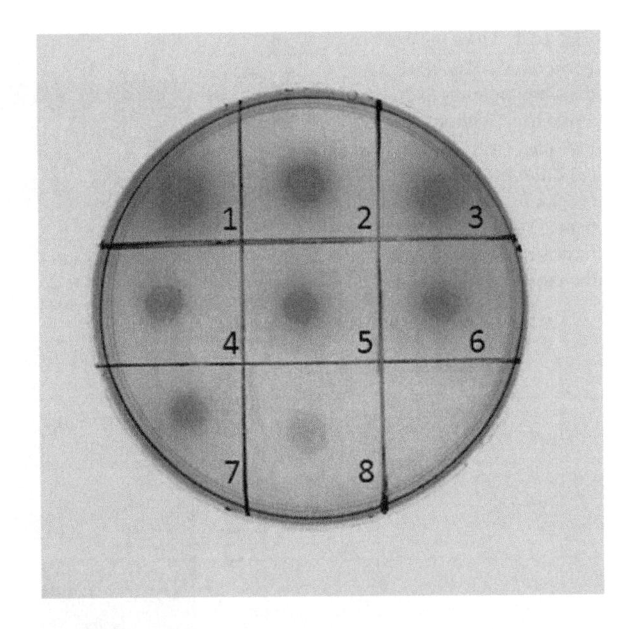

Fig. 14.1 Acetaldehyde production by *Streptococcus gordonii* on the PBB-Schiff's agar. (1) V2016wt; (2) *ΔadhA*; (3) *ΔadhB*; (4) *ΔadhAB*; (5) *ΔadhE*; (6) *ΔadhAE*; (7) *ΔadhBE*; (8) *ΔadhABE*

14.2.3 Acetaldehyde Production by S. gordonii *V2016* adh *Mutants*

The acetaldehyde production by the wild type and various gene-deletion mutants of V2016 was tested on the PBB-Schiff's agar plate. A positive reaction to acetaldehyde is indicated by the pinkish red color, while a negative reaction by white color. As shown in Fig. 14.1, the wild type strain displayed the strongest production, the single gene knockout mutants a slightly reduced production, double gene knockout mutants a much reduced production, and the triple knockout mutant a negative production of acetaldehyde.

14.2.4 Alcohol Dehydrogenases of S. gordonii *V2016*

Because existing enzyme activity gel assays (zymograms) were not sensitive enough to detect multiple ADH activities from crude bacterial samples, we developed a more sensitive zymogram method, which allowed us to detect multiple ADHs simultaneously on the same gel with crude bacterial lysates. ADH and ALDH activities of the testing bacteria were determined by a specific zymogram method improved from several methods described previously [11, 34, 35]. Nitroblue tetrazolium (NBT) in the presence of phenazine methosulfate (PMS) reacts with nicotinamide adenine dinucleotide phosphate (NADPH) produced by dehydrogenases to produce an insoluble blue-purple formazan. This new NBT-PMS detection method

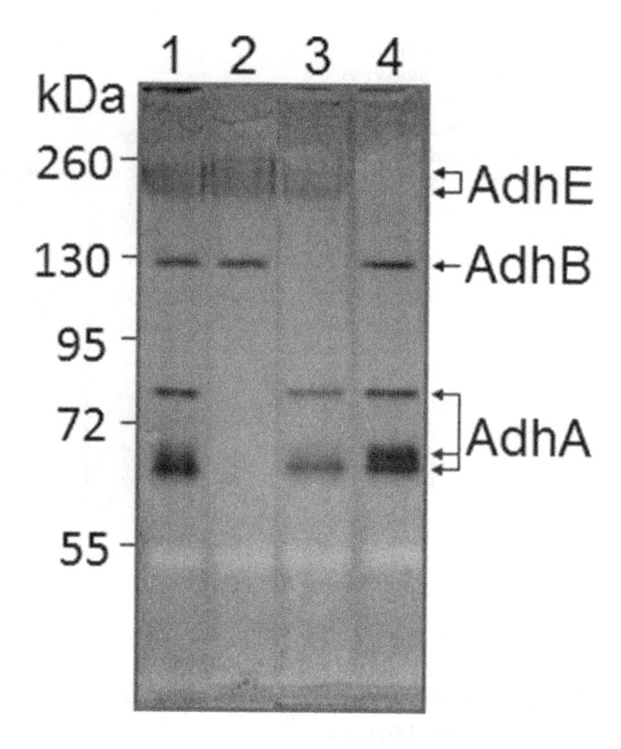

Fig. 14.2 *Streptococcus gordonii* V2016 ADH distribution analysis by zymogram without added Fe^{2+} and Zn^{2+}: (1) V2016wt; (2) *ΔadhA*; (3) *ΔadhB*; (4) *ΔadhE*. Note: missing band(s) of each *Δadh* mutant indicates the location(s) of the target ADH

was used to visualize ADH and ALDH in polyacrylamide gels [30]. Yeast ADH and ALDH (Sigma-Aldrich) were used as positive controls.

By knocking out three *adh* genes individually and in various combinations, we identified that *S. gordonii* V2016 has three primary ADHs, AdhA, AdhB, and AdhE (Fig. 14.2), which all recognized ethanol as a substrate with different activities (Fig. 14.3). Additionally, we have also identified a secondary ADH, S-AdhA, which specifically recognizes the secondary alcohol, 2-propanol, and a dehydrogenase specific for threonine (Fig. 14.4). These two dehydrogenases, however, do not recognize ethanol as their substrate (Figs. 14.2 and 14.3a), despite that AdhA recognizes both threonine and 2-propanol as its substrate (Fig. 14.4).

In addition to *S. gordonii* V2016, three other *S. gordonii* strains including V288, CHI, and 110-3, and five *S. sanguinis* laboratory strains including 133-79, S7, Blackburn, SK36 and ATCC 10556, and 11 oral *Streptococcus* isolates, including four strains produced only acetaldehyde, two strains produced both acetic acid and acetaldehyde, and five strains produced only acetic acid from ethanol (see legend to Fig. 14.5 for strain names), were also analyzed for both ADH and ALDH activities.

Among three ADHs in *S. gordonii* V2016, a cross regulation of their activities may exist. As shown in Fig. 14.2, when the activity of AdhE is weak due to the lack of its cofactor Fe^{2+} or missing due to gene deletion, the activities of AdhA and AdhB were relatively strong. However, when Fe^{2+} and Zn^{2+} were supplemented to the growth medium, the AdhE activity became substantially increased (Fig. 14.3a, Lanes 1–4) but the activities of both AdhA and AdhB were suppressed, possibly by

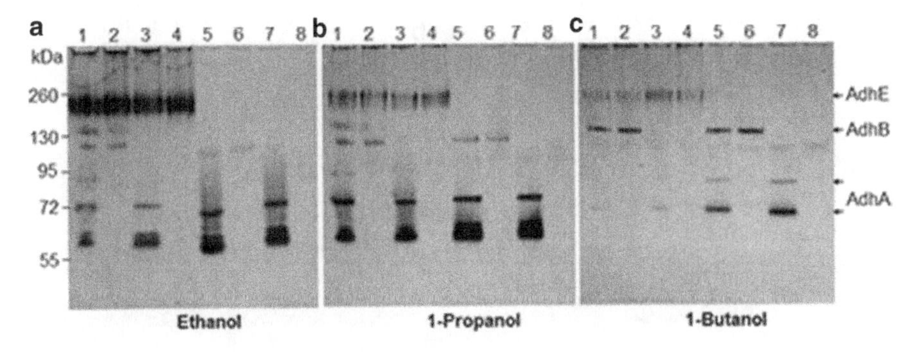

Fig. 14.3 Analysis of substrate preference of *Streptococcus gordonii* V2016 ADH. Fe^{2+} and Zn^{2+} were added to the growth medium and zymogram detection solution. (1) V2016wt; (2) *ΔadhA*; (3) *ΔadhB*; (4) *ΔadhAB*; (5) *ΔadhE*; (6) *ΔadhAE*; (7) *ΔadhBE*; (8) *ΔadhABE*. *Note*: in the wild type strain, AdhE prefers ethanol, AdhA prefers 1-propanol and AdhB prefers 1-butanol

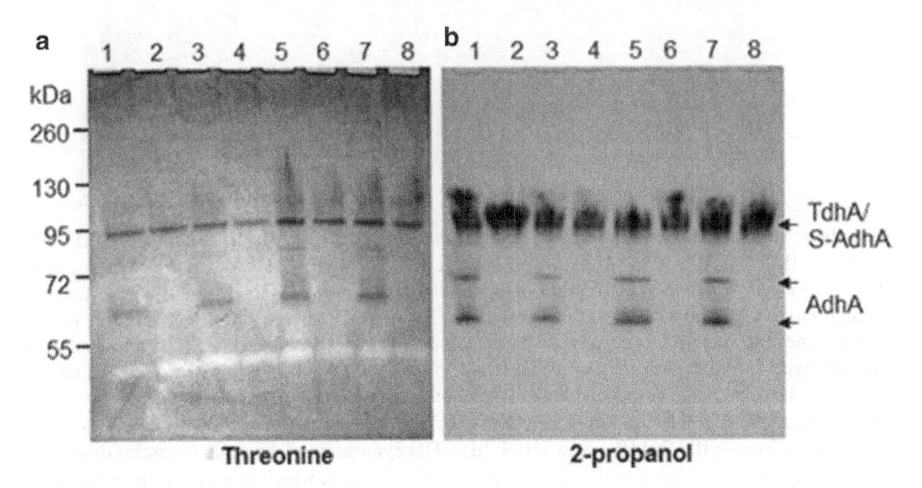

Fig. 14.4 Identification of two novel dehydrogenases in *S. gordonii*, the threonine dehydrogenase (TdhA) and the secondary alcohol dehydrogenase (S-AdhA) by zymograms. (1) V2016wt; (2) *ΔadhA*; (3) *ΔadhB*; (4) *ΔadhAB*; (5) *ΔadhE*; (6) *ΔadhAE*; (7) *ΔadhBE*; (8) *ΔadhABE*. *Note*: AdhA reacted with both threonine and 2-propanol because mutants with *ΔadhA* did not show these bands

the increased activity of AdhE. However, when the *adhE* gene was deleted, the activity of AdhA was increased (Fig. 14.3a, Lanes 5, 7), but not AdhB, which appeared to be relatively independent from AdhE regulation. A similar scenario was also observed when 1-propanol (Fig. 14.3b) and 1-butanol were used as substrates (Fig. 14.3c). These results suggest that AdhE may be the major ADH in *S. gordonii*. When its activity is upregulated, the activities of other ADHs, especially the AdhA, are reduced.

The ADH activities of V2016 and its various *adh* mutants were analyzed with the NBT-PMS zymogram method. First, the approximate size of each ADH was

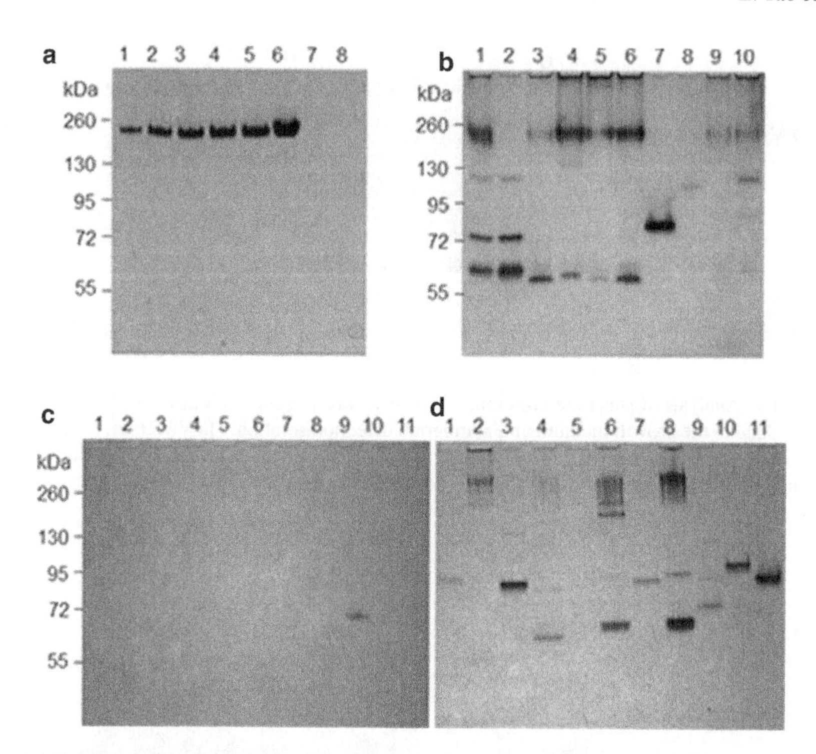

Fig. 14.5 (**a**) ALDH zymogram: 1-6, *Saccharomyces sereviciae* ALDH controls: 1, 0.1U; 2, 0.25U; 3, 0.5U; 4, 0.75U; 5, 1 U; 6, 3 U; 7, blank; 8, *S. gordonii* V2016wt. (**b**) ADH zymogram: 1-3, *S. gordonii* V2016wt, V2016*ΔadhE* and V288; 4 and 5, *S. sanguinis* S7 and Blackburn; 6 and 7, *S. gordonii* CHI and 110-3; 8-10, *S. sanguinis* 133-79, SK36 and ATCC 10556 (ALDH zymogram of these strains was all negative; data not shown). (**c**) ALDH zymogram: 1-4 (produced only acetaldehyde from ethanol), *S. salivarius* 101-1; *S. sanguinis* 104-5; *S. salivarius* 109-2, and *S. sanguinis* 109-3; 5 and 6 (produced both acetaldehyde and acetic acid from ethanol), *S. oralis* 108 and *S. mitis* 110-5; 7-11 (produced only acetic acid from ethanol), *S. salivarius* 101-7, *S. mitis* 104-4, *S. salivarius* 107-2, 110-1, and 110-4. (**d**) ADH zymogram of the same 11 strains displayed in (**c**). Note: only *S. salivarius* 107-2 displayed an ALDH band, which is different from AdhE. *S. oralis* 108 did not show detectable ADH

estimated by testing each *adh* gene-deletion mutant against the wild type on an ADH zymogram. As shown in Fig. 14.2, the wild type V2016 displays all three functional ADH enzymes (Lane 1). The Δ*adhA* mutant lacks three bands between 55 and 72 kDa (Lane 2). The Δ*adhB* mutant misses a single band near 130 kDa (Lane 3). The Δ*adhE* mutant misses top two bands around 260 kDa (Lane 4). Because the actual molecular sizes of these enzymes cannot be determined by the native polyacrylamide gel the sizes and shapes of these enzymes can only be estimated. For example, since three bands are related AdhA, this enzyme may take three different forms (e.g., monomer, dimer, and/or trimer).

The DNA sequences of *adhA* and *adhB* both show a zinc-binding domain, but only AdhA showed enhanced ADH activity after zinc supplementation and only in the absence of AdhE (Fig. 14.3a). Supplementing iron significantly enhanced AdhE activity (Fig. 14.3a, Lanes 1–4) suggesting that the AdhE protein is an iron-dependent ADH. However, in mutants with *adhE* inactivation (Fig. 14.3, Lanes 5 and 7), the activity of AdhA is enhanced, but the activity of AdhB is not. The AdhB protein has apparent one (Fig. 14.2) or two conformations (Fig. 14.3). The second AdhB band showed up only when zinc was supplemented into the growth medium and when AdhE is present.

14.2.5 Absence of ALDH in S. gordonii *V2016*

The *S. gordonii* genomic data [31] showed that this bacterium has a gene (*acdH,* SGO_0113) encoding the putative dual functional ALDH/ADH (AdhE). It has homology to the ALDH/ADH dual function AdhE in other bacteria [36]. Therefore, it is important to test whether *S. gordonii* V2016 AdhE also has dehydrogenase activity for acetaldehyde. As shown in Fig. 14.5a, we tested *S. gordonii* V2016 with the optimized NBT-PMS zymogram. However, *S. gordonii* V2016 did not show any detectable ALDH activity (Fig. 14.5a, Lane 8). To make sure that this method is sensitive enough to detect microbial ALDH, we used *S. cerevisiae* ALDH as a positive control. The zymogram detected ALDH activity as low as 0.1 U. This method has also detected ALDH from another oral *Streptococcus* strain, *S. salivarius* 107-2 (Fig. 14.5c). Therefore, the zymogram method should be reliable and the negative result indicated that *S. gordonii* V2016, as well as other tested oral *Streptococcus* strains, did not have a detectable ALDH. Because the *adhE* gene of these oral streptococci is highly homologous to *adhE* in other bacteria [36] that encodes a bifunctional ALDH/ADH, there might be a mutation(s) in its ALDH domain. This finding, together with the finding of oral *Neisseria* [11], indicates that genetic polymorphisms in ALDH in bacteria may exist similar to that seen in humans [21, 37]. Because most tested oral streptococcal strains showed multiple ADH, but no ALDH, the enzyme distribution bias may contribute to their excessive production of acetaldehyde from ethanol.

In the East Asian population of humans, a rather high percentage (up to 30 %) carries a defective ALDH2, which is caused by a point mutation resulting in a Glu to Lys substitution at the amino acid position 487, and is referred to as *ALDH2*487Lys* (previous symbol: *ALDH2**2) [38, 39]. In this study, we observed that in most strains of oral *Streptococcus* tested, the AdhE protein has only ADH but no ALDH activity. This is also true in *Neisseria* [11]. This indicates that the *adhE* gene of these bacteria might have lost its ability to express functional ALDH during the course of evolution. Based on a recent study on bacterial evolution [40], if a gene is nonessential for bacterial survival, more mutations can be accumulated in comparison with genes that are essential. Because *adhE* is nonessential, a random mutation in *adhE* could be allowed and be passed down to the offspring. The questions are how many bacterial species carry such a mutation in their *adhE* gene and which base substitution(s) may inactivate its ALDH activity.

Table 14.1 Substrate specificity of *S. gordonii* V2016 dehydrogenases

Substrate	AdhA	AdhB	AdhE	S-AdhA	TdhA
Acetaldehyde	–	–	–	–	–
Methanol	–	–	–	–	–
Ethanol	+[a]	+	+++	–	–
1-Propanol	+++	+	+	–	–
2-Propanol	+	–	–	+++	–
1-Butanol	+[a]	+++	+	±	–
tert-Butanol	–	–	–	–	–
Threonine	+	±	–	–	++

[a]When AdhE was present, AdhA activity was low, but when AdhE was absent, AdhA activity was high

14.2.6 *Substrate Specificities of* S. gordonii *ADHs*

In addition to ethanol, we also tested other alcohols, including methanol, 1-propanol, 2-propanol, 1-butanol, and tertiary-butanol, and the amino acid threonine with the NBT-PMS zymogram method. Except for methanol and tertiary-butanol which showed no activity all other tested substrates showed varied activities with these three primary ADHs (Figs. 14.3 and 14.4). The preferred substrates for AdhA, AdhB, and AdhE were 1-propanol, 1-butanol, and ethanol, respectively. Additionally, two new dehydrogenases for threonine and 2-propanol were observed. Insertion-inactivation study showed that the dehydrogenase encoded by the gene located at SGO_0440 was specific for threonine. We therefore named this gene *tdhA*, encoding the threonine dehydrogenase. To find the gene locus coding for 2-propanol dehydrogenase, we identified three genes with homologies to major dehydrogenases: SGO_0273, SGO_0440, and SGO_0841. However, none of the mutations in the three loci had inactivated the enzyme activity for 2-propanol. Therefore, the gene encoding the dehydrogenase specific for the secondary alcohol remains to be determined. The specificities of five dehydrogenases to various tested substrates are listed in Table 14.1. The substrate specificity analysis showed that three ADHs of *S. gordonii* V2016 all recognize a broad range of substrates besides ethanol, but the activities of S-AdhA and TdhA were quite specific to their preferred substrates. It appears to be disadvantageous for a bacterium to have multiple ADHs that all produce the toxic metabolite from ethanol. Having multiple different ADHs may offer the bacterium competitive growth advantage in the environment due to the capability of utilizing multiple different nutrient substrates.

14.2.7 *ADH/ALDH Profiles Vary Among Strains of Oral* Streptococcus

As shown in Fig. 14.5b, among four *S. gordonii* strains tested, only V2016 showed all three ADHs (AdhA, AdhB, and AdhE). *S. gordonii* V288 and CHI showed AdhA and AdhE, but no AdhB. However, *S. gordonii* 110-3 showed only one ADH

Table 14.2 Furin expression associated with *Streptococcus* adherence to immortalized oral keratinocytes (hTERT HOK)

Cell	% of Cells express furin[d]	% of Cell with ≥10 bacteria adhered to surface[d]
Untreated hTERT HOK control	22	ND
101-7+1 % ethanol[a]	45	75
101-7+1 % ethanol+CMK[b]	28	75
101-7	42	72
101-1+1 % ethanol[c]	66	74
101-1+1 % ethanol+CMK	46	31
101-1	37	43

Note: *S. salivarius* is a common oral bacterium of the tongue. Furin was detected with a primary rabbit polyclonal antibody and a secondary anti-rabbit antibody (FITC)

[a]*S. salivarius* 101-7 (does not produce acetaldehyde only acetic acid from ethanol) exposed to 1 % ethanol (added 50×10^3/well) and incubated for 3 h after 1 h exposure to ethanol

[b]CMK=UIGI-1 (5-FAM-Lys-Arg-Val-Lys-Arg-CMK) added to HOK before *S. salivarius* labeled with 5-Carboxy-di-O-acetylfluorescein *N*-succinimidyl ester

[c]*S. salivarius* 101-1 (produces acetaldehyde from ethanol) exposed to 1 % ethanol (added 50×10^3/ well and incubated for 3 h after 1 h exposure to ethanol

[d]Estimated % of cells (counted 3 × 100 cells/well in 6-well plates)

similar to AdhA. Among five *S. sanguinis* strains tested, S7 and Blackburn showed only AdhA and AdhE. 133-79 showed only a weak AdhB. SK36 showed only a weak AdhE. Although ATCC 10556 showed four bands representing AdhA, AdhB, and AdhE, their activities are relatively weak. As shown in Figs. 14.5c, d, three groups of oral *Streptococcus* strains included four strains produced only acetaldehyde, two strains produced both acetic acid and acetaldehyde, and five strains produced only acetic acid from ethanol were tested for both ALDH and ADH. Although most strains showed one or more ADHs, only one strain, *S. salivarius* 107-2 showed an ALDH activity band, which is significantly smaller than AdhE (Fig. 14.5c). Because its size is not within the range of AdhE, it may be a novel ALDH (Table 14.2).

With a broad range of substrate preferences and varied ADH profiles, these bacteria may metabolize ethanol differently. Because several acetic acid producers did not show ALDH activity bands, these bacteria may have either very weak ALDH or use different mechanisms to produce acetic acid from ethanol. In addition to enzymatic pathways, ethanol can also be oxidized by non-enzymatic free radical pathways to produce acetaldehyde [41, 42]. This might explain *S. oralis* 108 that showed no detectable ADH activity (Fig. 14.5d) but still produced excessive acetaldehyde from ethanol.

AdhE is highly conserved and may have multiple functions depending upon different bacterial species. For example, in *Leuconostoc*, AdhE is a bifunctional ALDH/ADH [36]. In *Escherichia coli* [43] and *Streptococcus bovis* [44], AdhE has three distinct enzymatic activities: ADH, acetaldehyde-CoA dehydrogenase, and pyruvate formate-lyase (PFL) deactivase. In *Listeria*, AdhE is also a major adhesion protein (named LAP, stands for *Listeria*-adhesion protein) and is located on the cell surface [45]. In *Thermoanaerobacter mathranii,* AdhE is a bifunctional ALDH/ADH responsible for ethanol production [46].

14.2.8 *Effect of* adh *Deletions on Bacterial Growth in Medium Containing Ethanol or Acetaldehyde*

The V2016 wild type and its seven mutants, *ΔadhA, ΔadhB, ΔadhE, ΔadhAB, ΔadhAE, ΔadhBE*, and *ΔadhABE*, were analyzed for growth in THY broth (control) and THY broth supplemented with 1 % ethanol or 1 % acetaldehyde. Each strain was grown in three tubes of 5 mL THY broth overnight with serial diluted inoculations. In the second morning, the culture at mid-exponential phase was transferred with a 1:100 dilution to the three different testing media and incubated at 37 °C. The optical density at 600 nm was measured every 30 min with a Genesys 20 Spectrophotometer. To better present the bacterial growth data, optical density readings as a function of time in the logarithmic growth phase were converted to doubling time (Fig. 14.6).

Deletion of any of these three *adh* genes did not show apparent difference in the growth doubling times in THY without supplemented ethanol. However, when ethanol was supplemented at 1 %, the growth became slowed with significantly longer doubling times for the wild type and the *ΔadhA* and/or *ΔadhB* mutants. The four mutants containing *adhE* deletion had largely the same doubling times between growth in THY alone and THY supplemented with 1 % ethanol. In comparison with the wild type, these four mutants showed significantly shorter doubling times in THY supplemented with 1 % ethanol. All eight strains displayed significantly longer doubling times in THY supplemented with 1 % acetaldehyde.

The significant increase in bacterial doubling time of all eight strains indicates that acetaldehyde is more toxic than ethanol. A similar effect is also reported in a study with yeast [47]. Therefore, mutants that lack the enzyme for the production of acetaldehyde can be more tolerant to ethanol than the wild type [5]. The growth study (Fig. 14.6) showed that all four mutants containing *ΔadhE* when grown in THY containing 1 % ethanol had no significant increase in doubling times comparing with growth in control THY. However, the *ΔadhA* and/or *ΔadhB* mutants showed increased doubling times like the wild type when grown in THY containing 1 % ethanol. This suggests that AdhA and AdhB may be less involved in acetaldehyde production from ethanol than AdhE in *S. gordonii* V2016.

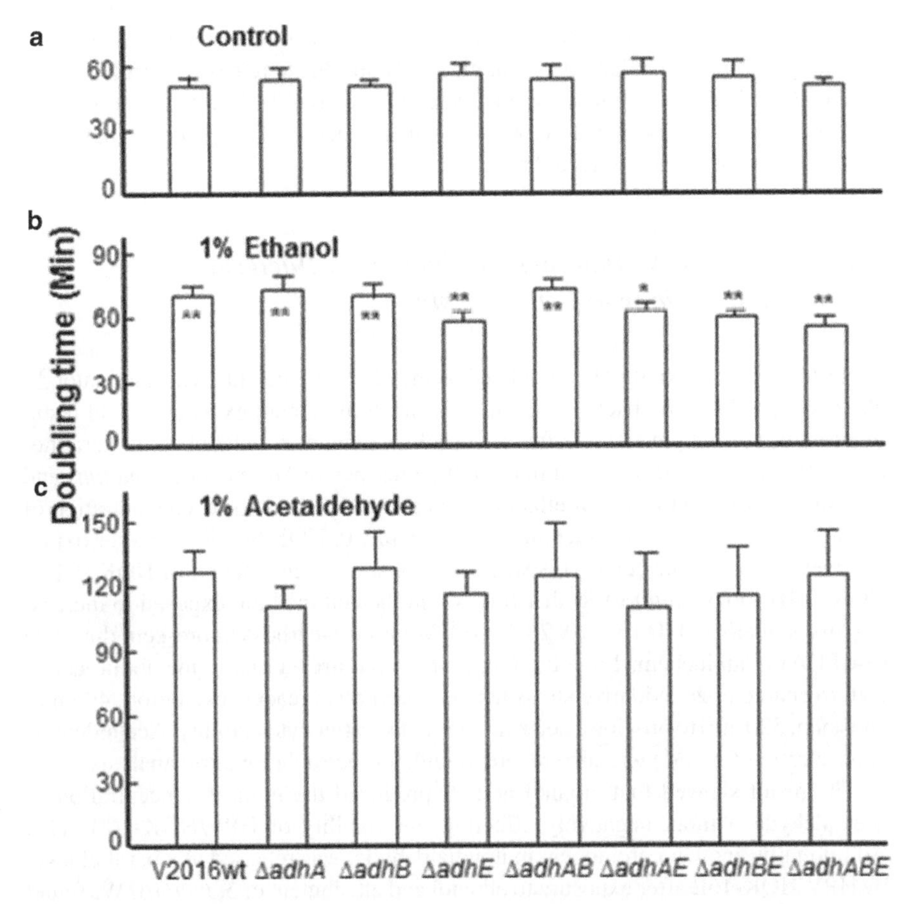

Fig. 14.6 Doubling times of *Streptococcus gordonii* V2016wt and seven *Δadh* mutants grown in THY or THY containing 1 % ethanol or 1 % acetaldehyde. *Asterisk* represents statistic difference by the Student's *t*-test (*$p<0.05$; **$p<0.01$). When *asterisk* is on top of the data bar, it represents significant difference between the doubling time of this strain and its wild type growing in the same medium. When *asterisk* is inside the data bar, it represents significant difference between the same strain growing in THY and THY containing 1 % ethanol. All strains growing in THY containing 1 % acetaldehyde had significantly longer doubling times than grown in other media. Each data bar represents the average of five measurements plus standard deviation

14.3 Ethanol Metabolism by Oral Streptococci Increases Bacterial Adhesion, HPV Entry, and HPV-Mediated Malignant Transformation of Oral Keratinocytes

Because both streptococci and HPV are commonly found in the oropharynx at relatively high levels and both independently produce damages to HOK, we hypothesized that bacteria and HPV may cooperate to affect oral mucosal cells and increase the risk for malignant transformation. When encountering environmental stress, such as exposure to ethanol, bacteria may increase attachment to the oral mucosa. This may disrupt the wellbeing of oral mucosal cells due to release of virulence factors and/or ethanol metabolites by the attached bacteria. As a result, oral mucosal cells may become vulnerable to viral infection or undergo virus-mediated malignant changes for cells already infected by a virus.

14.3.1 Ethanol Metabolites Production by Bacteria and Toxicity to Keratinocytes

We tested the cytotoxicity of ethanol and its metabolites, acetaldehyde and malondialdehyde, produced by bacteria. Even a low level of ethanol exposure could result in a loss of normal physiologic function such as oxidation metabolism in keratinocytes. This cell feature was examined in the presence of *Streptococcus mutans* and exposure to different levels of ethanol. We expected to find a low concentration of ethanol that enhanced the attachment of *S. mutans* to HOK but did not alter oxidative metabolism of the cells. This was accomplished by incubation of HOK (HPV/HOK-16B) and *S. mutans* together for 24 h in the cell medium exposed to increasing levels of ethanol (0.1, 1, 10, 20%, v/v). We used 5-carboxyfluorescein diacetate (5-CFDA) (Calbiochem, La Jolla, CA), which requires a viable metabolic active cell, to characterize oxidative stress using a microplate reader (excitation, 495 nm; emission, 520 nm) to provide a determination of keratinocyte viability. Acetaldehyde concentration (in mM) was also recorded with a spectrophotometric analysis.

The result showed that ethanol at 1 % produced the highest concentration of acetaldehyde without negatively affecting the viability of HPV/HOK-16B cells. Additionally, another carcinogen, malondialdehyde, was released at a parallel level by HPV/HOK-16B after exposure to ethanol and attachment of *S. mutans*. We found that 24 h of exposure of *S. mutans* and HPV/HOK-16B to ethanol produced significantly ($p<0.05$) higher levels of malondialdehyde (67 ± 5.5 mM/L) in comparison with exposure to ethanol with HPV/HOK-16B (malondialdehyde, 38 ± 42.0 mM/L) and with HPV/HOK-16B cells alone (malondialdehyde, 5 ± 5.4 mM/L).

Therefore, *S. mutans* co-incubation with HPV/HOK-16B and exposure to ethanol produced significantly increased release of two carcinogens, acetaldehyde and malondialdehyde, in comparison with ethanol exposure by only HPV/HOK-16B cells. Likewise, other oral streptococcus species, such as *S. gordonii* and *S. salivarius*, also produced similar results.

14.3.2 Ethanol Exposure Promotes Attachment of Streptococci to HOK

To evaluate effect of ethanol on the interaction among bacteria, viruses and host cells, we tested changes in *Streptococcus* adherence to HPV-infected and non-HPV-infected HOK cells [48]. The normal human foreskin keratinocyte (HFK) purchased from BioWittaker/Clonetics (Walkersville, MD) was used as a non-oral cell control

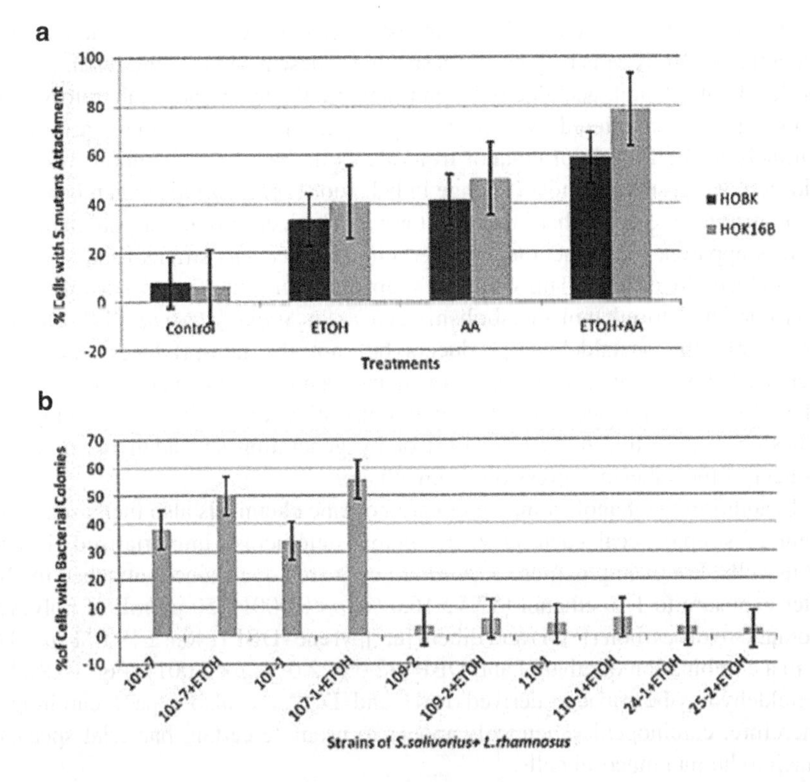

Fig. 14.7 (**a**) Percent of human oral buccal keratinocytes (HOBK) or HPV/HOK-16B cells with *S. mutans* attachment after exposure to ethanol (ETOH) and/or acetaldehyde (AA). After exposure to 1 % ethanol and/or acetaldehyde (3 h), *S. mutans* attachment to fresh HOBK and HPV/HOK-16B cells was significantly higher ($p<0.003$) than controls (*S. mutans* not exposed to ethanol with either human cell). (**b**) Identification of acetaldehyde producer *S. salivarius* by green fluorescent dye to detect attachment (with 20× magnification). The *S. salivarius* strain that produced acetaldehyde (101-7) and the strain that only produced acetic acid (109-2) were observed for attachment after exposure to 1 % ethanol. There was a significant level ($p<0.001$) of attachment by *S. salivarius* strains (101-7 and 107-1) without or in the presence of ethanol in comparison to controls. Controls included *S. salivarius* strains (109-2 and 110-1) that do not produce acetaldehyde and *L. rhamnosus* strains (1, 2) without or in the presence of ethanol and HPV/HOK-16B cells

and *Lactobacillus rhamnsus* was used as a non-streptococcus bacterial control. The plasmid pMHPV-16d was transfected into 70 % confluent normal HOK to produce the cell line HPV/HOK-16B (obtained from Dr. No Hee Park, University of California, Los Angeles). Cells were cultured in Dulbecco's modified Eagle's medium (DMEM) and Ham F-12 with 10 % fetal bovine serum (FBS). Assays were performed without penicillin and streptomycin, unless otherwise specified.

Certain *Streptococcus* strains, namely *S. mutans* LT11, *S. salivarius* 101-7 and 107-1, after low-level ethanol exposure (1 %) released carcinogens acetaldehyde and malondialdehyde. This occurred along with an increased adherence to the cell surfaces of HPV/HOK-16B (Fig. 14.7). However, other *S. salivarius* srains, such as 101-1 and 109-2, and two *Lactobacillus rhamnosus* strains, 24-1 and 25-2, did not show significant adherence. This suggested that the ethanol-promoted bacterial adherence to HOK cells may be strain specific. These results are interesting because the levels of ethanol, acetaldehyde, and malondialdehyde under laboratory conditions can be encountered in oral or oropharyngeal tissues. Humans are exposed routinely to high levels of ethanol from alcoholic beverages or mouth washes. In addition, we observed a novel change in behavior for *S. mutans*, which is normally a colonizer of the dental hard tissue but not a colonizer of oral mucosal membrane.

It is apparent that attachment by certain *Streptococcus* strains to keratinocytes was induced by ethanol. This could be simply by increasing expression of an adhesion unrelated to ethanol metabolism such as in *Staphylococcus* [49]. However, since only the acetaldehyde-producers but not the non-producers showed an increased attachment upon exposure to ethanol, a housekeeping enzyme involved in ethanol metabolism, such as AdhE, could also play a role either by serving as an adhesion protein like in *Listeria* [45] or by generating acetaldehyde, a stronger inducer of the adhesin expression, from ethanol.

In addition to ethanol, some other carcinogenic chemicals also increased attachment of streptococcal bacteria to the human telomerase immortalized (hTERT) HOK cells. For example, more *S. gordonii* cells (red dye stained) attached to HOK after exposure to 1 % ethanol (57.5 ± 16.0 %; $p < 0.0001$), 50 μg/mL of polycyclic aromatic hydrocarbon (PAH) and dibenz[a, l]pyrene (DBP) (40.4 ± 19.0 %, $p < 0.01$) or to a combination of ethanol and DBP (82.6 ± 22.0 %; $p < 0.001$) (Fig. 14.8). Like acetaldehyde, the tobacco-derived PAH and DBP are also type I carcinogens. Therefore, carcinogenic chemicals appear to promote certain bacterial species to attach to human mucosal cells.

14.3.3 Ethanol Metabolism by Oral Streptococci Promotes HPV 16 Entry into HOK

293TT cells of an adenovirus transformed human embryonic kidney cell line with a stably integrated SV40 genome with high levels of large T antigen were cultured in DMEM+ Enhanced GLU (GIBCO, Life Technologies, Grand Island, NY) supplemented with 10 % fetal bovine serum (Sigma-Aldrich, St. Louis MO). The HPV pseudovirus (PsV) was produced with an Optiprep purification or a maturation method using overnight incubation of crude cell lysate at 37 °C. HPV 16 PsV packaging plasmids (p16L1-GFP and pfwB) and the expression vector for luciferase (pCLucf) driven by the cytomegalovirus (CMV) promoter were used [50]. This system relies upon a co-propagation of L1/L2 expression plasmid together with a reporter plasmid [green fluorescent protein (GFP)] to generate high titers of mature PsV stocks for visualization of viral entry.

PsV particles (50 μL) were placed into wells in a 6-well plate coated with Type IV collagen containing hTERT HOK cells at 50–60 % confluence (5×10^5 cells) after inhibitor chemicals, and/or bacterial strains were added during a 24 h incubation and washing with DMEM + Enhance GluMax. Infection was monitored at least

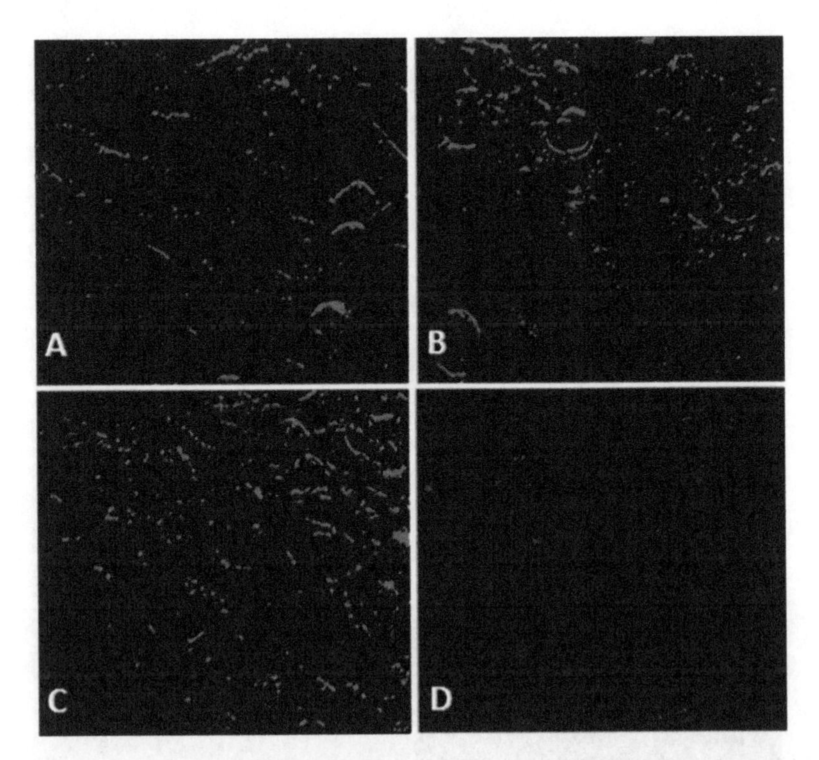

Fig. 14.8 Effect of environmental DNA damaging agents on attachment by *S. gordonii* to HPV/HOK-16B cells. (**a**) *S. gordonii* V2016 was exposed to 1 % ethanol (1 h), washed, stained and then added (50×10^3 CFU/well) to HPV/HOK-16B cells (3 h). *S. gordonii* was stained with a fluorescent red dye (LIVE/DEAD® BacLight™ Bacterial Viability Kit, Molecular Probe) after killing with isopropanol. (**b**), *S. gordonii* V2016 was exposed to the poly-cyclic aromatic hydrocarbon (PAH) and dibenz[a,l]pyrene (DBP), 50 μg/mL each, 1 h; washed, stained and added to HPV/HOK-16B cultures as above. (**c**), *S. gordonii* V2016 was exposed to 1 % ethanol and DBP (50 μg/mL) for 1 h, washed, stained, and then added to HPV/HOK-16B cells (3 h). Results show an increased attachment by *S. gordonii* after treatment with ethanol, PAH and/or DBP identified by the red dye detection. (**d**) Control shows a lack of attachment to HPV/HOK-16B cells by stained *S. gordonii* V2016 without treatment

7 days for GFP expression as a consequence of plasmid replication in hTERT HOK cells. Before addition of PsV for experimental studies we conducted a titration assay (200, 100, 50, 25, 10, and 5 μL) for each cell line to determine the maximum expression of GFP in a period of 7 days [51].

The result demonstrated that addition of streptococci and ethanol affected HPV 16 entry in a strain specific manner. When compared to control (no bacteria; 12.0 ± 10.0 %) *S. gordonii* V2016 wt (ADH positive) promoted HPV 16 entry into hTERT HOK cells (55.0 ± 20 %; $p < 0.01$), while the ADH null strain of *S. gordonii* V2016 *ΔadhABE* showed no significant increase (15.5 ± 11.0 %; $p > 0.5$) after exposure to 1 % ethanol (Fig. 14.9).

Because HPV entry requires furin activity, we tested if furin inhibitors could suppress the streptococcal adherence promoted furin expression. The furin inhibitor

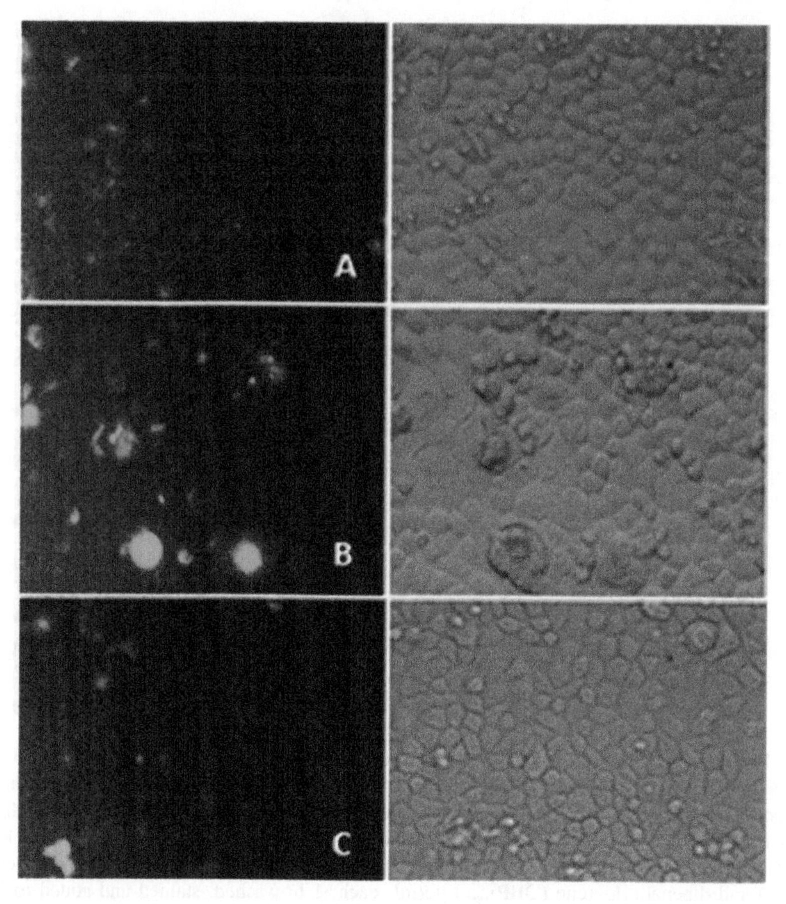

Fig. 14.9 HPV 16 PsV entry following exposure to ethanol treated *S. gordonii* strains. A GFP-tagged HPV 16 was used to track the viral entry in hTERT HOK cells. The dark field microscopy on the left was to distinguish fluorescence showing viral entry, while the light field one the right was to show the location of cells. (**a**) hTERT HOK cells displayed low levels of HPV 16 entry after incubation with 50×10^2 PsV of HPV 16 without prior exposure to *S. gordonii* cells. (**b**) Increased viral entry illustrated by augmented green fluorescence after hTERT HOK cells were treated with *S. gordonii* V2016, which was exposed to 1 % ethanol (1 h), washed and added to hTERT HOK cells (3 h) and then removed with a gentle wash. (**c**) A marked reduction in GFP-tagged cells with PsV replication and HPV 16 entry with cells treated with the ADH-negative strain *S. gordonii ΔadhABE* exposed to 1 % ethanol

peptide UIGI-1 (5-FAM-Lys-Arg-Val-Lys-Arg-CMK) (CMK) was added to HOK before exposure to *S. gordonii* or *S. salivarius* ADH expressing strains. We were able to visualize and detect this inhibition using 5-Carboxy-di-*O*-acetylfluorescein *N*-succinimidyl ester labeled CMK. Prior treatment with the CMK reduced attachment (red dye) of *S. salivarius* 101-1 (acetaldehyde producer from ethanol) and *S. salivarius* 101-7 (acetaldehyde non-producer from ethanol) to HOK, while in control HOK (not exposed to bacteria), furin expression was almost completely

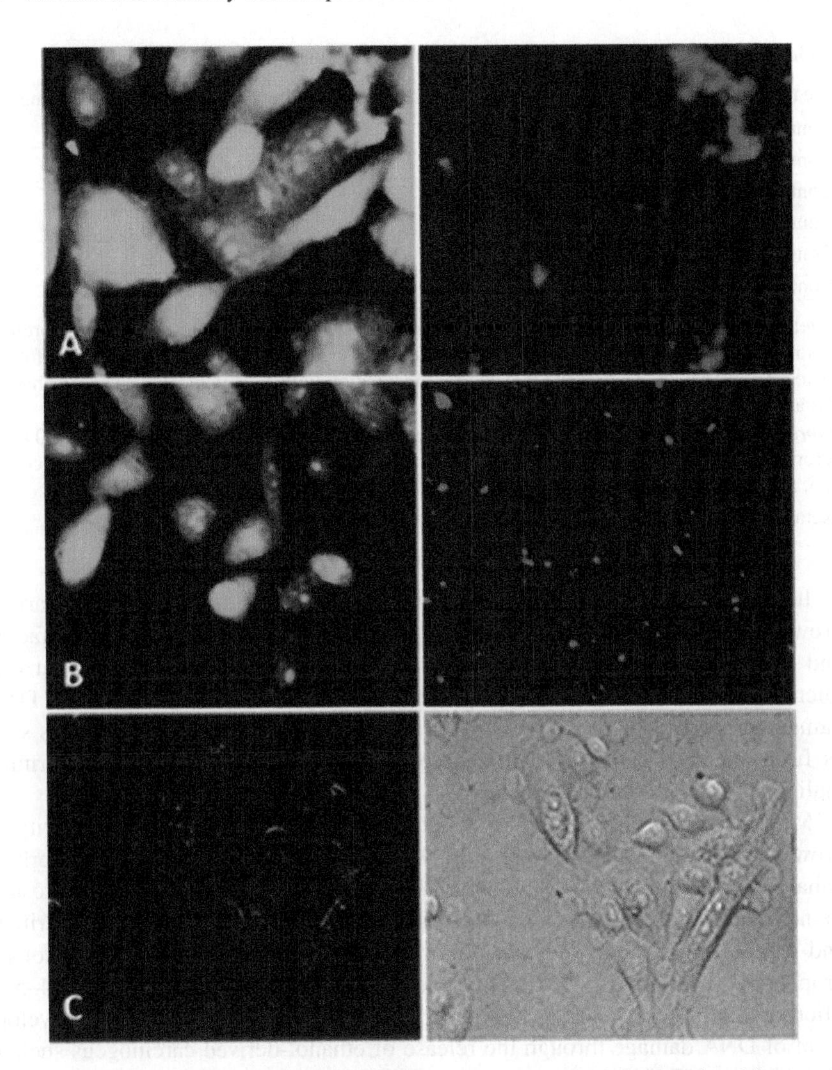

Fig. 14.10 *S. salivarius* attachment to HOK following exposure to 1 % ethanol and fluorescein-tagged CMK. One the *left panel*, the green fluorescence illustrates furin expression in HOK cells, while on the *right panel* the red fluorescence indicates adherent oral streptococci. (**a**) HOK cells were exposed to *S. salivarius* 107-1 (non-producer of acetaldehyde from ethanol) and CMK. (**b**) HOK cells were exposed to *S. salivarius* 101-1 (producer of acetaldehyde from ethanol) and CMK. (**c**) HOK cells were exposed to CMK only without bacteria. Note: furin is expressed in HOK cells while they were exposed to both CMK and streptococci, but furin is not detectable when HOK cells were exposed to only CMK without bacteria

suppressed (Fig. 14.10). The quantitative effect was shown in Table 14.3. It discloses that a loss of furin expression is associated with a decreased bacterial attachment. A similar result was also seen with *S. gordonii* wt and *ΔadhABE* strains (data not shown).

Table 14.3 Detection in 96-well basement membrane extract with cell invasion assay

Treatment[a]	Relative fluorescent units	No. of cells/well (mean)
Control (HPV/HOK-16B cells)	136	5,650
Control + *S. mutans* (5×10^3)	208	8,508
Control + *S. mutans* + 1 % ethanol	684	24,726
Control + *S. mutans* + 1 % ethanol + HS	685	24,778
Control + BS + 1 % ethanol	662	18,533
Control + BS + HS	342	14,020

Note: The percentage of HPV/HOK-16B cells that migrated through basement membrane proteins was recorded after exposure to *S. mutans* plus ethanol (1 h), heparin sulfate (100 U), or filtered bacterial supernatant. The results indicate that incubation with *S. mutans* plus ethanol enhanced this activity ($p < 0.001$)

Abbreviations: *HS* heparan sulfate (100 U) incubation (1 h) with HPV/HOK-16B cells (50×10^3) before placement into well, *BS* bacterial supernatant filtered (0.22-ìm syringe filter) from culture of *S. mutans* after incubation with ethanol for 3 h

[a]Keratinocytes (5×10^4) were plated into triplicate wells in a 96-well plate

It is recognized that furin-like convertase active domains are found in various growth factor receptors such as epithelial growth factor, transforming growth factor, and insulin growth factor receptors [9, 52]. Therefore, it is suggested that oral microbial attachment following exposure to ethanol/acetaldehyde or other DNA damaging agents (Figs. 14.7 and 14.8) can affect membrane related enzymes such as furin and also growth factor receptors critical for expansion of transforming malignant clones of epithelial cells [53].

Moreover, as proliferative dysregulation occurs from inappropriate activity of growth factors and weakening of normal epithelial intracellular bridge network an enhanced opportunity for HPV 16 entry to stem-cells like basal keratinocytes results to increase the risk for carcinoma development. For HPV 16 entry, timing is critical and to a degree specific sites are preferred, such as more common in the tonsil-oropharynx than in the oral cavity like lateral border of the tongue [54–56]. Microbial-biofilm changes at these critical sites can further influence the development of DNA damage through the release of ethanol-derived carcinogens such as acetaldehyde [57, 58].

14.3.4 Malignant Transformation of HOK Cells

Human oral buccal keratinocytes (donor sources had no history of HPV) were harvested for short-term (ethanol 10 mM/L; 24 h) incubation (5 % CO_2, 99 % humidity) by use of an oral brush harvest technique. A soft bristle brush was used after a 20-s oral rinse of phosphate-buffered saline (PBS), which reduced the numbers of surface non-nucleated exfoliated epithelial cells. A second oral brush harvest of keratinocytes occurred with another soft bristle brush with at least 20 brushing back-and-forth motions but without induction of any bleeding.

Harvested buccal keratinocytes were immediately placed into the cell culture medium (10^6 cells/10 μL), and then placed into tissue chamber slides for short-term culture (24 h).

Several characteristics associated with malignant transformation, including nuclear damage, acetaldehyde adduct formation, abnormal cell proliferation [monitored by 5′-bromo-2-deoxyuridine (BrdU) incorporation], and enhanced migration through integrin-coated basement membrane by HPV/HOK-16B, were analyzed [48]. To determine the presence of nuclear damage, nucleation and viability were assessed with a propidium iodide nuclear stain (80–90 % nucleated cells observed) with visualization of nuclear material by use of a 480-nm wavelength excitation. The observed physiologic modifications in keratinocytes suggest a transformation toward a malignant cell type because of an increase in DNA-acetaldehyde bulky adduct formation (Fig. 14.11), increased proliferation (Fig. 14.12), and migration through an integrin-laden basement membrane (Table 14.3) by HPV/HOK-16B. Although this is a cell laboratory study, a heretofore unrecognized contributor for malignant keratinocyte phenotype was suggested with *Streptococcus* species and bacterial release of acetaldehyde and malondialdehyde, whereas *L. rhamnosus* did not show such changes.

These results revealed that a complex interaction among oral *Streptococcus*, HPV-infected HOK and an environment containing ethanol leads to malignant changes in the keratinocytes. This suggested that poor oral hygiene, often associated with high presence of *Streptococcus* species and HPV, combined with the use of ethanol may increase the risk of cancer in the head and neck. We observed that under the influence of ethanol, certain oral *Streptococcus* species contributed to DNA damage, abnormal proliferation and migration activities in HPV/HOK-16B cells. The base of the tongue, hypopharynx and floor of the mouth are common sites for HPV and streptococcus exposures and squamous cell carcinoma. Further studies are required to observe these changes in clinical populations at risk for oral and oropharyngeal cancers.

14.4 Possible Study Limitations

First, expression of bacterial enzymes, namely ADH and ALDH, could be affected by the atmospheric conditions, such as aerobic, microaerophilic, and strict anaerobic [59]. Our enzyme analysis data were obtained from *Streptococcus* strains grown only in candle jars with a carbon dioxide-rich and oxygen-poor atmosphere (microaerophilic) condition. The enzyme activities might be different if the bacteria were grown under different atmospheric conditions. Second, we used only one concentration for both ethanol and acetaldehyde to study their difference in growth inhibitions against *S. gordonii* V2016. In real life, the ethanol level is high in the digestive tract only transiently when ethanol is first consumed. Once absorbed, the ethanol level in the saliva is much lower (0.02–0.1 %) and so is acetaldehyde (10–150 μM) [10]. We have not tested growth inhibitions of ethanol and acetaldehyde at such low concentrations. Third, our methods for determining the enzyme activities and productions of

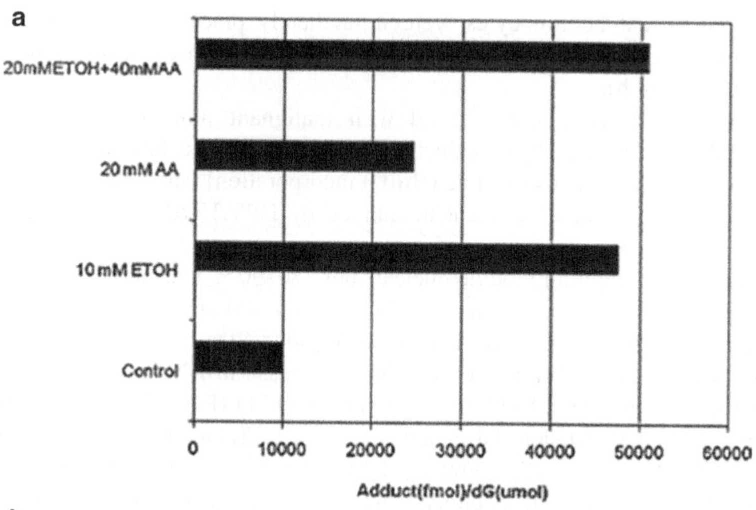

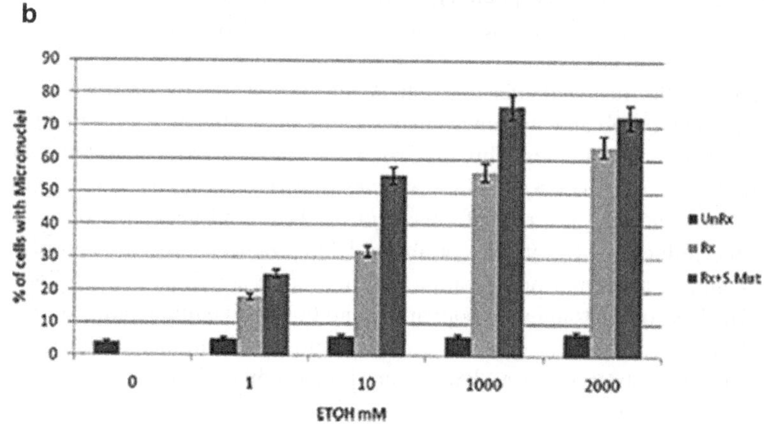

Fig. 14.11 (**a**) acetaldehyde adduct formation in HPV/HOK-16B cells after exposure to ethanol and/or acetaldehyde. DNA detection for N2-ethyldeoxyguanosine bulky adduct (fmol/dG μmol/L) in HPV/HOK-16B cells after exposure to ethanol [10–20 mmol/L (1–2 %), 24 h] or acetaldehyde [20–40 mmol/L (1–2 %), 24 h] was undertaken by use of liquid chromatography–electrospray ionization–tandem mass spectrometry for detection. Bulky adduct formation after exposure to ethanol and/or acetaldehyde was 2× or 4.5× the levels observed in controls. (**b**) Micronuclei observed after treatment with ethanol and ethanol + *S. mutans*. We obtained inverted-phase micrographs of HPV/HOK-16B exposed or not exposed to ETOH or *S. mutans* followed by propidium iodide nuclear stain to detect micronuclei. Micronuclei were counted in HPV/HOK-16B cells with ethanol treatment (Rx) (0.5 h: 0.1, 1, 10, and 20 %; 24 h) and co-cultured with *S. mutans*. Significant numbers of micronuclei ($p < 0.001$) were counted after increases in ethanol and/or presence of *S. mutans* compared with a lack of ethanol exposure (UnRx, untreated)

acetaldehyde and acetic acid from ethanol were either qualitative or semi-quantitative. Finally, our study about the interactions among *Streptococcus*, HPV and HOK after exposure to ethanol was under *in vitro* conditions. In the future, quantitative analyses of acetaldehyde and in vivo studies of malignant changes in HOK will be needed to confirm these observations.

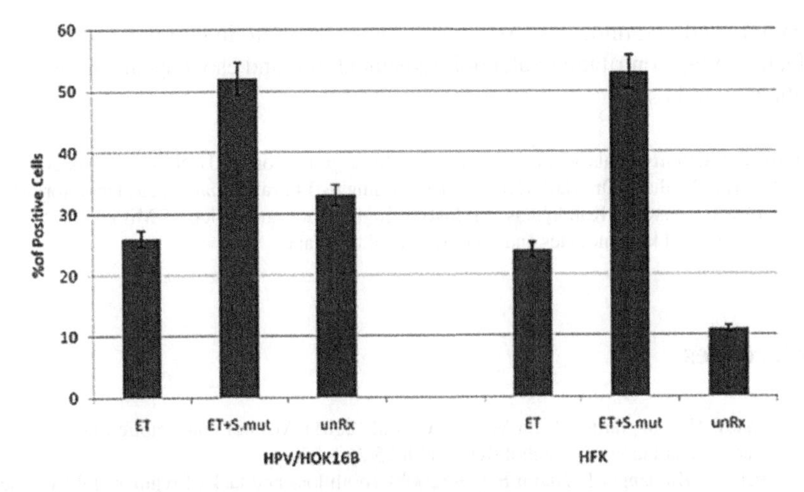

Fig. 14.12 HPV/HOK-16B or foreskin keratinocyte (HFK) cells were exposed to ethanol (ET) (10 mmol/L, 1 %) (5×10^3) and *S. mutans* (S. Mut), and immunohistochemistry was used to determine levels of expression for BrdU. Exposure to ethanol and *S. mutans* significantly increased ($p<0.001$) levels of BrdU as compared with untreated (unRx) cells

14.5 Summary

We have screened 52 oral *Streptococcus* strains for their capacity of acetic acid and acetaldehyde production from ethanol. Only 17 strains produced acetic acid from ethanol, while 19 produced acetaldehyde. Among all the strains tested, *S. gordonii* V2016 produced the most abundant acetaldehyde without acetic acid. Therefore, V2016 was selected for further study of its enzymes involved in acetaldehyde production from ethanol. We found that this bacterium displayed three different ADHs, AdhA, AdhB, and AdhE, that all oxidize ethanol to acetaldehyde, although their preferred substrates were 1-propanol, 1-butanol, and ethanol, respectively. Among the three enzymes, AdhA and AdhB are zinc-dependent, while AdhE is iron-dependent and is the major ADH in *S. gordonii* for ethanol metabolism because ethanol is its preferred substrate. Inactivation of *adhE*, but not *adhA* and *adhB*, resulted in a greater tolerance of the bacterium to ethanol due to reduction of the more toxic acetaldehyde production. *S. gordonii* V2016 did not show a detectable ALDH. Analysis of 19 additional strains of *S. gordonii, S. mitis, S. oralis, S. salivarius*, and *S. sanguinis* all showed similarly varied enzyme profiles of ADHs without detectable ALDH except one strain even though the *Streptococcus* carries the gene *adhE* that encodes a putative ALDH. The finding that a defective ALDH occurs frequently in bacteria is of evolutionary interest in biology because a significant number of humans in the East Asian population also express a defective ALDH. Therefore, activities of multiple ADHs but no ALDH in most oral streptococci may contribute to the excessive production of acetaldehyde from ethanol. Also, exposure to ethanol enhanced the attachment of acetaldehyde-producing *Streptococcus* strains to HOK and promoted HPV entry and malignant changes in

HPV-infected keratinocytes. As a result, these bacteria, in combination with HPV infection, may contribute to alcohol-associated oral and esophageal carcinogenesis in the human host.

Acknowledgements This work was supported by a grant from NIH National Cancer Institute (CA162537). We thank Dr. Mark Herzberg for sending us 14 oral *Streptococcus* laboratory strains. We thank Drs. Antonia Kolokythas, Mulokozi Lugakingira, and Michael Miloro for collecting human oral buccal keratinocytes and other technical assistance.

References

1. Bagnardi V, Blangiardo M, La Vecchia C et al (2001) Alcohol consumption and the risk of cancer: a meta-analysis. Alcohol Res Health 25:263–270
2. Abnet CC, Kamangar F, Islami F et al (2008) Tooth loss and lack of regular oral hygiene are associated with higher risk of esophageal squamous cell carcinoma. Cancer Epidemiol Biomarkers Prev 17:3062–3068
3. Homann N (2001) Alcohol and upper gastrointestinal tract cancer: the role of local acetaldehyde production. Addict Biol 6:309–323
4. Homann N, Tillonen J, Rintamäki H et al (2001) Poor dental status increases acetaldehyde production from ethanol in saliva: a possible link to increased oral cancer risk among heavy drinkers. Oral Oncol 37:153–158
5. Brown LM, Check DP, Devesa SS (2011) Oropharyngeal cancer incidence trends: diminishing racial disparities. Cancer Causes Control 22:753–763
6. Brown SD, Guss AM, Karpinets TV et al (2011) Mutant alcohol dehydrogenase leads to improved ethanol tolerance in *Clostridium thermocellum*. Proc Natl Acad Sci U S A 108: 13752–13757
7. Meurman JH, Uittamo J (2008) Oral micro-organisms in the etiology of cancer. Acta Odontol Scand 66:321–326
8. Bassi DE, Fu J, Lopez de Cicco R et al (2005) Proprotein convertases: "master switches" in the regulation of tumor growth and progression. Mol Carcinog 44:151–161
9. Thomas G (2002) Furin at the cutting edge: from protein traffic to embryogenesis and disease. Nat Rev Mol Cell Biol 3:753–766
10. Homann N, Jousimies-Somer H, Jokelainen K et al (1997) High acetaldehyde levels in saliva after ethanol consumption: methodological aspects and pathogenetic implications. Carcinogenesis 18:1739–1743
11. Muto M, Hitomi Y, Ohtsu A et al (2000) Acetaldehyde production by non-pathogenic *Neisseria* in human oral microflora: implications for carcinogenesis in upper aerodigestive tract. Int J Cancer 88:342–350
12. Nobbs AH, Lamont RJ, Jenkinson HF (2009) *Streptococcus* adherence and colonization. Microbiol Mol Biol Rev 73:407–450
13. Kurkivuori J, Salaspuro V, Kaihovaara P et al (2007) Acetaldehyde production from ethanol by oral streptococci. Oral Oncol 43:181–186
14. Väkeväinen S, Tillonen J, Blom M et al (2001) Acetaldehyde production and other ADH-related characteristics of aerobic bacteria isolated from hypochlorhydric human stomach. Alcohol Clin Exp Res 25:421–426
15. Woutersen RA, Appelman LM, Van Garderen-Hoetmer A et al (1986) Inhalation toxicity of acetaldehyde in rats. III. Carcinogenicity study. Toxicology 41:213–231
16. Obe G, Anderson D (1987) International Commission for Protection against Environmental Mutagens and Carcinogens. ICPEMC Working Paper No. 15/1. Genetic effects of ethanol. Mutat Res 186:177–200

17. Vaca CE, Fang JL, Schweda EK (1995) Studies of the reaction of acetaldehyde with deoxynucleosides. Chem Biol Interact 98:51–67
18. Espina N, Lima V, Lieber CS et al (1988) *In vitro* and *in vivo* inhibitory effect of ethanol and acetaldehyde on O^6-methylguanine transferase. Carcinogenesis 9:761–766
19. Secretan B, Straif K, Baan R et al (2009) A review of human carcinogens. Part E. Tobacco, areca nut, alcohol, coal smoke, and salted fish. Lancet Oncol 10:1033–1034
20. Rehm J, Kanteres F, Lachenmeier DW (2010) Unrecorded consumption, quality of alcohol and health consequences. Drug Alcohol Rev 29:426–436
21. Druesne-Pecollo N, Tehard B, Mallet Y et al (2009) Alcohol and genetic polymorphisms: effect on risk of alcohol-related cancer. Lancet Oncol 10:173–180
22. Marttila E, Bowyer P, Sanglard D et al (2013) Fermentative 2-carbon metabolism produces carcinogenic levels of acetaldehyde in *Candida albicans*. Mol Oral Microbiol 28:281–291
23. Uittamo J, Siikala E, Kaihovaara P et al (2009) Chronic candidosis and oral cancer in APECED-patients: production of carcinogenic acetaldehyde from glucose and ethanol by *Candida albicans*. Int J Cancer 124:754–756
24. Dewhirst FE, Chen T, Izard J et al (2010) The human oral microbiome. J Bacteriol 192:5002–5017
25. Turnbaugh PJ, Ley RE, Hamady M (2007) The human microbiome project. Nature 449:804–810
26. O'Hara AM, Shanahan F (2006) The gut flora as a forgotten organ. EMBO Rep 7:688–693
27. Salaspuro MP (2003) Acetaldehyde, microbes, and cancer of the digestive tract. Crit Rev Clin Lab Sci 40:183–208
28. Väkeväinen S, Tillonen J, Agarwal DP et al (2000) High salivary acetaldehyde after a moderate dose of alcohol in ALDH2-deficient subjects: strong evidence for the local carcinogenic action of acetaldehyde. Alcohol Clin Exp Res 24:873–877
29. Väkeväinen S, Tillonen J, Salaspuro M (2001) 4-Methylpyrazole decreases salivary acetaldehyde levels in ALDH2-deficient subjects but not in subjects with normal ALDH2. Alcohol Clin Exp Res 25:829–834
30. Pavlova SI, Jin L, Gasparovich SR et al (2013) Multiple alcohol dehydrogenases but no functional acetaldehyde dehydrogenase causing excessive acetaldehyde production from ethanol by oral streptococci. Microbiology 159:1437–1446
31. Vickerman MM, Iobst S, Jesionowski AM et al (2007) Genome-wide transcriptional changes in *Streptococcus gordonii* in response to competence signaling peptide. J Bacteriol 189: 7799–7807
32. Lau PC, Sung CK, Lee JH et al (2002) PCR ligation mutagenesis in transformable streptococci: application and efficiency. J Microbiol Methods 49:193–205
33. Sambrook J, Fritsch EF, Maniatis T (1989) Molecular cloning: a laboratory manual, 2nd edn. Cold Spring Harbor Laboratory, Cold Spring Harbor
34. Gabriel O (1971) Locating enzymes on gels. Methods Enzymol 22:578–604
35. Grell EH, Jacobson KB, Murphy JB (1968) Alterations of genetics material for analysis of alcohol dehydrogenase isozymes of *Drosophila melanogaster*. Ann N Y Acad Sci 151: 441–455
36. Koo OK, Jeong DW, Lee JM et al (2005) Cloning and characterization of the bifunctional alcohol/acetaldehyde dehydrogenase gene (adhE) in *Leuconostoc mesenteroides* isolated from Kimchi. Biotechnol Lett 27:505–510
37. Hiyama T, Yoshihara M, Tanaka S et al (2007) Genetic polymorphisms and esophageal cancer risk. Int J Cancer 121:1643–1658
38. Lewis SJ, Smith GD (2005) Alcohol, ALDH2, and esophageal cancer: a meta-analysis which illustrates the potentials and limitations of a Mendelian randomization approach. Cancer Epidemiol Biomarkers Prev 14:1967–1971
39. Yokoyama A, Muramatsu T, Ohmori T et al (1998) Alcohol-related cancers and aldehyde dehydrogenase-2 in Japanese alcoholics. Carcinogenesis 19:1383–1387
40. Martincorena I, Seshasayee AS, Luscombe NM (2012) Evidence of non-random mutation rates suggests an evolutionary risk management strategy. Nature 485:95–98

41. Reinke LA, Rau JM, McCay PB (1994) Characteristics of an oxidant formed during iron (II) autoxidation. Free Radic Biol Med 16:485–492
42. Welch KD, Davis TZ, Aust SD (2002) Iron autoxidation and free radical generation: effects of buffers, ligands, and chelators. Arch Biochem Biophys 397:360–369
43. Nnyepi MR, Peng Y, Broderick JB (2007) Inactivation of *E. coli* pyruvate formate-lyase: role of AdhE and small molecules. Arch Biochem Biophys 459:1–9
44. Asanuma N, Yoshii T, Hino T (2004) Molecular characteristics and transcription of the gene encoding a multifunctional alcohol dehydrogenase in relation to the deactivation of pyruvate formate-lyase in the ruminal bacterium *Streptococcus bovis*. Arch Microbiol 181:122–128
45. Jagadeesan B, Koo OK, Kim KP et al (2010) LAP, an alcohol acetaldehyde dehydrogenase enzyme in *Listeria*, promotes bacterial adhesion to enterocyte-like Caco-2 cells only in pathogenic species. Microbiology 156:2782–2795
46. Yao S, Mikkelsen MJ (2010) Identification and overexpression of a bifunctional aldehyde/alcohol dehydrogenase responsible for ethanol production in *Thermoanaerobacter mathranii*. J Mol Microbiol Biotechnol 19:123–133
47. Brendel M, Marisco G, Ganda I et al (2010) DNA repair mutant *pso2* of *Saccharomyces cerevisiae* is sensitive to intracellular acetaldehyde accumulated by disulfiram-mediated inhibition of acetaldehyde dehydrogenase. Genet Mol Res 9:48–57
48. Schwartz J, Pavlova S, Kolokythas A et al (2013) Streptococci-human papilloma virus interaction with ethanol exposure leads to keratinocyte damage. J Oral Maxillofac Surg 70:1867–1879
49. Knobloch JK, Horstkotte MA, Rohde H et al (2002) Alcoholic ingredients in skin disinfectants increase biofilm expression of *Staphylococcus epidermidis*. J Antimicrob Chemother 49:683–687
50. Buck CB, Thompson CD (2007) Production of papillomavirus-based gene transfer vectors. Curr Protoc Cell Biol, Chapter 26 (Unit 26 21)
51. Richards RM, Lowry DR, Schiller JT et al (2006) Cleavage of the papillomavirus minor capsid protein, L2 at a furin consensus site is necessary for infection. Proc Natl Acad Sci U S A 103:1522–1527
52. AbeY OM, Inagaki F et al (1998) Disulfide bond structure of human epidermal growth factor receptor. J Biol Chem 273:11150–11157
53. Gschwind A, Zwick E, Prenzel N et al (2001) Cell communication networks: epidermal growth factor receptor transactivation as the paradigm for intereceptor signal transmission. Oncogene 30:1594–1600
54. Poling JS, Ma XJ, Bui S et al (2014) Human papillomavirus (HPV) status of non-tobacco related squamous cell carcinoma of the lateral tongue. Oral Oncol 50:306–310
55. Pytynia KB, Dahlstrom KR, Sturgis EM (2014) Epidemiology of HPV-associated oropharynx cancer. Oral Oncol 50:380–386
56. Termine N, Panzarella V, Falaschini S et al (2008) HPV in oral squamous cell carcinoma vs head and neck squamous cell carcinoma biopsies: a meta-analysis (1988-2007). Ann Oncol 19:1681–1690
57. Smith EM, Rubenstein LM, Haugen TH et al (2010) Tobacco and alcohol use increases the risk of both HPV-associated and HPV-independent head and neck cancers. Cancer Causes Control 21:1369–1378
58. Smith EM, Rubenstein LM, Haugen TH et al (2012) Complex etiology underlies risk and survival in head and neck cancer human papillomavirus, tobacco, and alcohol: a case for multifactor disease. J Oncol 2012:571862. doi:10.1155/2012/571862
59. Salaspuro V, Nyfors S, Heine R et al (1999) Ethanol oxidation and acetaldehyde production in vitro by human intestinal strains of Escherichia coli under aerobic, microaerobic, and anaerobic conditions. Scand J Gastroenterol 34:967–973
60. Park NH, Min BM, Li SL et al (1991) Immortalization of normal human oral keratinocytes with type 16 human papillomavirus. Carcinogenesis 12:1627–1631
61. Ranhand JM (1974) Simple, inexpensive procedure for the disruption of bacteria. J Appl Microbiol 28:66–69

Chapter 15
Genetic Polymorphisms of Alcohol Dehydrogense-1B and Aldehyde Dehydrogenase-2, Alcohol Flushing, Mean Corpuscular Volume, and Aerodigestive Tract Neoplasia in Japanese Drinkers

Akira Yokoyama, Takeshi Mizukami, and Tetsuji Yokoyama

Abstract Genetic polymorphisms of alcohol dehydrogenase-1B (ADH1B) and aldehyde dehydrogenase-2 (ALDH2) modulate exposure levels to ethanol/acetaldehyde. Endoscopic screening of 6,014 Japanese alcoholics yielded high detection rates of esophageal squamous cell carcinoma (SCC; 4.1 %) and head and neck SCC (1.0 %). The risks of upper aerodigestive tract SCC/dysplasia, especially of multiple SCC/dysplasia, were increased in a multiplicative fashion by the presence of a combination of slow-metabolizing *ADH1B*1/*1* and inactive heterozygous *ALDH2*1/*2* because of prolonged exposure to higher concentrations of ethanol/acetaldehyde. A questionnaire asking about current and past facial flushing after drinking a glass (≈ 180 mL) of beer is a reliable tool for detecting the presence of inactive ALDH2. We invented a health-risk appraisal (HRA) model including the flushing questionnaire and drinking, smoking, and dietary habits. Esophageal SCC was detected at a high rate by endoscopic mass-screening in high HRA score persons. A total of 5.0 % of 4,879 alcoholics had a history of (4.0 %) or newly diagnosed (1.0 %) gastric cancer. Their high frequency of a history of gastric cancer is partly explained by gastrectomy being a risk factor for alcoholism because of altered ethanol metabolism, e.g., by blood ethanol level overshooting. The combination of *H. pylori-associated* atrophic gastritis and *ALDH2*1/*2* showed the greatest risk of gastric cancer in alcoholics. High detection rates of advanced colorectal adenoma/carcinoma were found in alcoholics, 15.7 % of 744 immunochemical fecal occult blood test (IFOBT)-negative alcoholics and 31.5 % of the 393 IFOBT-positive alcoholics.

A. Yokoyama, M.D. (✉) • T. Mizukami, M.D.
National Hospital Organization Kurihama Medical and Addiction Center,
5-3-1 Nobi, Yokosuka, Kanagawa 239-0841, Japan
e-mail: a_yokoyama@kurihama1.hosp.go.jp

T. Yokoyama, M.D.
Department of Health Promotion, National Institute of Public Health, Saitama, Japan

V. Vasiliou et al. (eds.), *Biological Basis of Alcohol-Induced Cancer*,
Advances in Experimental Medicine and Biology 815,
DOI 10.1007/978-3-319-09614-8_15

Macrocytosis with an MCV ≥ 106 fl increased the risk of neoplasia in the entire aerodigestive tract of alcoholics, suggesting that poor nutrition as well as ethanol/acetaldehyde exposure plays an important role in neoplasia.

Keywords Acetaldehyde • Alcohol dehydrogenase • Alcoholic • Aldehyde dehydrogenase • Colorectal neoplasia • Esophageal cancer • Head and neck cancer • Mean corpuscular volume • Stomach cancer

15.1 Endoscopy and Esophageal Iodine Staining of Screening Japanese Alcoholic Men for Upper Aerodigestive Tract Neoplasia

Alcohol consumption, tobacco smoking, inadequate intake of fruits and vegetables, and low body mass index (BMI) are risk factors for squamous cell carcinoma (SCC) of the upper aerodigestive tract (UADT, i.e., the oral cavity, pharynx, larynx, and esophagus), and many alcoholic patients have all of these risk factors. We introduced an endoscopic screening program at Kurihama Medical and Addiction Center in 1993 [1], and by 2010 initial screening of 6,014 Japanese alcoholic men by endoscopy combined with head and neck inspection and esophageal iodine staining had detected esophageal SCC in 243 (4.0 %) of them and head and neck SCC in 65 (1.1 %) of them. Barrett adenocarcinoma was detected in only two patients. Technical improvements in endoscopes and a growing understanding of the endoscopic findings of early SCC in the UADT [2, 3] have enabled very early detection of SCC in the UADT. Treatment of early SCC in UADT by endoscopic or endoscope-guided mucosectomy has become a widespread practice in Japan and succeeded in improving the outcome of this high-mortality cancer [3, 4].

15.2 ALDH2 Genotype and Alcohol Metabolism

Genetic polymorphism of aldehyde dehydrogenase-2 (ALDH2, rs671) modulates exposure levels to acetaldehyde after drinking (Fig. 15.1). A mutant *ALDH2*2* allele encoding an inactive subunit of ALDH2 was carried by Han Chinese as they spread throughout East Asia [5]. About 40 % of Japanese, 7 % being homozygotes and 35 % heterozygotes [6], have the inactive *ALDH2*2* allele which acts in a dominant manner. After drinking a small amount of alcohol, people with inactive ALDH2 tend to experience a flushing response that includes facial flushing, palpitations, nausea, and drowsiness [7]. People with inactive ALDH2 [8] or with inactive-ALDH2-associated facial flushing [9] tend to experience a hangover in the morning after drinking a smaller amount of alcohol than people without either of them. Thus presence of inactive ALDH2 in some East Asians tends to prevent them from drinking heavily [6]. Because of their very intense flushing responses *ALDH2*2/*2* homozygotes are

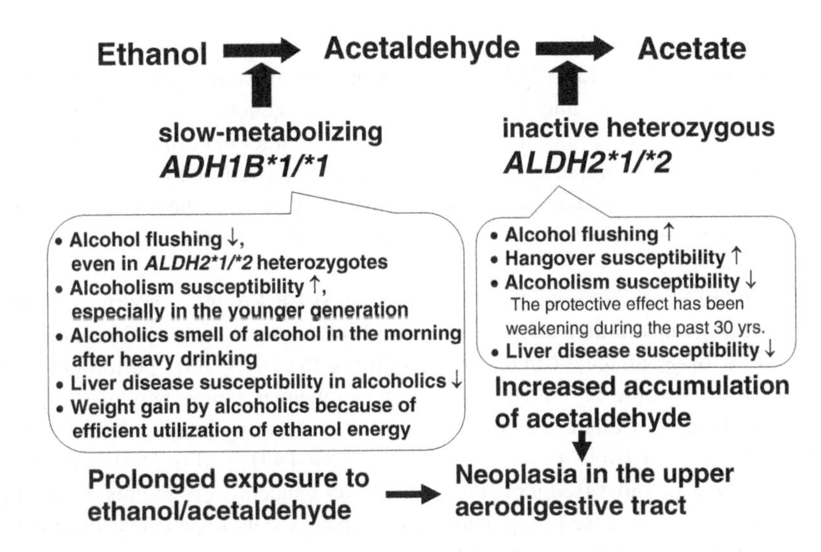

Fig. 15.1 Phenotypes of the *alcohol dehydrogenase-1B*1/*1* (*ADH1B*1/*1*) and *aldehyde dehydrogenase-2*1/*2* (*ALDH2*1/*2*) genotypes and their relationships to ethanol and acetaldehyde exposure and accumulation

usually nondrinkers or occasional drinkers. However, the inhibitory effect of being heterozygous for inactive ALDH2 on heavy drinking is influenced by sociocultural factors. The proportion of Japanese alcoholics with the *ALDH2*1/*2* genotype increased dramatically from 2.5 % in 1979, to 8.0 % in 1986, and to 13.0 % in 1992 [10], and the proportion continued to increase from 13.0 % in 1996–2000, to 14.0 % in 2001–2005, and to 15.4 % in 2006–2010 [11].

15.3 ALDH2 Genotype and Squamous Cell Neoplasia in the UADT

In 1996, we first reported that being heterozygous for inactive ALDH2 is a strong risk factor for esophageal cancer in daily drinkers and alcoholics [12]. Since then, epidemiological studies have almost consistently demonstrated a strong association between the *ALDH2*1/*2* genotype and the risk of UADT cancer in East-Asian drinkers [13]. A meta-analysis of Asian case–control studies of esophageal cancer showed that the *ALDH2*1/*2*-associated odds ratio (95 % confidence interval) [OR (95 % CI)] for esophageal cancer was 3.12 (1.95–5.02) in moderate drinkers, 5.64 (1.57–20.25) in ex-drinkers, and 7.12 (4.67–10.86) in heavy drinkers [14]. The *ALDH2*1/*2* genotype has been found to be a very strong risk factor for esophageal cancer among Taiwan Chinese and Japanese drinkers (OR = 4.74–6.21 in moderate drinkers and 9.21–9.75 in heavy drinkers), but the OR associated with the *ALDH2*1/*2* genotype in the high incidence regions of esophageal cancer of Mainland China was not so high (OR = 1.98 in moderate drinkers and 1.31 in heavy drinkers). The meta-analysis confirmed that being homozygous for the inactive allele, i.e., having the *ALDH2*2/*2* genotype greatly increased the risk of esophageal cancer in drinkers, and that the heterozygous genotype, i.e., *ALDH2*1/*2* increased the risk of esophageal cancer in a similar manner in both men and women. A meta-analysis of 6 Japanese case–control studies showed that the *ALDH2*1/*2-associated* risk for head and neck cancer was also greater in heavy drinkers [OR (95 % CI) = 3.57 (1.21–2.77)] than in moderate drinkers [OR (95 % CI) = 1.68 (1.22–2.27)] [15].

The United States National Cancer Institute's SEER Program reported synchronous multiple cancers and metachronous multiple cancers in only 2 % and 3 %, respectively, of esophageal cancer patients during the 1973–2003 period [16]. The prevalence of multiple organ cancers among esophageal cancer patients treated at the National Cancer Center of Japan increased at an alarming rate between 1969 and 1996 from 6.3 % in 1969–1980, to 22.2 % in 1981–1991, and to 39.0 % in 1992–1996 [17], and the most frequent sites of the other cancers were the head and neck and stomach. The increased proportion of multiple cancers in Japanese esophageal cancer patients may partly explained by a dramatic increase in the proportion of *ALDH2*1/*2* heterozygotes among Japanese heavy drinkers during the same period [10]. The *ALDH2*1/*2* genotype has consistently been demonstrated to be a strong determinant of the risk of synchronous and metachronous SCC of the UADT in Japanese drinkers [13].

The ORs (95 % CI) associated with the *ALDH2*1/*2* genotype in Japanese alcoholics has been found to increase for the very early stages of the esophageal neoplasia, from 2.88 (1.81–4.57) for low-grade intraepithelial neoplasia, to 5.14 (2.87–9.19) for high-grade intraepithelial neoplasia, and to 4.07 (1.97–8.40) for invasive SCC [18]. The presence of multiple esophageal iodine-unstained lesions or a large esophageal dysplasia has been found to be a strong predictor of the development of multiple cancers in the UADT in Japanese [13], and to be associated with the *ALDH2*1/*2* genotype [18], p53 alteration [19], and telomere shortening in the esophagus [20] in Japanese alcoholics. Thus, acetaldehyde, which is an established human carcinogen, plays a critical role in the multicentric development of neoplasia throughout the UADT.

15.4 The Simple Flushing Questionnaire

The discovery of ALDH2-associated cancer susceptibility emphasizes the importance of developing screening tests for inactive ALDH2 based on alcohol flushing. We devised a flushing questionnaire to identify person with inactive ALDH2 (Fig. 15.2) [21, 22]. The simple flushing questionnaire consists of two questions: (A) Do you have a tendency to flush in the face immediately after drinking a glass ($\approx$180 mL) of beer? (B) Did you have a tendency to flush in the face immediately after drinking a glass of beer during the first to second year after you started drinking? The results are used to classify the subjects as a current flusher, former flusher, or

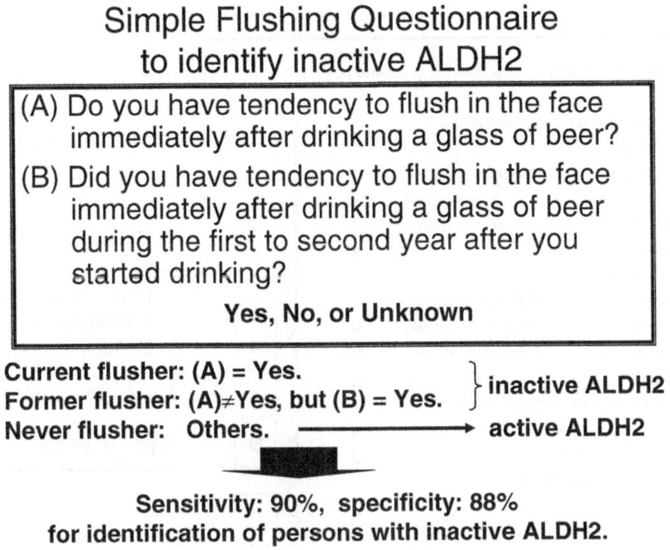

Simple Flushing Questionnaire
to identify inactive ALDH2

(A) Do you have tendency to flush in the face immediately after drinking a glass of beer?

(B) Did you have tendency to flush in the face immediately after drinking a glass of beer during the first to second year after you started drinking?

Yes, No, or Unknown

Current flusher: (A) = Yes.
Former flusher: (A)≠Yes, but (B) = Yes. } **inactive ALDH2**
Never flusher: Others. → **active ALDH2**

Sensitivity: 90%, specificity: 88%
for identification of persons with inactive ALDH2.

Fig. 15.2 The simple flushing questionnaire to identify persons with inactive ALDH2. Sensitivity and specificity are both approximately 90 % in Japanese 40 years of age and over, regardless of gender

never flusher. When current flushers or former flushers were assumed to have inactive ALDH2, the simple flushing questionnaire had 90 % sensitivity and 88 % specificity when evaluated in 610 Japanese men 40 years of age or older [22] and 88 % sensitivity and 92 % specificity when evaluated in 381 Japanese women 40 years of age or older [23]. The individuals' risks of UADT cancer estimated on the basis of the replies to the flushing questionnaire were slightly lower than but essentially comparable to their risks estimated on the basis of ALDH2 genotyping [21–23].

15.5 Mass-Screening for SCC of the UADT by Means of Health-Risk Appraisal Models

Based on the results of a case–control study [24], we devised health-risk appraisal (HRA) models for esophageal cancer that include ALDH2 genotype or alcohol flushing [25]. The total risk score is calculated by adding the scores A–E (Fig. 15.3). If a person's risk score is 11 or more according to the HRA-Flushing model, that person's risk of esophageal cancer is in the top 10 % of the study population. A cross-validation study predicted that approximately 60 % of the esophageal and hypopharyngeal SCCs in the entire population could be detected by examining only people whose risk scores were in the top 10 % in the HRA models. Follow-up endoscopy with esophageal iodine staining (median follow-up period: 5.0 years)

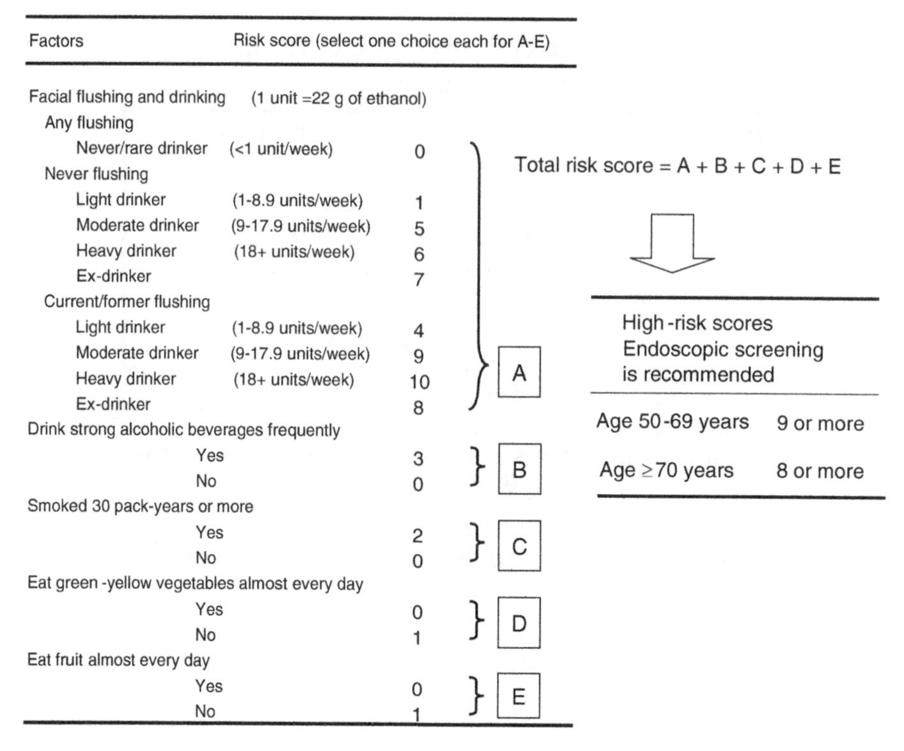

Fig. 15.3 Health-risk appraisal model for esophageal cancer combined with the simple flushing questionnaire. The questionnaire enables makes it possible for people to easily determine their risk of esophageal cancer, and public awareness campaigns that use the questionnaire will help persuade high-risk persons to undergo endoscopic screening or enable them to change their lifestyle to prevent esophageal cancer

was performed on 404 cancer-free controls and resulted in the diagnosis of six esophageal SCCs and two pharyngeal SCCs [26]. The risk scores of six of these eight cancer patients at baseline were in the top 10 % according to the HRA-Flushing model. The cancer detection rate per 100 person–years in the top 10 % risk group of the cancer-free controls was 2.3 for esophageal cancer and 3.5 for esophageal or pharyngeal cancer. We applied this HRA-Flushing questionnaire to endoscopic mass-screening programs of 2,221 Japanese men during 2008 and 2009 at five cancer screening facilities, and esophageal cancer was diagnosed in 19 persons as a result [27]. The HRA-Flushing score of 5 % of the examinees was 11 or greater, and esophageal cancer was detected in 4.3 % of them, as opposed to in 0.7 % of the other examinees. A receiver operating characteristic curve analysis showed that when we used an cutoff point of an HRA score of ≥9 in the 50–69 age group and of ≥8 in the 70–89 age group to select individuals with a high risk for esophageal cancer, the sensitivity and false-positive rate was 52.6 % and 15.2 %, respectively, and cancer was detected in 2.91 % of the examinees in the high-risk group, as opposed to 0.48 % in the other group. Although the cutoff values for high-risk groups should be changed to achieve better performance of the HRA model according to the population targeted, these figures encouraged using our questionnaire to screen larger populations of Japanese men.

Our simple questionnaire makes it possible for many people to identify their risk of UADT cancer very easily, and use of the questionnaire in public awareness campaigns will help persuade high-risk persons to undergo endoscopic screening or enable them to change their lifestyle to prevent UADT cancer. Our HRA model that includes ALDH2 genotype yielded a slightly better positive predictive value [25] and a slightly higher cancer detection rate [26] than the HRA-Flushing model. Genotyping ALDH2 would entail an initial cost, but genotyping needs to be performed only once in a lifetime, the data are always available, and the unit cost would be greatly discounted if a huge number of samples are analyzed.

15.6 ADH1B Genotype, Alcohol Metabolism, and SCC of the UADT

Alcohol dehydrogenase-1B (ADH1B, previously called ADH2) also has a functional genetic polymorphism (rs1229984; Fig. 15.1). The *ADH1B*1/*1* genotype, which results in expression of slow-metabolizing ADH1B, is prevalent in Caucasians (present in approximately 90 %), but present in only a small fraction of East Asians (e.g., 7 % of Japanese) [6]. The presence of slow-metabolizing ADH1B is a stronger risk factor for alcoholism and SCC of the UADT in East Asians than the presence of the fast-metabolizing ADH1B because of having the *ADH1B*2* allele. Approximately 30 % of Japanese alcoholics have the slow-metabolizing ADH1B, and the slow-metabolizing ADH1B has been found to be more frequent in the younger generations of Japanese alcoholics, probably because the *ADH1B*1/*1* genotype accelerates the progression of alcohol dependence [11].

Alcohol challenge tests have failed to demonstrate any associations between ADH1B genotype and the blood ethanol or acetaldehyde concentrations after ingestion of small to moderate doses of ethanol [13]. However, the slow-metabolizing ADH1B has an approximately 40 times lower Vmax in vitro than the fast-metabolizing ADH1B [28], and an experiment in which a clamping technique and intravenous alcohol infusion were used showed a modestly but significantly lower ethanol elimination rate (11–18 %) among Jews with the slow-metabolizing ADH1B than with the fast-metabolizing ADH1B [29]. At the time of their first visit to our Addiction Center, we evaluated associations between ADH1B and ALDH2 genotypes and the blood ethanol levels of 805 Japanese alcoholic men in the morning after drinking within the previous 34 h [30]. The results showed no significant differences in age-adjusted usual alcohol consumption according to ADH1B or ALDH2 genotypes. Higher blood ethanol levels persisted for longer periods in the group with the slow-metabolizing ADH1B who were *ADH1B*1/*1* carriers ($n=246$) than in the group with the fast-metabolizing ADH1B who were *ADH1B*2* carriers ($n=559$), and blood ethanol levels ≥ 0.3 mg/mL (criterion for drunk driving according to Japanese law) after a 12.1–18-h interval since the last drink were observed in a significantly higher proportion of the *ADH1B*1/*1* carriers than of the *ADH1B*2* carriers (40 % vs. 14–17 %, $p<0.0001$). Multivariate analyses showed that the ethanol levels were 0.500 mg/mL higher in the group with the *ADH1B*1*1* genotype, and the OR (95 % CI) for an ethanol level ≥0.3 mg/mL in the presence of the *ADH1B*1/*1* genotype was 3.44 (2.34–5.04). There were no significant differences in blood ethanol levels according to ALDH2 genotype.

We evaluated associations between ADH1B and ALDH2 genotypes and the body weight and BMI of 1,301 Japanese alcoholic men on the day of their first visit [31]. There were no significant differences in usual caloric intake in the form of alcoholic beverages according to ADH1B genotype in any of the age brackets, but the presence of the slow-metabolizing ADH1B was more strongly associated with weight gain in all age brackets. This result links the slower ethanol elimination by the *ADH1B*1*/*1* alcoholics with their more efficient utilization of ethanol as an energy source. No effects of ALDH2 genotype on body weight or BMI were observed.

In a study in which we simultaneously measured the blood and salivary ethanol and acetaldehyde levels of Japanese alcoholics in the morning on the day of their first visit [32, 33], we found that ethanol and acetaldehyde remained in the blood and saliva for much longer periods and at much higher levels in the group with the slow-metabolizing ADH1B than in the group with the fast-metabolizing ADH1B, even after adjusting for age, body weight, the amount of alcohol consumed, and interval since the previous drink. Chronic heavy drinking by alcoholics may amplify the modest effect of *ADH1B*1/*1* on ethanol metabolism and lead to clear prolongation of the presence of ethanol in the body, including in the UADT. The blood and salivary ethanol levels of the subjects were similar, but the acetaldehyde levels in their saliva were much higher than in their blood because of acetaldehyde production by oral microorganisms. [32, 33]

ADH1B genotype markedly affects alcohol flushing [22, 34]. Alcohol flushing in fast-metabolizing ADH1B carriers is triggered by a rapid initial rise in the blood acetaldehyde concentration, whereas the slow initial rise in the blood acetaldehyde in slow-metabolizing ADH1B carriers may weaken the alcohol flushing [22, 34]. The results of our flushing questionnaire showed that despite the presence of inactive ALDH2 in heterozygotes, 25 % of the slow-metabolizing ADH1B carriers were never flushers and 38 % were former flushers, and thus there was a clear association between alcohol consumption by inactive ALDH2 heterozygotes and their facial flushing categories [22].

The above findings provide clues as to why slow-metabolizing ADH1B increases the risk of both alcoholism and UADT cancer. First, slow-metabolizing ADH1B diminishes the intensity of facial flushing, thereby accounting for the greater susceptibility to heavy drinking. Second, chronic heavy drinking amplifies the modest effect of slow-metabolizing ADH1B on ethanol elimination, which leads to much longer exposure to ethanol and, in turn, results in increasing the risk of developing alcoholism. When ethanol lingers in the body, the UADT is exposed to high levels of acetaldehyde as a result of acetaldehyde production in saliva, and that creates a condition that increases the risk of UADT cancer.

15.7 The *ADH1B*1/*1* and *ALDH2*1/*2* Genotype Combination and SCC in the UADT

The risk of SCC of the head and neck and esophagus of Japanese and Taiwanese drinkers has been found to be extremely increased in a multiplicative fashion by the combination of slow-metabolizing ADH1B in the *ADH1B*1/*1* genotype and inactive ALDH2 in the *ALDH2*1/*2* genotype (OR = 22–122 for esophageal SCC) [13, 18, 24, 35–37]. In Japanese alcoholics, the OR (95 % CI) with the *ADH1B*1/*1* and *ALDH2*1/*2* genotype combination has been found to increase for the very early stages of the esophageal neoplasia, from 4.53 (2.17–9.47) for low-grade intraepithelial neoplasia, to 10.4 (4.34–24.7) for high-grade intraepithelial neoplasia, and 21.7 (7.96–59.3) for invasive SCC [18]. When the two genotypes were combined with other risk factors, based on the multivariate OR for each risk factor the ORs (95 % CI) of Japanese for esophageal SCC increased enormously, by 248 times when combined with consumption of 198–395 g ethanol/week and by 414 times when combined with consumption of ≥396 g ethanol/week [24], in Taiwanese by 382 (47–3,085) times when combined with consumption of >30 g ethanol/day [35], and in Japanese by 189 (95–377) times when combined with both consumption of >96.5 g ethanol/week and smoking [36] and by 357 (105–1,210) times when combined with both drinking and smoking [37].

15.8 ADH1B and ALDH2 Genotype and Liver Disease in Japanese Alcoholic Men

Contrary to results of investigations of the relationships between the ADH1B and ALDH2 genotypes and susceptibility to UADT cancer, a cross-sectional survey of 1,902 Japanese alcoholic men showed that liver cirrhosis was associated with the presence of the *ADH1B*2* allele and *ALDH2*1/*1* genotype [38]. When non-cirrhotic patients with no or only mild liver fibrosis as controls, the results showed that the OR (95 % CI) of the *ADH1B*2* allele increased according to the severity of their liver disease, from 1.67 (1.32–2.11) in the non-cirrhotic group with liver fibrosis, to 1.81 (1.24–2.63) in the Child-Pugh class A cirrhosis group, 2.97 (1.79–4.93) in the Child-Pugh class B cirrhosis group, and 4.32 (1.48–12.6) in the Child-Pugh class C cirrhosis group. Since age-adjusted daily alcohol consumption did not differ according to ADH1B/ALDH2 genotypes in the alcoholics, their ADH1B/ALDH2-associated increase in risk of liver disease cannot be explained by the levels of ethanol and acetaldehyde exposure.

15.9 Gastric Cancer in Japanese Alcoholic Men

The lifetime drinking profiles of Japanese alcoholic men have shown that gastrectomy increases susceptibility to alcoholism [39]. Gastrectomy results in swift passage of ethanol from the small intestine into the systemic circulation. Because of this dynamic change in ethanol delivery, overshoot of blood ethanol levels and subsequent high ethanol exposure have been observed after gastrectomy [40], and they lead to rapid development of alcohol dependence. A large survey of 4,879 Japanese alcoholic men demonstrated that a high proportion of them had a history of gastrectomy, although the proportion decreased from 13.3 to 7.8 % during the 1996–2010 period [11]. Many alcoholic men with a history of gastrectomy had changed their drinking pattern after the gastrectomy and had became alcoholics after a shorter period of heavy drinking and after a lower cumulative alcohol intake than alcoholics with no history of gastrectomy [39]. There were more frequent blackouts in the gastrectomy group, and that may have reflected the sharper rise in their blood ethanol level. Since a history of gastrectomy increases the risk of alcohol dependence, this acquired risk factor increases susceptibility to alcohol dependence in the absence of the alcoholism-susceptibility genotype *ADH1B*1/*1*, and that explains the lower frequency of *ADH1B*1/*1* that was found in a gastrectomy group of Japanese alcoholic men than in a non-gastrectomy group [11].

The prevalence of gastric cancer is extremely high among Japanese alcoholic men. A study of 4,879 Japanese male alcoholic patients revealed that 187 had a history of gastrectomy for gastric cancer and ten had a history of mucosectomy for gastric cancer, and 47 were diagnosed with gastric cancer during the initial endoscopic screening [11]. A total of 244 (5.0 %) of the patients had a history of gastric cancer or were newly diagnosed with gastric cancer.

Inactive ALDH2, macrocytosis, and simultaneous presence of UADT cancer as well as *H. pylori*-associated atrophic gastritis have been found to be associated with the risk of gastric cancer detected by endoscopic screening of Japanese alcoholic men [41]. This finding partly explains why gastric, esophageal, and head and neck cancers are often concurrent in Japanese alcoholic men. However, the frequency of the *ALDH2*1/*2* genotype in alcoholics with a history of gastrectomy for gastric cancer was found to be as low as in alcoholics without a history of gastrectomy [11]. These findings suggest different causal associations between alcoholism and each group of gastric cancer.

The risk of metachronous gastric cancer is high in Japanese with esophageal SCC, especially among alcoholic men, suggesting a common cause of both cancers. Endoscopic follow-up (median, 47 months) after the initial diagnosis of esophageal SCC was performed in 99 Japanese alcoholic men [42]. A serum pepsinogen test showed a higher seroprevalence of severe chronic atrophic gastritis among the esophageal SCC cases than among age-matched alcoholic controls, whereas their *H. pylori* status was similar. The accelerated progression of severe chronic atrophic gastritis observed in Japanese alcoholic men with esophageal SCC suggests the existence of a common mechanism by which both esophageal SCC and *H. pylori-related* severe chronic atrophic gastritis develop in the alcoholics. Metachronous gastric adenocarcinoma was diagnosed in 11 of the 99 gastric cancer-free patients in the same study, and the cumulative rate of metachronous gastric cancer within 5 years was estimated to be 15 %. The hazard ratio [HR (95 % CI)] of metachronous gastric cancer was 7.87 (1.43–43.46) in the group with severe chronic atrophic gastritis in comparison with the group without chronic atrophic gastritis. Inactive heterozygous

ALDH2 was not associated with an increased risk of metachronous gastric cancer. Accelerated development of severe chronic atrophic gastritis at least partially explained the very high frequency of development of metachronous gastric cancer in this population of Japanese men with an initial diagnosis of SCC of the esophagus.

15.10 Colonoscopic Screening of Japanese Alcoholic Men for Colorectal Neoplasia

The results of colonoscopic screening of Japanese alcoholic men for colorectal neoplasia yielded an extremely high rate of advanced colorectal neoplasia: 15.7 % in the group of 744 subjects with a negative immunochemical fecal occult blood test (IFOBT) and 31.6 % in the group of 393 subjects with a positive IFOBT [43]. Advanced colorectal neoplasia has been reported to have been detected in 2.6 % of an IFOBT-negative group and 16.0 % of an IFOBT-positive group in the Japanese general population [44]. Advanced colorectal neoplasia includes adenomas ≥10 mm, villous and tubulovillous adenomas, high-grade dysplasia, carcinoma-in-situ, and invasive cancers. Thus, screening alcoholic men by the IFOBT alone is inadequate, and colonoscopy should be recommended to the patients. There were no significant associations between ALDH2 genotypes and the risk of advanced colorectal neoplasia [43].

15.11 Association Between a High Mean Corpuscular Volume and Increased Risk of Aerodigestive Tract Neoplasia in Japanese Alcoholic Men

Epidemiological evidence indicates that the red cell mean corpuscular volume (MCV) of inactive ALDH2 carriers is increased by exposure to acetaldehyde [45–51]. Alcoholism, severe acetaldehyde exposure because of the presence of inactive ALDH2, smoking, low BMI, and folate deficiency are associated with both increased MCV and increased risk of aerodigestive tract cancer. The simultaneous presence of a high MCV of 106 or more, *ALDH2*1/*2* genotype, and *ADH1B*1/*1* genotype in Japanese alcoholic men synergistically increase their risk of esophageal SCC [OR (95 % CI)=320 (27–>1,000)] [47]. An endoscopic follow-up study of cancer-free Japanese alcoholics revealed that cancer of the UADT developed much more frequently among alcoholics with a high MCV of 106 or more [HR (95 % CI)=2.52 (1.22–5.22)] [51]. An MCV of 106 or more in Japanese alcoholic men was found to increase their risks of head and neck SCC [OR (95 % CI)=2.71 (1.42–5.16)] [52], esophageal SCC [OR (95 % CI)=3.68 (1.96–6.93)] [48], and gastric cancer [OR (95 % CI)=2.5 (1.2–5.2)] [41], and to increase their risk of advanced colorectal neoplasia in an IFOBT-negative group [ORs (95 % CI)=1.65 (1.02–2.64)] and an IFOBT-positive group [2.83 (1.15–6.93)] [43] (Table 15.1).

Table 15.1 Age-adjusted odds ratios for aerodigestive neoplasia and high MCV in Japanese alcoholic men

MCV ≥ 106 fl	Controls	Head and neck cancer [52]	Controls	Esophageal cancer [48]
	N=215	N=43	N=206	N=65
Frequency	23 %	50 %	17 %	43 %
OR (95 % CI)	1 reference	2.71 (1.42–5.16)	1 reference	3.68 (1.96–6.93)
MCV ≥ 106 fl	Controls	Stomach cancer [41]	Controls	Advanced colorectal neoplasia [43]
	N=281	N=45	N=400	N=241
Frequency	20 %	38 %	22 %	32 %
OR (95 % CI)	1 reference	2.5 (1.2–5.2)	1 reference	1.65 (1.02–2.64)[a]
				2.83 (1.15–6.93)[b]

[a]Immunochemical fecal occult blood test negative subjects

[a]Immunochemical fecal occult blood test positive subjects

15.12 Conclusions

- The *ADH1B*1/*1* genotype and *ALDH2*1/*2* genotype are associated with an increased risk of UADT neoplasia in East-Asian drinkers.
- Prolonged exposure to high ethanol and acetaldehyde concentrations and inefficient degradation of acetaldehyde in the UADT explains the high risk of UADT neoplasia in people with these genotypes.
- Health-risk appraisal models that include alcohol flushing or ALDH2 genotype are useful tools for mass-screening for UADT neoplasia.
- The ADH1B genotype of Japanese alcoholic men affects their body weight and susceptibility to liver disease.
- Gastric cancer and advanced colorectal neoplasia are more common in Japanese alcoholic men than in nonalcoholic Japanese men.
- A high MCV in Japanese alcoholic men increases their risk of aerodigestive tract neoplasia.

References

1. Yokoyama A, Ohmori T, Makuuchi H et al (1995) Successful screening for early esophageal cancer in alcoholics using endoscopy and mucosa iodine staining. Cancer 76:928–934
2. Muto M, Nakane M, Katada C et al (2004) Squamous cell carcinoma in situ at oropharyngeal and hypopharyngeal mucosal sites. Cancer 101:1375–1381
3. Makuuchi H (2001) Endoscopic mucosal resection for mucosal cancer in the esophagus. Gastrointest Endosc Clin N Am 11:445–458
4. Sato Y, Omori T, Yokoyama A et al (2006) Treatment of superficial carcinoma in the pharynx and the larynx. Shokaki Naishikyo 18:1407–1416 (in Japanese with English abstract)

5. Li H, Borinskaya S, Yoshimura K et al (2009) Refined geographic distribution of the Oriental ALDH2*504Lys (nee 487Lys) variant. Ann Hum Genet 73:335–345
6. Higuchi S, Matsushita S, Murayama M et al (1995) Alcohol and aldehyde dehydrogenase polymorphisms and the risk for alcoholism. Am J Psychiatry 152:1219–1221
7. Harada S, Agarwal DP, Goedde HW (1981) Aldehyde dehydrogenase deficiency as cause of facial flushing reaction to alcohol in Japanese. Lancet 2(8253):982
8. Yokoyama M, Yokoyama A, Yokoyama T et al (2005) Hangover susceptibility in relation to aldehyde dehydrogenase-2 genotype, alcohol flushing, and mean corpuscular volume in Japanese workers. Alcohol Clin Exp Res 29:1165–1171
9. Yokoyama M, Suzuki N, Yokoyama T et al (2012) Interactions between migraine and tension-type headache and alcohol drinking, alcohol flushing, and hangover in Japanese. J Headache Pain 13:137–145
10. Higuchi S, Matsushita S, Imazeki H et al (1994) Aldehyde dehydrogenase genotypes in Japanese alcoholics. Lancet 343:741–742
11. Yokoyama A, Yokoyama T, Matsui T et al (2013) Trends in gastrectomy and ADH1B and ALDH2 genotypes in Japanese alcoholic men and their gene-gastrectomy, gene-gene and gene-age interactions for risk of alcoholism. Alcohol Alcohol 48:146–152
12. Yokoyama A, Muramatsu T, Ohmori T et al (1996) Esophageal cancer and aldehyde dehydrogenase-2 genotypes in Japanese males. Cancer Epidemiol Biomarkers Prev 5:99–102
13. Yokoyama A, Omori T, Yokoyama T (2010) Alcohol and aldehyde dehydrogenase polymorphisms and a new strategy for prevention and screening for cancer in the upper aerodigestive tract in East Asians. Keio J Med 59:115–130
14. Yang SJ, Yokoyama A, Yokoyama T et al (2010) Relationship between genetic polymorphisms of ALDH2 and ADH1B and esophageal cancer risk: a meta-analysis. World J Gastroenterol 16:4210–4220
15. Boccia S, Hashibe M, Galli P et al (2009) Aldehyde dehydrogenase 2 and head and neck cancer: a meta-analysis implementing a Mendelian randomization approach. Cancer Epidemiol Biomarkers Prev 18:248–254
16. Hayat MJ, Howlader N, Reichman ME et al (2007) Cancer statistics, trends, and multiple primary cancer analyses from the Surveillance, Epidemiology, and End Results (SEER) Program. Oncologist 12:20–37
17. Watanabe H (1998) Present status and management of multiple primary esophageal cancer associated with head and neck cancer. J Jpn Bronchoesophagol Soc 49:151–155 (in Japanese)
18. Yokoyama A, Hirota T, Omori T et al (2012) Development of squamous neoplasia in esophageal iodine-unstained lesions and the alcohol and aldehyde dehydrogenase genotypes of Japanese alcoholic men. Int J Cancer 130:2949–2960
19. Yokoyama A, Tanaka Y, Yokoyama T et al (2011) p53 protein accumulation, iodine-unstained lesions, and alcohol dehydrogenase-1B and aldehyde dehydrogenase-2 genotypes in Japanese alcoholic men with esophageal dysplasia. Cancer Lett 308:112–117
20. Aida J, Yokoyama A, Shimomura N et al (2013) Telomere shortening in the esophagus of Japanese alcoholics: relationships with chromoendoscopic findings, ALDH2 and ADH1B genotypes and smoking history. PLoS One 8:e63860
21. Brooks PJ, Enoch MA, Doldman D et al (2009) The alcohol flushing response: an unrecognized risk factor of esophageal cancer from alcohol consumption. PLoS Med 6:e50
22. Yokoyama T, Yokoyama A, Kato H et al (2003) Alcohol flushing, alcohol and aldehyde dehydrogenase genotypes, and risk for esophageal squamous cell carcinoma in Japanese men. Cancer Epidemiol Biomarkers Prev 12:1227–1233
23. Yokoyama A, Kato H, Yokoyama T et al (2006) Esophageal squamous cell carcinoma and aldehyde dehydrogenase-2 genotypes in Japanese females. Alcohol Clin Exp Res 30:491–500
24. Yokoyama A, Kato H, Yokoyama T et al (2002) Genetic polymorphisms of alcohol and aldehyde dehydrogenases and glutathione S-transferees M1 and drinking, smoking, and diet in Japanese men with esophageal squamous cell carcinoma. Carcinogenesis 23:1851–1859
25. Yokoyama T, Yokoyama A, Kumagai Y et al (2008) Health Risk Appraisal Models for Mass Screening of Esophageal Cancer in Japanese Men. Cancer Epidemiol Biomarkers Prev 17:2846–2854

26. Yokoyama A, Kumagai Y, Yokoyama T, Omori T, Kato H, Igaki H, Tsujinaka T, Muto M, Yokoyama M, Watanabe H (2009) Health risk appraisal models for mass screening for esophageal and pharyngeal cancer: an endoscopic follow-up study of cancer-free Japanese men. Cancer Epidemiol Biomarkers Prev 18:651–655
27. Yokoyama A, Oda J, Iriguchi Y et al (2013) A health-risk appraisal model and endoscopic mass screening for esophageal cancer in Japanese men. Dis Esophagus 26:148–153
28. Yin SJ, Bosron WF, Magnes LJ et al (1984) Human liver alcohol dehydrogenase: Purification and kinetic characterization of the $\beta_2\beta_2$, $\beta_2\beta_1$, $\alpha\beta_2$ and $\beta_2\gamma_1$ 'Oriental' isozymes. Biochemistry 23:5847–5853
29. Neumark YD, Friedlander Y, Durst R et al (2004) Alcohol dehydrogenases polymorphisms influence alcohol-elimination rates in a male Jewish population. Alcohol Clin Exp Res 28:10–14
30. Yokoyama A, Yokoyama T, Mizukami T et al (2014) Blood ethanol levels of nonabstinent Japanese alcoholic men in the morning after drinking and their ADH1B and ALDH2 genotypes. Alcohol Alcohol 49(1):31–37
31. Yokoyama A, Yokoyama T, Matsui T et al (2013) Alcohol dehydrogenase-1B genotype (rs1229984) is a strong determinant of the relationship between body weight and alcohol intake in Japanese alcoholic men. Alcohol Clin Exp Res 37:1123–1132
32. Yokoyama A, Tsutsumi E, Imazeki H et al (2007) Contribution of the alcohol dehydrogenase-1B genotype and oral microorganisms to high salivary acetaldehyde concentrations in Japanese alcoholic men. Int J Cancer 121:1047–1054
33. Yokoyama A, Tsutsumi E, Imazeki H et al (2010) Polymorphisms of alcohol dehydrogenase-1B and aldehyde dehydrogenase-2 and the blood and salivary ethanol and acetaldehyde concentrations of Japanese alcoholic men. Alcohol Clin Exp Res 34:1246–1256
34. Yokoyama A, Yokoyama T, Omori T (2010) Past and current tendency for facial flushing after a small dose of alcohol is a marker for increased risk of upper aerodigestive tract cancer in Japanese drinkers. Cancer Sci 101:2497–2498
35. Lee CH, Lee JM, Wu DC et al (2008) Carcinogenetic impact of ADH1B and ALDH2 genes on squamous cell carcinoma risk of the esophagus with regard to the consumption of alcohol, tobacco and betel quid. Int J Cancer 122:1347–1356
36. Cui R, Kamatani Y, Takahashi A et al (2009) Functional variants in ADH1B and ALDH2 coupled with alcohol and smoking synergistically enhance esophageal cancer risk. Gastroenterology 137:1768–1775
37. Tanaka F, Yamamoto K, Suzuki S et al (2010) Strong interaction between the effects of alcohol consumption and smoking on oesophageal squamous cell carcinoma among individuals with ADH1B and/or ALDH2 risk alleles. Gut 59:1457–1464
38. Yokoyama A, Mizukami T, Matsui T et al (2013) Genetic polymorphisms of alcohol dehydrogenase-1B and aldehyde dehydrogenase-2 and liver cirrhosis, chronic calcific pancreatitis, diabetes mellitus, and hypertension among Japanese alcoholic men. Alcohol Clin Exp Res 37:1391–1401
39. Yokoyama A, Takagi T, Ishii H et al (1995) Gastrectomy enhances vulnerability to the development of alcoholism. Alcohol 12:213–216
40. Caballeria J, Frezza M, Hernández-Muñoz R et al (1989) Gastric origin of the first-pass metabolism of ethanol in humans: effect of gastrectomy. Gastroenterology 97:1205–1209
41. Yokoyama A, Yokoyama T, Omori T et al (2007) Helicobacter pylori, chronic atrophic gastritis, inactive aldehyde dehydrogenase-2, macrocytosis and multiple upper aerodigestive tract cancers and the risk for gastric cancer in alcoholic Japanese men. J Gastroenterol Hepatol 22:210–217
42. Yokoyama A, Omori T, Yokoyama T et al (2009) Chronic atrophic gastritis and metachronous gastric cancer in Japanese alcoholic men with esophageal squamous cell carcinoma. Alcohol Clin Exp Res 33:898–905
43. Mizukami T, Yokoyama A, Yokoyama T et al (2012) Immunochemical fecal occult blood test and colonoscopic screening in Japanese alcoholic men. Alcohol Clin Exp Res 36(Suppl):94A (Abstract No. P009

44. Morikawa T, Kato J, Yamaji Y et al (2005) A comparison of the immunochemical fecal occult blood test and total colonoscopy in the asymptomatic population. Gastroenterology 129:422–428
45. Nomura F, Itoga S, Tamura M et al (2000) Biological markers of alcoholism with respect to genotypes of low-Km aldehyde dehydrogenase (ALDH2) in Japanese subjects. Alcohol Clin Exp Res 24(4 Suppl):30S–33S
46. Hashimoto Y, Nakayama T, Futamura A et al (2002) Erythrocyte mean cell volume and genetic polymorphism of aldehyde dehydrogenase 2 in alcohol drinkers. Blood 99:3487–3488
47. Yokoyama M, Yokoyama A, Yokoyama T et al (2003) Mean corpuscular volume and the aldehyde dehydrogenase-2 genotype in male Japanese workers. Alcohol Clin Exp Res 27: 1395–1401
48. Yokoyama A, Yokoyama T, Muramatsu T et al (2003) Macrocytosis, a new predictor for esophageal squamous cell carcinoma in Japanese men. Carcinogenesis 24:1773–1778
49. Yokoyama A, Yokoyama T, Kumagai Y et al (2005) Mean corpuscular volume, alcohol flushing and the predicted risk of squamous cell carcinoma of the esophagus in cancer-free Japanese men. Alcohol Clin Exp Res 29:1877–1883
50. Yokoyama T, Saito K, Lwin H et al (2005) Epidemiological evidence that acetaldehyde plays a significant role in the development of decreased serum folate concentration and elevated mean corpuscular volume in alcohol drinkers. Alcohol Clin Exp Res 29:622–630
51. Yokoyama A, Omori T, Yokoyama T et al (2006) Risk of squamous cell carcinoma of the upper aerodigestive tract in cancer-free alcoholic Japanese men: an endoscopic follow-up study. Cancer Epidemiol Biomarkers Prev 15:2209–2215
52. Yokoyama A, Omori T, Yokoyama T et al (2010) Risk factors of squamous cell carcinoma in the oropharynx, hypopharynx, and epilarynx of Japanese alcoholic men: A case-control study based on an endoscopic screening program. Stomach Intestine (Tokyo) 45:180–189 (In Japanese with English abstract, tables, and figures)

Chapter 16
Acetaldehyde and Retinaldehyde-Metabolizing Enzymes in Colon and Pancreatic Cancers

S. Singh, J. Arcaroli, D.C. Thompson, W. Messersmith, and V. Vasiliou

Abstract Colorectal cancer (CRC) and pancreatic cancer are two very significant contributors to cancer-related deaths. Chronic alcohol consumption is an important risk factor for these cancers. Ethanol is oxidized primarily by alcohol dehydrogenases to acetaldehyde, an agent capable of initiating tumors by forming adducts with proteins and DNA. Acetaldehyde is metabolized by ALDH2, ALDH1B1, and ALDH1A1 to acetate. Retinoic acid (RA) is required for cellular differentiation and is known to arrest tumor development. RA is synthesized from retinaldehyde by the retinaldehyde dehydrogenases, specifically ALDH1A1, ALDH1A2, ALDH1A3, and ALDH8A1. By eliminating acetaldehyde and generating RA, ALDHs can play a crucial regulatory role in the initiation and progression of cancers. ALDH1 catalytic activity has been used as a biomarker to identify and isolate normal and cancer stem cells; its presence in a tumor is associated with poor prognosis in colon and pancreatic cancer. In summary, these ALDHs are not only biomarkers for CRC and pancreatic cancer but also play important mechanistic role in cancer initiation, progression, and eventual prognosis.

Keywords Acetaldehyde • ALDH • Biomarker • Colorectal cancer • Pancreatic cancer • Retinaldehyde • Stem cells

S. Singh
Department of Pharmaceutical Sciences, University of Colorado Anschutz Medical Campus, Mail Stop C238-P20, 12850 E Montview Blvd, Aurora, CO 80045, USA

J. Arcaroli • W. Messersmith
Division of Medical Oncology, University of Colorado School of Medicine, Mail Stop C238-P20, 12850 E Montview Blvd, Aurora, CO 80045, USA

D.C. Thompson
Department of Clinical Pharmacy, University of Colorado School of Medicine, Mail Stop C238-P20, 12850 E Montview Blvd, Aurora, CO 80045, USA

V. Vasiliou (✉)
Department of Environmental Health Sciences, Yale School of Public Health, New Haven, CT, USA
e-mail: Vasilis.Vasiliou@ucdenver.edu

V. Vasiliou et al. (eds.), *Biological Basis of Alcohol-Induced Cancer*, Advances in Experimental Medicine and Biology 815,
DOI 10.1007/978-3-319-09614-8_16

16.1 Introduction

Colorectal cancer (CRC) and pancreatic cancer represent serious health concerns because of their very high morbidity and mortality. Each year, more than one million new CRC cases are diagnosed and over 500,000 deaths are associated with this condition worldwide [1]. In the USA, CRC is the fourth most commonly diagnosed cancer and second leading cause of cancer-related death. Pancreatic cancer ranks tenth in incidence but is disproportionately fatal in being the fourth largest cause of cancer-related deaths in the USA. The American Cancer Society estimated diagnosis of 142,820 new cases of CRC in the USA during 2013; of these, approximately 50,830 people are expected to die. The estimated incidence of pancreatic cancer is 45,220 with 38,460 deaths [2]. Although the exact mechanisms that promote CRC remain obscure, there is increasing evidence suggesting the involvement of lifestyle-related factors in addition to genetic predisposition. These factors include waist circumference, folate, and multivitamins in the diet, high fat and high energy diet, physical exercise, tobacco smoking, and alcohol consumption [3, 4].

According to dose–response meta-analysis and pooled results from cohort studies, chronic daily consumption of approximately 50 g of alcohol increases the relative risk for colon cancer by 40 % [5, 6]. Alcohol and its primary metabolite, acetaldehyde, have also been linked with pancreatic cancer [7]. Various theories have been advanced regarding the mechanism by which alcohol induces cancer. For example, ethanol may enhance mucosal penetration of a carcinogen by serving as a solvent. In addition, ethanol induces cytochrome P4502E1 (CYP2E1), an enzyme capable of generating reactive oxygen species (Fig. 16.1). However, the most well-accepted theory regarding ethanol-induced cancer involves acetaldehyde acting as a carcinogen [8, 9]. Ethanol is metabolized to acetaldehyde by alcohol dehydrogenase (ADH), CYP2E1, and catalase [10, 11] (Fig. 16.1). Acetaldehyde is a molecule capable of forming adducts with DNA which is considered an initial step in carcinogenesis [12]. In 2009, the International Agency for Research on Cancer (IARC) designated acetaldehyde (as associated with alcohol consumption) to be a group I human carcinogen [13]. Acetaldehyde is metabolized to acetate, a process catalyzed by aldehyde dehydrogenase (ALDH) 2, ALDH1B1, and ALDH1A1 (Fig. 16.1) [14]. The ability of these ALDHs to repress cellular acetaldehyde levels is consistent with a role for ALDHs in colon and pancreatic cancers and is supported strongly by the association of ALDH2 deficiency with high incidence of CRC and pancreatic cancer in heavy ethanol drinkers [15, 16]. In addition to metabolizing acetaldehyde, ALDH1 isozymes are the primary enzymes involved in the metabolism of retinaldehyde to retinoic acid (RA), a signaling molecule that plays a crucial role in cellular proliferation and differentiation [11]. Given their ability to affect cellular RA levels, it is likely that RA-generating ALDHs have a role in modulating carcinogenesis. Several other observations lend support to the notion that ALDHs are implicated in cancer. First, ALDH activity has been used to identify and isolate normal and cancer stem cells (CSCs) of various lineages [17–19] (Table 16.1). Second, high ALDH expression has been found to be associated with poor clinical outcome in leukemia [20],

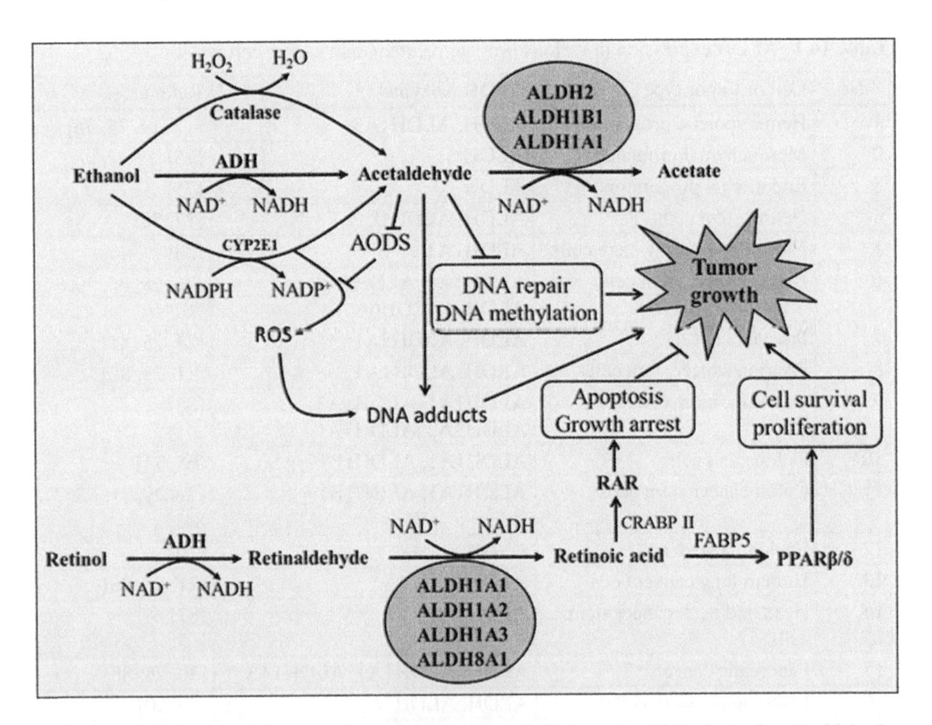

Fig. 16.1 ALDHs modulate carcinogenesis by metabolizing acetaldehyde and retinaldehyde. Ethanol is metabolized by alcohol dehydrogenase (ADH), catalase, and CYP2E1 to acetaldehyde. Acetaldehyde can interfere with antioxidative defense systems (AODS) and generate reactive oxygen species (ROS); inhibits DNA repair and methylation; and forms DNA and protein adducts to promote tumor growth. Acetaldehyde is metabolized to acetate primarily by ALDH2, ALDH1B1, and ALDH1A1. Retinaldehyde, formed from retinol by ADH, is converted to retinoic acid (RA) by retinaldehyde-metabolizing ALDHs. RA exerts anticarcinogenic activity by binding to cellular retinoic acid-binding proteins (CRBPII) and activating the RA receptor (RAR). When RA binds to fatty acid-binding protein 5 (FABP5), it activates orphan nuclear receptor peroxisome proliferator-activated receptor (PPAR)β/δ and acts as procarcinogenic agent. *ALDH* aldehyde dehydrogenase, *NAD*+ NAD(P), nicotinamide adenine dinucleotide (phosphate), H_2O_2 hydrogen peroxide

ovarian [21–23], prostate [24, 25], breast [26–28], colorectal [29], and pancreatic cancer [15, 30]. Third, ALDH+ cells (cells with very high ALDH expression) exhibit a greater tumorigenic capacity, as reflected in colony-forming capability in vitro and in xenograft-induced tumor formation in vivo [31]. We have found very strong up-regulation of ALDH1B1 expression in an animal model of colon polyps, specifically adenomatous polyposis coli multiple intestinal neoplasia (*Apc*(*Min*)/+) mice (our unpublished data). These mice have point mutation in *Apc*, a tumor suppressor gene which when mutated leads to dysregulation of the Wnt-signaling pathway and results in up-regulation of oncogenes like *c-Myc* [32]. Overexpression of ALDH1B1 in polyps from these mice is suggestive of a possible relationship between Wnt-signaling and ALDH1B1 expression, a consideration that warrants further study.

Table 16.1 ALDH expression in various progenitor, stem, and cancer cell types

S.No.	Cell or tumor type	ALDH isozyme(s)[a]	Reference
1	Hematopoietic progenitor	ALDH, ALDH1A3	[17–19, 75, 76]
2	Mesenchymal progenitors	ALDH	[75]
3	Endothelial progenitors	ALDH	[75]
4	Neural stem cells	ALDH, ALDH1L1	[77, 78]
5	Normal mammary stem cells	ALDH1A1	[26]
6	Breast cancer stem cells	ALDH1A1, ALDH1A3, ALDH2, ALDH6A1,	[26, 28, 76, 79, 80]
7	Prostate cancer	ALDH, ALDH7A1	[24, 25, 81]
8	Ovarian cancer stem cells	ALDH, ALDH1A1	[21–23, 82]
9	Ovarian cancer cells	ALDH1A1, ALDH1A3, ALDH3A2, ALDH7A1	[83]
10	Colon stem cells	ALDH1A1, ALDH1B1	[65, 71]
11	Colon cancer stem cells	ALDH1A1, ALDH1B1	[21, 29, 31, 65, 71, 84]
12	Leukemia stem cells	ALDH	[20]
13	Human lung cancer cells	ALDH1A1	[21, 85, 86]
14	Head and neck cancer stem cells	ALDH1A1	[87]
15	Pancreatic cancer	ALDH, ALDH1A1, ALDH1A3	[30, 76, 88]
16	Liver cancer stem cells	ALDH, ALDH1A1	[89, 90]

[a]ALDH is designated for studies in which ALDH+ cells were identified and isolated using the ALDEFLUOR™ assay

A causal relationship exists between alcohol consumption and CRC or pancreatic cancer and this may be mediated, at least in part, by acetaldehyde [12, 15]. The significance of retinaldehyde and acetaldehyde in tumor formation, and very high expression of the ALDHs in colorectal and pancreatic cancer are suggestive of a crucial role for acetaldehyde- and retinaldehyde-metabolizing ALDHs in these cancers. Lack of ALDH2 activity and resultant high acetaldehyde levels are linked with colon cancer initiation. By contrast high ALDH1 activity (primarily ALDH1A1 and ALDH1B1) is required for the stemness and tumorigenic potential of CSCs.

16.2 Acetaldehyde: A carcinogen

Acetaldehyde is categorized as "carcinogenic to humans" and "reasonably anticipated to be a human carcinogen" according to IARC regulations and US National Toxicology Program (NTP), respectively [13, 33]. Acetaldehyde has been shown to be a highly toxic, mutagenic, and carcinogenic compound in a variety of in vitro and in vivo studies. Its effects range from damaging antioxidant defenses [11] to interfering with DNA methylation and repair mechanisms through formation of adducts with DNA and proteins (Fig. 16.1) [10, 12]. In the colon, acetaldehyde is primarily produced from ethanol by resident bacteria and, to a lesser extent, by mucosal alcohol

Table 16.2 Affinity of ALDHs for acetaldehyde and retinaldehyde

S.No.	ALDH isozyme(s)	Substrate	K_m (μM)	Reference
1	ALDH1A1	Acetaldehyde	180	[14]
		All-*trans* retinaldehyde	11.6–26.8	[91]
		9-*cis* retinaldehyde	3.59	Jackson et al., under preparation
2	ALDH1A2	All-*trans* retinaldehyde	0.66	[92]
		9-*cis* retinaldehyde	0.62	
3	ALDH1A3	All-*trans* retinaldehyde	0.2	[93]
4	ALDH1B1	Acetaldehyde	55	[14]
		Retinaldehyde	24.9	Jackson et al., under preparation
5	ALDH2	Acetaldehyde	3.2	[14]
6	ALDH8A1	9-*cis* retinaldehyde	3.15	[46]

dehydrogenase (ADH). As a result of metabolism by intra-colonic microbes, large quantities (nine-fold higher than normal) of acetaldehyde accumulate in the rat colon 2 h after intraperitoneal injection of ethanol [34]. Human colon mucosal cells harbor ADH1, ADH3, and ADH5, with the ADH1 and ADH3 isozymes being most active [35]. In an in vitro experiment, human colon contents were able to generate 60–250 μM acetaldehyde when incubated with concentration of ethanol (10–100 mg%), which is known to be attained during normal ethanol drinking [36]. The high levels of acetaldehyde attained in the colon after drinking ethanol likely underlies the correlation between chronic, heavy ethanol consumption, and CRC in humans. In ethanol-treated rats, a high concentration of acetaldehyde (50–350 μM) in the colon mucosa has been shown to correlate positively with hyper-proliferation of the colon crypt cells. Such a phenomenon would be anticipated to favor the development of CRC [37, 38].

Acetaldehyde is metabolized primarily by mitochondrial ALDH2 and ALDH1B1 and, to lesser extent, by cytosolic ALDH1A1 (Table 16.2) [14]. The most convincing evidence for a role of acetaldehyde in CRC initiation emanates from studies involving Asians who possess a polymorphism in their ALDH2 enzyme known as ALDH2*2. These subjects possess a single nucleotide polymorphism (SNP) that leads to a lysine to glutamate substitution at residue 487 that renders the enzyme functionally inactive [39, 40]. Approximately 40 % of the Asian population carry an ALDH2*2 allele; this compromises their ability to metabolize acetaldehyde and increases their colon cancer risk 3.4 times [16].

16.3 Opposing Effects of Retinoic Acid on Cancer Cell Proliferation

Retinoids exert many physiologically important and diverse functions in relation to cellular proliferation and differentiation of normal and cancer cells. For example, retinoids are crucial for embryonic development and adult tissue remodeling. The retinoids comprise all of the derivatives of retinol, including all-*trans*-, 9-*cis*-, and

13-*cis*- retinoic acid (RA). Retinol is oxidized to retinaldehyde by retinol dehydrogenases. The resultant retinaldehydes are further metabolized to their corresponding RA by retinaldehyde dehydrogenases which include RALDH1 (ALDH1A1), RALDH2 (ALDH1A2), RALDH3 (ALDH1A3), and RALDH4 (ALDH8A1) (Table 16.2) [41–46]. Among the RAs, all-*trans*-RA (ATRA) is the most biologically potent retinoid. Abnormally low levels of ALDH1A2 have been observed in breast and prostate cancers [47, 48]. Impaired RA formation and high levels of CYP26A1 (an RA-metabolizing enzyme) in human breast cancer are consistent with a protective role for RA in this cancer [47–49]. The physiological actions of the retinoids are mediated through binding of the RA receptor (RAR) and retinoid X receptor (RXR) heterodimer to the regulatory region of retinoid-responsive genes, known as RA response elements [50]. RARs and RXRs are ligand-dependent transcription factors and exist as α, β, or γ isoforms. RAR isoforms interact with both ATRA and 9-*cis* RA, whereas RXR isoforms interacts only with 9-*cis* RA [51, 52]. The binding of RA with the RAR/RXR dimer recruits co-activator proteins and initiates transcriptional activation of the retinoid-responsive genes [50]. Retinoids have been found to be effective for the treatment of acute promyelocytic leukemia and prevention of liver, lung, breast, prostate, skin, and colon cancers [53–55]. In vivo studies involving rats have revealed that retinoids added to the diet reduced colon cancer cell proliferation and prevented azoxymethane-induced aberrant crypt foci (putative precancerous lesions in colon) and colon tumor formation [55, 56]. An RXR-selective retinoid, AGN194204, has been found to inhibit the proliferation of human pancreatic cancer cells, an effect that can be reversed by an RXR-selective antagonist [57]. In addition to inhibiting the growth of pancreatic cancer cells, RA increases the sensitivity of pancreatic adenocarcinoma cells to the antineoplastic drugs gemcitabine and cisplatin [58].

In contrast to the antiproliferative and anti-survival role of RA in cancer cells, dietary ATRA has been shown to enhance initiation and growth of intestinal tumors in the *Apc*(*Min*)/+ mouse model in vivo [59]. RA can promote cell survival and hyperplasia in cells expressing high levels of fatty acid-binding protein 5 (FABP5) by activating an orphan nuclear receptor, peroxisome proliferator-activated receptor (PPAR)β/δ [60]. PPARβ/δ mediates antiapoptotic properties partly by inducing the PDK1/Akt survival pathway [61]. RA binds to intracellular lipid-binding proteins (iLBPs), including cellular retinoic acid-binding proteins (CARBPII and FABP5). CARBPII and FABP5 are selective for nuclear receptors RARα and PPARβ/δ, respectively [60]. Hence, RA induces CARBPII- or FABP5- mediated activation of RAR or PPARβ/δ (respectively), depending on the ratio of FABP5/CRBPII in the cells [60]. Human CRC cell lines (specifically, T84, COLO205, SW620, SW480, HCT116, and DLD-1) express ~30-fold higher levels of FABP5 relative to normal colorectal cells (CCD18-Co), suggesting the possibility of pro-proliferative and antiapoptotic roles for RA in these cells [60, 62]. However, the expression levels of PPARβ/δ in CRC cells and its role in tumorigenesis are unresolved in various cancers, including CRC [63].

RA inhibits the proliferation and increases chemosensitivity of pancreatic cancer cells. However, the involvement of RA in CRC is less clear, with opposing findings suggesting pro- or antiproliferative roles.

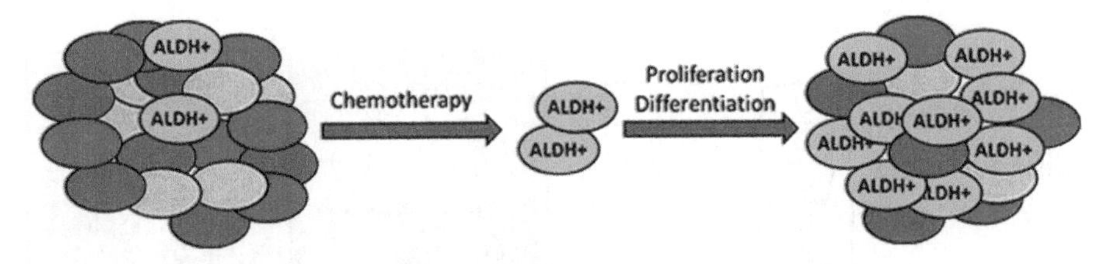

Fig. 16.2 ALDH-expressing cells are responsible for chemoresistance and relapse of many tumors after chemotherapy. Most current chemotherapy drugs are effective against the bulk of the tumor cells. However, the high ALDH-expressing (ALDH+) cancer stem cells are resistant to these treatments. As a result, during chemotherapy, the ALDH+ cells proliferate and promote tumor growth. The resultant tumors contain an increased proportion of ALDH+ cells making them more resistant to chemotherapy than the original tumor

16.4 ALDHs and Cancer Stem Cells

In the gastrointestinal (GI) tract, tissue-specific stem cells are at the top of the cellular hierarchy and play a critical role in regulating tissue homeostasis. These specialized epithelial cells are characterized by their ability to self-renew and differentiate into a variety of cellular populations that perform specific functions within the GI tract. Currently, it is believed that these tissue-specific stem cells (or progenitor cells), when oncogenically transformed, become CSCs or tumor-initiating cells (TICs) since they functionally possess the capacity to form tumors and maintain tumor growth. Accumulating evidence also suggests that CSCs are responsible for chemotherapeutic/radiation resistance and tumor recurrence (Fig. 16.2). ALDH catalytic activity has been identified in many human cancers [28] and, as such, is used as a marker of CSCs, including colorectal and pancreatic cancer. The pathophysiological function of ALDHs in CSCs remains unresolved. Intense research of ALDH enzymes is underway in order to elucidate the role of these proteins in the development and progression of cancer as well as drug resistance.

16.4.1 Colorectal Cancer

Although earlier stages of CRC are highly curable, therapeutic interventions in advanced disease have proven to be poorly effective at increasing the 5-year survival rate. Recent drug development has focused on targeting the CSC population as a potential therapy. In normal colon, CSCs reside at the bottom of the crypt and generate upward, migrating and differentiating transit amplifying cells (in the middle of the crypt) which become terminally differentiated cells as they move upward and eventually shed into the lumen (Fig. 16.3a) [64]. In CRC, several different molecules, including the cell surface markers CD133 and CD44 as well as ALDH activity, have been proposed as biomarkers for identification and isolation of the CSC

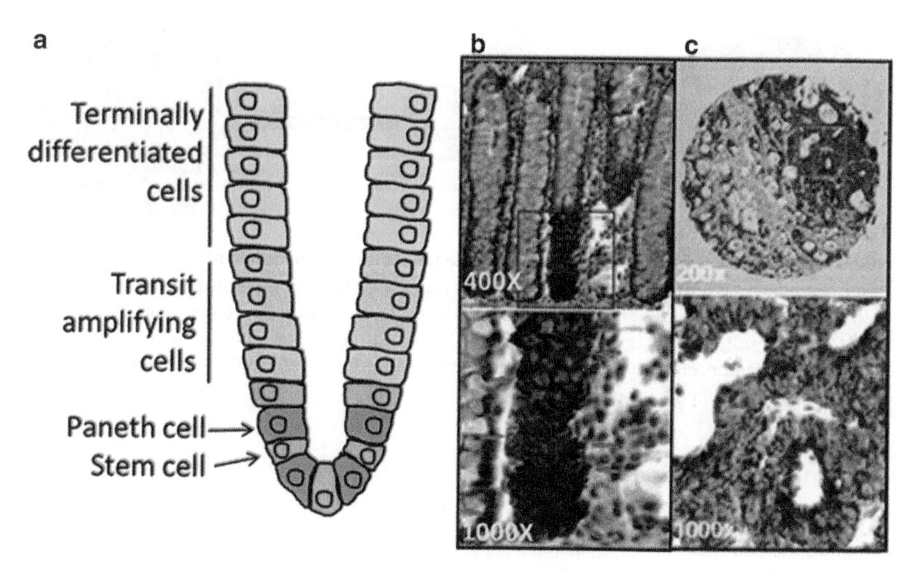

Fig. 16.3 ALDH1B1 expression pattern in normal colon and colon adenocarcinoma. Location of various cell types in normal colon (**a**). ALDH1B1 expression (*red arrows*) is strictly localized to stem-like cells at the base of crypts in the normal human colon (**b**). ALDH1B1 is expressed at extremely high levels throughout all cells of human colon adenocarcinomas (**c**). In figures (**b**, **c**) (reproduced from Chen et al. [71]), *lower panels* are higher magnification of areas identified by *squares* in the *upper panel*

population [31, 65–68]. CD133+ colon cancer cells were initially shown to be tumorigenic [67, 68]. However, subsequent studies identified that both CD133+ and CD133− cells possess tumorigenic potential [69]. CD44+ (either with or without epithelial-specific antigen (ESA+)) was demonstrated to be a marker in colon CSCs [69]. However, additional studies showed that CD44+ cells reside throughout the entire crypt, including the proliferative compartment, suggesting that the CD44+ colon cells are not necessarily stem-like [65]. We have examined CD44 and ALDH together in one of our CRC patient-derived tumor xenograft (PDTX) models to determine if CD44+ cells had tumorigenic properties [70]. Despite ALDH+/CD44+ cells showing some tumorigenic growth, ALDH+/CD44− cells exhibited a higher incidence and faster growing tumors. In this same PDTX model, isolation and injection of ALDH+ and ALDH− cells in mice showed a significant difference with respect to tumor growth [70]. ALDH+ cells produced fast growing and large tumors when compared to ALDH− cells that either produced very small tumors or no tumors in five separate PDTX models. Importantly, all ALDH+ tumors looked morphologically the same as the original tumor. Several other studies have shown that injection of ALDH+ cells from colitis and colon cancer patients facilitated spheroid formation (in vitro three-dimensional spheroid cell culture that more closely resembles the in vivo environment) and tumor growth in a xenograft model, while ALDH− cells were incapable of tumor growth [31, 65]. These studies demonstrate that ALDH catalytic activity appears to be a robust marker of CSCs in CRC.

Given the apparent promise of ALDH activity as a potential biomarker of CSCs, many investigations are currently exploring the role of ALDH in CSC function. In particular, a great deal of focus is being placed on which ALDH isoform(s) mediate the catalytic activity in the CSCs. In normal colon stem cells, ALDH1 has been demonstrated to be primarily expressed at the bottom of the crypt compartment in the colon (where colon-specific stem cells are located) and ALDH1 levels are significantly elevated in the development and progression of CRC [65]. Interestingly, ALDH1 protein levels are elevated in the colon of patients with ulcerative colitis (a risk factor for colon cancer) compared to normal colon cells; such expression may be important in the transformation from colitis to colon cancer [31]. We have shown that ALDH1B1 protein is 5.6-fold higher when compared to ALDH1A1 in CRC patients and may be a potential biomarker in CRC (Fig. 16.3b, c) [71]. Similarly, very high expression of ALDH1B1 was found in the colon polyps of *Apc(Min)/+* mice (our unpublished data). While these studies indicate elevations in individual ALDH isoforms in CRC, the contribution of these enzymes to the progression of CSCs and CRC remain to be clarified.

A common problem associated with standard chemotherapeutic regimens in CRC is treatment resistance. Although chemotherapy is effective at reducing tumor burden, many CRC patients will experience disease recurrence and ultimately succumb to their disease. CSCs are thought to be responsible for chemotherapy resistance and disease recurrence [72]. Therefore, therapeutic elimination of this population would be predicted to reduce tumor recurrence and ultimately improve survival. In our CRC PDTX model, the effects of an inhibitor of the Notch pathway (considered to be important for self-renewal of colon stem cells) in combination with irinotecan was investigated on the ALDH+ cell population [70]. The combination therapy was effective at reducing the number of ALDH+ cells as well as tumor recurrence, even after treatment was discontinued when compared to single agent Notch pathway inhibition and irinotecan. Administration of the combination therapy for 28 days prevented tumor growth in the ALDH+ cell xenograft model; this protection continued for 3 months after combination treatment was completed [70]. These data indicate that the ALDH+ population has the ability to self-renew, and significantly reducing this population of cells delays tumor recurrence (Fig. 16.2). Whether specific ALDH isozymes contribute to chemotherapy resistance remains to be determined.

16.4.2 Pancreatic Cancer

Despite considerable research, the 5-year survival rate for pancreatic cancer still remains extremely poor. A concerted effort is underway to delineate pathways that are dysregulated in the CSC population of this disease and thereby identify novel potential therapeutic targets.

In pancreatic cancer, ALDH+ cells have been shown to possess stem cell features, as evidenced by enhanced clonogenicity in vitro and tumorigenic growth in mice [73]. These cells also have greater tumorigenic potential than CD133+ cells [73].

Interestingly, ALDH+ cells from pancreatic cancers have been demonstrated to: (1) express many genes of the mesenchymal phenotype, (2) have an increased capacity to migrate and invade, and (3) be more numerous in metastatic lesions [30]. Furthermore, in pancreatic cancer patients, expression of ALDH in tumors is associated with a worse survival rate than those tumors that do not express ALDH [30]. ALDH1A1 expression has been linked to resistance to chemotherapy in a pancreas PDTX model. In this context, treatment with gemcitabine was shown to enhance gene and protein expression of ALDH. Inclusion of an inhibitor of hedgehog (a pathway important for stem cell regulation in the pancreas) with gemcitabine resulted in decreased expression of ALDH [74]. These studies suggest that ALDH+ cells are stem-like cells in pancreas cancer and may be important contributors in disease progression and chemoresistance; therefore, contribute to the negative outcomes in patients with pancreatic cancer.

16.5 Summary

There is accumulating evidence that supports a role for ALDHs in cancer development and progression. The exact mechanisms by which ALDHs influence tumorigenesis remain to be defined. Certainly, metabolism of acetaldehyde and/or the generation of retinoic acid represent modalities by which ALDHs could influence CRC and pancreatic cancer. ALDH catalytic activity appears to be an excellent biomarker that can be utilized for the isolation and characterization of the CSC population in tumors obtained from patients with CRC or pancreatic cancer. It is becoming apparent that the various ALDH isozymes may have different roles in tumorigenesis (from metabolism of the carcinogen to modulation of the proliferation-regulating retinoids) and that the timing and cellular localization of isozyme expression may be critical factors that influence how ALDHs modulate cancer development and progression. Further studies are needed that identify (1) the importance of ALDH catalytic activity in modulation of tumorigenesis, (2) the specific ALDH isozymes involved (and that regulate CSCs), and (3) the signaling pathways that regulate tumor-associated ALDH expression. The results obtained from such studies should lead to the development of novel therapies that may more effectively treat these devastating diseases.

Acknowledgments We would like to thank our colleagues for critically reviewing this manuscript. This work was supported, in part, by the following NIH grants; AA022057 and EY11490.

References

1. Parkin DM, Bray F, Ferlay J, Pisani P (2005) Global cancer statistics, 2002. CA Cancer J Clin 55:74–108
2. Siegel R, Naishadham D, Jemal A (2013) Cancer statistics, 2013. CA Cancer J Clin 63:11–30

3. Wei EK, Wolin KY, Colditz GA (2010) Time course of risk factors in cancer etiology and progression. J Clin Oncol 28:4052–4057
4. Kaneko R, Sato Y, An Y et al (2010) Clinico-epidemiologic study of the metabolic syndrome and lifestyle factors associated with the risk of colon adenoma and adenocarcinoma. Asian Pac J Cancer Prev 11:975–983
5. Cho E, Smith-Warner SA, Ritz J et al (2004) Alcohol intake and colorectal cancer: a pooled analysis of 8 cohort studies. Ann Intern Med 140:603–613
6. Fedirko V, Tramacere I, Bagnardi V et al (2011) Alcohol drinking and colorectal cancer risk: an overall and dose-response meta-analysis of published studies. Ann Oncol 22:1958–1972
7. Wilson JS, Apte MV (2003) Role of alcohol metabolism in alcoholic pancreatitis. Pancreas 27:311–315
8. Brennan P, Boffetta P (2004) Mechanistic considerations in the molecular epidemiology of head and neck cancer. IARC Sci Publ 393–414
9. Seitz HK, Stickel F (2007) Molecular mechanisms of alcohol-mediated carcinogenesis. Nat Rev Cancer 7:599–612
10. Seitz HK, Stickel F (2010) Acetaldehyde as an underestimated risk factor for cancer development: role of genetics in ethanol metabolism. Genes Nutr 5:121–128
11. Singh S, Brocker C, Koppaka V et al (2013) Aldehyde dehydrogenases in cellular responses to oxidative/electrophilic stress. Free Radic Biol Med 56:89–101
12. Yu HS, Oyama T, Isse T et al (2010) Formation of acetaldehyde-derived DNA adducts due to alcohol exposure. Chem Biol Interact 188:367–375
13. Secretan B, Straif K, Baan R et al (2009) A review of human carcinogens—Part E: tobacco, areca nut, alcohol, coal smoke, and salted fish. Lancet Oncol 10:1033–1034
14. Stagos D, Chen Y, Brocker C et al (2010) Aldehyde dehydrogenase 1B1: molecular cloning and characterization of a novel mitochondrial acetaldehyde-metabolizing enzyme. Drug Metab Dispos 38:1679–1687
15. Kanda J, Matsuo K, Suzuki T et al (2009) Impact of alcohol consumption with polymorphisms in alcohol-metabolizing enzymes on pancreatic cancer risk in Japanese. Cancer Sci 100: 296–302
16. Yokoyama A, Muramatsu T, Ohmori T et al (1998) Alcohol-related cancers and aldehyde dehydrogenase-2 in Japanese alcoholics. Carcinogenesis 19:1383–1387
17. Storms RW, Trujillo AP, Springer JB et al (1999) Isolation of primitive human hematopoietic progenitors on the basis of aldehyde dehydrogenase activity. Proc Natl Acad Sci U S A 96:9118–9123
18. Hess DA, Meyerrose TE, Wirthlin L et al (2004) Functional characterization of highly purified human hematopoietic repopulating cells isolated according to aldehyde dehydrogenase activity. Blood 104:1648–1655
19. Armstrong L, Stojkovic M, Dimmick I et al (2004) Phenotypic characterization of murine primitive hematopoietic progenitor cells isolated on basis of aldehyde dehydrogenase activity. Stem Cells 22:1142–1151
20. Ran D, Schubert M, Pietsch L et al (2009) Aldehyde dehydrogenase activity among primary leukemia cells is associated with stem cell features and correlates with adverse clinical outcomes. Exp Hematol 37:1423–1434
21. Deng S, Yang X, Lassus H et al (2010) Distinct expression levels and patterns of stem cell marker, aldehyde dehydrogenase isoform 1 (ALDH1), in human epithelial cancers. PLoS One 5:e10277
22. Silva IA, Bai S, McLean K et al (2011) Aldehyde dehydrogenase in combination with CD133 defines angiogenic ovarian cancer stem cells that portend poor patient survival. Cancer Res 71:3991–4001
23. Wang YC, Yo YT, Lee HY et al (2012) ALDH1-bright epithelial ovarian cancer cells are associated with CD44 expression, drug resistance, and poor clinical outcome. Am J Pathol 180:1159–1169
24. van den Hoogen C, van der Horst G, Cheung H et al (2010) High aldehyde dehydrogenase activity identifies tumor-initiating and metastasis-initiating cells in human prostate cancer. Cancer Res 70:5163–5173

25. van den Hoogen C, van der Horst G, Cheung H et al (2011) The aldehyde dehydrogenase enzyme 7A1 is functionally involved in prostate cancer bone metastasis. Clin Exp Metastasis 28:615–625
26. Ginestier C, Hur MH, Charafe-Jauffret E et al (2007) ALDH1 is a marker of normal and malignant human mammary stem cells and a predictor of poor clinical outcome. Cell Stem Cell 1:555–567
27. Charafe-Jauffret E, Ginestier C, Iovino F et al (2010) Aldehyde dehydrogenase 1-positive cancer stem cells mediate metastasis and poor clinical outcome in inflammatory breast cancer. Clin Cancer Res 16:45–55
28. Marcato P, Dean CA, Pan D et al (2011) Aldehyde dehydrogenase activity of breast cancer stem cells is primarily due to isoform ALDH1A3 and its expression is predictive of metastasis. Stem Cells 29:32–45
29. Langan RC, Mullinax JE, Ray S et al (2012) A pilot study assessing the potential role of non-CD133 colorectal cancer stem cells as biomarkers. J Cancer 3:231–240
30. Rasheed ZA, Yang J, Wang Q et al (2010) Prognostic significance of tumorigenic cells with mesenchymal features in pancreatic adenocarcinoma. J Natl Cancer Inst 102:340–351
31. Carpentino JE, Hynes MJ, Appelman HD et al (2009) Aldehyde dehydrogenase-expressing colon stem cells contribute to tumorigenesis in the transition from colitis to cancer. Cancer Res 69:8208–8215
32. McCart AE, Vickaryous NK, Silver A (2008) Apc mice: models, modifiers and mutants. Pathol Res Pract 204:479–490
33. Zakhari S, Vasiliou V, Guo QM (eds) (2011) Alcohol and cancer, 1st edn. Springer, New York
34. Visapaa JP, Jokelainen K, Nosova T, Salaspuro M (1998) Inhibition of intracolonic acetaldehyde production and alcoholic fermentation in rats by ciprofloxacin. Alcohol Clin Exp Res 22:1161–1164
35. Jelski W, Zalewski B, Chrostek L, Szmitkowski M (2004) The activity of class I, II, III, and IV alcohol dehydrogenase isoenzymes and aldehyde dehydrogenase in colorectal cancer. Dig Dis Sci 49:977–981
36. Salaspuro M (1997) Microbial metabolism of ethanol and acetaldehyde and clinical consequences. Addict Biol 2:35–46
37. Seitz HK, Simanowski UA, Garzon FT et al (1990) Possible role of acetaldehyde in ethanol-related rectal cocarcinogenesis in the rat. Gastroenterology 98:406–413
38. Simanowski UA, Suter P, Russell RM et al (1994) Enhancement of ethanol induced rectal mucosal hyper regeneration with age in F344 rats. Gut 35:1102–1106
39. Chen YC, Peng GS, Tsao TP et al (2009) Pharmacokinetic and pharmacodynamic basis for overcoming acetaldehyde-induced adverse reaction in Asian alcoholics, heterozygous for the variant ALDH2*2 gene allele. Pharmacogenet Genomics 19:588–599
40. Jackson BC, Holmes RS, Backos DS et al (2013) Comparative genomics, molecular evolution and computational modeling of ALDH1B1 and ALDH2. Chem Biol Interact 202:11–21
41. Duester G (2000) Families of retinoid dehydrogenases regulating vitamin A function: production of visual pigment and retinoic acid. Eur J Biochem 267:4315–4324
42. Yoshida A, Hsu LC, Dave V (1992) Retinal oxidation activity and biological role of human cytosolic aldehyde dehydrogenase. Enzyme 46:239–244
43. Suzuki R, Shintani T, Sakuta H et al (2000) Identification of RALDH-3, a novel retinaldehyde dehydrogenase, expressed in the ventral region of the retina. Mech Dev 98:37–50
44. Niederreither K, Fraulob V, Garnier JM et al (2002) Differential expression of retinoic acid-synthesizing (RALDH) enzymes during fetal development and organ differentiation in the mouse. Mech Dev 110:165–171
45. Sima A, Parisotto M, Mader S, Bhat PV (2009) Kinetic characterization of recombinant mouse retinal dehydrogenase types 3 and 4 for retinal substrates. Biochim Biophys Acta 1790:1660–1664
46. Lin M, Napoli JL (2000) cDNA cloning and expression of a human aldehyde dehydrogenase (ALDH) active with 9-cis-retinal and identification of a rat ortholog, ALDH12. J Biol Chem 275:40106–40112

47. Kim H, Lapointe J, Kaygusuz G et al (2005) The retinoic acid synthesis gene ALDH1a2 is a candidate tumor suppressor in prostate cancer. Cancer Res 65:8118–8124
48. Mira YLR, Zheng WL, Kuppumbatti YS et al (2000) Retinol conversion to retinoic acid is impaired in breast cancer cell lines relative to normal cells. J Cell Physiol 185:302–309
49. Touma SE, Perner S, Rubin MA et al (2009) Retinoid metabolism and ALDH1A2 (RALDH2) expression are altered in the transgenic adenocarcinoma mouse prostate model. Biochem Pharmacol 78:1127–1138
50. Tang XH, Gudas LJ (2011) Retinoids, retinoic acid receptors, and cancer. Annu Rev Pathol 6:345–364
51. Chambon P (1996) A decade of molecular biology of retinoic acid receptors. FASEB J 10:940–954
52. Shimizu M, Takai K, Moriwaki H (2009) Strategy and mechanism for the prevention of hepatocellular carcinoma: phosphorylated retinoid X receptor alpha is a critical target for hepatocellular carcinoma chemoprevention. Cancer Sci 100:369–374
53. Boone CW, Kelloff GJ, Malone WE (1990) Identification of candidate cancer chemopreventive agents and their evaluation in animal models and human clinical trials: a review. Cancer Res 50:2–9
54. Warrell RP Jr, Frankel SR, Miller WH Jr et al (1991) Differentiation therapy of acute promyelocytic leukemia with tretinoin (all-trans-retinoic acid). N Engl J Med 324:1385–1393
55. Zheng Y, Kramer PM, Lubet RA et al (1999) Effect of retinoids on AOM-induced colon cancer in rats: modulation of cell proliferation, apoptosis and aberrant crypt foci. Carcinogenesis 20:255–260
56. Wargovich MJ, Jimenez A, McKee K et al (2000) Efficacy of potential chemopreventive agents on rat colon aberrant crypt formation and progression. Carcinogenesis 21:1149–1155
57. Balasubramanian S, Chandraratna RA, Eckert RL (2004) Suppression of human pancreatic cancer cell proliferation by AGN194204, an RXR-selective retinoid. Carcinogenesis 25:1377–1385
58. Pettersson F, Colston KW, Dalgleish AG (2001) Retinoic acid enhances the cytotoxic effects of gemcitabine and cisplatin in pancreatic adenocarcinoma cells. Pancreas 23:273–279
59. Mollersen L, Paulsen JE, Olstorn HB et al (2004) Dietary retinoic acid supplementation stimulates intestinal tumour formation and growth in multiple intestinal neoplasia (Min)/+mice. Carcinogenesis 25:149–153
60. Schug TT, Berry DC, Shaw NS et al (2007) Opposing effects of retinoic acid on cell growth result from alternate activation of two different nuclear receptors. Cell 129:723–733
61. Di-Poi N, Tan NS, Michalik L et al (2002) Antiapoptotic role of PPARbeta in keratinocytes via transcriptional control of the Akt1 signaling pathway. Mol Cell 10:721–733
62. Koshiyama A, Ichibangase T, Imai K (2013) Comprehensive fluorogenic derivatization-liquid chromatography/tandem mass spectrometry proteomic analysis of colorectal cancer cell to identify biomarker candidate. Biomed Chromatogr 27:440–450
63. Peters JM, Shah YM, Gonzalez FJ (2012) The role of peroxisome proliferator-activated receptors in carcinogenesis and chemoprevention. Nat Rev Cancer 12:181–195
64. McDonald SA, Preston SL, Lovell MJ et al (2006) Mechanisms of disease: from stem cells to colorectal cancer. Nat Clin Pract Gastroenterol Hepatol 3:267–274
65. Huang EH, Hynes MJ, Zhang T et al (2009) Aldehyde dehydrogenase 1 is a marker for normal and malignant human colonic stem cells (SC) and tracks SC overpopulation during colon tumorigenesis. Cancer Res 69:3382–3389
66. Dalerba P, Dylla SJ, Park IK et al (2007) Phenotypic characterization of human colorectal cancer stem cells. Proc Natl Acad Sci U S A 104:10158–10163
67. Ricci-Vitiani L, Lombardi DG, Pilozzi E et al (2007) Identification and expansion of human colon-cancer-initiating cells. Nature 445:111–115
68. O'Brien CA, Pollett A, Gallinger S, Dick JE (2007) A human colon cancer cell capable of initiating tumour growth in immunodeficient mice. Nature 445:106–110
69. Shmelkov SV, Butler JM, Hooper AT et al (2008) CD133 expression is not restricted to stem cells, and both CD133+ and CD133- metastatic colon cancer cells initiate tumors. J Clin Invest 118:2111–2120

70. Arcaroli JJ, Powell RW, Varella-Garcia M et al (2012) ALDH+ tumor-initiating cells exhibiting gain in NOTCH1 gene copy number have enhanced regrowth sensitivity to a gamma-secretase inhibitor and irinotecan in colorectal cancer. Mol Oncol 6:370–381
71. Chen Y, Orlicky DJ, Matsumoto A et al (2011) Aldehyde dehydrogenase 1B1 (ALDH1B1) is a potential biomarker for human colon cancer. Biochem Biophys Res Commun 405:173–179
72. Dean M, Fojo T, Bates S (2005) Tumour stem cells and drug resistance. Nat Rev Cancer 5:275–284
73. Kim MP, Fleming JB, Wang H et al (2011) ALDH activity selectively defines an enhanced tumor-initiating cell population relative to CD133 expression in human pancreatic adenocarcinoma. PLoS One 6:e20636
74. Jimeno A, Feldmann G, Suarez-Gauthier A et al (2009) A direct pancreatic cancer xenograft model as a platform for cancer stem cell therapeutic development. Mol Cancer Ther 8:310–314
75. Gentry T, Foster S, Winstead L et al (2007) Simultaneous isolation of human BM hematopoietic, endothelial and mesenchymal progenitor cells by flow sorting based on aldehyde dehydrogenase activity: implications for cell therapy. Cytotherapy 9:259–274
76. Marcato P, Dean CA, Giacomantonio CA, Lee PW (2011) Aldehyde dehydrogenase: its role as a cancer stem cell marker comes down to the specific isoform. Cell Cycle 10:1378–1384
77. Corti S, Locatelli F, Papadimitriou D et al (2006) Identification of a primitive brain-derived neural stem cell population based on aldehyde dehydrogenase activity. Stem Cells 24:975–985
78. Foo LC, Dougherty JD (2013) Aldh1L1 is expressed by postnatal neural stem cells in vivo. Glia 61:1533–1541
79. Tanei T, Morimoto K, Shimazu K et al (2009) Association of breast cancer stem cells identified by aldehyde dehydrogenase 1 expression with resistance to sequential Paclitaxel and epirubicin-based chemotherapy for breast cancers. Clin Cancer Res 15:4234–4241
80. Charafe-Jauffret E, Ginestier C, Iovino F et al (2009) Breast cancer cell lines contain functional cancer stem cells with metastatic capacity and a distinct molecular signature. Cancer Res 69:1302–1313
81. Le Magnen C, Bubendorf L, Rentsch CA et al (2013) Characterization and clinical relevance of ALDH bright populations in prostate cancer. Clin Cancer Res 19:5361–5371
82. Landen CN Jr, Goodman B, Katre AA et al (2010) Targeting aldehyde dehydrogenase cancer stem cells in ovarian cancer. Mol Cancer Ther 9:3186–3199
83. Saw YT, Yang J, Ng SK et al (2012) Characterization of aldehyde dehydrogenase isozymes in ovarian cancer tissues and sphere cultures. BMC Cancer 12:329
84. Dylla SJ, Beviglia L, Park IK et al (2008) Colorectal cancer stem cells are enriched in xenogeneic tumors following chemotherapy. PLoS One 3:e2428
85. Jiang F, Qiu Q, Khanna A et al (2009) Aldehyde dehydrogenase 1 is a tumor stem cell-associated marker in lung cancer. Mol Cancer Res 7:330–8
86. Zhang Q, Taguchi A, Schliekelman M et al (2011) Comprehensive proteomic profiling of aldehyde dehydrogenases in lung adenocarcinoma cell lines. Int J Proteomics 2011:145010
87. Chen YC, Chen YW, Hsu HS et al (2009) Aldehyde dehydrogenase 1 is a putative marker for cancer stem cells in head and neck squamous cancer. Biochem Biophys Res Commun 385:307–13
88. Kim SK, Kim H, Lee DH et al (2013) Reversing the intractable nature of pancreatic cancer by selectively targeting ALDH-high, therapy-resistant cancer cells. PLoS One 8:e78130
89. Ma S, Chan KW, Lee TK et al (2008) Aldehyde dehydrogenase discriminates the CD133 liver cancer stem cell populations. Mol Cancer Res 6:1146–53
90. Colombo F, Baldan F, Mazzucchelli S et al (2011) Evidence of distinct tumour-propagating cell populations with different properties in primary human hepatocellular carcinoma. PLoS One 6:e21369
91. Gagnon I, Duester G, Bhat PV (2003) Enzymatic characterization of recombinant mouse retinal dehydrogenase type 1. Biochem Pharmacol 65:1685–90
92. Gagnon I, Duester G, Bhat PV (2002) Kinetic analysis of mouse retinal dehydrogenase type-2 (RALDH2) for retinal substrates. Biochim Biophys Acta 1596:156–62
93. Graham CE, Brocklehurst K, Pickersgill RW, Warren MJ (2006) Characterization of retinaldehyde dehydrogenase 3. Biochem J 394:67–75

Chapter 17
Alcohol, Carcinoembryonic Antigen Processing and Colorectal Liver Metastases

Benita McVicker, Dean J. Tuma, Kathryn E. Lazure, Peter Thomas, and Carol A. Casey

Abstract It is well established that alcohol consumption is related to the development of alcoholic liver disease. Additionally, it is appreciated that other major health issues are associated with alcohol abuse, including colorectal cancer (CRC) and its metastatic growth to the liver. Although a correlation exists between alcohol use and the development of diseases, the search continues for a better understanding of specific mechanisms. Concerning the role of alcohol in CRC liver metastases, recent research is aimed at characterizing the processing of carcinoembryonic antigen (CEA), a glycoprotein that is associated with and secreted by CRC cells. A positive correlation exists between serum CEA levels, liver metastasis, and alcohol consumption in CRC patients, although the mechanism is not understood. It is known that circulating CEA is processed primarily by the liver, first by nonparenchymal Kupffer cells (KCs) and secondarily, by hepatocytes via the asialoglycoprotein receptor (ASGPR). Since both KCs and hepatocytes are known to be significantly impacted by alcohol, it is hypothesized that alcohol-related effects to these liver cells will lead to altered CEA processing, including impaired asialo-CEA degradation, resulting in changes to the liver microenvironment and the metastatic potential of CRC cells. Also, it is predicted that CEA processing will affect cytokine production in the alcohol-injured liver, resulting

B. McVicker, Ph.D. (✉)
Research Service (151), Veterans Affairs Nebraska-Western Iowa Health Care System, 4101 Woolworth Avenue, Omaha, NE 68105, USA
e-mail: bmcvicker@unmc.edu

D.J. Tuma • K.E. Lazure • C.A. Casey
Research Service (151), Veterans Affairs Nebraska-Western Iowa Health Care System, 4101 Woolworth Avenue, Omaha, NE 68105, USA

The Department of Internal Medicine University of Nebraska Medical Center, Omaha, NE USA
e-mail: dtuma@unmc.edu; kelazure@unmc.edu; ccasey@unmc.edu

P. Thomas
Department of Surgery & Biomedical Sciences, Creighton University School of Medicine, Omaha, NE, USA
e-mail: peterthomas@creighton.edu

V. Vasiliou et al. (eds.), *Biological Basis of Alcohol-Induced Cancer*, Advances in Experimental Medicine and Biology 815,
DOI 10.1007/978-3-319-09614-8_17

in pro-metastatic changes such as enhanced adhesion molecule expression on the hepatic sinusoidal endothelium. This chapter examines the potential role that alcohol-induced liver cell impairments can have in the processing of CEA and associated mechanisms involved in CEA-related colorectal cancer liver metastasis.

Keywords Carcinoembryonic antigen • Colorectal cancer • Liver metastases • Alcoholic liver disease • Asialoglycoprotein receptor • Kupffer cells • Carcinoembryonic antigen receptor • Pro-inflammatory cytokines • Liver microenvironment

17.1 Introduction

It is estimated that 65 % of all patients with colorectal cancer (CRC) will develop distant metastases with the liver being the most common site [1]. Despite advancements in diagnosis, surgical interventions and chemotherapeutics, the mechanisms involved in CRC liver metastasis remain uncharacterized. The metastatic ability of cancers varies widely; however, the concentration of carcinoembryonic antigen (CEA), a cell surface glycoprotein associated with carcinomas, has been shown to be predictive of the metastatic potential of colon tumors [2]. Additionally, the consumption of alcohol has been linked to elevated serum CEA levels and increases in liver metastasis in CRC patients [3]. It is proposed that key parameters involved in CRC metastases to the liver, especially in an alcohol-injured organ, may involve the specific processing of CEA by liver cells. In particular, the binding and degradation of circulating CEA occurs primarily in the liver by both nonparenchymal and parenchymal cells. CEA is removed from the circulation by Kupffer cells (KCs) that leads to the production of inflammatory cytokines that are known to facilitate tumor cell adhesion and survival [2, 4, 5]. Ultimately, CEA levels are controlled by hepatocellular degradation of the CEA that is processed (desialylated) by and released from KCs followed by hepatocellular endocytosis via the parenchymal cell-specific asialoglycoprotein receptor (ASGPR) [6].

It has been established that both KCs and hepatocytes are significantly affected by alcohol consumption and thus linked to the development and progression of liver disease. As the resident macrophage in the liver, KCs have been shown to be important producers of ethanol-induced reactive oxygen species (ROS) and inflammatory cytokines, especially tumor necrosis factor-alpha (TNF-α), that are implicated in the development of alcoholic liver injury [7, 8]. As the ethanol metabolizing cell and direct recipient of ethanol-induced injury (i.e., damage elicited by KC-produced cytokines), the hepatocyte has been shown to be extremely susceptible to the effects of alcohol. Many potential mechanisms and defects in hepatocellular function in response to alcohol have been documented [9]. Among the prominent and consistent defects identified in ethanol-injured hepatocytes were significant alterations observed in hepatocellular protein trafficking events, including marked defects in the activity and function of the ASGPR [10, 11]. Considering such extensive and cell-specific effects of ethanol in the liver, it is proposed that alcohol consumption may also increase the susceptibility of the liver to the colonization of tumor cells, especially those coming from a primary CRC lesion. Although the general mechanisms of tumorigenesis including processes of metastases have been largely identified, the biochemical mechanisms involved in CRC liver metastases remains uncharacterized. Importantly, the identification of specific changes in the liver microenvironment that support the development of liver metastases from CRC in the alcoholic could lead to the generation of better therapeutic options for cancer patients.

17.2 Disease Impact: CRC and Liver Metastases

17.2.1 Colon Carcinoma, Alcohol Comorbidity, and Metastasis

CRC is the third most common cancer worldwide, and in the United States alone is expected to account for over 50,000 deaths annually [1]. In recent years, progress has been made in reducing the incidence rate of CRC primarily through the use of screening techniques and the removal of noncancerous polyps. In spite of these efforts, survival disparities still exist within various groups, especially between patients diagnosed with localized disease compared to those with CRC spread to a distant site (e.g., liver). Exact reasons for the differences in CRC morbidity and mortality rates are unknown, but certainly may be related to CRC risk factors such as family history, obesity, consumption of red or processed meats, and moderate-to-heavy alcohol drinking.

CRC is a complex disease that typically develops slowly over a period of 10–15 years, demonstrating why favorable outcomes are seen with early detection and removal of adenomatous polyps. The vast majority of CRCs develop from alterations in the glandular tissue forming adenocarcinomas. CRC can expand in the lining and wall of the colon or rectum, spread to nearby lymph nodes and travel in blood vessels to distant locations, primarily the liver. The development of CRC has been linked to several risk factors including eating habits and the consumption of alcohol. Reported estimates state that individuals with a lifetime average intake of 2–4 drinks per day have a 23 % increased risk of CRC compared to those who consume less than one drink per day [12]. Multiple research studies are actively in search of mechanisms that contribute to CRC development in chronic and excessive alcohol users. Areas of investigation include the characterization of P450 cytochrome induction, effect of acetaldehyde or other alcohol metabolites on apoptosis and DNA repair, DNA methylation defects, and the role of ROS in the transformation of normal cells into cancerous cells. In addition to research focused on defining how ethanol influences carcinogenesis in the colon and rectum, another important and emerging area of study is aimed at deciphering the effect of alcohol on the biology of CRC spread to the liver.

17.2.2 Colorectal Liver Metastases

Although studies have provided evidence that chronic alcohol use is involved in the promotion of epithelial cell cancers including CRC, several works have also shown that alcohol consumption is associated with more aggressive courses and poorer prognosis of CRC [13]. More specifically, the spread of CRC to distant sites is considered a key contributing factor of poorer outcomes in patients with an alcohol use history. In a recent retrospective study, more alcohol consumption was reported in patients with colorectal liver metastasis (CLM) compared to CRC patients with no evidence of liver involvement [14]. In another study, alcohol consumption was identified as an independent risk factor of CRC liver metastasis [3]. A significant correlation was observed between alcohol consumption and synchronous liver metastasis with an increased trend observed in metachronous CLM that was detected later in the course of CRC disease [3]. To date, the mechanisms by which alcohol promotes CLM has not been determined. Due to the varied response of alcohol in humans, the primary mechanisms for CLM in alcohol consumers may be complex, but it has been shown that the duration as well as amount of alcohol consumed is related to impaired liver function and parameters associated with CRC metastatic growth such as impaired natural killer cell activity and gastrin release [15–18]. Additionally, consequences of alcohol metabolism may be involved in multiple steps of the complex metastatic process from the arrest of CRC cells in the liver sinusoids, adhesion to the liver epithelium, and establishment in the hepatic parenchyma. Overall, the study of CLM in conjunction with clinicopathological factors of alcohol-mediated liver injury is vitally important for the betterment of CRC patients and therefore warrants continued efforts in the search and identification of contributing mechanisms.

The liver is the most common and often only site of CRC metastasis as well as the most frequent site of recurrence [19]. Considering this, liver involvement of CRC is considered a major determining factor of survival. Sadly, half of CRC patients are expected to develop CLM at either the time of CRC diagnosis (synchronous) or later in the course of disease. The overall median survival is 6–12 months if left untreated leaving surgical resection as a primary treatment. Since alcohol consumption has been identified as a risk factor for CLM, it is important to determine mechanisms involved in CRC metastasis in the liver since alcohol may affect not only the development of CLM, but also the recurrence of disease following surgical interventions. In studying the metastatic process, it is important to note that the location and unique architecture of the liver are well suited to facilitate the metastatic potential of CRC. Circulating cancer cells enter the liver via the portal vein system and travel within the sinusoids where encounters with the various specialized liver cells occur determining the fate and metastatic potential of the CRC cells. During the initial phase of CRC cell colonization from a primary lesion in the colon to the liver, the cancer cells must survive mechanical stresses and escape elimination by the immune system in order to proceed. Although the liver is poised to efficiently eliminate incoming cancer cells, the inflammatory responses induced by the invading cells can elicit mechanisms that actually facilitate tumor cell arrest and growth. The overall promotion of metastasis occurs through the pathways of arrest, extravasation, vascularization, and ultimately placement and growth of CRC cells in the liver tissue [20]. There is evidence indicating that an important factor in the success of cancer cell invasion into the liver depends on cross-talk events that occur between various liver cells and the CRC metastatic cells [21]. The cells of the hepatic sinusoid appear to be especially crucial in the early steps of the metastatic process where the tumor cells adhere and migrate into the hepatic parenchyma [22–24]. During the time tumor cells are trying to survive immune surveillance or pro-death signaling mechanisms, the environment in the sinusoids is rich in inflammatory associated cytokines produced and released from the liver sinusoidal endothelial cells (LSECs) and resident KCs. In response to the cytokine-enriched environment, there is a multidimensional response to the tumor cells that includes an increase in the expression and secretion of adhesion molecules that can enhance the adhesion of the CRC cells to the endothelial lining of the sinusoids. A number of adhesion molecules have been identified that are thought to play prominent roles in metastatic adhesion of CRC cells such as E-selectin, intercellular adhesion molecule-1 (ICAM-1), and the CEA [22, 23]. Also, the activity of liver KCs appears to be crucial during this phase of the metastatic process which can be significantly affected by CEA-induced cytokine release from KCs and related increases in LSEC adhesion molecule expression [6, 24]. It is theorized that the characterization of events at the early phases of metastasis may yield the identification of better therapeutic targets compared to events that occur later in CLM such as during the extravasation and establishment of CRC cells in the liver parenchyma. Thus, the role of KCs during the arrest and adhesion phases of metastases is of current interest, especially in understanding the interplay of KCs with CEA, the CRC associated adhesion molecules, and changes resulting from CEA processing that affect the liver microenvironment and potential of CRC tumors to colonize.

17.3 The Carcinoembryonic Antigen, Colon Adenocarcinoma, and Liver Metastasis

17.3.1 CEA and Colorectal Cancer

The CEA is a 180 kDa glycoprotein that belongs to the immunoglobulin gene family that codes for several adhesion proteins. CEA has been studied extensively for half a century, initially described as an antigen found in embryonic colon tissue as well as in colon adenocarcinomas [25] and then to a lesser extent in certain benign tissues [26]. Elevations in serum CEA levels were soon found to be associated with the presence of various malignancies ultimately designating CEA as one of the most widely used tumor markers worldwide. Due to the heterogeneity of CEA expression during the presence of various cancers as well as nonspecific closely related proteins, the use of serum CEA levels as a screening tool proved unreliable [27]. However, the significant overexpression of CEA associated with colorectal carcinomas has allowed CEA to remain as an excellent and useful marker for CRC. Positive correlations to serum CEA concentrations have been shown for colorectal tumor stage, grade and site. Serum CEA was found to be increased with increasing disease stage [28], with well-differentiated tumors [29] and with tumors presenting on the left side of the colon [28]. A significant benefit of using CEA serum measures was also found in assessing prognosis in CRC patients. Several studies have shown that patients with high preoperative CEA serum concentrations (>5 ng/mL) had worse outcomes [30–32] resulting in the ranking of preoperative CEA as a category I prognostic marker for CRC [33]. Additionally, it was determined that postoperative serum CEA levels correlated to the frequency of disease recurrence [34]. Moreover and notably, CEA measures proved to be a sensitive and specific prognostic marker for liver metastasis. It is well established that the spread of CRC to distant organs occurs primarily to the liver which is associated with adverse outcomes (20–50 % 5-year survival rates posthepatic resection) [35]. CEA has been shown to be increased in >80 % of patients with distant metastasis and it was found that serial CEA measures were greater than 95 % accurate in detecting liver metastatic disease [36, 37]. Despite this evidence, the monitoring of CEA as a suitable biomarker continues to be debated since low numbers of CRC patients (less than 5 %) are able to benefit from hepatic surgical resection, and because of studies that question the prognostic value of CEA imaging [38, 39]. However, serum CEA concentrations remain the preferred marker of CRC liver involvement and continued research into the biological processing of CEA in the liver may aid in the treatment, use of CEA as a biomarker, and importantly better outcomes for CRC patients.

17.3.2 CEA and Liver Metastasis

Although, it is clear that CEA is a major determinant of CRC tumor progression and metastasis, characterization of the biological function and processing of CEA remains incomplete, especially during liver metastatic disease. Over the last 30 years, the development of molecular techniques and appropriate animal models has significantly contributed to our understanding with evidence presented that CEA has important biologic function related to tumorigenicity and CRC metastatic spread. In early studies, CEA was found to function as an intercellular adhesion molecule facilitating contact with epithelial cell membranes [40, 41]. In other works, CEA has been shown to be involved in stimulating the survival of tumor cells through the inhibition of apoptotic or anoikis mechanisms [42, 43]. In addition to its role as an adhesion molecule, CEA was also shown to be related to multiple pro-metastatic immunosuppressive events. It was demonstrated that CEA is involved in (1) the stimulation of anti-inflammatory factor release from lymphocytes [44]; (2) inhibition of natural killer cell activity [45]; and (3) suppression of dendritic cells resulting in weak T cell response to tumor cells [46]. Altogether, it became clear that CEA is intimately involved in CRC metastasis by affecting immunity and by promoting CRC colonization in the liver [47] and/or attachment in the hepatic sinusoids [41, 48]. In further seminal investigations, it was demonstrated that CEA could facilitate liver colonization of CRC cells by affecting the liver microenvironment. Specifically, it was shown that in contrast to the direct adhesion mechanisms of CEA-bearing tumor cells, the metabolism of CEA was found to be actively involved in the colonization of CRC cells in the liver. Work from the laboratories of Thomas and Jessup showed that soluble CEA enhanced liver metastasis which involved the increased release of cytokines from CEA-stimulated Kupffer cells, the resident liver macrophage [49, 50]. It was determined that KCs have a specific CEA receptor (CEAR) and that CEA–CEAR interaction leads to pro-inflammatory cytokine secretion that subsequently stimulates the up-regulation of adhesion molecules and ultimately the adhesion of colorectal cells [51, 52]. Furthermore, in addition to pro-metastatic effects induced by CEA-mediated KC cytokine release, it was also determined that the metabolism of CEA is a multistep process that also involves hepatocytes, the parenchymal cell of the liver. It was shown that subsequent to internalization via CEAR, CEA is partially degraded in the KC and released as a desialylated asialo-CEA molecule that can then be recognized by the hepatocyte-specific asialoglycoprotein receptor and ultimately degraded [53]. From these works, it is appreciated that understanding the combined processing of CEA by both populations of liver cells could provide key information concerning the mechanistic role of CEA in promoting CRC liver metastatic disease.

17.3.3 Contribution of Liver Cell CEA Degradation in CRC Liver Metastasis

The liver is the major site of uptake and eventual clearance of circulating CEA that involves the processing of CEA by both nonparenchymal (Kupffer Cell) and parenchymal (hepatocyte) cells [53].

17.3.3.1 Kupffer Cell Metabolism of CEA

It is known that Kupffer cell binding of CEA is highly conserved allowing comparative studies between rodent models and human tissue [54]. Additionally, the details of CEA liver uptake in animal models has been established through in vitro as well as in vivo studies. As previously noted, CEA belongs to a family of CEA-related cell adhesion molecules of the immunoglobulin (Ig) supergene family. The differentiation between the cell adhesion family members was shown to involve variations in the extracellular domain of the molecule. All the CEA family members are highly glycosylated proteins that typically have one variable N-domain and up to six Ig domains in the extracellular region that function as adhesion molecules or receptors [55]. What identifies the family members is the number of Ig domains that stem from the anchorage point in the cell membrane. Of the best characterized CEA cell adhesion family members, CEA has been shown to bear six of the Ig domains whereas other closely related family members such as the biliary glycoprotein and the nonspecific cross-reacting antigen bear 2–3 domains. The number of Ig domains has been shown to mediate interactions with other cellular proteins and is associated with the diversity between their functions. For CEA, the interaction has been characterized to occur with the 80 kDa CEA receptor (CEAR), a heterogeneous nuclear ribonucleoprotein M that is found on both human and rodent cells [56]. Specifically, CEA is recognized and internalized by CEAR via a specific peptide sequence, Pro-Glu-Leu-Pro-Lys (PELPK), which is located between the N domain and the first Ig domain in the extracellular portion of CEA. The binding of CEA to CEAR on Kupffer cells initiates a series of signaling events that leads to the tyrosine phosphorylation of at least two intracellular proteins followed by the induction and release of several cytokines including the interleukins (IL-1α, IL-6, and IL-10), and TNF-α. The production of these cytokines has been shown to affect the up-regulation of adhesion molecules on the hepatic sinusoidal endothelium, the protection of tumor cells against cytotoxicity by nitric oxide and other reactive oxygen radicals, and the action of pro-angiogenic factors that enhance cancer cell survival [49, 56, 57]. Overall, it is thought that the responses of KCs to CEA is integral for the remodeling of the microenvironment that takes place allowing tumor cell establishment and CRC growth, especially during metastatic disease in the liver.

17.3.3.2 Hepatocellular CEA Processing

The uptake and eventual clearance of CEA in the liver occurs through a multistep process that involves two distinct liver cell types, macrophages (KCs) and hepatocytes. Following CEA binding, internalization, and stimulation of KCs as described, CEA is processed further by KCs resulting in the production and exportation of desialylated CEA. The carbohydrate side chain of CEA contains mannose, galactose, *N*-acetylglucosamine, fucose, and sialic acid [53]. Subsequent to internalization, KCs modify CEA by removing sialic acid residues producing asialo-CEA. The asialo-CEA protein that now expresses exposed terminal galactose residues is exocytosed by the KC into the hepatic sinusoid where it is endocytosed by liver parenchymal cells via the asialoglycoprotein receptor (ASGPR). The asialoglycoprotein receptor, also termed the hepatic binding protein, was discovered four decades ago and described as a hepatocellular surface carbohydrate that binds glycoproteins lacking terminal sialic acid residues (asialoglycoproteins with exposed sugars such as galactose) [58, 59]. Proteins that are internalized via the ASGPR follow the well-characterized process of receptor-mediated endocytosis (RME) and are ultimately degraded into constituent amino acids and sugars. The fate of CEA processing in the hepatocyte was confirmed as asialo-CEA was shown to be endocytosed by hepatocytes via the ASGPR in clathrin-coated pits that develop into transport vesicles which ultimately fuse with the lysosomes where degradation of CEA takes place [53]. The contribution of asialo-CEA processing by hepatocytes in CRC hepatic metastasis remains to be determined. It is known that the carbohydrate-lectin interaction between asialo-CEA and the ASGPR is different from the protein–protein interaction used by the CEAR for the KC processing of CEA. Thus, the specific recognition mechanisms used for CEA and asialo-CEA processing by liver cells most likely formulates the responses in the liver and the potential of CRC cells to seed and grow. It is speculated that defects in asialo-CEA degradation by hepatocytes could be involved in enhanced CRC metastasis. It is hypothesized that altered asialo-CEA clearance by hepatocytes could lead to further stimulation of KCs or other liver cells and the secretion of products that induce pro-metastatic changes such as the up-regulation of adhesion molecules (i.e., ICAM-1), growth factors (i.e., vascular endothelial growth factor), and modulating enzymes (i.e., matrix metalloproteinases). Also, asialo-CEA that is not properly removed by hepatocellular degradation has the potential to directly contribute to CRC cell survival by interacting with and stimulating anti-apoptotic mechanisms within tumor cells. Clearly, more information is needed to define the role of asialo-CEA in livers with hepatocellular damage and particularly in situations in which the ASGPR is impaired.

17.4 The Role of Alcohol in CEA Processing and Potential to Promote CRC Liver Metastasis

17.4.1 Effect of Alcohol on Liver Cells

The clinical manifestations of alcoholic liver disease (ALD) include the development of steatosis, alcoholic hepatitis, fibrosis, and cirrhosis [60]. Increasing evidence suggests that the liver is sensitized to particular triggers (e.g., oxidative stress and endotoxins) in earlier phases of ALD such as steatosis that ultimately results in hepatocyte injury and the subsequent development of more advanced disease [9, 61]. Accompanying these events are striking changes to various liver cells, especially Kupffer cells and hepatocytes. Ongoing work in our laboratory is examining how such ethanol-mediated effects to liver cells can impact CEA processing and the enhanced potential of CRC to metastasize in the alcohol-injured liver (Fig. 17.1). We suggest that alcohol consumption significantly affects the direct processing of CEA by KCs, the subsequent release of cytokines, and the role of those cytokines in facilitating the expression of pro-metastatic changes in the liver. Additionally, we believe that the processing of asialo-CEA is markedly altered by ethanol-mediated impairments to the hepatocyte ASGPR resulting in enhanced metastatic potential of CRC cells in the alcohol-injured liver.

17.4.1.1 Effect of Alcohol on Kupffer Cells

Kupffer cells are known to play an important role in ethanol and/or its metabolite's detrimental effects on liver function. Therefore, the activation of liver macrophages and related consequences are of central importance. Kupffer cell (KC) activation results in the secretion of inflammatory cytokines and cytotoxic products including

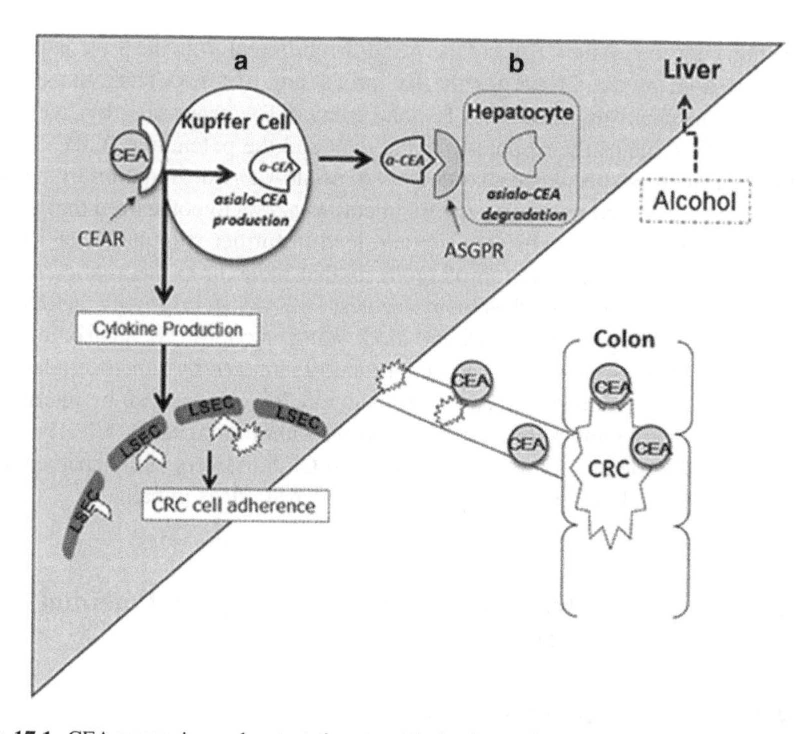

Fig. 17.1 CEA processing and metastatic potential of colorectal cancer cells (CRC) in the alcohol-injured liver. (**a**) The potential effects of alcohol on the direct processing of CEA by Kupffer Cells (KCs), the release of cytokines, and the role of those factors in facilitating the expression of pro-metastatic adhesion molecules on liver sinusoidal endothelial cells (LSECs). (**b**) The indirect processing of CEA by KCs and hepatocytes and the potential ethanol-elicited impairments to this process can enhance KC activation and pro-metastatic changes in the liver. *Key*: *a-CEA* asialo-CEA, *ASGPR* hepatic asialoglycoprotein receptor, *CEA* carcinoembryonic antigen, *CEAR* CEA receptor

LPS-stimulated TNF-α production through signaling mediated by the KC macrophage endotoxin receptor, CD14. The evidence for KC involvement in ALD was confirmed when enhanced measures of injury (aminotransferase levels, steatosis, and cell death) were alleviated by gadolinium chloride-induced elimination of KCs in the liver [62]. In addition to ethanol-induced endotoxin-mediated activation of KCs and the consequential secretion of cytokines, KCs play multiple roles in the injured liver including cell adhesion, phagocytosis, and as a mediator of inflammation [63]. However, the contribution of liver macrophages in disease states is extremely complex due to the knowledge that activated KCs release different chemokines, reactive oxygen intermediates, and cytokines in response to ethanol or other stimuli. In support of our hypothesis that the effects of alcohol would facilitate enhanced CEA-mediated KC responses, it is important to note that the cytokine TNF-α has been linked to pro-metastatic changes of the microenvironment. It was demonstrated that TNF-α serum concentrations correlated to the increased risk of metastasis of certain tumors through the up-regulation of adhesion molecules on the endothelium [64]. In the setting of alcohol, the production of TNF-α is one of the

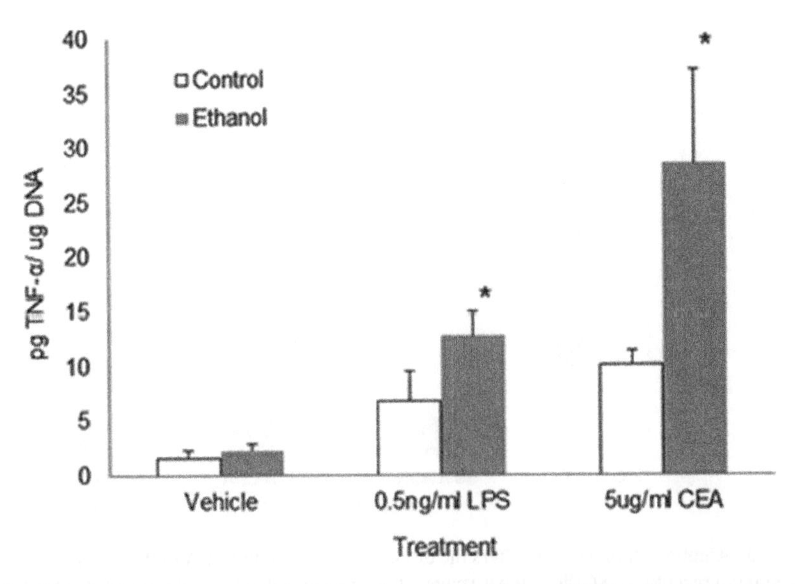

Fig. 17.2 Alcohol administration results in enhanced production of TNF-α from KCs in response to CEA. Kupffer cells (KCs) were isolated from the livers of rats chronically fed control or alcohol-containing liquid diet. TNF-α production was measured in the media of cultured KCs after 4 h of incubation with vehicle, LPS, or CEA as indicated. Values significant from control-fed animals are indicated (*$p<0.05$)

earliest responses to damage implicating these events as key players in the pathogenesis of ethanol-induced liver injury [8, 65]. Notably, in recent studies we have determined that alcohol administration results in the enhanced KC production of TNF-α in response to CEA (Fig. 17.2). In this work, KCs were isolated from rats fed the Lieber–DeCarli control or ethanol-containing diets for up to 6 weeks, a well-established rodent model of alcohol-induced liver injury [66]. The obtained KCs were used in an in vitro assay in which they were stimulated in the presence of either human CEA or low concentrations of LPS as a positive control for KC stimulation, followed by analysis of the media for the presence of TNF-α. The results demonstrate that KCs from both control and ethanol-fed rats were stimulated to produce TNF-α in response to CEA as well as LPS as expected. Importantly, the KC response from alcohol-fed animals was found to be enhanced over that produced by KCs obtained from the pair-fed control animals. This data supports our hypothesis that the alcohol-mediated effects on liver macrophages contributes to increases in TNF-α expression, a cytokine known to potentiate metastatic processes.

17.4.1.2 Alcohol-Mediated Defects to the Hepatocyte ASGPR and Asialo-CEA Processing

Alcohol consumption can lead to a variety of pathological consequences including alcohol-induced alterations to the essential resident cell of the liver parenchyma, the hepatocyte [67]. Studies by our Liver Study Group reported that particular processes are highly susceptible to the detrimental effects of ethanol as defects were

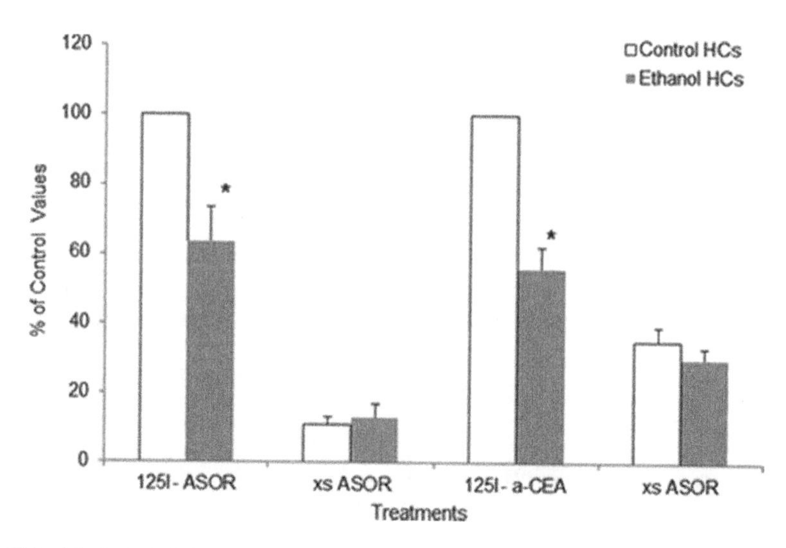

Fig. 17.3 Alcohol administration alters the binding of desialylated CEA to the hepatocyte asialoglycoprotein receptor (ASGPR). Hepatocytes (HCs) were isolated from control or ethanol-fed rats and incubated at 4 °C for 60 min with either radiolabeled asialoorosomucoid (^{125}I ASOR) or asialo-CEA (^{125}I a-CEA). Nonspecific binding was assessed by the inclusion of 100-fold excess of unlabeled ligand (xASOR). Values significant from control-fed HCs are indicated (*$p<0.05$)

identified in protein trafficking pathways, including the cellular process of RME [68–70]. Those works demonstrated that ethanol administration significantly impacted the function of the abundant hepatocyte ASGPR that was involved in the observed RME defects. It was shown that ethanol feeding to animals resulted in impairments to the ASGPR from initial binding of ligands and internalization of receptor–ligand complexes to its subsequent sorting and delivery to lysosomes for degradation [10, 11]. Several sites of alcohol-mediated alterations in the ASGP receptor were recorded that included a significant decrease in number of surface receptors and the ability to internalize and degrade ligands. The hepatocellular RME process is thought to be especially important in the regulation of CEA in the liver. In particular, the asialo-CEA that is produced by KCs can be recognized, bound to, and internalized via the ASGPR which facilitates the ultimate degradation of the CEA glycoprotein in the hepatocyte. In our current work, we examined how ethanol-mediated impairments to the ASGPR would affect the clearance of asialo-CEA by performing binding and degradation assays using radiolabeled desialylated CEA (asialo-CEA) and a known ligand for the ASGPR, asialoorosomucoid (ASOR) (Fig. 17.3). As predicted from our previous work, we found that ASOR binding to the surface of hepatocytes was impaired in cells from ethanol-fed animals compared to controls and that the binding was specific for ASGPR (inhibited by excess unlabeled ASOR). Importantly, we determined that asialo-CEA binding was impaired in hepatocytes from ethanol-fed animals and furthermore, was inhibited by unlabeled ASOR, demonstrating specificity to the ASGPR. Additionally, the terminal

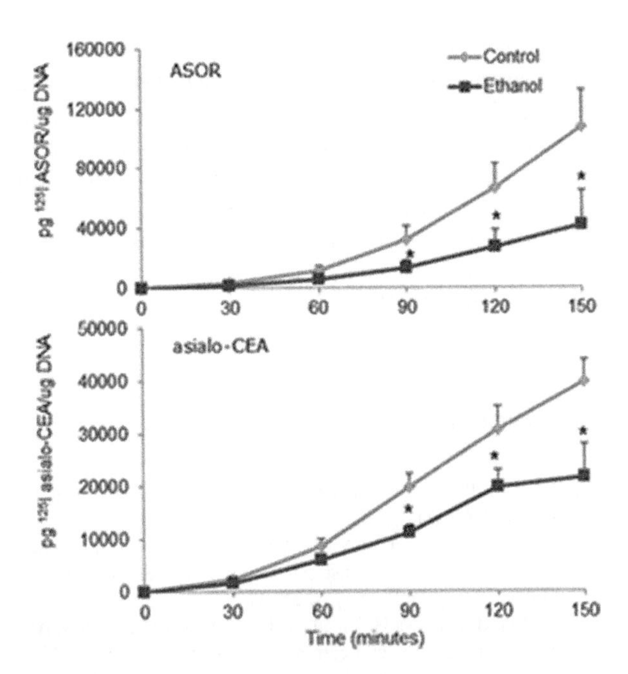

Fig. 17.4 Degradation of asialo-CEA is impaired in hepatocytes isolated from ethanol-fed rats. Hepatocytes isolated from rats chronically fed control or alcohol-containing diets were incubated over a 150 min time course with radiolabeled desialylated CEA (^{125}I asialo-CEA) or a specific ligand for the hepatocyte asialoglycoprotein receptor, asialoorosomucoid (^{125}I ASOR). The degradation of the iodinated ligands was determined by the acid-soluble radioactivity measured in the cell culture media. Values significant from control-fed animals are indicated (*$p<0.05$)

degradation of asialo-CEA in normal and ethanol-injured hepatocytes was measured (Fig. 17.4). Again, classic ethanol-induced reductions were observed in the degradation of ASOR, a well-characterized ligand for the ASGPR. In a similar fashion, the degradation of asialo-CEA was also found to be significantly reduced in hepatocytes obtained from chronically fed animals compared to controls. It is evident from these initial studies that ethanol-induced derangements of hepatocellular endocytosis can disrupt the regulation of CEA in the liver by altering effective clearance mechanisms.

17.5 Summary

Alcohol consumption has been shown to be a major etiologic factor of both acute and chronic diseases affecting many other critical organ systems (e.g., liver, pancreas, brain, and lung). In addition, it is clear that alcohol is a recognized carcinogen and that chronic and/or heavy consumption is a significant risk factor for the development of cancers as well as metastasis to secondary organs. Sadly, the development of liver metastasis in CRC patients has been correlated to alcohol consumption. Due to this correlation, alcohol consuming CRC patients require intensive examination and follow-up with respect to liver metastasis since most recurrences to the liver are not resectable leading to poor survival rates. Even though it is known that alcohol consumption correlates with liver metastasis and poor outcomes in

CRC patients, the mechanism(s) by which alcohol participates in, or provides an environment supportive of CRC metastases is unknown. Some of the current research in this field is aimed at characterizing the role of CEA, a hallmark predictive factor of metastatic colon cancer. The relationships between serum CEA levels in alcoholics, ethanol-mediated alterations to liver cells and the consequences of alcohol-related CEA processing in the liver are being investigated. Understanding CEA processing in the liver has the potential to significantly impact the field and lead to the development of therapies for early detection and/or prevention of liver involvement in CRC disease.

References

1. Jemal A, Bray F, Center MM, Ferlay J, Ward E et al (2011) Global cancer statistics. CA Cancer J Clin 61:69–90
2. Wagner HE, Toth CA, Steele GD Jr, Thomas P (1992) Metastatic potential of human colon cancer cell lines: relationship to cellular differentiation and carcinoembryonic antigen production. Clin Exp Metastasis 10:25–31
3. Maeda M, Nagawa H, Maeda T, Koike H, Kasai H (1998) Alcohol consumption enhances liver metastasis in colorectal carcinoma patients. Cancer 83:1483–1488
4. Thomas P, Hayashi H, Zimmer R, Forse RA (2004) Regulation of cytokine production in carcinoembryonic antigen stimulated Kupffer cells by beta-2 adrenergic receptors: implications for hepatic metastasis. Cancer Lett 209:251–257
5. Thomas P, Zamcheck N, Rogers AE, Fox JG (1982) Plasma clearance of carcinoembryonic antigen and asialo carcinoembryonic antigen by the liver of the nutritionally deficient rhesus monkey. Clin Lab Med 2:459–467
6. Thomas P, Forse RA, Bajenova O (2011) Carcinoembryonic antigen (CEA) and its receptor hnRNP M are mediators of metastasis and the inflammatory response in the liver. Clin Exp Metastasis 28:923–932
7. Thakur V, Pritchard MT, McMullen MR, Wang Q, Nagy LE (2006) Chronic ethanol feeding increases activation of NADPH oxidase by lipopolysaccharide in rat Kupffer cells: role of increased reactive oxygen in LPS-stimulated ERK1/2 activation and TNF-alpha production. J Leukoc Biol 79:1348–1356
8. Thurman RG (1998) II. Alcoholic liver injury involves activation of Kupffer cells by endotoxin. Am J Physiol 275:G605–G611
9. Tsukamoto H, Lu SC (2001) Current concepts in the pathogenesis of alcoholic liver injury. FASEB J 15:1335–1349
10. Casey CA, Kragskow SL, Sorrell MF, Tuma DJ (1987) Chronic ethanol administration impairs the binding and endocytosis of asialo-orosomucoid in isolated hepatocytes. J Biol Chem 262:2704–2710
11. Tworek BL, Tuma DJ, Casey CA (1996) Decreased binding of asialoglycoproteins to hepatocytes from ethanol-fed rats. Consequence of both impaired synthesis and inactivation of the asialoglycoprotein receptor. J Biol Chem 271:2531–2538
12. Ferrari P, Jenab M, Norat T, Moskal A, Slimani N et al (2007) Lifetime and baseline alcohol intake and risk of colon and rectal cancers in the European prospective investigation into cancer and nutrition (EPIC). Int J Cancer 121:2065–2072
13. Seitz HK, Stickel F (2007) Molecular mechanisms of alcohol-mediated carcinogenesis. Nat Rev Cancer 7:599–612
14. Dong H, Tang J, Li LH, Ge J, Chen X et al (2013) Serum carbohydrate antigen 19-9 as an indicator of liver metastasis in colorectal carcinoma cases. Asian Pac J Cancer Prev 14:909–913

15. Ciccotosto GD, McLeish A, Hardy KJ, Shulkes A (1995) Expression, processing, and secretion of gastrin in patients with colorectal carcinoma. Gastroenterology 109:1142–1153
16. Hanjal FFM, Valenzuela JE (1988) Impaired gastrin release after a meal and wine in chronic alcoholics. Gastroenterology 94
17. Thomas P, Gangopadhyay A, Steele G Jr, Andrews C, Nakazato H et al (1995) The effect of transfection of the CEA gene on the metastatic behavior of the human colorectal cancer cell line MIP-101. Cancer Lett 92:59–66
18. Yirmiya R, Ben-Eliyahu S, Gale RP, Shavit Y, Liebeskind JC et al (1992) Ethanol increases tumor progression in rats: possible involvement of natural killer cells. Brain Behav Immun 6:74–86
19. Ismaili N (2011) Treatment of colorectal liver metastases. World J Surg Oncol 9:154
20. Van den Eynden GG, Majeed AW, Illemann M, Vermeulen PB, Bird NC et al (2013) The multifaceted role of the microenvironment in liver metastasis: biology and clinical implications. Cancer Res 73:2031–2043
21. Bird NC, Mangnall D, Majeed AW (2006) Biology of colorectal liver metastases: a review. J Surg Oncol 94:68–80
22. Auguste P, Fallavollita L, Wang N, Burnier J, Bikfalvi A et al (2007) The host inflammatory response promotes liver metastasis by increasing tumor cell arrest and extravasation. Am J Pathol 170:1781–1792
23. Brodt P, Fallavollita L, Bresalier RS, Meterissian S, Norton CR et al (1997) Liver endothelial E-selectin mediates carcinoma cell adhesion and promotes liver metastasis. Int J Cancer 71:612–619
24. Paschos KA, Majeed AW, Bird NC (2010) Role of Kupffer cells in the outgrowth of colorectal cancer liver metastases. Hepatol Res 40:83–94
25. Gold P, Freedman SO (1965) Specific carcinoembryonic antigens of the human digestive system. J Exp Med 122:467–481
26. Boucher D, Cournoyer D, Stanners CP, Fuks A (1989) Studies on the control of gene expression of the carcinoembryonic antigen family in human tissue. Cancer Res 49:847–852
27. Duffy MJ (2001) Carcinoembryonic antigen as a marker for colorectal cancer: is it clinically useful? Clin Chem 47:624–630
28. Wanebo HJRB, Pinsky CM, Hoffman RG, Streams M, Schwartz MK et al (1978) The use of preoperative carcinoembryonic antigen level as a prognostic indicator to complement pathological staging. N Engl J Med 299:448–451
29. Bhatnagar J, Tewari HB, Bhatnagar M, Austin GE (1999) Comparison of carcinoembryonic antigen in tissue and serum with grade and stage of colon cancer. Anticancer Res 19: 2181–2187
30. LA Carriquiry PA (1999) Should carcinoembryonic antigen be used in the management of patients with colorectal cancer. Dis Colon Rectum 42:921–929
31. Chu DZ, Erickson CA, Russell MP, Thompson C, Lang NP et al (1991) Prognostic significance of carcinoembryonic antigen in colorectal carcinoma. Serum levels before and after resection and before recurrence. Arch Surg 126:314–316
32. Harrison LE, Guillem JG, Paty P, Cohen AM (1997) Preoperative carcinoembryonic antigen predicts outcomes in node-negative colon cancer patients: a multivariate analysis of 572 patients. J Am Coll Surg 185:55–59
33. Compton CC, Fielding LP, Burgart LJ, Conley B, Cooper HS et al (2000) Prognostic factors in colorectal cancer. College of American Pathologists Consensus Statement 1999. Arch Pathol Lab Med 124:979–994
34. Filella XMR, Pique JM, Grau JJ, Garcia-Valdecasas JC, Biete A et al (1994) CEA as a prognostic factor in colorectal cancer. Anticancer Res 14:795–798
35. Gupta GP, Massague J (2006) Cancer metastasis: building a framework. Cell 127:679–695
36. Aldulaymi B, Bystrom P, Berglund A, Christensen IJ, Brunner N et al (2010) High plasma TIMP-1 and serum CEA levels during combination chemotherapy for metastatic colorectal cancer are significantly associated with poor outcome. Oncology 79:144–149

37. Bhattacharjya S, Aggarwal R, Davidson BR (2006) Intensive follow-up after liver resection for colorectal liver metastases: results of combined serial tumour marker estimations and computed tomography of the chest and abdomen - a prospective study. Br J Cancer 95:21–26
38. Ballantyne GH, Modlin IM (1988) Postoperative follow-up for colorectal cancer: who are we kidding? J Clin Gastroenterol 10:359–364
39. Jantscheff P, Terracciano L, Lowy A, Glatz-Krieger K, Grunert F et al (2003) Expression of CEACAM6 in resectable colorectal cancer: a factor of independent prognostic significance. J Clin Oncol 21:3638–3646
40. Benchimol S, Fuks A, Jothy S, Beauchemin N, Shirota K et al (1989) Carcinoembryonic antigen, a human tumor marker, functions as an intercellular adhesion molecule. Cell 57:327–334
41. Hostetter RB, Augustus LB, Mankarious R, Chi KF, Fan D et al (1990) Carcinoembryonic antigen as a selective enhancer of colorectal cancer metastasis. J Natl Cancer Inst 82:380–385
42. Eidelman FJ, Fuks A, DeMarte L, Taheri M, Stanners CP (1993) Human carcinoembryonic antigen, an intercellular adhesion molecule, blocks fusion and differentiation of rat myoblasts. J Cell Biol 123:467–475
43. Ordonez C, Screaton RA, Ilantzis C, Stanners CP (2000) Human carcinoembryonic antigen functions as a general inhibitor of anoikis. Cancer Res 60:3419–3424
44. Medoff JR, Jegasothy BV, Roche JK (1984) Carcinoembryonic antigen-induced release of a suppressor factor from normal human lymphocytes in vitro. Cancer Res 44:5822–5827
45. Heiskala MK, Stenman UH, Koivunen E, Carpen O, Saksela E et al (1988) Characteristics of soluble tumour-derived proteins that inhibit natural killer activity. Scand J Immunol 28:19–27
46. van Gisbergen KP, Aarnoudse CA, Meijer GA, Geijtenbeek TB, van Kooyk Y (2005) Dendritic cells recognize tumor-specific glycosylation of carcinoembryonic antigen on colorectal cancer cells through dendritic cell-specific intercellular adhesion molecule-3-grabbing nonintegrin. Cancer Res 65:5935–5944
47. Jessup JM, Petrick AT, Toth CA, Ford R, Meterissian S et al (1993) Carcinoembryonic antigen: enhancement of liver colonisation through retention of human colorectal carcinoma cells. Br J Cancer 67:464–470
48. Jessup JM (1998) Growth potential of human colorectal carcinoma in nude mice; association with the preoperative serum concentration of carcinoembryonic antigen in paitents. Cancer Res 48:1689–1692
49. Edmiston KH, Gangopadhyay A, Shoji Y, Nachman AP, Thomas P et al (1997) In vivo induction of murine cytokine production by carcinoembryonic antigen. Cancer Res 57:4432–4436
50. Jessup JM, Ishii S, Mitzoi T, Edmiston KH, Shoji Y (1999) Carcinoembryonic antigen facilitates experimental metastasis through a mechanism that does not involve adhesion to liver cells. Clin Exp Metastasis 17:481–488
51. Gangopadhyay A, Lazure DA, Thomas P (1998) Adhesion of colorectal carcinoma cells to the endothelium is mediated by cytokines from CEA stimulated Kupffer cells. Clin Exp Metastasis 16:703–712
52. Laguinge L, Bajenova O, Bowden E, Sayyah J, Thomas P et al (2005) Surface expression and CEA binding of hnRNP M4 protein in HT29 colon cancer cells. Anticancer Res 25:23–31
53. Thomas P, Toth CA, Saini KS, Jessup JM, Steele G Jr (1990) The structure, metabolism and function of the carcinoembryonic antigen gene family. Biochim Biophys Acta 1032:177–189
54. Kammerer R, Zimmermann W (2010) Coevolution of activating and inhibitory receptors within mammalian carcinoembryonic antigen families. BMC Biol 8:12
55. Beauchemin N, Arabzadeh A (2013) Carcinoembryonic antigen-related cell adhesion molecules (CEACAMs) in cancer progression and metastasis. Cancer Metastasis Rev 32(3–4): 643–671
56. Gangopadhyay A, Bajenova O, Kelly TM, Thomas P (1996) Carcinoembryonic antigen induces cytokine expression in Kuppfer cells: implications for hepatic metastasis from colorectal cancer. Cancer Res 56:4805–4810
57. Jessup JM, Laguinge L, Lin S, Samara R, Aufman K et al (2004) Carcinoembryonic antigen induction of IL-10 and IL-6 inhibits hepatic ischemic/reperfusion injury to colorectal carcinoma cells. Int J Cancer 111:332–337

58. Ashwell G, Kawasaki T (1978) A protein from mammalian liver that specifically binds galactose-terminated glycoproteins. Methods Enzymol 50:287–288
59. Ashwell G, Morell AG (1974) The role of surface carbohydrates in the hepatic recognition and transport of circulating glycoproteins. Adv Enzymol Relat Areas Mol Biol 41:99–128
60. French SW, Nash J, Shitabata P, Kachi K, Hara C et al (1993) Pathology of alcoholic liver disease. VA Cooperative Study Group 119. Semin Liver Dis 13:154–169
61. Cederbaum AI (2001) Introduction-serial review: alcohol, oxidative stress and cell injury. Free Radic Biol Med 31:1524–1526
62. Adachi Y, Bradford BU, Gao W, Bojes HK, Thurman RG (1994) Inactivation of Kupffer cells prevents early alcohol-induced liver injury. Hepatology 20:453–460
63. Wheeler MD (2003) Endotoxin and Kupffer cell activation in alcoholic liver disease. Alcohol Res Health 27:300–306
64. Okahara H, Yagita H, Miyake K, Okumura K (1994) Involvement of very late activation antigen 4 (VLA-4) and vascular cell adhesion molecule 1 (VCAM-1) in tumor necrosis factor alpha enhancement of experimental metastasis. Cancer Res 54:3233–3236
65. Tilg H, Diehl AM (2000) Cytokines in alcoholic and nonalcoholic steatohepatitis. N Engl J Med 343:1467–1476
66. Lieber CS, DeCarli LM (1982) The feeding of alcohol in liquid diets: two decades of applications and 1982 update. Alcohol Clin Exp Res 6:523–531
67. Tuma DJ, Sorrell MF (1988) Effects of ethanol on protein trafficking in the liver. Semin Liver Dis 8:69–80
68. Casey CA, Kragskow SL, Sorrell MF, Tuma DJ (1989) Ethanol-induced impairments in receptor-mediated endocytosis of asialoorosomucoid in isolated rat hepatocytes: time course of impairments and recovery after ethanol withdrawal. Alcohol Clin Exp Res 13:258–263
69. Casey CA, Wiegert RL, Tuma DJ (1993) Chronic ethanol administration impairs ATP-dependent acidification of endosomes in rat liver. Biochem Biophys Res Commun 195:1127–1133
70. McVicker BL, Casey CA (1999) Effects of ethanol on receptor-mediated endocytosis in the liver. Alcohol 19:255–260

Chapter 18
Alcohol Consumption and Antitumor Immunity: Dynamic Changes from Activation to Accelerated Deterioration of the Immune System

Hui Zhang, Zhaohui Zhu, Faya Zhang, and Gary G. Meadows

Abstract The molecular mechanisms of how alcohol and its metabolites induce cancer have been studied extensively. However, the mechanisms whereby chronic alcohol consumption affects antitumor immunity and host survival have largely been unexplored. We studied the effects of chronic alcohol consumption on the immune system and antitumor immunity in mice inoculated with B16BL6 melanoma and found that alcohol consumption activates the immune system leading to an increase in the proportion of IFN-γ-producing NK, NKT, and T cells in mice not injected with tumors. One outcome associated with enhanced IFN-γ activation is inhibition of melanoma lung metastasis. However, the anti-metastatic effects do not translate into increased survival of mice bearing subcutaneous tumors. Continued growth of the subcutaneous tumors and alcohol consumption accelerates the deterioration of the immune system, which is reflected in the following: (1) inhibition in the expansion of memory $CD8^+$ T cells, (2) accelerated decay of Th1 cytokine-producing cells, (3) increased myeloid-derived suppressor cells, (4) compromised circulation of B cells and T cells, and (5) increased NKT cells that exhibit an IL-4 dominant cytokine profile, which is inhibitory to antitumor immunity. Taken together, the dynamic effects of alcohol consumption on antitumor immunity are in two opposing phases: the first phase associated with immune stimulation is tumor inhibitory and the second phase resulting from the interaction between the effects of alcohol and the tumor leads to immune inhibition and resultant tumor progression.

Keywords Alcohol • Antitumor immunity • B16BL6 melanoma • NK cells • T cells • NKT cells • Metastasis • Survival

---

H. Zhang (✉) • Z. Zhu • F. Zhang • G.G. Meadows
Department of Pharmaceutical Sciences, College of Pharmacy, Washington State University, Spokane, WA 99210-1495, USA
e-mail: hzhang@wsu.edu

V. Vasiliou et al. (eds.), *Biological Basis of Alcohol-Induced Cancer*,
Advances in Experimental Medicine and Biology 815,
DOI 10.1007/978-3-319-09614-8_18

18.1 Introduction

Epidemiological and experimental data convincingly indicate that alcohol consumption increases the incidence of multiple types of cancer, most notably digestive system cancers in both genders, although more common in men, and breast cancer in women [1–6]. There is no doubt that alcohol and its metabolites such as acetaldehyde are carcinogenic in humans and animals. Therefore, the International Agency for Research on Cancer in 2012 listed alcoholic beverages and acetaldehyde associated with the consumption of alcoholic beverages as group 1 carcinogens, which are the substances, mixtures, and exposure circumstances that are carcinogenic to humans.

The research on alcohol consumption and cancer can be divided into two equally important areas. One of the areas is focused on the mechanistic studies related to the carcinogenic activity of alcohol and its metabolites. The other area involves the study of how alcohol consumption affects tumor progression and the survival of cancer patients. Compared to the extensive studies on the molecular mechanisms of how alcohol and its metabolites act as carcinogens, studies on how alcohol consumption affects tumor progression and survival of cancer patients are very limited. Most of the research on the survival of human alcoholics with cancer has come from epidemiological surveys. There are no studies that reveal the underlying mechanisms of how chronic alcohol consumption affects the survival of cancer patients with different types of cancer. The results often vary with the type of cancer. Epidemiological research consistently indicates that alcohol consumption decreases the survival of patients with oral cavity, pharyngeal, laryngeal, and esophageal cancer [7–11]. However, most studies indicate that chronic alcohol consumption, especially low and moderate alcohol consumption, does not significantly affect the survival of breast cancer patients [12–15]. Some research indicates that alcohol consumption decreases the survival of high intensity drinkers in postmenopausal women with breast cancer [16–18]; whereas, other studies indicate that low and moderate alcohol consumption benefits the survival of young breast cancer patients [12, 19]. A paradoxical observation is that while alcohol consumption is associated with a decrease in the incidence of Non-Hodgkin's Lymphoma (NHL) [20–22], it also decreases the survival of these patients [23–25]. In our mouse studies, chronic alcohol consumption inhibits lung metastasis of B16BL6 melanoma inoculated intravenously [26]; however, survival of mice bearing subcutaneous melanomas is decreased [27].

Multiple factors are involved in the regulation of cancer progression and patient survival. The immune system plays a key role. Tumor immunotherapy has become one of the most promising approaches to treat and cure cancers. A large body of research conducted in human alcoholics and experimental animals indicates that chronic alcohol consumption is immunomodulatory [28, 29]. Although it is well-known that alcohol is an immunosuppressant, it also is well documented that chronic alcohol consumption activates the immune system, especially T, NKT, and dendritic cells in human and experimental animals [30–33]. However, the effects on how chronic alcohol consumption affects tumor progression and antitumor immunity are largely unexplored. Using a murine chronic alcohol consumption and B16BL6 melanoma model, we have systematically studied how chronic alcohol consumption affects the antitumor immunity.

18.2 Animal Model of Chronic Alcohol Consumption and B16BL6 Melanoma Inoculation

18.2.1 Alcohol Administration

Female C57BL/6 mice at 6–7 weeks of age were purchased from Charles River laboratories (Wilmington, MA). After arrival, mice were single housed in plastic cages with micro-filter tops in the Wegner Hall Vivarium, which is fully accredited by the Association for the Assessment and Accreditation of Laboratory Animal Care. Mice were allowed ad libitum access to Purina 5001 rodent laboratory chow and sterilized Milli-Q water. After 1 week of acclimation, mice were randomly divided into two groups. One group as control was continuously provided with chow and Milli-Q water. The other group was provided with chow and 20 % (w/v) alcohol diluted from 190-proof Everclear (St. Louis, MO) with sterilized Milli-Q water. Mice consume at least 30 % of their caloric intake from alcohol [34]. Previous studies that incorporated a pair-fed group indicated no difference in energy intake between water-drinking and alcohol-consuming groups [35]. Mice were used for experiments 2–6 months after starting alcohol consumption, since during this time period the immune parameters induced by chronic alcohol consumption are relatively stable [36]. This chronic alcohol consumption model does not cause liver injury. All protocols involving mice were approved by the Institutional Animal Care and Use Committee at Washington State University.

18.2.2 Tumor Inoculation

B16BL6 melanoma was used in this study, and it is a well-established model to study tumor metastasis, tumor immunology, and immunotherapy [37, 38]. For the lung metastasis study, 5×10^4 B16BL6 cells in 200 μL of PBS were injected intravenously via the lateral tail vein. Mice were euthanized 3 weeks after tumor cell inoculation. Lungs was collected and fixed in phosphate buffered formalin. Lung colonies of melanoma were counted under a dissecting microscope. For survival and antitumor immune response experiments, tumor cells were inoculated subcutaneously into the right side of hip area with 2×10^5 cells in 200 μL PBS. Related antitumor immune parameters were determined from 5 to 28 days after tumor inoculation.

18.3 Chronic Alcohol Consumption Inhibits B16BL6 Melanoma Lung Metastasis in an IFN-γ Signaling Pathway-Dependent Fashion

Most cancer patients die of metastasis. Little clinical data are available regarding the effects of chronic alcohol consumption on tumor metastasis in human alcoholics. One clinical observation indicated that alcohol consumption significantly enhanced liver metastasis in colorectal cancer patients [39]. Two recent reports demonstrated that alcohol consumption did not significantly affect the metastasis of duodenal wall gastrinomas and laryngeal squamous cell cancers, which are two types of cancer that are related to alcohol consumption [40, 41]. Using the mouse chronic alcohol consumption and B16BL6 melanoma model, we found that alcohol consumption inhibited B16BL6 lung metastasis [26]. This inhibition is independent of perforin and granzyme-induced NK cell cytotoxic effects [42]. The anti-metastatic effect induced during alcohol consumption is maintained in γC knockout mice, which lack NK, B, and $CD8^+$ T cells (Fig. 18.1) [43]. While tumor metastasis is a complicated process, IFN-γ plays an important role in control of melanoma metastasis [44]. The inhibition of B16BL6 melanoma lung metastasis in mice chronically consuming alcohol was abrogated in IFN-γ knockout (KO) mice (Fig. 18.1) [43]. This suggests that the anti-metastatic effect induced by chronic alcohol consumption is dependent on the IFN-γ signaling pathway and that multiple cell types are involved in this process. It should be noted that in this system, melanoma cells were inoculated into the blood via the tail vein. These cells circulate in the blood and then seed the lung. The survival of these cells in the blood determines their fate and ability to form lung metastases. The tumor cells are cleared from the blood within 48 h after inoculation. Therefore, the activated immune cells in the blood play a key role in the clearance of the tumor cells. Chronic alcohol consumption decreases metastasis of B16BL6 melanoma into the lung in an IFN-γ-dependent fashion,

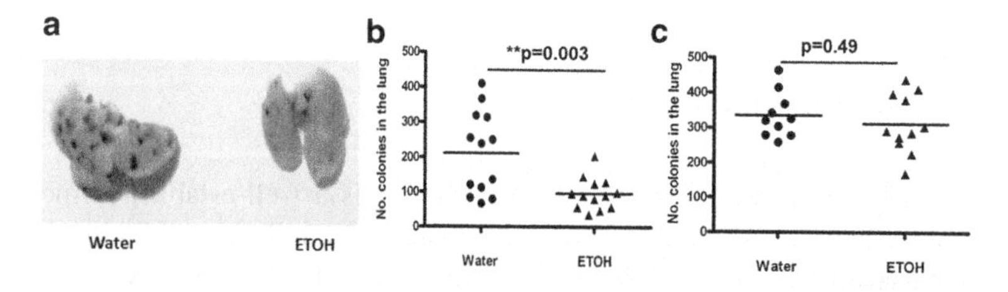

Fig. 18.1 Chronic alcohol consumption-induced inhibition on B16BL6 melanoma metastasis is IFN-γ-dependent. (**a**) Images showing colonies of melanoma (*black dots*) on the lung of water-drinking mice and alcohol-consuming mice. (**b, c**) Histogram showing melanoma lung colony number in γC KO (**b**) and IFN-γ KO (**c**) mice drinking water or consuming alcohol for 3 months. Each group contained 10–13 mice. Two-tailed Student-*t* test was used to test the difference between the two groups. The difference was defined as significant when *p* value was less than 0.05. With kind permission from Springer Science + Business Media, this figure is adapted from *Experimental and Clinical Metastasis*, IFN-γ is essential for the inhibition of B16BL6 melanoma lung metastasis in chronic alcohol drinking mice, 28(3), 2011, 301–307, Zhang H, Zhu Z, McKinley JM, and Meadows, GG, Fig. 1 and Fig. 2d, 2e

suggesting that alcohol increases activated immune cells, especially those producing IFN-γ. Indeed, we found that chronic alcohol consumption increases IFN-γ-producing T cells, NK cells, and NKT cells in the blood.

18.4 Effects of Chronic Alcohol Consumption on T Cells: Induction of T Cell Activation Through Homeostatic Proliferation in the Steady State and Acceleration of T Cell Dysfunction in Melanoma-Bearing Mice

T cells are important cells of the adaptive immune response. T cells can be divided into $CD4^+$ T cells, also called helper T (Th) cells, and $CD8^+$ T cells, known as cytotoxic T lymphocytes (CTL). Based on their T cell receptor (TCR) activation status, these cells can be divided further into naïve cells, which are T cells that have not contacted antigen, and memory cells, which are the T cells that have been activated by antigen. Upon activation memory T cells produce more cytokines and exhibit stronger functional response compared to naïve T cells. Based on the function, especially cytokine production, $CD4^+$ T cells are divided into different subtypes. Th1 cells produce inflammatory cytokines such as IFN-γ, which will activate macrophage, dendritic cells, NK cells and $CD8^+$ T cells. Th1 cells play important roles in antitumor immunity. Th2 cells produce IL-4, IL-5, IL-10, IL-13, etc. cytokines. These cytokines inhibits Th1 response and antitumor immunity. Th2 cells play important roles in allergy and humoral immune response. $CD4^+CD25^+FoxP3^+$ regulatory T (Treg) cells produce IL-10 and TGF-β. Treg cells inhibit $CD8^+$ T cell and NK cell function and facilitate tumor progression [45]. Th17 cells produce IL-17 family cytokines. These cells play critical roles in autoimmune diseases, but exhibit controversial function in antitumor immunity [46, 47]. $CD8^+$ T cells are the key effector cells in antitumor immunity [48]. Upon activation, $CD8^+$ T cells produce Th1 cytokines and the cytotoxic effector molecules perforin and granzymes to kill target cells. IFN-γ produced by memory and tumor-specific T cells play the crucial role in the control of tumor progression, metastasis, and host survival [49]. Compared to naïve T cells, memory T cells are more efficient and are potent producers of IFN-γ. This effect could be associated with the low threshold for demethylation in the promoter region of the IFN-γ gene in the memory T cells [50–52]. Memory $CD8^+$ T cells are also the important cell population that provides the early source of IFN-γ before T cell receptor activation [53]. Alcohol consumption increases activated $CD8^+$ T cells and IFN-γ-producing $CD8^+$ T cells in human alcoholics [54]. We and others found that chronic alcohol consumption in the steady state increases the percentage of $CD8^+$ T cells exhibiting the memory phenotype and also $CD8^+$ T cells producing IFN-γ [31, 33]. We further showed that chronic alcohol consumption increases memory T cells in the steady state through the induction of T cell homeostatic proliferation [33]. These increased memory and IFN-γ-producing $CD8^+$ T cells in alcohol-consuming mice could play an important protective role in the antitumor immune response.

Although alcohol consumption increases steady state levels of memory and IFN-γ-producing CD8+ T cells in mice not injected with tumors, it inhibits memory and tumor-specific CD8+ T cell expansion in the melanoma-bearing mice (Fig. 18.2) and also accelerates CD8+ T cell dysfunction, which is reflected in the repaid decline in cytokine-producing cells (Fig. 18.3) [55]. Tumor cells produce factors that induce immune inhibitory cells, which inhibit CD8+ T cell function. These inhibitory cells include tumor associated macrophages (TAM), myeloid-derived suppressor cells (MDSC), regulatory T cells (Treg), inhibitory B cells ($CD1d^{hi}CD5^{+}$), and NKT cells. We found that chronic alcohol consumption does not alter the percentage of

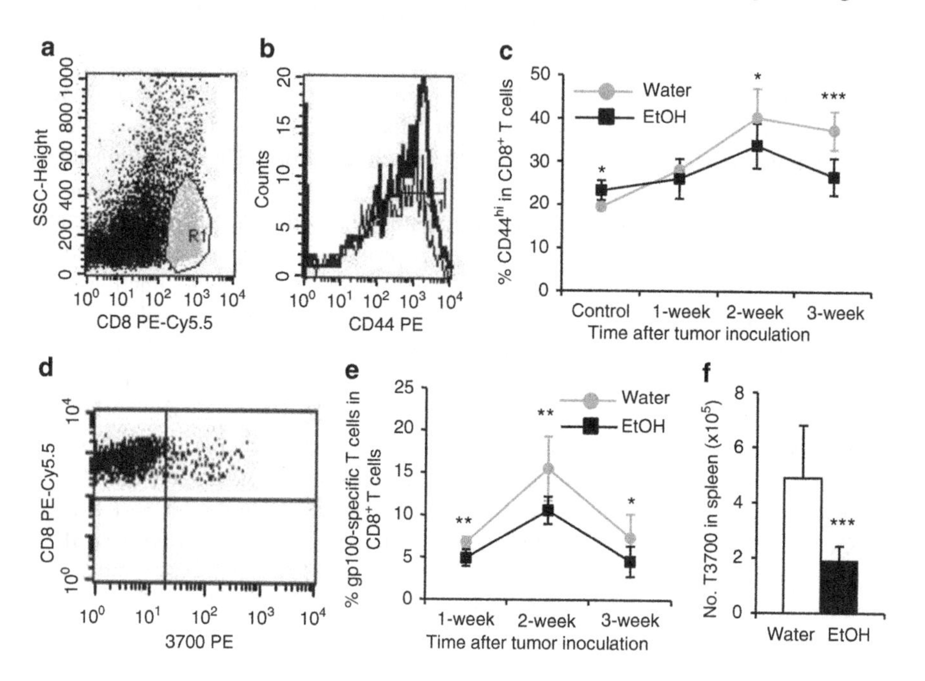

Fig. 18.2 Chronic alcohol consumption inhibits memory and tumor-specific CD8+ T cell expansion. (**a**) *Dot plot* showing gated CD8+ T cells (R1). (**b**) *Histogram* showing $CD44^{hi}$ memory CD8+ T cells in the spleen of melanoma-bearing mice. (**c**) Percentage of $CD8^{+}CD44^{hi}$ cells in CD8+ splenocytes from non-tumor injected mice (control) and melanoma-bearing mice at the indicated time points after tumor inoculation. (**d**) *Dot plot* showing gp100-tetramer+cells (3700 PE) in the gated splenic CD8+ T cells of melanoma-bearing mice. (**e**) Percentage of gp-100-specific CD8+ T cells in splenic CD8+ T cells at the indicated time points after tumor inoculation. (**f**) Number of gp100-specific CD8+ T cells in the spleen of tumor-bearing mice 3-week after tumor inoculation. Water = water-drinking mice, ETOH = alcohol-consuming mice. $*p<0.05$, $**p<0.01$, $***p<0.001$. With kind permission from Springer Science + Business Media, this figure is adapted from *Cancer Immunology, Immunotherapy*, Chronic alcohol consumption enhances myeloid-derived suppressor cells (MDSC) in B16BL6 melanoma-bearing mice, 59 (8), 2010, 1151–1159, Zhang, H and Meadows, GG, Fig. 1 and 2

Fig. 18.3 (continued) 2015,1070–1080, Zhang, H and Meadows GG, Fig. 7. With kind permission from Springer Science+Business Media, Fig. 18.3b is adapted from *Cancer Immunology, Immunotherapy*, Chronic alcohol consumption enhances MDSC in B16BL6 melanoma-bearing mice, 59 (8), 2010, 1151–1159, Zhang, H and Meadows, GG, Fig. 3

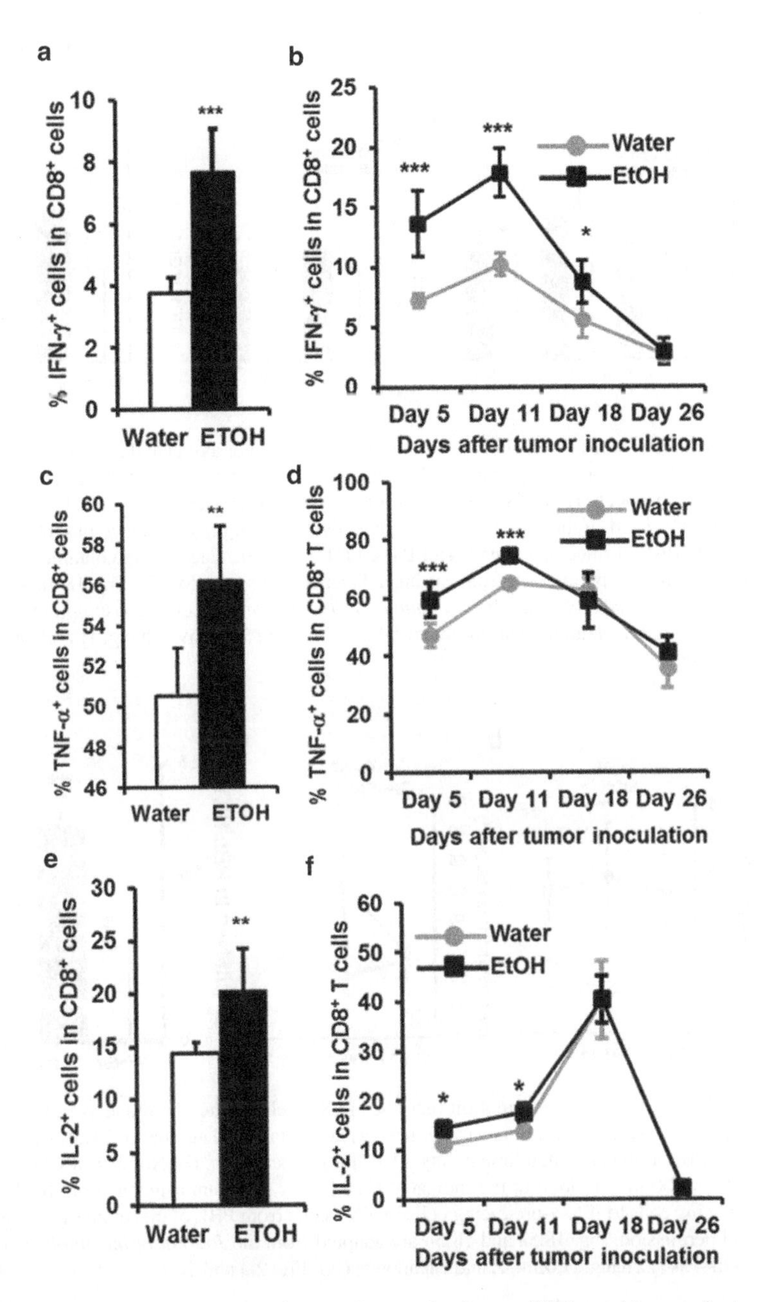

Fig. 18.3 Chronic alcohol consumption accelerates the decay of Th1 cytokine-producing CD8+ T cells in the melanoma-bearing mice. Mice were given alcohol for 2 months (non-tumor injected mice) to 3 months (tumor-bearing mice). IFN-γ (**a**), TNF-α (**c**) and IL-2 (**e**)-producing cells in splenic CD8+ T cells of non-tumor injected mice and melanoma-bearing mice (**b**, **d**, **f**, respectively) at the indicated time points after tumor inoculation as determined by intracellular staining. Each group contained ten mice. Two-tailed Student-*t* test was used to test the difference between the two groups. The difference was defined as significant when *p* value was less than 0.05. $*p<0.05$, $**p<0.01$, $***p<0.001$. With kind promession, Fig. 18.3a (**a**) is adapted from the *Journal of Leukocyte Biology*, Chronic alcohol consumption in mice increases the proportion of peripheral memory T cells by homeostatic proliferation, 78(5),

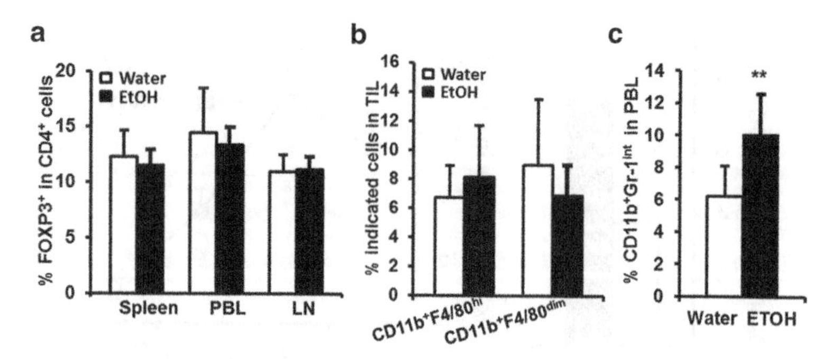

Fig. 18.4 Effects of chronic alcohol consumption on Treg, tumor associated macrophages (TAM) and MDSC in melanoma-bearing mice. (**a**) Percentage of FoxP3⁺CD4⁺ regulatory T cells in spleen, PBL, and lymph nodes (LN) of melanoma-bearing mice. (**b**) Percentage of $CD11b^{+}F4/80^{hi}$ and $CD11b^{+}F4/80^{dim}$ TAM in the tumor infiltrated leukocytes (TIL). (**c**) Percentage of $CD11b^{+}Gr\text{-}1^{int}$ MDSC in the PBL of mice inoculated with B16BL6 for 1-week. Each group contained ten mice. $^{**}p<0.01$. With kind permission from Springer Science+Business Media, Fig. 18.4a and 18.4c are adapted from *Cancer Immunology, Immunotherapy*, Chronic alcohol consumption enhances MDSC in B16BL6 melanoma-bearing mice, 59 (8), 2010, 1151–1159, Zhang, H and Meadows, GG, Fig. 4 and 5

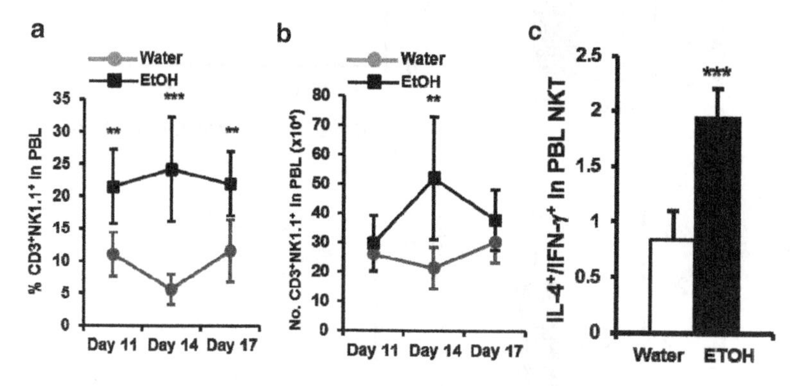

Fig. 18.5 Chronic alcohol consumption increases iNKT cells in the blood and skews iNKT cell cytokine profile toward Th2-dominant cytokines. (**a**) Percentage and number of $CD3^{+}NK1.1^{+}$ NKT cells in the PBL at the indicated time points after tumor inoculation. (**b**) Number of $CD3^{+}NK1.1^{+}$ NKT cells in 700 μL of blood at the indicated time points after tumor inoculation. (**c**) Ratio of IL-4-producing cells to IFN-γ-producing cells in NKT cells from PBL of melanoma-bearing mice. With kind permission, Fig. 18.5a and 18.5b are adapted from the *Journal of Immunology* 189(3), 2012, 1340–1348, Zhang, H, Zhu, Z and Meadows, GG, Fig. 2D and 2I

Treg and TAM, but decreases inhibitory B cells, increases MDSC (Fig. 18.4), and increases NKT cells (Fig. 18.5a, b) in melanoma-bearing mice. These results suggest that MDSC and NKT cells could be the major immunoregulatory cells that inhibit $CD8^{+}$ T cell function in the alcohol-consuming, melanoma-bearing mice. One of the signaling pathways employed by MDSC to inhibit $CD8^{+}$ T cell function is IL-13/IL13R/iNOS-arginase/arginine [56]. IL-13 activates MDSC to upregulate the expression of iNOS and arginase 1. These enzymes metabolize arginine to decrease the availability of this amino acid, which in turn (1) block the translation of CD3ζ, one of the important components of TCR complex; (2) inhibit T cell proliferation; and (3) promote T cell apoptosis [57]. We found that chronic alcohol consumption significantly upregulates the expression of IL-4/IL-13 receptor α-chain (CD124) on MDSC [55], and decreases the concentration of arginine in the plasma of melanoma-bearing mice [35]. The increased MDSC could be associated with the inhibition of the memory and tumor-specific $CD8^+$ T cell proliferation and dysfunction in the tumor-bearing mice. The dysfunction of $CD8^+$ T cells will promote tumor progression.

In summary, the increase in memory and IFN-γ-producing T cells in mice due to alcohol consumption leads to inhibition of lung metastasis after intravenous melanoma inoculation and may even inhibit tumor growth at the early stage after subcutaneous tumor inoculation. However, tumor progression is facilitated by the continued presence of and interaction between melanoma and alcohol, which accelerates $CD8^+$ T cell dysfunction, negating any established early antitumor activity. Thus, the net outcome is no increase or even the decrease in survival of melanoma-bearing mice.

18.5 Effects of Chronic Alcohol Consumption on NK Cells: Impaired NK Cell Release from the Bone Marrow and Decreased Mature NK Cells in the Periphery

NK cells are innate immune cells. More than 95 % of these cells originate, develop, and mature in the bone marrow. Around 5 % of NK cells develop and mature in the thymus [58]. Bone marrow-derived NK cells have strong cytolytic activity, but are weak in cytokine production. Thymus-derived NK cells exhibit strong cytokine production, but exhibit weak cytolytic activity. Upon maturation these cells circulate to peripheral tissues and organs such as blood, spleen, lymph nodes, and liver. Unlike T cells and B cells, NK cells do not have a rearranged antigen specific receptor. The receptors governing NK cell function are Ly-49 family C-type lectin receptors in mice and immunoglobulin-like receptors in human. Most of these receptors are inhibitory. Their ligands are MHC molecules [59]. Most virus-infected cells and transformed cancer cells lose the expression of MHC molecules on their cell surface, which decreases the inhibitory signals in NK cells, thus leading to activation and production of perforin and granzymes to kill viral infected cells or cancer cells. Activated NK cells also produce Th1 cytokines to induce and enhance $CD8^+$ T cell and the Th1 cell immune response. Therefore NK cells play important roles in cancer surveillance and antitumor immunity [60]. Chronic alcohol consumption decreases the numbers of NK cells and compromises their cytolytic activity in the blood of human alcoholics [61]. Chronic alcohol consumption in mice also decreases NK cells in the peripheral organs, and inhibits NK cell cytolytic activity in the blood and spleen [62, 63]. We found that chronic alcohol consumption compromises

NK cell release from the bone barrow and this contributes to the decrease in mature NK cells in the spleen and blood [36]. Due to the decrease of bone marrow-derived mature NK cells, the portion of thymus-derived NK cells, which are a population of NK cells that express IL-7Rα (CD127) and produce large amount IFN-γ upon activation, is increased in the periphery [36]. Chronic alcohol consumption inhibits NK cell migration to lymph nodes through the downregulation of CD62L expression [64]. NK cells play an important role in preventing B16 melanoma metastasis into lymph nodes [65]. We found that chronic alcohol consumption increases B16BL6 melanoma metastasis into the draining lymph nodes in mice bearing subcutaneous tumor [64].

18.6 Effects of Chronic Alcohol Consumption on B Cells: Impaired B Cell Circulation in B16BL6 Melanoma-Bearing Mice

B cells are the largest population of lymphocytes in the spleen of mice and also the largest population of antigen presenting cells. The major function of B cells is to produce antibodies and orchestrate the humoral immune response. However, the effects of B cells in antitumor immunity have not been studied fully. Using gene mutation and cell depletion, it was found that depletion of B cells enhances antitumor immunity, suggesting that B cells have an inhibitory function on antitumor immunity [66]. A group of B cells that are $CD19^{+}CD1d^{hi}CD5^{+}$ produce IL-10 and inhibit T cells function [67]. Depletion of these inhibitory B cells enhances antitumor immunity [68]. Mature B cells play important roles in antitumor immunity through enhancing T cell activation and cytokine production, which depend on the antigen presenting function of B cells [69]. Therefore, two opposite functions of B cells in antitumor immunity have been identified. The first is an inhibitory function through IL-10 production. The second is an antitumor function through presentation of antigen to T cells to enhance T cell activation and cytokine production. We found that in the steady state chronic alcohol consumption does not significantly affect the B cell phenotype and nor their distribution in blood and lymph nodes; however, B cell numbers decrease in the spleen [33]. In melanoma-bearing mice B cells decrease around fourfold in the blood of alcohol-consuming mice with prolonged tumor growth compared to their water-drinking counterparts [70]. The cells that decrease are mature $CD23^{+}$ B cells. The decrease of mature B cells in the blood results from impaired B cell circulation associated with a compromised sphingosine-1-phosphate (S1P)/S1PR1 signaling pathway [70]. Effective circulation of B cells is important in order to capture antigen and for T cell activation. Therefore the impaired circulation of mature B cells would be expected to negatively affect T cell function in the melanoma-bearing, alcohol-consuming mice.

18.7 Effects of Chronic Alcohol Consumption on iNKT Cells: Increased Mature iNKT Cells That Produce an IFN-γ-Dominant Th1 Cytokine Profile in the Steady State, and an IL-4 Dominant Th2-Cytokine Profile in Melanoma-Bearing Mice

NKT cells are a unique population of T cells that recognize lipid antigens presented by the MHC I-like molecule, CD1d, but do not recognize peptide antigens presented by MHC molecules. Therefore, these cells are CD1-restricted T cells. Around 80 % of these cells express an invariant TCR α chain: Vα14Jα18 in mice and Vα24Jα18 in humans. They are called invariant NKT cells or iNKT cells. One of the most important features of iNKT cells is that these cells rapidly produce large amount of and a broad spectrum of cytokines once activated. These cytokines include Th1 cytokines such as IFN-γ, Th2 cytokines such as IL-4, IL-10, and IL-13, and Th17 cytokines such as IL-17A and IL-9. Due to the broad spectrum of cytokines produced, iNKT cells function more like immune regulatory cells than effector cells. iNKT cells play important roles in the regulation of antitumor immunity. The balance of Th1 and Th2 cytokines will shape the downstream antitumor immune response. When iNKT cells produce Th1-dominant cytokines such as IFN-γ, they will activate dendritic cells to produce IL-12. IL-12 synergizes with IFN-γ to further activate NK, iNKT, and T cells to produce more IFN-γ, which induces a strong Th1 immune response that inhibits Th2 immune responses and inhibits tumor progression. When iNKT cells produce a Th2-dominant cytokine profile these cytokines will not only directly inhibit the Th1 immune response, but also induce regulatory dendritic cells to produce IL10 and inhibit NK and $CD8^+$ T cell function (Fig. 18.6). These Th2 cytokines also induce MDSC, TAM to further enhance the Th2 immune response, which will facilitate tumor progression (Fig. 18.6). Under normal conditions iNKT cells produce Th1-dominant cytokines and favor antitumor immunity. Repeat activation of iNKT cells induces iNKT cell anergy [71]. The anergic iNKT cells produce Th2-dominant cytokines which inhibit antitumor immunity and favor tumor progression [71].

Since the ligands of iNKT cell receptors are lipids, and since alcohol consumption alters the metabolism of lipids, it is highly possible that alcohol consumption will affect iNKT cells through inducing iNKT cell activation. Indeed, we found that in the steady state chronic alcohol consumption significantly increases iNKT cells in the thymus and liver, but not in the spleen, blood, or bone marrow. The increased iNKT cells are $NK1.1^+$ mature cells. These cells are potent IFN-γ-producing cells. Upon activation under normal conditions, iNKT cells produce an IFN-γ-dominant Th1 cytokine profile, which favors antitumor immune responses. However, in melanoma-bearing mice, alcohol consumption not only increases iNKT cells in the thymus and liver, but also significantly increases iNKT cells in the blood

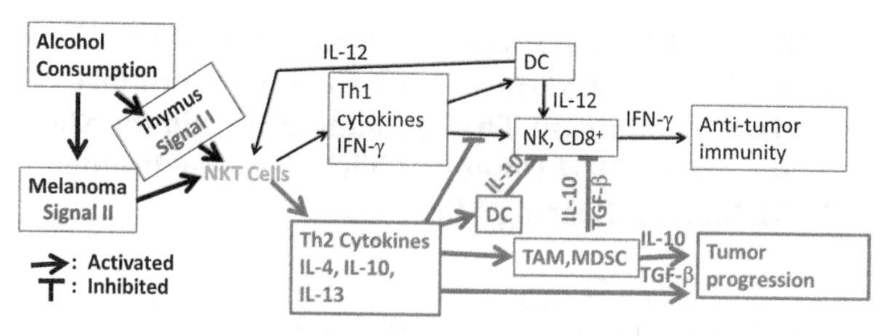

Fig. 18.6 A mechanistic scheme for NKT cell regulation of tumor immune responses and the hypothetical interaction between alcohol consumption and melanoma in modulating NKT cell anti- and pro-tumor (in *bold*) immunity. Once activated, NKT cells produce Th1 and Th2 cytokines. If the cytokine profile is dominated by the Th1 cytokine, IFN-γ, this will activate DC to produce IL-12, which in turn activates NK and CD8$^+$ T cells in addition to further activation of NKT cells to produce additional IFN-γ. IFN-γ produced by NKT cells activate NK and CD8$^+$ T cells to produce IFN-γ. This positive feedback loop forms a strong antitumor immune response. If the NKT cell cytokine profile is dominated by Th2 cytokines such as IL-4, IL-10, and IL-13, these cytokines will inhibit the Th1 response. IL-13 can activate MDSC and TAM. These cells produce TGF-β and IL-10, which can inhibit NK and CD8$^+$ T cell activation and attenuate antitumor immunity. Th2 cytokines produced by NKT, TAM, and MDSC also facilitate tumor growth and progression. Alcohol consumption induces thymocytes to produce lipids which work as Signal I to simulate NKT cell activation and proliferation. Alcohol consumption interacts with melanoma to induce/increase lipids which work as Signal II to stimulate NKT cells. The continuous stimulation of Signal I and Signal II induces NKT cell anergy leading to a Th2-dominant cytokine profile that inhibits antitumor immunity and promotes tumor progression

(Fig. 18.5a, b). More importantly, once activated, the iNKT cells produce an IL-4 dominant cytokine profile (Fig. 18.5c). These results indicate that in the melanoma-bearing mice alcohol interacts with the tumor to not only activate iNKT cells, but also to reverse the cytokine profile from Th1-dominant to Th2-dominant. This Th2-dominant cytokine profile favors tumor progression.

In summary, chronic alcohol consumption induces a signal to enhance iNKT cell activation and maturation in the steady state. We designate this signal as Signal I. This signal induces iNKT cells to produce and maintain a Th1-dominant cytokine profile. Activation of these iNKT cells will induce NK and CD8$^+$ T cells to generate Th1 immune responses, which favor antitumor immunity. In the tumor-bearing mice, alcohol interacts with melanoma cells to induce another signal that also activates iNKT cells. We designate this signal as Signal II. We suggest that the continuous activation of iNKT cells by these two signals induces iNKT cell anergy to produce a Th2-dominant cytokine profile that enhances MDSC and TAM function, but inhibits NK and CD8$^+$ T cell function. The net outcome would facilitate tumor progression (Fig. 18.6).

18.8 Immunological Basis of Chronic Alcohol Consumption on Tumor Surveillance, Progression, and the Survival of Cancer Patients

If chronic alcohol consumption activates the immune system, why does chronic alcohol consumption also increase the incidence of multiple types of cancer? We provide the following explanation based on the current knowledge regarding tumor immunosurveillance and tumor immunoediting. Normal cells have intrinsic functions, such as tumor suppressor genes, active DNA repair mechanisms, and apoptosis, to prevent carcinogen-induced cellular transformation [72]. Once some cells are transformed into cancer cells, they are controlled by the surveillance of immune system. This process is called tumor immunoediting, which includes three stages: elimination, equilibrium, and escape [73]. In the elimination stage, both innate and adaptive immune systems are involved in the elimination of transformed cells. Most of the transformed cells will be eliminated at this stage. Some of the transformed cells may survive the immunosurveillance, and enter the equilibrium stage. At the equilibrium stage, the adaptive immune system will keep the transformed cells in check and continue eliminating the transformed cells. Under the selection pressure of the immune system, the tumor cells will change their tumor antigens to avoid immunosurveillance. At this stage the tumor cells survive, but the tumor is clinically invisible. This stage may last for years, possibly even decades. Once tumor cells escape the control of the immune system, they will grow quickly and form a clinically visible tumor. At this escape stage, the tumor will generate multiple factors including inhibitory cells to inhibit immune system function and cause immune system exhaustion. Chronic alcohol consumption generates some carcinogens, such as acetaldehyde. These carcinogens accumulate in some specific organs, such as mammary gland and the upper aerodigestive tract, to induce cell transformation in these organs. Although alcohol consumption activates the immune system, which may help eliminate the transformed cells, the continuous stimulation of carcinogens from chronic alcohol consumption will induce a high frequency of cellular transformation. The frequent tumor cell transformation will accelerate immune system exhaustion and increase the chance of tumor escape. Once tumor cells escape the immune system control, they can interact with alcohol to induce immune inhibitory factors, such as MDSC and iNKT, as we found in the B16BL6 melanoma model, to accelerate the dysfunction of the immune system. The outcome would be the decreased host survival. As an example, the reduction in the incidence of NHL in alcoholics could be associated with an activated immune system. Because the NHL cancer cells are circulating, this will increase the chance that they will encounter activated immune cells and be eliminated, thus diminishing the incidence of NHL. Indeed, it was found that the occurrence of NHL is correlated with the Th1 cytokine level in the blood [74]. However, once NHL cancer cells escape immunosurveillance, this will quickly lead to immune system exhaustion resulting in a decreased survival in the alcoholics.

18.9 Prospective and Possible Strategies for Tumor Immunotherapy in Alcoholics

The ultimate goal of tumor immunology research is to find optimized approaches for tumor immunotherapy. Our research in the mouse model of alcohol and tumor immunology suggests the following strategies of tumor immunotherapy in alcohol-consuming and melanoma-bearing mice that could ultimately be translated to humans. Targeting iNKT cells and $CD8^+$ T cells should be good candidates to recover antitumor immunity in alcoholics. For example, we found in the B16BL6 melanoma model that alcohol consumption increases iNKT cells, and these cells are tumor inhibitory in the steady state and at the early stage of tumor growth. However, with the progression of tumor growth, the crosstalk between alcohol and melanoma induces iNKT cell anergy and reverses the iNKT cell cytokine profile from Th1 dominant to Th2 dominant. Therefore, developing immunotherapeutic strategies to prevent iNKT cell anergy through blocking the negative crosstalk between alcohol and melanoma cells could greatly enhance the antitumor immune response and therapeutic outcome in alcoholics. Blockade of the PD-1/PD-L1 signaling pathway can prevent activation-induced iNKT cell anergy and return IFN-γ production in iNKT cells to its former state [75]. Blockade of NKG2A, a receptor that inhibits activation of iNKT cells and that is increased in alcohol-consuming mice, can break the IFN-γ-induced negative regulation feedback loop [76, 77]. Antibodies or siRNA could be used to block PD-1/PD-L1 or/and NKG2A signaling pathway to prevent iNKT cell dysfunction and enhance antitumor immunity via activating iNKT cells in alcohol-consuming, tumor-bearing mice.

An additional approach to prevent iNKT cell anergy in alcoholics with cancer would be to manipulate lipid metabolism (modulate Signal I and Signal II).

MDSC and iNKT cells are the key inhibitory cells in alcohol-consuming mice bearing melanoma that modulate $CD8^+$ T cell function. Targeting MDSC and iNKT cells could result in recovery of $CD8^+$ T cell function. As discussed above, blockade of PD-1/PD-L1 and NKG2A could prevent iNKT cell anergy and enhance IFN-γ production of iNKT cells, which should enhance $CD8^+$ T cell function. In addition, anergic $CD8^+$ T cells express PD-1 and NKG2A [78, 79]. Blockade of PD-1 and NKG2A will not only prevent iNKT cell anergy, but also directly overcome the inhibitory signals in $CD8^+$ T cells. IL-15/IL-15Rα immunotherapy could be another option to boost $CD8^+$ T cell function in alcohol-consuming, tumor-bearing mice. We previously found that alcohol consumption decreases IL-15-producing dendritic cells, which are critical for memory T cell and NK cell proliferation and survival [80]. Therefore, IL-15/IL-15Rα combined with PD-1 and NKG2A blockade could be an additional approach to recover $CD8^+$ T cells function that is lost as a result of the alcohol/melanoma tumor interaction. Experimental and clinical data indicate that a combination of different therapies produce more efficient and powerful therapeutic outcome than single therapy [81].

18.10 Overall Summary and Concluding Remarks

Our research using the mouse chronic alcohol consumption and B16BL6 melanoma model clearly demonstrates that the dynamic effects of alcohol consumption on antitumor immunity occur in two opposing phases. The first phase associated with immune stimulation is tumor inhibitory and the second phase resulting from the interaction between the effects of alcohol and the tumor leads to immune inhibition and tumor progression. These finding may provide some mechanistic explanation for the controversial effects of alcohol consumption on the survival of different type of cancers, such as the reduction in NHL risk and the decrease in survival of NHL patients. Chronic alcohol consumption modulates T cells, NK cells, NKT cells, and B cells in non-tumor injected mice and in tumor-bearing mice by different mechanisms. Chronic alcohol consumption in mice not injected with tumor (1) increases memory and IFN-γ-producing T cells through the induction of T cell homeostatic proliferation, (2) increases IFN-γ-producing NK cells via the increase in the proportion of thymus-derived NK cells, and (3) increases iNKT cells and their activation and maturation through the induction of a lipid signal, Signal I. In melanoma-bearing mice, chronic alcohol consumption inhibits memory- and tumor-specific $CD8^+$ T expansion and accelerates $CD8^+$ T and NK cell dysfunction through the induction of MDSC as well as the induction of a Th2-skewed cytokine profile in iNKT cells. In addition alcohol also interacts with melanoma cells to induce another iNKT cell activation signal, Signal II. The continuous presence of Signal I and Signal II induces not only skews the iNKT cell cytokine profile from Th1 toward Th2, but also induces iNKT cell anergy. Moreover, alcohol consumption impairs the circulation of mature B cells through modulation of the S1P/S1PR1 signaling pathway to impair T cell activation and antitumor cytokine production. These findings provide new insights regarding the selection of targets for cancer immunotherapy in alcoholics.

18.11 Future Directions of Research in Alcohol and Tumor Immunology

Although research in this area so far has defined the effects of alcohol consumption on the antitumor immune response at the cellular level, the underlying molecular mechanism is largely unknown. Once the precise molecular basis is identified, it will be possible to select more efficient targets for tumor immunotherapy in alcoholics. Major research directions required to further understand the role of alcohol in regulation of antitumor immunology include determining the following: (1) the molecular mechanism of how alcohol interacting with tumor induces iNKT cell anergy, (2) the molecular mechanism of how alcohol consumption induces MDSC in the melanoma-bearing mice, (3) the effects of impaired mature B cell circulation on antitumor immunity and the survival of tumor-bearing mice, and (4) identification of the molecular basis underlying how alcohol interacting with tumor modulates the S1P/S1PR1 signaling pathway and compromises the circulation of B cells. In addition to these basic immunology studies, it is also important to study the dynamic effects of different immunization regimens on the antitumor immune response in the alcohol-consuming, tumor-bearing mice. With continued research in these and other areas, it is highly likely that new effective immunotherapeutic approaches will become available for the treatment of alcoholics with cancer.

Acknowledgement This project was supported by K05AA017149 and by funds provided for medical and biological research by the State of Washington Initiative Measure No. 171.

References

1. Nelson DE et al (2013) Alcohol-attributable cancer deaths and years of potential life lost in the United States. Am J Public Health 103(4):641–648
2. Boffetta P, Hashibe M (2006) Alcohol and cancer. Lancet Oncol 7(2):149–156
3. Haas SL, Ye W, Lohr JM (2012) Alcohol consumption and digestive tract cancer. Curr Opin Clin Nutr Metab Care 15(5):457–467
4. Berstad P et al (2008) Alcohol intake and breast cancer risk among young women. Breast Cancer Res Treat 108(1):113–120
5. Longnecker MP et al (1995) Risk of breast cancer in relation to lifetime alcohol consumption. J Natl Cancer Inst 87(12):923–929
6. Brooks PJ, Zakhari S (2013) Moderate alcohol consumption and breast cancer in women: from epidemiology to mechanisms and interventions. Alcohol Clin Exp Res 37(1):23–30
7. Thrift AP et al (2012) The influence of prediagnostic demographic and lifestyle factors on esophageal squamous cell carcinoma survival. Int J Cancer 131(5):E759–E768
8. Mayne ST et al (2009) Alcohol and tobacco use prediagnosis and postdiagnosis, and survival in a cohort of patients with early stage cancers of the oral cavity, pharynx, and larynx. Cancer Epidemiol Biomarkers Prev 18(12):3368–3374
9. Jerjes W et al (2012) The effect of tobacco and alcohol and their reduction/cessation on mortality in oral cancer patients: short communication. Head Neck Oncol 4:6
10. Wu IC et al (2012) Substance use (alcohol, areca nut and cigarette) is associated with poor prognosis of esophageal squamous cell carcinoma. PLoS One 8(2):e55834
11. Wang JB et al (2012) Attributable causes of esophageal cancer incidence and mortality in China. PLoS One 7(8):e42281
12. Newcomb PA et al (2013) Alcohol consumption before and after breast cancer diagnosis: associations with survival from breast cancer, cardiovascular disease, and other causes. J Clin Oncol 31(16):1939–1946
13. Kwan ML et al (2012) Postdiagnosis alcohol consumption and breast cancer prognosis in the after breast cancer pooling project. Cancer Epidemiol Biomarkers Prev 22(1):32–41
14. Harris HR, Bergkvist L, Wolk A (2012) Alcohol intake and mortality among women with invasive breast cancer. Br J Cancer 106(3):592–595
15. Flatt SW et al (2010) Low to moderate alcohol intake is not associated with increased mortality after breast cancer. Cancer Epidemiol Biomarkers Prev 19(3):681–688
16. McDonald PA et al (2002) Breast cancer survival in African American women: is alcohol consumption a prognostic indicator? Cancer Causes Control 13(6):543–549
17. Weaver AM et al (2013) Alcohol intake over the life course and breast cancer survival in Western New York exposures and breast cancer (WEB) study: quantity and intensity of intake. Breast Cancer Res Treat 139(1):245–253

18. Holm M et al (2013) Pre-diagnostic alcohol consumption and breast cancer recurrence and mortality: results from a prospective cohort with a wide range of variation in alcohol intake. Int J Cancer 132(3):686–694
19. Barnett GC et al (2008) Risk factors for the incidence of breast cancer: do they affect survival from the disease? J Clin Oncol 26(20):3310–3316
20. Morton LM et al (2005) Alcohol consumption and risk of non-Hodgkin lymphoma: a pooled analysis. Lancet Oncol 6(7):469–476
21. Tramacere I et al (2012) Alcohol drinking and non-Hodgkin lymphoma risk: a systematic review and a meta-analysis. Ann Oncol 23(11):2791–2798
22. Gapstur SM et al (2012) Alcohol intake and the incidence of non-Hodgkin lymphoid neoplasms in the cancer prevention study II nutrition cohort. Am J Epidemiol 176(1):60–69
23. Geyer SM et al (2010) Smoking, alcohol use, obesity, and overall survival from non-Hodgkin lymphoma: a population-based study. Cancer 116(12):2993–3000
24. Talamini R et al (2008) The impact of tobacco smoking and alcohol drinking on survival of patients with non-Hodgkin lymphoma. Int J Cancer 122(7):1624–1629
25. Battaglioli T et al (2006) Cigarette smoking and alcohol consumption as determinants of survival in non-Hodgkin's lymphoma: a population-based study. Ann Oncol 17(8):1283–1289
26. Meadows GG et al (1993) Alcohol consumption suppresses metastasis of B16-BL6 melanoma in mice. Clin Exp Metastasis 11:191–199
27. Blank SE, Meadows GG (1996) Ethanol modulates metastatic potential of B16BL6 melanoma and host responses. Alcohol Clin Exp Res 20:624–628
28. Szabo G, Mandrekar P (2009) A recent perspective on alcohol, immunity, and host defense. Alcohol Clin Exp Res 33(2):220–232
29. Cook RT (1998) Alcohol abuse, alcoholism, and damage to the immune system—a review. Alcohol Clin Exp Res 22(9):1927–1941
30. Cook RT et al (1991) Activated CD-8 cells and HLA DR expression in alcoholics without overt liver disease. J Clin Immunol 11:246–253
31. Song K et al (2002) Chronic ethanol consumption by mice results in activated splenic T cells. J Leukoc Biol 72:1109–1116
32. Laso FJ et al (2007) Chronic alcohol consumption is associated with changes in the distribution, immunophenotype, and the inflammatory cytokine secretion profile of circulating dendritic cells. Alcohol Clin Exp Res 31(5):846–854
33. Zhang H, Meadows GG (2005) Chronic alcohol consumption in mice increases the proportion of peripheral memory T cells by homeostatic proliferation. J Leukoc Biol 78(5):1070–1080
34. Abdallah RM, Starkey JR, Meadows GG (1988) Toxicity of chronic high alcohol intake on mouse natural killer cell activity. Res Commun Chem Pathol Pharmacol 59:245–258
35. Núñez NP, Carter PA, Meadows GG (2002) Alcohol consumption promotes body weight loss in melanoma-bearing mice. Alcohol Clin Exp Res 26:617–626
36. Zhang H, Meadows GG (2008) Chronic alcohol consumption perturbs the balance between thymus-derived and bone marrow-derived natural killer cells in the spleen. J Leukoc Biol 83(1):41–47
37. Weber JS et al (1987) Immunotherapy of a murine tumor with interleukin 2. Increased sensitivity after MHC class I gene transfection. J Exp Med 166(6):1716–1733
38. Winter H et al (2001) Immunotherapy of melanoma: a dichotomy in the requirement for IFN-gamma in vaccine-induced antitumor immunity versus adoptive immunotherapy. J Immunol 166(12):7370–7380
39. Maeda M et al (1998) Alcohol consumption enhances liver metastasis in colorectal carcinoma patients. Cancer 83(8):1483–1488
40. Oksuzler O et al (2009) Investigation of the synergism between alcohol consumption and herpes simplex virus in patients with laryngeal squamous cell cancers. Eur Arch Otorhinolaryngol 266(12):1977–1982
41. Wilson SD et al (2011) Zollinger-Ellison syndrome associated with a history of alcohol abuse: coincidence or consequence? Surgery 150(6):1129–1135

42. Spitzer JH et al (2000) The modulation of B16BL6 melanoma metastasis is not directly mediated by cytolytic activity of natural killer cells in alcohol-consuming mice. Alcohol Clin Exp Res 24(6):837–844
43. Zhang H et al (2011) IFN-gamma is essential for the inhibition of B16BL6 melanoma lung metastasis in chronic alcohol drinking mice. Clin Exp Metastasis 28(3):301–307
44. Dunn GP, Koebel CM, Schreiber RD (2006) Interferons, immunity and cancer immunoediting. Nat Rev Immunol 6(11):836–848
45. Savage PA, Malchow S, Leventhal DS (2013) Basic principles of tumor-associated regulatory T cell biology. Trends Immunol 34(1):33–40
46. Martin F, Apetoh L, Ghiringhelli F (2012) Controversies on the role of Th17 in cancer: a TGF-beta-dependent immunosuppressive activity? Trends Mol Med 18(12):742–749
47. Ye J, Livergood RS, Peng G (2013) The role and regulation of human Th17 cells in tumor immunity. Am J Pathol 182(1):10–20
48. DuPage M et al (2012) Expression of tumour-specific antigens underlies cancer immunoediting. Nature 482(7385):405–409
49. Shankaran V et al (2001) IFNgamma and lymphocytes prevent primary tumour development and shape tumour immunogenicity. Nature 410(6832):1107–1111
50. Kersh EN et al (2006) Rapid demethylation of the IFN-gamma gene occurs in memory but not naive CD8 T cells. J Immunol 176(7):4083–4093
51. Fitzpatrick DR, Shirley KM, Kelso A (1999) Cutting edge: stable epigenetic inheritance of regional IFN-gamma promoter demethylation in CD44highCD8+ T lymphocytes. J Immunol 162(9):5053–5057
52. Winders BR, Schwartz RH, Bruniquel D (2004) A distinct region of the murine IFN-gamma promoter is hypomethylated from early T cell development through mature naive and Th1 cell differentiation, but is hypermethylated in Th2 cells. J Immunol 173(12):7377–7384
53. Kambayashi T et al (2003) Memory CD8+ T cells provide an early source of IFN-gamma. J Immunol 170(5):2399–2408
54. Laso FJ et al (1999) Chronic alcoholism is associated with an imbalanced production of Th-1/Th-2 cytokines by peripheral blood T cells. Alcohol Clin Exp Res 23(8):1306–1311
55. Zhang H, Meadows GG (2010) Chronic alcohol consumption enhances myeloid-derived suppressor cells in B16BL6 melanoma-bearing mice. Cancer Immunol Immunother 59(8):1151–1159
56. Highfill SL et al (2010) Bone marrow myeloid-derived suppressor cells (MDSCs) inhibit graft-versus-host disease (GVHD) via an arginase-1-dependent mechanism that is up-regulated by interleukin-13. Blood 116(25):5738–5747
57. Zhu X et al (2014) The central role of arginine catabolism in T-cell dysfunction and increased susceptibility to infection after physical injury. Ann Surg 259(1):171–178
58. Vosshenrich CA et al (2006) A thymic pathway of mouse natural killer cell development characterized by expression of GATA-3 and CD127. Nat Immunol 7(11):1217–1224
59. Vivier E et al (2008) Functions of natural killer cells. Nat Immunol 9(5):503–510
60. Vivier E et al (2012) Targeting natural killer cells and natural killer T cells in cancer. Nat Rev Immunol 12(4):239–252
61. Perney P et al (2003) Specific alteration of peripheral cytotoxic cell perforin expression in alcoholic patients: a possible role in alcohol-related diseases. Alcohol Clin Exp Res 27(11):1825–1830
62. Meadows GG et al (1992) Ethanol induces marked changes in lymphocyte populations and natural killer cell activity in mice. Alcohol Clin Exp Res 16:474–479
63. Blank SE et al (1993) Ethanol-induced changes in peripheral blood and splenic NK cells. Alcohol Clin Exp Res 17(3):561–565
64. Zhang H, Zhu Z, Meadows GG (2011) Chronic alcohol consumption decreases the percentage and number of NK cells in the peripheral lymph nodes and exacerbates B16BL6 melanoma metastasis into the draining lymph nodes. Cell Immunol 266(2):172–179
65. Chen S et al (2005) Suppression of tumor formation in lymph nodes by L-selectin-mediated natural killer cell recruitment. J Exp Med 202(12):1679–1689

66. Inoue S et al (2006) Inhibitory effects of B cells on antitumor immunity. Cancer Res 66(15): 7741–7747
67. Yanaba K et al (2008) A regulatory B cell subset with a unique CD1dhiCD5+ phenotype controls T cell-dependent inflammatory responses. Immunity 28(5):639–650
68. Horikawa M et al (2011) Regulatory B cell production of IL-10 inhibits lymphoma depletion during CD20 immunotherapy in mice. J Clin Invest 121(11):4268–4280
69. DiLillo DJ, Yanaba K, Tedder TF (2010) B cells are required for optimal CD4+ and CD8+ T cell tumor immunity: therapeutic B cell depletion enhances B16 melanoma growth in mice. J Immunol 184(7):4006–4016
70. Zhang H, Zhu Z, Meadows GG (2012) Chronic alcohol consumption impairs distribution and compromises circulation of B cells in B16BL6 melanoma-bearing mice. J Immunol 189(3):1340–1348
71. Parekh VV et al (2005) Glycolipid antigen induces long-term natural killer T cell anergy in mice. J Clin Invest 115(9):2572–2583
72. Hanahan D, Weinberg RA (2011) Hallmarks of cancer: the next generation. Cell 144(5): 646–674
73. Schreiber RD, Old LJ, Smyth MJ (2011) Cancer immunoediting: integrating immunity's roles in cancer suppression and promotion. Science 331(6024):1565–1570
74. Saberi Hosnijeh F et al (2010) Plasma cytokines and future risk of non-Hodgkin lymphoma (NHL): a case-control study nested in the Italian European Prospective Investigation into Cancer and Nutrition. Cancer Epidemiol Biomarkers Prev 19(6):1577–1584
75. Parekh VV et al (2009) PD-1/PD-L blockade prevents anergy induction and enhances the antitumor activities of glycolipid-activated invariant NKT cells. J Immunol 182(5):2816–2826
76. Kawamura T et al (2009) NKG2A inhibits invariant NKT cell activation in hepatic injury. J Immunol 182(1):250–258
77. Ota T et al (2005) IFN-gamma-mediated negative feedback regulation of NKT-cell function by CD94/NKG2. Blood 106(1):184–192
78. Fourcade J et al (2010) Upregulation of Tim-3 and PD-1 expression is associated with tumor antigen-specific CD8+ T cell dysfunction in melanoma patients. J Exp Med 207(10): 2175–2186
79. Gooden M et al (2011) HLA-E expression by gynecological cancers restrains tumor-infiltrating CD8(+) T lymphocytes. Proc Natl Acad Sci U S A 108(26):10656–10661
80. Mortier E et al (2009) Macrophage- and dendritic-cell-derived interleukin-15 receptor alpha supports homeostasis of distinct CD8+ T cell subsets. Immunity 31(5):811–822
81. Ott PA, Hodi FS, Robert C (2013) CTLA-4 and PD-1/PD-L1 blockade: new immunotherapeutic modalities with durable clinical benefit in melanoma patients. Clin Cancer Res 19(19): 5300–5309

Chapter 19
A Perspective on Chemoprevention by Resveratrol in Head and Neck Squamous Cell Carcinoma

Sangeeta Shrotriya, Rajesh Agarwal, and Robert A. Sclafani

Abstract Head and neck squamous cell carcinoma (HNSCC) accounts for around 6 % of all cancers in the USA. Few of the greatest obstacles in HNSCC include development of secondary primary tumor, resistance and toxicity associated with the conventional treatments, together decreasing the overall 5-year survival rate in HNSCC patients to ≤50 %. Radiation and chemotherapy are the conventional treatment options available for HNSCC patients at both early and late stage of this cancer type malignancy. Unfortunately, patients response poorly to these therapies leading to relapsed cases, which further, emphasizes the need of additional strategies for the prevention/intervention of both primary and the secondary primary tumors post-HNSCC therapy. In recent years, growing interest has focused on the use of natural products or their analogs to reduce the incidence and mortality of cancer, leading to encouraging results. Resveratrol, a component from grape skin, is one of the well-studied agents with a potential role in cancer chemoprevention and other health benefits. As an anticancer agent, resveratrol suppresses metabolic activation of pro-carcinogens to carcinogens by modulating the metabolic enzymes responsible for their activation, and induces phase II enzymes, thus, further detoxifying the effect of pro-carcinogens. Resveratrol also inhibits cell growth and induces cell death in cancer cells by targeting cell survival and cell death regulatory pathways. Growing evidence also suggest that resveratrol directly binds to DNA and RNA, activates antioxidant enzymes, prevents inflammation, and stimulates DNA damage checkpoint kinases affecting genomic integrity more specifically in malignant cells.

S. Shrotriya
Department of Pharmaceutical Sciences, University of Colorado Denver, Aurora, CO, USA

R. Agarwal
Department of Pharmaceutical Sciences, University of Colorado Denver, Aurora, CO, USA

University of Colorado Cancer Center, University of Colorado Denver, Aurora, CO, USA

R.A. Sclafani (✉)
University of Colorado Cancer Center, University of Colorado Denver, Aurora, CO, USA

Department of Biochemistry and Molecular Genetics, University of Colorado Denver, Aurora, CO, USA
e-mail: Robert.Sclafani@ucdenver.edu

V. Vasiliou et al. (eds.), *Biological Basis of Alcohol-Induced Cancer*, Advances in Experimental Medicine and Biology 815,
DOI 10.1007/978-3-319-09614-8_19

Abbreviations

4NQO	4-Nitroquinoline 1-oxide
ADH	Alcohol dehydrogenase
ALDH	Aldehyde dehydrogenase
AP-1	Activator protein 1
ARE	Antioxidant response element
ATM	Ataxia telangiectasia mutated
ATR	Ataxia telangiectasia-Rad3-related
B[A]P	Benzo[a]pyrene
BER	Base excision repair
Brca1	Breast cancer gene 1
Cdc25C	Cell division cycle 25C
CDKs	Cyclin dependent kinases
Chk1/2	Checkpoint kinase 1/2
COX2	Cyclooxygenase 2
CYP450	Cytochrome P450s
DMBA	7,12-Dimethylbenz[α]anthracene
DSBs	Double strand break
EGFR	Epidermal growth factor receptor
Egr 1	Early growth response 1
EMT	Epithelial mesenchymal transition
FA	Fanconi anemia
GSH	Glutathione
HNSCC	Head and neck squamous cell carcinoma
HPV	Human papilloma virus
HRR	Homologous recombination repair
IGFR	Insulin-like growth factor receptor
iNOS	Inducible nitric oxide synthetase
JNK	c-Jun N-terminal kinase
MAPK	Mitogen-activated protein kinases
MCM	Minichromosome maintenance
MMPs	Metalloproteinases
MMR	Mismatch repair
mTOR	Mammalian target of rapamycin
NER	Nucleotide excision repair
NF-kB	Nuclear factor kappa B
NHEJ	Non-homologous end joining
NNK	4-Methylnitrosamine-1-(3-pyridyl)-1-butanone
P53	Protein 53
PARP	Poly (ADP-ribose) polymerases
PCNA	Proliferating cell nuclear antigen
RA	Retinoic acid
Rb	Retinoblastoma protein

Res	Resveratrol
ROS/RNS	Reactive oxygen species/reactive nitrogen species
SPT	Secondary primary tumor
SSBs	Single strand break
STAT3	Signal transducer and activator of transcription 3

19.1 Introduction

This chapter aims to summarize recent development in the etiology and treatment of head and neck squamous cell carcinoma (HNSCC). Alcohol and its by-products have been reported to be strong possible carcinogens in different types of malignancies including HNSCC. In this perspective, chemopreventive agents are revealed to exert their chemopreventive efficacy either by blocking the activity of carcinogens or its metabolism or by targeting various cell survival pathways in tumor cells [69]. Further in-depth mechanistic studies revealed that nutraceutical resveratrol may inhibit carcinogenesis by affecting the molecular events in the initiation, promotion, and progression stages. Therefore, our goal in this chapter is to briefly describe the molecular mechanism involved in alcohol-induced oral cancer and to propose that these oral cancers can be prevented through the use of the nontoxic natural product resveratrol.

19.1.1 Head and Neck Squamous Cell Carcinoma

HNSCC is a devastating disease worldwide accounting for 650,000 new cases and 350,000 deaths every year [14]. According to the American Cancer Society, in the year 2013, approximately 53,640 new cases and 11,520 deaths are projected to occur, in the USA alone [59]. HNSCC is mostly curable with conventional treatment therapy when is diagnosed in early stage (I or II); unfortunately most of the HNSCC patients are diagnosed at advanced stage of the disease (III or IV), and survival rate is below 50 %. Epidemiological data suggest that several behavioral, environmental, viral, and genetic factors have been associated with the development of HNSCC. Tobacco and alcohol consumption accounts for ≥80 % of the risk factors associated with HNSCC [48, 56]. For individuals using both tobacco and alcohol, there is a synergistic effect, accelerating the risk of both oral and pharyngeal cancer by nearly 35-fold [65]. One cohort study with nearly 0.5 million participants focused on investigating the relationship between HNSCC risk and alcohol consumption, suggested that moderate drinkers (up to one drink a day) showed reduced risk of HNSCC compared to nondrinkers and heavy drinkers [17]. Reflecting this linearity, alcohol consumption is one of the major risk factors for nonsmoker. This was further supported by a population-based survey of 1,090 oral or pharyngeal cancer patients where the risk of secondary primary tumors (SPTs) was documented to be ≥50 % who continue with their drinking behavior after treatment [10]. Hence, considering concurrent risk of cancer development among alcoholics, understanding the relationship between alcohol consumption and cancer development, specifically HNSCC, is crucial for therapeutics and preventive purposes.

19.1.2 Major Molecular and Genetic Predispositions in HNSCC

Molecular and genetic analyses have revealed that multiple pathways are compromised in head and neck cancer [34]. Pathways that are critically altered in HNSCC include protein 53 (p53), retinoblastoma protein (Rb), epidermal growth factor receptor, signal transducer and activator of transcription, vascular endothelial growth factor, DNA repair regulators, and mammalian target of rapamycin; these pathways are thus also identified as potential therapeutic targets [18, 26]. Overexpression of dominant negative p53 and cyclin D1, together with increased telomerase activity (≥80 %), confer deregulated cell cycle and resistance to DNA damage stimulators in HNSCC [7]. Viral proteins E6 and E7 encoded by human papilloma virus (HPV) bind and inactivate p53 and Rb, respectively, disrupting the cell cycle regulation in HPV+tumors [29]. Most HPV+tumors are observed among young nonsmokers and nondrinkers, and are usually present at an advanced stage at the time of diagnosis [16, 22]. In addition, a clinical study with HNSCC patients ($n=37$) and healthy individuals ($n=35$) showed that tumor cells are more sensitive to irradiation-mediated DNA damage and display impaired DNA repair, indicating the crucial role of DNA repair mechanism in HNSCC treatment [8, 54]. Another study with archival human head and neck cancer tumor specimens revealed that Ku80, a DNA repair protein, was overexpressed and correlated with increased drug resistance, thus indicating this pathway as an attractive therapeutic target [8, 37, 41]. In relation to this linearity, patients with Fanconi Anemia (FA), a genetic syndrome with defective DNA repair mechanism, have been associated with an early lymph node metastases with poor clinical outcome [68]. DNA repair defects in Fanconi patients lead to pre-cancerous cells with increased levels of DNA damage and mutations. Although there is no evidence for genetic mutations in Fanc/Brca pathway in HNSCCs, loss or reduced expression of Fanc/Brca pathway genes has been reported in sporadic HNSCCs [36]. Likewise, amplification of several oncogenes [e.g., c-myc, Ras, EGFR, erbB2, nitric oxide synthetase, and cyclooxygenase 2 (COX2)] has been observed in HNSCC and has been associated with poor prognosis [28, 51]. The direct downstream target of tyrosine kinase receptors (EGFR, IGF-1R) and the PI3K/Akt signaling pathway is mammalian target of rapamycin (mTOR), which is found to be activated in 90–100 % of HNSCC [43]. Together, all these oncogenic pathways interlinked cellular pathways, deregulated in HNSCC, serve as potential therapeutic and preventive targets for HNSCC [5, 34, 45]. All these pathways directly or indirectly involved in cancer initiation, promotion, or progression are reported as possible targets of alcohol or its metabolites [12, 57]. Herein, identification of these cellular and molecular processes that are disrupted by exposure to alcohol is necessary to consider for the therapeutic and preventive intervention of cancer included HNSCC [26].

19.1.3 *Alcohol: Carcinogen or Co-carcinogen in HNSCC*

As mentioned earlier in this chapter, heavy alcohol consumption is directly associated with increased risk of cancer included HNSCC. In notion to this, International Agency for Cancer has classified alcohol as carcinogen to human, as it may influence cancer incidence by modulating different stages in cancer development [57]. In this chapter, we have briefly summarized cellular and molecular processes altered by alcohol.

19.1.3.1 Alcohol Metabolism and Metabolic Enzymes

The bacterial microfloras present in the oral cavity or esophagus convert ethanol to acetaldehyde with the help of alcohol dehydrogenase (ADHs), cytochrome P4502E1 (CYP2E1), and catalase [17]. Acetaldehyde so formed is further processed to acetate in the presence of aldehyde dehydrogenase (ALDH). Acetaldehyde binds covalently with DNA forming DNA adducts and interferes with DNA synthesis and repair, thereby initiating multistage carcinogenesis process after continuous exposure [57]. There are ample evidences showing that exposure to acetaldehyde produce mucosal lesions and adenocarcinomas in the nasal mucosa in rats [32, 70]. A study by Balbo and colleague have reported that N^2-ethyl-2′-deoxyguanosine, a major acetaldehyde derived-DNA adduct was increased by several-fold from baseline after alcohol use in humans [2]. Genetic evidence suggests that individuals having fast metabolites alleles variants for ADHs [ADH1B*2, ADH1C*1] and the null allele for ALDH 2 [ALDH2*2] have increased acetaldehyde levels and inefficient alcohol metabolism, thereby increasing susceptibility to cancer after alcohol consumption [9, 57, 7]. Similarly, the activities of ADH and ALDH are shown to be significantly higher in cancerous than in healthy tissues [30], further suggesting the importance of these enzymes in alcohol metabolism and aggressive cancer. Furthermore, chronic alcohol consumption leads to induction of CYP2E1, which metabolizes alcohol to acetaldehyde, as well as generates reactive oxygen species (ROS) and reactive nitrogen species (RNS) as by-products [12]. These formed ROS and/or RNS can directly damage DNA, generate lipid peroxidation products, and increase inflammation. Moreover, prolonged ethanol exposure leads to decreased levels of endogenous antioxidants, thereby reducing cellular defense mechanisms [12]. Together, this supports the key role of alcohol metabolites in causing cancer including HNSCC in humans through different endogenous mechanisms.

19.1.3.2 Alcohol: DNA Damage and DNA Repair

Reactive oxygen- or nitrogen-containing molecules that are generated during alcohol metabolism can result in different types of DNA damage included single-base lesions, single-strand breaks (SSBs), and double-strand breaks (DSBs). To counteract the DNA damage, cells have elaborated DNA repair mechanisms, thus protecting genomic integrity [24]. SSBs and single-base lesions are repaired by base-excision repair (BER), nucleotide-excision repair (NER), and mismatch repair (MMR); and DSBs are repaired through non-homologous end joining (NHEJ) and homologous repair (HR) mechanisms. A prolonged consumption of alcohol results in higher levels of oxidative DNA damage, lipid peroxidation adducts, and acetaldehyde-DNA adducts, and thus overwhelming the relevant DNA repair mechanism and impairing genome function [12]. In recent years, polymorphisms in DNA repair have been directly linked with increased risk to DNA damaging agents with a likelihood of oncogenic transformation [19].

19.1.3.3 Alcohol Interacts with Oncogenes and Tumor Suppressor Pathways

Apart from mutagenic potential of alcohol metabolites through DNA adducts formation, alcohol is also shown to disrupt cellular and molecular pathways in the multistage carcinogenesis process. It has been suggested that acetaldehyde abrogates a cell's ability to repair DNA damage [57]. Similarly, alcohol has been reported to either directly interacting with cell membrane proteins or modulating cellular function [57]. Long-term alcohol exposure preferentially causes *K*-ras mutation, impairs p16INK4A protein expression, and induces mutation in both retinoblastoma (Rb) and p53 proteins thereby, triggering cancer promotion [51]. Aberration of all these genes promote cell survival, evade apoptosis, and stimulate cell proliferation of cancer cells [49]. A recent study has reported that ethanol in cancer cells is ultimately converted to acetyl-CoA (a high-energy mitochondrial fuel) that can be used to synthesize ketone bodies, fueling tumor cell growth via oxidative mitochondrial metabolism (OXPHOS) [55]. Alcohol consumption has been associated with disruption of retinoid metabolism; retinoic acid (RA)-receptor signaling pathways involved in regulating cell proliferation, differentiation, lipid metabolism, and inflammation in many alcohol-associated malignancies [25]. Different natural products are shown to activate retinoid receptors expressed in cancer cell types initiating redifferentiation in cancer cells [27].

Recently, a connection between alcohol as a carcinogen and cancer has been made in Fanconi anemia (FA) patients, who are at risk for leukemia and HNSCC [20, 23]. Consistent with this, a very recent study in humans have suggested a compromised FA pathway which likely leads to an increased accumulation of aldehyde-induced DNA damage in hematopoietic stem cells, resulting in p53/p21 mediated cell death or senescence [23]. Earlier, a similar finding was observed in a preclinical model in which the hematopoietic stem cells in ALDH2/FANCD2 double knockout mice accumulate more DNA damage than hematopoietic stem cells in either of the single knockout mice [20]. In summary, loss of the ALDH2 isozyme in FA humans and knockout mice results in accumulation of acetaldehyde, which then acts as a carcinogen by producing DNA crosslinks that are not repaired due to the loss of the FA/BRCA DNA repair pathway. These studies may provide a rationale to explain why alcohol is a factor in the etiology of non-FA HNSCC. In this idea, continual exposure to alcohol may result in enough acetaldehyde being produced to cause an accumulation of mutations and increased risk for HNSCC.

19.1.3.4 Alcohol and Nutrition

Various studies have shown that chronic alcohol abuse may alter the way body processes nutrients, consequently changing its carcinogenic potential. This has been supported by the studies reporting that alcoholics have reduced levels of zinc, iron, vitamin E, and vitamin B [17, 57]. Once nutrients are absorbed, alcohol can prevent their utilization in the body by altering their transport, storage, and excretion [51].

19.1.4 Chemoprevention

19.1.4.1 Biologic Basis of Chemoprevention

Molecular and genetic analyses have revealed that various pathways are compromised in different malignancies included head and neck cancer [34], and that these pathways are being targeted by chemotherapeutic agents to improve life quality of patients, but have limited success [1]. It is, thus, inevitable that additional alternative strategies are required to significantly enhance the therapeutic index of conventional treatment modalities. In this regard, dietary components are reported to inhibit cancer development, progression, and metastasis by modulating different mechanisms of cancer development, under a modality known as "cancer chemoprevention."

19.1.5 Resveratrol as Chemopreventive Agent

Resveratrol, a grape-derived stilbene, is one of the most widely investigated chemopreventive agents retaining a wide variety of health-beneficial activities, including anticancer properties [42]. It has been shown to decelerate carcinogenesis process via direct and indirect multiple targets, mechanisms involved in the survival of cancer cells, and accelerating cell death. As detailed review of chemopreventive potential of resveratrol is beyond the scope of this chapter, we will briefly summarize the underlying molecular mechanism involved in chemopreventive efficacy of resveratrol in multistage carcinogenesis process.

19.1.5.1 Resveratrol in Xenometabolism

Under continuous exposure to pro-carcinogenic agent, normal cells undergo DNA mutations leading to initiation of the carcinogenesis process. Generally, carcinogens such as 7,12-dimethylbenzo[a]anthracene, benzanthracene, benzopyrene [B(a) P], and 4-nitroquinolone-1-oxide are converted to active metabolites in the presence of Phase I metabolic enzymes. These metabolites interact with genomic DNA forming adducts that directly correlate with cancer initiation, promotion, and progression stages. Resveratrol is shown to be effective in suppressing cancer initiation process by inhibiting these metabolic enzymes that play an important role in activation of pro-carcinogen [21]. Resveratrol modulates the activities of CYP1A1, CYP1B1, CYP3A, glutathione-*S*-transferase, UDP-glucuronosyl transferase (UGT), aryl hydrocarbon receptor (AhR), *O*-acetyl transferase, and sulfotransferase in both preclinical and clinical studies, thereby changing the risk of carcinogen undergoing metabolism [21, 69]. Resveratrol also activates phase II enzymes gene expression through modulation of mitogen-activated protein kinase pathways [21]. Resveratrol is also shown to differentially induce NAD(P)H quinone reductase (NQO), glutathione (GSH), glutathione reductase, and hemoxygenase 1 (HO-1), subsequently inhibiting DNA damage [21]. In contrast, in some of the cancer cells, resveratrol exerts pro-oxidant effect in the presence of transition metals such as Cu (II) depending on its concentration and time of exposure, leaving normal cells unharmed [40].

19.1.5.2 Resveratrol: As Anti-inflammatory Agent

In general, excessive generation of ROS and nitrogen species plays a pivotal role during inflammation [35]. Other two important enzymes involved in triggering inflammatory responses include cyclooxygenase and lipoxygenase, promoting tumor growth, progression and metastasis [52]. Among cyclooxygenase enzymes, COX1 plays a key role in prostaglandin synthesis, and COX2 is involved during inflammation [52]. Both of these enzymes are known to promote DNA synthesis, cell proliferation, invasion, and metastasis [27]. In this context, resveratrol is known to inhibit de novo synthesis of iNOS and COX2 via inhibition of NF-kB pathway as well the upstream kinases activating this pathway [51]. Resveratrol also targets nitric oxide production and lipoxygenase-stimulated inflammatory responses [7]. In conclusion, resveratrol counteracts ROS- and/or RNS-associated DNA damage and inflammatory responses as a part of its potential cancer chemopreventive efficacy.

19.1.5.3 Resveratrol: Cell Growth and Death Regulatory Pathways

Resveratrol exerts anti-proliferative activity via induction of cell growth inhibition, and induces apoptosis by modulating the major survival pathways present in the cancer cells [20]. Phosphatidylinositol 3-kinase (PI3K)/AKT pathway is considered as one of the major regulators of cell survival. Resveratrol is reported to suppress the activation of mitogen-activated protein kinases (MAPKs), c-Jun N-terminal kinase (JNK), MEK/ERK/NF-kB, cyclins, and cyclin dependent kinases (CDKs) leading to cell cycle arrest at specific stage and reversal of epithelial-to-mesenchymal transition (EMT) [4]. For instance, resveratrol has been shown to inhibit cell growth of different cancer cells at G0/G1 phase by altering cyclin D1/CDK4; in addition, resveratrol also causes S and G2/M phase cell cycle arrest resulting in increased cyclin A and E levels [62, 66, 67]. Moreover, a study by Shi and colleagues has shown that resveratrol exerts synergistic effects, when given in combination with the cytotoxic agent tamoxifen, on inhibiting the growth of both MCF-7/TR cells and metastasis [4]. Regarding the apoptosis induction, it is reported that resveratrol activates TNF-related apoptosis ligand, modulates the expression of pro-(Bax) and anti-(Bcl2 and Bcl-xL) apoptotic proteins, disrupts mitochondrial membrane potential, and causes caspases 9, 3 and poly (ADP-ribose) polymerase (PARP) cleavage, thus facilitating apoptotic cell death both in culture and mouse xenograft models [6, 50, 53, 62, 66]. As an upstream trigger to this cellular function, resveratrol is also shown to stimulate DNA damage response molecules and their downstream targets facilitating DNA damage response, cell cycle arrest, and apoptosis in various cancer models including HNSCC [11, 66]. Collectively, detailed mechanistic studies in preclinical cancer models have revealed that resveratrol targets PI3K/Akt/mTOR, NF-kB, MAPKs and apoptotic pathways, cell cycle regulatory molecules, different cellular receptors, and angiogenic pathways, which attribute towards its cancer chemopreventive efficacy.

19.1.5.4 Resveratrol: An Antioxidant

Antioxidant property of resveratrol is considered one of the crucial mechanisms involved in its chemopreventive and anticancer efficacy. In support to this idea, resveratrol is shown to prevent the generation of ROS following exposure to oxidizing agents, e.g., chemical carcinogens generated from tobacco and alcohol, hydrogen peroxide, etc. [21]. Resveratrol strongly increases the activity of antioxidant response element (ARE)/Nrf-2, leading to enhanced expression of phase II-detoxifying enzymes and antioxidant enzymes including NAD(P)H: quinone oxidoreductase 1(NQO 1) and hemoxygenase 1(HO1) via modulation of p38/PI3K/Akt pathways [21, 60]. Resveratrol has also been shown to enhance histone/protein deacetylase SIRT1 and adenosine monophosphate-activated protein kinase (AMPK) leading to metabolic changes in the cancer cells [62]. Likewise, resveratrol exhibits protective effect against ROS-mediated lipid peroxidation and DNA damage in normal cells, versus toxic effect in malignant cells including head and neck cancer [66]. More recently, Bishayee and colleagues have reported that resveratrol suppresses 4-(methylnitrosamino)-1-(3-pyridyl)-1-butanone (NNK)- and benzo[a]pyrene (B[a]P)-mediated precancerous changes in the human epithelial breast cancer cell MCF10A [6, 30].

19.1.5.5 Resveratrol: Interaction with DNA Polymerase, Topoisomerase, and Telomerase

Various preclinical studies have revealed that resveratrol directly binds to DNA and RNA through H-bond and stabilizes the double-helical structure, and this might serve as a molecular basis for anti-mutagenic effect of resveratrol [21, 39]. Resveratrol also impedes DNA replication by specifically interacting with DNA polymerase α and δ, resulting in DNA damage [20]. DNA polymerases are the enzymes required for de novo synthesis of DNA, and are equally crucial in protecting cells against the effects of DNA damage, and for initiating DNA repair [47]. Resveratrol is also reported to result in DNA damage through pro-oxidant effect in the presence of transition metal [40]. This might raise doubts of its beneficial function in normal cells versus cancer cells in general. However, it has been reported that pro-oxidant effect of these phytochemicals are more specific to cancer cells, for the fact that cancer cells have higher levels of transition metals, produce more ROS, etc. [63]. Similarly, Topoisomerase activity is particularly higher by several folds in cancer cells including squamous cell carcinoma, and has been targeted by many anticancer drugs [15]. Topoisomerase is essential for DNA replication, facilitating chromosome tangling, condensation, and mitotic segregation [21]. Three different types of topoisomerases are described in mammalian cells (type IA, type IB, and type II). Among all these, molecular modeling has confirmed that resveratrol specially binds to topoisomerase II and stabilizes the cleavage complexes of DNA and topoisomerase compromising DNA topology in cancer cells [3]. Similarly, resveratrol is shown to inhibit telomerase, a cellular reverse transcriptase that catalyzes the synthesis and extension of telomeric DNA, further assisting the cells to proliferate and delay the development of cell senescence [33]. Recently, several studies have also shown that resveratrol downregulates telomerase activity as well as inhibits the nuclear localization of human telomerase reverse transcriptase (hTERT), a subunit of telomerase [21, 33]. As hTERT is regulated by upstream protein kinase C, NF-kB, MAP kinases, and effect of resveratrol on hTERT might be indirect due to inhibition of these upstream kinases as well.

19.1.5.6 Resveratrol: DNA Damage and DNA Damage Repair

As mentioned earlier in this chapter that chemopreventive and anticancer activities of resveratrol have been implicated to involve DSBs. After DSBs, DNA repair molecules, for instance Mre11, Rad50, and Nbs1 (MRN complex), are stimulated at the site of DSBs, and recruits PI3K-related kinases such as ataxia telangiectasia mutated (ATM)/ATR to the DNA break site. Once activated, ATM phosphorylates several key DSB repair and checkpoint control factors like H2AX, p53, Nbs1, BRCA1, SMC1, Chk2, and Chk1 [19]. H2A.X is critical for facilitating the assembly of specific DNA-repair complexes on damaged DNA site [44]. Resveratrol is reported to induce S-phase arrest through activation of an ATM/ATR–Chk1/Chk2–Cdc25C pathway [67]. A study by Galicia and colleagues have shown that resveratrol at higher concentrations also triggers the downregulation of several genes of MMR, DNA replication, homologous recombination (HR), and cell growth inhibition [36]. Similarly, resveratrol modulates another set of genes up-regulated in different cancers, namely mini-chromosome maintenance [46] genes family, that are recruited to sites of DNA replication and facilitate DNA replication via helicase activity [38, 39]. In one recent study, resveratrol was shown to possess an anti-MCM effect by inducing a significant decrease in MCM2–MCM7 and MCM10 gene expression [36]. Interestingly, in the same study, the authors have found that resveratrol downregulated the expression of Rad51, BRCA1, and BRCA2 genes in MCF-7 breast cancer cells [36]. In contrast to this finding, study from our group has shown that resveratrol at lower concentrations (5–50 μmol/L) induces DNA damage in HNSCC cells but not in normal human epidermal keratinocytes, and causes S-phase arrest together with induction of Brca1 and gamma-H2AX foci in both in vitro and in vivo models [66]. This discrepancy may be due to the fact that effect of resveratrol in different systems depends upon the phenotype of cancer cells. It should be noted that BRCA1 plays an important role in the maintenance of genome stability, specifically through the HR pathway for double-strand DNA repair [61]. The results from a study suggest that resveratrol reduces the expression of these targets, blocks DNA repair mechanisms, which in turn leads to cell death [36].

19.1.5.7 Resveratrol: Clinical Studies

According to the US National Institute of Health, at present, several clinical trials are being conducted assessing the potential effects of resveratrol alone or in combination with other agents on human diseases, including cancer (http://clinicaltrials.gov). Most of these clinical studies are designed to assess the therapeutic efficacy of resveratrol as an antioxidant, anti-hypertensive, anti-diabetic, anti-inflammatory, cardiac problems, anti-erythema, metabolic syndrome, and anticancer agent [62]. Red wine rich in resveratrol, has long been thought to be beneficial in preventing heart disease by increasing the levels of "good" cholesterol and protecting against artery damage. In one of the randomized double blinded placebo controlled studies, 50 healthy adult smokers were given 500 mg^{-1} resveratrol or placebo for 30 days, followed by a wash-out period of 4 weeks to determine the effect of resveratrol on endothelial function. The investigators have reported that 4 weeks of resveratrol treatment, significantly reduced triglyceride concentration and increased antioxidant status [4]. In another study resveratrol was evaluated against systemic oxidative stress, which revealed that it significantly increases antioxidant properties [6]. Similarly, high level of resveratrol is shown to inhibit CYPs and interact with transporters which modify the CYPs-mediated metabolism [72]. In another randomized control trial by De Groote et al., the investigator evaluated the change in the gene expression after resveratrol (150 mg/day for 28 days) ingestion. The results from this study have demonstrated that genes affected include antioxidants, inflammatory, stress-response and those regulating cell growth [11]. Another published study with colon cancer patients receiving either 500 or 1,000 mg of resveratrol has shown that it decreases the expression of Ki-67 proliferation marker. In contrary, another study with same dose of resveratrol did not exhibit any significant difference [62]. In a study by Howells et al., hepatic cancer patients ($n=3$ receiving 5 g/day for 10–21 days) receiving resveratrol prior to surgery, have shown that apoptosis marker (cleaved caspase 3) was significantly increased in malignant hepatic tissues; however, there was no effect in the hepatic tissue levels of IGF-1, Ki-67, phospho-Akt, phospho-GSK, GSK3, phospho-ERK, ERK, β-catenin, survivin, Bcl-1, Bax, or PARP [27]. There are many mechanisms and pathways which are proven to be regulated by resveratrol in both preclinical and clinical studies, and to summarize all of these studies is beyond the scope of this chapter.

19.1.6 Conclusions and Prospect of Resveratrol in HNSCC

In HNSCC, the acute toxicity associated with conventional therapies is observed in 30–40 % of cases, thereby preventing from timely completion of therapy, affecting overall survival rates of the patients [64]. The results from different studies have revealed that only 10–20 % of patients are benefited from single targeted therapies, and remaining exhibit intrinsic resistance, thereby, limiting their clinical use [13, 31, 58]. Furthermore, >50 % of the head and neck patients undergoing surgical resection of primary tumors develop loco-regional recurrence, and the molecular genesis driving this phenomenon of secondary tumor development is largely unknown. Hence, from translational viewpoint, natural agents like resveratrol have demonstrated promising anti-tumor and cancer chemopreventive efficacy against different malignancies included head and neck cancer, by exerting pleiotropic effects on different signaling pathways as summarized in Fig. 19.1. Similarly, resveratrol is also shown to bind to DNA, inhibit DNA polymerase α/δ, modulate topoisomerase and telomerase activity, induce DNA damage, and alter DNA repair pathways (schematically shown in Fig. 19.1). Considering the importance of DNA repair mechanism in developing resistance to DNA damaging agents, often limiting the therapeutic efficacy of chemotherapy agents, it is suggestive that resveratrol might help to overcome drug resistance, reduce drug toxicity, and enhance efficacy when used in combination with other therapeutic agents. Furthermore, considering the selective toxicity to cancer cells, resveratrol might be a promising chemopreventive agent in a wide-range of HNSCC patients including smokers, alcoholics and FA patients; however, this assumption needs rigorous pre-clinical followed by clinical studies before resveratrol could "really" be beneficial to the identified cancer patient population.

Grant Support: Supported by grants to R. Sclafani from the Fanconi Anemia Research Foundation and the University of Colorado Cancer Center.

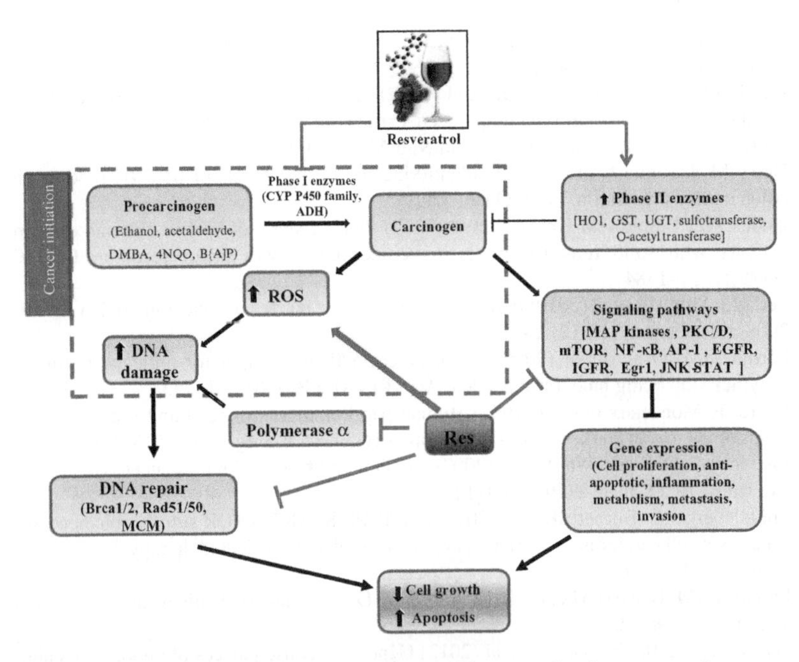

Fig. 19.1 Schematic representation illustrating the different molecular targets of resveratrol in head and neck squamous cell cancer

References

1. Aggarwal BB, Shishodia S (2006) Molecular targets of dietary agents for prevention and therapy of cancer. Biochem Pharmacol 71(10):1397–1421
2. Balbo S, Meng L et al (2012) Kinetics of DNA adduct formation in the oral cavity after drinking alcohol. Cancer Epidemiol Biomarkers Prev 21(4):601–608
3. Basso E, Fiore M et al (2013) Effects of resveratrol on topoisomerase II-alpha activity: induction of micronuclei and inhibition of chromosome segregation in CHO-K1 cells. Mutagenesis 28(3):243–248
4. Bo S, Ciccone G et al (2013) Anti-inflammatory and antioxidant effects of resveratrol in healthy smokers a randomized, double-blind, placebo-controlled, cross-over trial. Curr Med Chem 20(10):1323–1331
5. Brown GC, Borutaite V (2011) There is no evidence that mitochondria are the main source of reactive oxygen species in mammalian cells. Mitochondrion 12(1):1–4
6. Buonocore D, Lazzeretti A et al (2012) Resveratrol-procyanidin blend: nutraceutical and anti-aging efficacy evaluated in a placebocontrolled, double-blind study. Clin Cosmet Investig Dermatol 5:159–165
7. Choi S, Myers JN (2008) Molecular pathogenesis of oral squamous cell carcinoma: implications for therapy. J Dent Res 87(1):14–32
8. Connell PP, Jayathilaka K et al (2006) Pilot study examining tumor expression of RAD51 and clinical outcomes in human head cancers. Int J Oncol 28(5):1113–1119
9. Cui R, Kamatani Y et al (2009) Functional variants in ADH1B and ALDH2 coupled with alcohol and smoking synergistically enhance esophageal cancer risk. Gastroenterology 137(5): 1768–1775

10. Day GL, Blot WJ et al (1994) Second cancers following oral and pharyngeal cancers: role of tobacco and alcohol. J Natl Cancer Inst 86(2):131–137
11. De Groote D, Van Belleghem K et al (2012) Effect of the intake of resveratrol, resveratrol phosphate, and catechin-rich grape seed extract on markers of oxidative stress and gene expression in adult obese subjects. Ann Nutr Metab 61(1):15–24
12. Derry MM, Raina K et al (2013) Identifying molecular targets of lifestyle modifications in colon cancer prevention. Front Oncol 3:119
13. Dudek AZ, Lesniewski-Kmak K et al (2009) Phase I study of bortezomib and cetuximab in patients with solid tumours expressing epidermal growth factor receptor. Br J Cancer 100(9):1379–1384
14. Ferlay J, Shin HR et al (2010) Estimates of worldwide burden of cancer in 2008: GLOBOCAN 2008. Int J Cancer 127(12):2893–2917
15. Fortune JM, Osheroff N (2000) Topoisomerase II as a target for anticancer drugs: when enzymes stop being nice. Prog Nucleic Acid Res Mol Biol 64:221–253
16. Fouret P, Monceaux G et al (1997) Human papillomavirus in head and neck squamous cell carcinomas in nonsmokers. Arch Otolaryngol Head Neck Surg 123(5):513–516
17. Freedman ND, Schatzkin A et al (2007) Alcohol and head and neck cancer risk in a prospective study. Br J Cancer 96(9):1469–1474
18. Freudlsperger C, Burnett JR et al (2010) EGFR-PI3K-AKT-mTOR signaling in head and neck squamous cell carcinomas: attractive targets for molecular-oriented therapy. Expert Opin Ther Targets 15(1):63–74
19. Furgason JM, Bahassi M et al (2012) Targeting DNA repair mechanisms in cancer. Pharmacol Ther 137(3):298–308
20. Garaycoechea JI, Crossan GP et al (2012) Genotoxic consequences of endogenous aldehydes on mouse haematopoietic stem cell function. Nature 489(7417):571–575
21. Gatz SA, Wiesmuller L (2008) Take a break—resveratrol in action on DNA. Carcinogenesis 29(2):321–332
22. Gillison ML, Koch WM et al (2000) Evidence for a causal association between human papillomavirus and a subset of head and neck cancers. J Natl Cancer Inst 92(9):709–720
23. Hira A, Yabe H et al (2013) Variant ALDH2 is associated with accelerated progression of bone marrow failure in Japanese Fanconi anemia patients. Blood 122(18):3206–3209
24. Hoeijmakers JH (2001) Genome maintenance mechanisms for preventing cancer. Nature 411(6835):366–374
25. Hong WK, Endicott J et al (1986) 13-cis-retinoic acid in the treatment of oral leukoplakia. N Engl J Med 315(24):1501–1505
26. Howard JD, Lu B et al (2012) Therapeutic targets in head and neck squamous cell carcinoma: identification, evaluation, and clinical translation. Oral Oncol 48(1):10–17
27. Howells LM, Berry DP et al (2011) Phase I randomized, double-blind pilot study of micronized resveratrol (SRT501) in patients with hepatic metastases–safety, pharmacokinetics, and pharmacodynamics. Cancer Prev Res (Phila) 4(9):1419–1425
28. Irish JC, Bernstein A (1993) Oncogenes in head and neck cancer. Laryngoscope 103(1 Pt 1): 42–52
29. Ishiji T (2000) Molecular mechanism of carcinogenesis by human papillomavirus-16. J Dermatol 27(2):73–86
30. Jelski W, Kutylowska E et al (2011) Alcohol dehydrogenase (ADH) and aldehyde dehydrogenase (ALDH) as candidates for tumor markers in patients with pancreatic cancer. J Gastrointestin Liver Dis 20(3):255–259
31. Kundu SK, Nestor M (2012) Targeted therapy in head and neck cancer. Tumour Biol 33(3):707–721
32. Lachenmeier DW, Monakhova YB (2011) Short-term salivary acetaldehyde increase due to direct exposure to alcoholic beverages as an additional cancer risk factor beyond ethanol metabolism. J Exp Clin Cancer Res 30:3
33. Lanzilli G, Fuggetta MP et al (2006) Resveratrol down-regulates the growth and telomerase activity of breast cancer cells in vitro. Int J Oncol 28(3):641–648

34. Leemans CR, Braakhuis BJ et al (2011) The molecular biology of head and neck cancer. Nat Rev Cancer 11(1):9–22
35. Lenaz G (2001) The mitochondrial production of reactive oxygen species: mechanisms and implications in human pathology. IUBMB Life 52(3–5):159–164
36. Leon-Galicia I, Diaz-Chavez J et al (2013) Resveratrol induces downregulation of DNA repair genes in MCF-7 human breast cancer cells. Eur J Cancer Prev 22(1):11–20
37. Maacke H, Opitz S et al (2000) Over-expression of wild-type Rad51 correlates with histological grading of invasive ductal breast cancer. Int J Cancer 88(6):907–913
38. Maiorano D, Lutzmann M et al (2006) MCM proteins and DNA replication. Curr Opin Cell Biol 18(2):130–136
39. Markus MA, Marques FZ et al (2011) Resveratrol, by modulating RNA processing factor levels, can influence the alternative splicing of pre-mRNAs. PLoS One 6(12):e28926
40. Martins LA, Coelho BP et al (2014) Resveratrol induces pro-oxidant effects and time-dependent resistance to cytotoxicity in activated hepatic stellate cells. Cell Biochem Biophys 68(2):247–257
41. Mitra A, Jameson C et al (2009) Overexpression of RAD51 occurs in aggressive prostatic cancer. Histopathology 55(6):696–704
42. Nassiri-Asl M, Hosseinzadeh H (2009) Review of the pharmacological effects of Vitis vinifera (Grape) and its bioactive compounds. Phytother Res 23(9):1197–1204
43. Nguyen SA, Walker D et al (2012) mTOR inhibitors and its role in the treatment of head and neck squamous cell carcinoma. Curr Treat Options Oncol 13(1):71–81
44. O'Neil N, Rose A (2006) DNA repair. WormBook, pp 1–12
45. Ohba S, Fujii H et al (2009) Overexpression of GLUT-1 in the invasion front is associated with depth of oral squamous cell carcinoma and prognosis. J Oral Pathol Med 39(1):74–78
46. Paglin S, Hollister T et al (2001) A novel response of cancer cells to radiation involves autophagy and formation of acidic vesicles. Cancer Res 61(2):439–444
47. Pavlov YI, Shcherbakova PV et al (2006) Roles of DNA polymerases in replication, repair, and recombination in eukaryotes. Int Rev Cytol 255:41–132
48. Petti S (2009) Lifestyle risk factors for oral cancer. Oral Oncol 45(4–5):340–350
49. Poeta ML, Manola J et al (2007) TP53 mutations and survival in squamous-cell carcinoma of the head and neck. N Engl J Med 357(25):2552–2561
50. Radhakrishnan S, Reddivari L et al (2011) Resveratrol potentiates grape seed extract induced human colon cancer cell apoptosis. Front Biosci (Elite Ed) 3:1509–1523
51. Radojicic J, Zaravinos A et al (2012) HPV, KRAS mutations, alcohol consumption and tobacco smoking effects on esophageal squamous-cell carcinoma carcinogenesis. Int J Biol Markers 27(1):1–12
52. Rakoff-Nahoum S (2006) Why cancer and inflammation? Yale J Biol Med 79(3–4):123–130
53. Roy S, Kaur M et al (2007) p21 and p27 induction by silibinin is essential for its cell cycle arrest effect in prostate carcinoma cells. Mol Cancer Ther 6(10):2696–2707
54. Rusin P, Olszewski J et al (2009) Comparative study of DNA damage and repair in head and neck cancer after radiation treatment. Cell Biol Int 33(3):357–363
55. Sanchez-Alvarez R, Martinez-Outschoorn UE et al (2013) Ethanol exposure induces the cancer-associated fibroblast phenotype and lethal tumor metabolism: implications for breast cancer prevention. Cell Cycle 12(2):289–301
56. Sankaranarayanan R, Masuyer E et al (1998) Head and neck cancer: a global perspective on epidemiology and prognosis. Anticancer Res 18(6B):4779–4786
57. Seitz HK, Stickel F (2007) Molecular mechanisms of alcohol-mediated carcinogenesis. Nat Rev Cancer 7(8):599–612
58. Sharafinski ME, Ferris RL et al (2010) Epidermal growth factor receptor targeted therapy of squamous cell carcinoma of the head and neck. Head Neck 32(10):1412–1421
59. Siegel R, Naishadham D et al (2013) Cancer statistics, 2013. CA Cancer J Clin 63(1):11–30
60. Signorelli P, Ghidoni R (2005) Resveratrol as an anticancer nutrient: molecular basis, open questions and promises. J Nutr Biochem 16(8):449–466

61. Smith J, Tho LM et al (2010) The ATM-Chk2 and ATR-Chk1 pathways in DNA damage signaling and cancer. Adv Cancer Res 108:73–112
62. Tome-Carneiro J, Larrosa M et al (2013) Resveratrol and clinical trials: the crossroad from in vitro studies to human evidence. Curr Pharm Des 19(34):6064–6093
63. Trachootham D, Alexandre J et al (2009) Targeting cancer cells by ROS-mediated mechanisms: a radical therapeutic approach? Nat Rev Drug Discov 8(7):579–591
64. Tsao AS, Garden AS et al (2006) Phase I/II study of docetaxel, cisplatin, and concomitant boost radiation for locally advanced squamous cell cancer of the head and neck. J Clin Oncol 24(25):4163–4169
65. Turati F, Garavello W et al (2010) A meta-analysis of alcohol drinking and oral and pharyngeal cancers. Part 2: results by subsites. Oral Oncol 46(10):720–726
66. Tyagi A, Gu M et al (2011) Resveratrol selectively induces DNA damage, independent of Smad4 expression, in its efficacy against human head and neck squamous cell carcinoma. Clin Cancer Res 17(16):5402–5411
67. Tyagi A, Singh RP et al (2005) Resveratrol causes Cdc2-tyr15 phosphorylation via ATM/ATR-Chk1/2-Cdc25C pathway as a central mechanism for S phase arrest in human ovarian carcinoma Ovcar-3 cells. Carcinogenesis 26(11):1978–1987
68. Wang M, Chu H et al (2013) Molecular epidemiology of DNA repair gene polymorphisms and head and neck cancer. J Biomed Res 27(3):179–192
69. Weng CJ, Yen GC (2012) Chemopreventive effects of dietary phytochemicals against cancer invasion and metastasis: phenolic acids, monophenol, polyphenol, and their derivatives. Cancer Treat Rev 38(1):76–87
70. Woutersen RA, Appelman LM et al (1986) Inhalation toxicity of acetaldehyde in rats. III. Carcinogenicity study. Toxicology 41(2):213–231
71. Yokoyama A, Muramatsu T et al (1998) Alcohol-related cancers and aldehyde dehydrogenase-2 in Japanese alcoholics. Carcinogenesis 19(8):1383–1387
72. Chow, H. H., L. L. Garland, et al. (2010). "Resveratrol modulates drug- and carcinogen-metabolizing enzymes in a healthy volunteer study." Cancer Prev Res (Phila) 3(9): 1168–1175

Chapter 20
The Effects of Alcohol and Aldehyde Dehydrogenases on Disorders of Hematopoiesis

Clay Smith, Maura Gasparetto, Craig Jordan, Daniel A. Pollyea, and Vasilis Vasiliou

Abstract Hematopoiesis involves the orderly production of millions of blood cells per second from a small number of essential bone marrow cells termed hematopoietic stem cells (HSCs). Ethanol suppresses normal hematopoiesis resulting in leukopenia, anemia, and thrombocytopenia and may also predispose to the development of diseases such as myelodysplasia (MDS) and acute leukemia. Currently the exact mechanisms by which ethanol perturbs hematopoiesis are unclear. The aldehyde dehydrogenase (ALDH) gene family plays a major role in the metabolism of reactive aldehydes derived from ethanol in the liver and other organs. At least one of the ALDH isoforms, ALDH1A1, is expressed at high levels in HSCs in humans, mice, and other organisms. Recent data indicate that ALDH1A1 and possibly other ALDH isoforms may metabolize reactive aldehydes in HSCs and other hematopoietic cells as they do in the liver and elsewhere. In addition, loss of these ALDHs leads to perturbation of a variety of cell processes that may predispose HSCs to disorders in growth and leukemic transformation. From these findings, we suggest a hypothesis that the cytopenias and possible increased risk of MDS and acute leukemia in heavy alcohol users is due to polymorphisms in genes responsible for metabolism of alcohol derived reactive aldehydes and repair of their DNA adducts in HSCs and other hematopoietic cells. In the article, we will summarize the biological properties of hematopoietic cells and diseases related to ethanol consumption, discuss molecular characteristics of ethanol metabolism, and describe a model to explain how ethanol derived reactive aldehydes may promote HSC damage.

C. Smith, M.D. (✉)
Division of Hematology, University of Colorado,
Mail Stop F704, 1665 Aurora Court, Room 4320, Aurora, CO 80045, USA

Department of Medicine, University of Colorado, Aurora, CO, USA
e-mail: Clayton.Smith@ucdenver.edu

M. Gasparetto • C. Jordan • D.A. Pollyea
Department of Medicine, University of Colorado, Aurora, CO, USA

V. Vasiliou
Department of Environmental Health Sciences, Yale School of Public Health,
New Haven, CT, USA

V. Vasiliou et al. (eds.), *Biological Basis of Alcohol-Induced Cancer*,
Advances in Experimental Medicine and Biology 815,
DOI 10.1007/978-3-319-09614-8_20

Keywords ALDH • Alcohol • Hematopoiesis • Leukemia • Myelodysplasia

Abbreviations

AML	Acute myeloid leukemia
CFU	Colony forming unit
HSC	Hematopoietic stem cell
MDS	Myelodysplasia
RBC	Red blood cell

20.1 Hematopoiesis

The human body has at any one time about 10^{12} bone marrow cells from which all the mature white blood cells (including lymphocytes and myeloid cells), red blood cells (RBCs), and platelets that comprise the hematopoietic system are generated [1]. The lymphocytes include T-cells, B-cells, and NK-cells all of which play critical roles in immunity, tumor surveillance, autoimmune diseases, immunodeficiencies, organ and bone marrow transplantation as well as other processes. The T-, B-, and NK-cell lineages are comprised of a wide range of subpopulations and ultimately individual clonal populations. These share specific T-cell receptors, immunoglobulins, and NK inhibitory receptors, respectively, leading to tremendous diversity in the immune repertoire. Myeloid cells include neutrophils, monocytes, eosinophils, and basophils, which also play critical roles in immunity, inflammation, and other processes. In addition, to being very diverse, the hematopoietic system is extremely dynamic so that specific cell populations respond rapidly to a particular demand and then quickly return to baseline when the demand no longer exists. For example, with significant bleeding, RBC production quickly increases and then when the hematocrit returns to normal levels, RBC production declines back to baseline levels. Similarly with a bacterial infection, neutrophil production rapidly increases and once the infection is under control, also returns to baseline levels. All of the blood cell types have limited life spans, which range from hours to days for neutrophils, a week for platelets, 3–4 months for RBCs, and up to decades for some T- and B-lymphocytes. To keep up with this constant loss of blood cells, millions of new blood cells are produced each second and this process typically lasts for the entirety of human life spans unless it is interrupted by a hematopoietic disorder.

The process by which all of these diverse blood cell lineages are produced in such a robust, dynamic, and durable fashion is termed hematopoiesis [2]. Most of what we know about hematopoiesis is derived from murine studies but there is some data from large animal and human studies to believe that hematopoiesis proceeds through relatively conserved and consistent processes in these different organisms. The current model of hematopoiesis is summarized in Fig. 20.1. In this model, rare cells termed hematopoietic stem cells (HSCs) reside primarily in the bone marrow

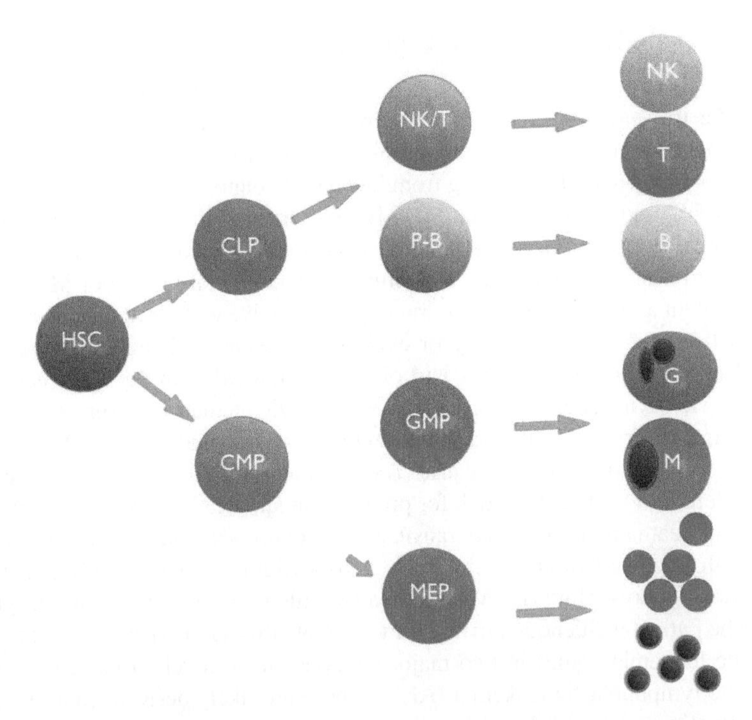

Fig. 20.1 Hematopoiesis. The current model of hematopoiesis is depicted below. In this model hematopoietic cells are derived from a small number of HSCs that progress through a series of progenitors that are characterized by progressive restriction of the ability to proliferate and generate different hematopoietic progenitors. *HSC* hematopoietic stem cells, *CLP* common lymphoid progenitor, *CMP* common myeloid progenitors, *G* granulocytes, *M* monocytes, *P-B* pre-B cells, *GMP* granulocyte macrophage progenitor, myelo-erythroid progenitor

and produce all of the various cell lineages through a successive series of intermediate cells termed progenitors [3]. The HSCs have tremendous replicative capacity and can produce all the progenitors, which ultimately yield lymphocytes, myeloid cells, RBCs, and platelets. In contrast, hematopoietic progenitors have progressively restricted capacity to divide and to yield multiple lineages. For example, there are common lymphoid progenitors that can produce T-, B-, and NK-cells but not myeloid cells. Downstream of these are more restricted T-cell progenitors that can produce T-cells but not B- or NK-cells. Ultimately, terminally differentiated cells are produced which can no longer divide and are pre-programmed to undergo apoptosis or be scavenged by various cells within the reticuloendothelial system.

Hematopoiesis is governed by a complex set of intrinsic coordinated gene expression processes that can be strongly influenced by a variety of environmental factors, which include cytokines, chemokines, direct cell contact with other cell types, and tissue oxygen levels. Particularly important is the bone marrow microenvironment, which is very complex with a wide array of growth, differentiation, migration, and other signals constantly bathing the HSCs and progenitors and influencing their growth and development [4].

20.2 Myelodysplasia and Acute Leukemia

Given the tremendous amount of cell turnover and the complex intracellular and environmental regulation of hematopoiesis, it is not surprising that there are a number of hematopoietic disorders ranging from low blood counts (cytopenias) to myelodysplasia (MDS) and acute leukemia. MDS is a spectrum of disorders of blood production, typically characterized by anemia, neutropenia, or thrombocytopenia and a tendency to transform into acute leukemia [5]. The natural history of MDS ranges from indolent and stable for years to rapidly progressive with death occurring within months from infections, bleeding or evolution to acute leukemia. Approximately 15,000 people, most in their 60s and older, develop MDS in the USA each year. The known environmental associations with MDS include exposure to benzene, radiation, and prior chemotherapy treatment of other malignancies but in most patients, there is not an obvious cause. The treatment of MDS depends on whether it is considered low risk or high risk for progression and death. Low risk patients may require no treatment or may need transfusions, growth factor administration, or treatment with hypomethylating agents such as azacytidine, which broadly effect gene expression patterns. High risk MDS is treated with allogeneic stem cell transplantation if the patient is fit enough to be able to tolerate this aggressive form of therapy.

Acute leukemia comes in two major versions, acute myeloid leukemia (AML) and acute lymphoblastic leukemia (ALL). These are likely derived from malignant transformation of myeloid and lymphoid hematopoietic progenitors, respectively, although some evidence indicates that in some cases, leukemic transformation may occur at the level of the HSC. Approximately 14,000 people develop AML each year and it typically effects people in their 60s or older although it can develop at any age. AML is also a wide range of diseases, driven by a variety of different molecular processes [6]. Most patients with AML are treated with chemotherapy and patients deemed at high risk for relapse but otherwise fit are further treated with an allogeneic stem cell transplant in an effort to reduce the chances of relapse. Overall about 30–40 % of patients with AML can be cured; however there is great variation depending on the genetic abnormalities in the AML and the age and fitness of the patient. In patients with high risk cytogenetic features, <5–10 % of patients can be cured [7]. In contrast, ALL occurs more commonly in children although it can occur in adults of any age as well [8]. Approximately 6,000 people are diagnosed per year with ALL, and in childhood the majority of patients are now cured with prolonged courses of multiple chemotherapy agents. However, in adults the cure rate is much lower, largely due to relapse of the disease. Patients deemed at high risk for relapse after treatment with chemotherapy alone are treated with allogeneic stem cell transplantation.

20.3 The Effects of Alcohol on Hematopoiesis

Heavy and prolonged alcohol use has long been known to cause cytopenias including anemia, leukopenia, and thrombocytopenia. In a group of heavy alcohol users, anemia occurred in around half and thrombocytopenia and leukopenia were also

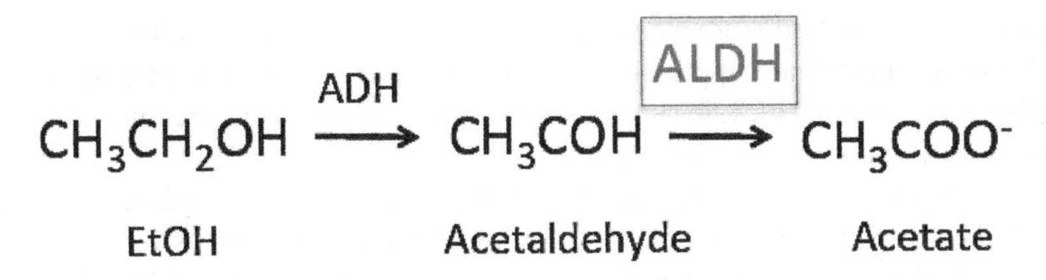

Fig. 20.2 Role of ALDHs in ethanol metabolism. *ALDH* aldehyde dehydrogenase, *ADH* alcohol dehydrogenase

common [9]. A variety of abnormalities in red cell size and structure have been observed including increased RBC cell volume and the development of abnormal RBCs including macrocytes, stomatocytes, and others. Morphologic analysis of bone marrow samples revealed vacuolization and other abnormalities indicative of toxic effects [9–11]. In addition to suppression of red cells, platelet, and neutrophils, alcohol may also suppress B-cell development and thereby impact immunity [12]. Ethanol and its primary reactive metabolite acetaldehyde (Fig. 20.2) also significantly suppress neutrophil and macrophage function and increase intracellular reactive oxygen species (ROS) [13].

The toxic effects of alcohol on hematopoiesis are compounded by other factors including concomitant medications and illnesses. For example, in a murine model of HIV, chronic administration of ethanol for 7 weeks resulted in a substantial reduction in erythroid and myeloid progenitors, which was worsened by expression of HIV proteins in hematopoietic cells or by treatment with the HIV drug azidothymidine (AZT). This study indicates that ethanol ingestion in HIV-1-infected individuals, particularly those on antiretroviral drugs, might increase bone marrow toxicity and contribute to HIV-1-associated hematopoietic impairment [14]. In part this may be due to the adverse effects of AZT on mitochondrial DNA replication that may be additive to the known deleterious effects on other aspects of mitochondrial function [15, 16]. Similarly, in mice exposed to ethanol and benzene, another marrow toxin, a variety of abnormalities in RBC production were noted [17].

The cytopenias from alcohol have been investigated in both animal models and in human in-vitro models of hematopoietic cell growth. Laboratory measurements of hematopoietic progenitor cell growth such as colony forming (CFU) assays demonstrate sensitivity to ethanol and an even more marked inhibition by acetaldehyde. Suppression of erythroid progenitors occurred more readily than suppression of myeloid progenitors [18], which may partially explain why anemia is more common than leukopenia in heavy alcohol users. In contrast, suppression of primitive multipotent progenitors was relatively minimal with concentrations of ethanol and acetaldehyde that suppressed more committed myeloid and erythroid progenitor cells. These findings indicated that suppression of hematopoiesis by alcohol occurred primarily at the level of committed hematopoietic progenitors, there was differential sensitivity of myeloid and erythroid progenitors to alcohol and that both ethanol and acetaldehyde played roles in the development of cytopenias following alcohol use [18]. The effects of ethanol specifically on platelet formation have also been investigated. Immature megakaryocyte progenitors as measured in CFU assays had a dose dependent suppression by ethanol. In contrast, acetaldehyde did not inhibit more mature megakaryocytes unless very high concentrations were used while they retained sensitivity to ethanol [19, 20].

In addition to suppressing baseline hematopoiesis, chronic ethanol may also block the dynamic response of hematopoiesis to demands such as infections. In a murine model, long-term ethanol exposure blocked the increase in blood granulocyte counts normally associated with intrapulmonary inoculation with *S. pneumoniae*. This suppression was associated with a significant decrease in bone marrow myeloid progenitor cell proliferation and was accompanied by increased STAT3 phosphorylation. In addition, ethanol increased G-CSF induced activation of the STAT3–p27^{Kip1} pathway in a mouse hematopoietic progenitor cell line while inhibiting growth proliferation and inducing cell cycle arrest [21]. Similarly, in a murine model of pneumonia, ethanol exposure impaired the pneumococcal-induced increase in neutrophil recruitment into the alveolar space, decreased bacterial clearance from the lung, and increased mortality from pneumonia. At least part of this was due to impaired proliferation of early marrow hematopoietic progenitors [22]. In addition to these direct effects of alcohol on hematopoiesis, indirect effects may occur as well. For example, ethanol has been shown to decrease the production of growth factors from T-cells which are important for normal hematopoiesis [23]. The exact molecular mechanisms by which ethanol and acetaldehyde suppress hematopoiesis remain unknown. Several lines of evidence implicate the formation of reactive aldehyde adducts with proteins, DNA, and other cellular constituents as part of the mechanism. In immunohistochemical analysis of erythroid cells from blood and marrow from patients with heavy ethanol use, acetaldehyde adducts occurred on both the cell membrane and intracellularly [24, 25].

Together, these findings demonstrate that ethanol and its metabolite acetaldehyde have wide ranging direct and indirect effects on hematopoiesis, immunity, and cell function that in part be mediated by adduct formation by acetaldehyde.

20.3.1 Alcohol and the Risk of MDS and Leukemia

Given the wide-spread deleterious effects of ethanol on the hematopoietic system, associations with the development of MDS and leukemia have been investigated in a number of epidemiologic and laboratory studies. In a case controlled Japanese study, alcohol consumption was associated with approximately a twofold increase in the risk of MDS and this risk was dose dependent [26]. An Italian population based case control study demonstrated that there may be an increased incidence of AML in people consuming high levels of alcohol [27]. A hospital case control study in Shanghai also demonstrated that alcohol may be a possible risk factor for the development of AML [28]. In contrast, in a US population based case control interview study, there was a mildly increased risk for MDS and leukemia in alcohol users, which did not reach statistical significance [29]. In patients in the USA with

AML and known pre-existing MDS, no association was found with alcohol use [30]. Similarly, in a large cohort study from the Netherlands no association with the development of AML was observed with ethanol while the association with MDS was not examined [31]. Together, these epidemiologic studies are inconsistent so that it currently remains unclear whether there is an association between alcohol use and development of MDS and leukemia.

In laboratory studies, there is more consistent and clear evidence that ethanol and acetaldehyde can lead to AML development. In mice genetically engineered to be deficient in the acetaldehyde metabolizing enzyme Aldh2 and the DNA repair enzyme Fancd2 (which is responsible in part for repairing acetaldehyde induced DNA crosslinks), ethanol exposure led to bone marrow failure and spontaneous AML development [32]. In follow-up studies, Aldh2 and Fancd2 deficient mouse HSCs were found to be the more sensitive to damage by reactive aldehydes than more committed progenitors. This further confirmed that reactive aldehydes from both exogenous and endogenous sources might play an important role in HSC biology. In addition, these studies indicated that congenital bone marrow disorders with a propensity to bone marrow failure and AML development, such as Fanconi's anemia (which is caused by deficiency of Fancd2), might also be caused by excess accumulation of reactive aldehydes [33]. In support of this possible explanation for human diseases of the marrow, Japanese Fanconi's anemia patients with polymorphisms in ALDH2 that affected enzymatic function had accelerated progression to bone marrow failure compared to their counterparts with normal aldehyde dehydrogenase (ALDH) activity [34]. These studies show that the affect of alcohol on hematopoietic cells including HSCs may be complex and at least in part be dependent on the presence or absence of deficiencies in DNA repair enzymes like FancD2 and acetaldehyde metabolizing enzymes including the ALDHs.

20.4 ALDH and HSCs

As discussed elsewhere in this volume, ALDHs play a key role in metabolizing alcohol (Fig. 20.2). ALDH activity is higher in HSCs compared to more mature progenitors and most differentiated hematopoietic cells. Based on this distinction, a fluorescent ALDH substrate termed Aldefluor has been developed that is useful in the isolation of human hematopoietic cells as well as stem cells in other tissues and species [35]. In studies investigating the biology of ALDHs in HSCs and other hematopoietic cells, Aldefluor was found to be primarily a substrate for the ALDH1A1 isoform and to a lesser extent the ALDH3A1 isoform while other studies have also found it is a substrate for ALDH2 as well [36, 37]. Surprisingly, however, loss of ALDH1A1 in hematopoietic cells did not lead to any clear phenotype [38]. However, loss of ALDH1A1 led to an increase in expression in the ALDH3A1 isoform, presumably as a compensatory mechanism [36]. When mice deficient in both ALDH1A1 and ALDH3A1 were examined, there were a wide range of hematopoietic defects under baseline conditions including a block in B-cell development as well as abnormalities in cell cycling, intracellular signaling, and gene expression along with reduced numbers of HSCs. Intriguingly, hematopoietic cells from these mice had increased ROS and reduced metabolism of the reactive aldehyde 4-hydroxynonenal and sensitivity to DNA damage from these agents [36]. Together these findings further support a role for ALDHs in HSCs in metabolizing endogenous and exogenous reactive aldehydes such as acetaldehyde derived from ethanol.

20.5 A Possible Model for Ethanol Induced Bone Marrow Damage and Development of MDS/AML

Clearly, the understanding of the effects of alcohol on hematopoiesis and the development of cytopenias, MDS, and AML is at a very early stage with much that is unclear. However it is possible that there is an important but complex association between ethanol derived reactive aldehydes and bone marrow disease states. One testable hypothesis that links the various findings discussed above is that ethanol derived acetaldehyde forms protein and DNA adducts but that these species are more prevalent (and therefore more damaging) in people with reduced ALDH activity or reduced DNA adduct repair activity (see Fig. 20.3). This would be consistent with the prior findings that hematopoietic progenitors appear to be more sensitive to ethanol and acetaldehyde than HSCs, possibly because HSCs normally have high levels of ALDH activity that effectively metabolize acetaldehyde to acetate so that DNA and protein adducts do not accumulate. Similarly with high levels of DNA repair activity in HSCs, any DNA adducts that form may be readily corrected. However people with insufficient ALDH or DNA repair activity in HSCs may be more susceptible to accumulating DNA damage from ethanol-derived acetaldehyde. Ultimately the DNA damage could lead to mutagenesis and the development of MDS and AML.

There are known polymorphisms in the DNA repair proteins that predispose to leukemia formation as well as polymorphisms in ALDHs that may increase risk of other cancers [39, 40]. If the above model is correct, then people with these polymorphisms could be specifically vulnerable to the effects of ethanol. These relatively small populations may not be obvious in large epidemiologic studies, where larger, less susceptible populations may dilute their contribution to the overall risk assessments. This hypothesis could be tested both in the laboratory and in future epidemiologic studies. For example, in the laboratory, it will be interesting to test whether mice engineered to have various combinations of ALDH deficiencies with and without DNA repair deficiencies are prone to MDS and AML development following exposure to ethanol. In population studies, it would be useful to test persons with MDS and AML for known polymorphisms in ALDHs and DNA repair genes along with ethanol use and look for associations. If these studies confirm that there are defined populations specifically susceptible to hematopoietic damage by alcohol, then perhaps simple clinical tests can be developed to readily identify these persons in order to help them mitigate their risks from ethanol use.

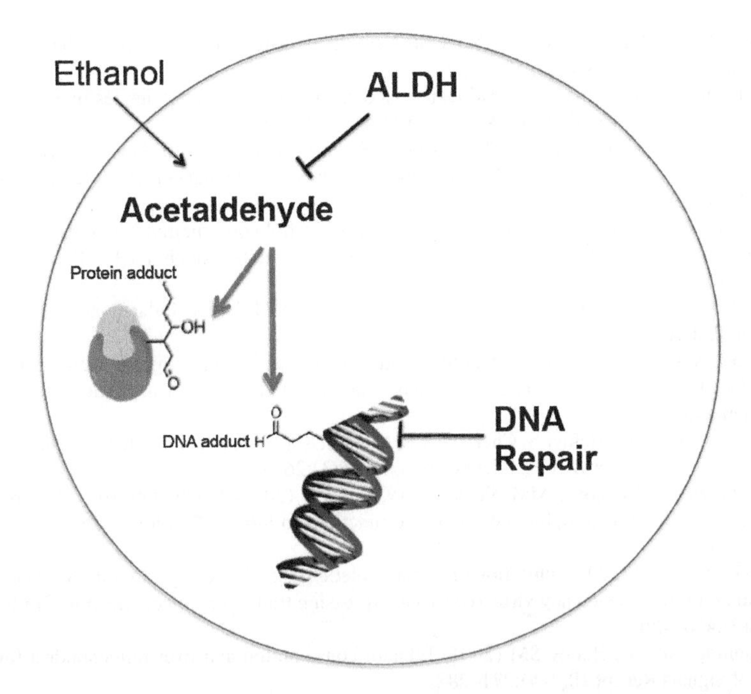

Fig. 20.3 Model for the possible role of ALDHs in hematopoietic progenitor cells. In this model, deficiencies in ALDH or DNA repair activity in individuals with either polymorphisms or some other abnormality in function will be more susceptible to DNA and protein damage leading to widespread cytopenias and a propensity to develop MDS and AML. These effects may be potentiated in HSCs where particularly high levels of ALDH activity are found which normally would protect these critical cells from acetaldehyde produced from ethanol

In summary, the intersection between alcohol, reactive aldehydes, HSCs, and hematopoiesis may be complex but improved understanding of these relationships could lead to new approaches to preventing alcohol related blood disorders as well as new insights into the biology of the blood system and stem cells.

References

1. Bianconi E et al (2013) An estimation of the number of cells in the human body. Ann Hum Biol 40(6):463–471
2. Doulatov S, Notta F, Laurenti E, Dick JE (2012) Hematopoiesis: a human perspective. Cell Stem Cell 10(2):120–136
3. Seita J, Weissman IL (2010) Hematopoietic stem cell: self-renewal versus differentiation. Wiley Interdiscip Rev Syst Biol Med 2(6):640–653
4. Urao N, Ushio-Fukai M (2013) Redox regulation of stem/progenitor cells and bone marrow niche. Free Radic Biol Med 54:26–39
5. Natelson EA, Pyatt D (2013) Acquired myelodysplasia or myelodysplastic syndrome: clearing the fog. Adv Hematol 2013:309637

6. Stein EM, Tallman MS (2012) Novel and emerging drugs for acute myeloid leukemia. Curr Cancer Drug Targets 12(5):522–530
7. O'Donnell MR (2013) Risk stratification and emerging treatment strategies in acute myeloid leukemia. J Natl Compr Canc Netw 11(5 Suppl):667–669
8. Mathisen MS, Kantarjian H, Thomas D, O'Brien S, Jabbour E (2013) Acute lymphoblastic leukemia in adults: encouraging developments on the way to higher cure rates. Leuk Lymphoma 54(12):2592–2600
9. Latvala J, Parkkila S, Niemela O (2004) Excess alcohol consumption is common in patients with cytopenia: studies in blood and bone marrow cells. Alcohol Clin Exp Res 28(4): 619–624
10. Ballard HS (1980) Alcohol-associated pancytopenia with hypocellular bone marrow. Am J Clin Pathol 73(6):830–834
11. Budde R, Hellerich U (1995) Alcoholic dyshaematopoiesis: morphological features of alcohol-induced bone marrow damage in biopsy sections compared with aspiration smears. Acta Haematol 94(2):74–77
12. Wang H, Zhou H, Mahler S, Chervenak R, Wolcott M (2011) Alcohol affects the late differentiation of progenitor B cells. Alcohol Alcohol 46(1):26–32
13. Vrsalovic M, Vrsalovic MM, Presecki AV, Lukac J (2007) Modulating role of alcohol and acetaldehyde on neutrophil and monocyte functions in vitro. J Cardiovasc Pharmacol 50(4): 462–465
14. Prakash O et al (2001) Inhibition of hematopoietic progenitor cell proliferation by ethanol in human immunodeficiency virus type 1 tat-expressing transgenic mice. Alcohol Clin Exp Res 25(3):450–456
15. Cunningham CC, Bailey SM (2001) Ethanol consumption and liver mitochondria function. Biol Signals Recept 10(3–4):271–282
16. Samuels DC (2006) Mitochondrial AZT metabolism. IUBMB Life 58(7):403–408
17. Baarson KA, Snyder CA (1991) Evidence for the disruption of the bone marrow microenvironment by combined exposures to inhaled benzene and ingested ethanol. Arch Toxicol 65(5):414–420
18. Meagher RC, Sieber F, Spivak JL (1982) Suppression of hematopoietic-progenitor-cell proliferation by ethanol and acetaldehyde. N Engl J Med 307(14):845–849
19. Cooper GW, Dinowitz H, Cooper B (1984) The effects of administration of ethyl alcohol to mice on megakaryocyte and platelet development. Thromb Haemost 52(1):11–14
20. Levine RF, Spivak JL, Meagher RC, Sieber F (1986) Effect of ethanol on thrombopoiesis. Br J Haematol 62(2):345–354
21. Siggins RW et al (2011) Alcohol suppresses the granulopoietic response to pulmonary Streptococcus pneumoniae infection with enhancement of STAT3 signaling. J Immunol 186(7):4306–4313
22. Raasch CE et al (2010) Acute alcohol intoxication impairs the hematopoietic precursor cell response to pneumococcal pneumonia. Alcohol Clin Exp Res 34(12):2035–2043
23. Imperia PS, Chikkappa G, Phillips PG (1984) Mechanism of inhibition of granulopoiesis by ethanol. Proc Soc Exp Biol Med 175(2):219–225
24. Balbo S et al (2012) Time course of DNA adduct formation in peripheral blood granulocytes and lymphocytes after drinking alcohol. Mutagenesis 27(4):485–490
25. Latvala J, Parkkila S, Melkko J, Niemela O (2001) Acetaldehyde adducts in blood and bone marrow of patients with ethanol-induced erythrocyte abnormalities. Mol Med 7(6):401–405
26. Ido M et al (1996) A case-control study of myelodysplastic syndromes among Japanese men and women. Leuk Res 20(9):727–731
27. Gorini G et al (2007) Alcohol consumption and risk of leukemia: a multicenter case-control study. Leuk Res 31(3):379–386
28. Wong O, Harris F, Yiying W, Hua F (2009) A hospital-based case-control study of acute myeloid leukemia in Shanghai: analysis of personal characteristics, lifestyle and environmental risk factors by subtypes of the WHO classification. Regul Toxicol Pharmacol 55(3): 340–352

29. Brown LM et al (1992) Alcohol consumption and risk of leukemia, non-Hodgkin's lymphoma, and multiple myeloma. Leuk Res 16(10):979–984
30. Crane MM, Keating MJ (1991) Exposure histories in acute nonlymphocytic leukemia patients with a prior preleukemic condition. Cancer 67(8):2211–2214
31. Heinen MM et al (2013) Alcohol consumption and risk of lymphoid and myeloid neoplasms: results of the Netherlands cohort study. Int J Cancer 133(7):1701–1712
32. Langevin F, Crossan GP, Rosado IV, Arends MJ, Patel KJ (2011) Fancd2 counteracts the toxic effects of naturally produced aldehydes in mice. Nature 475(7354):53–58
33. Garaycoechea JI et al (2012) Genotoxic consequences of endogenous aldehydes on mouse haematopoietic stem cell function. Nature 489(7417):571–575
34. Hira A et al (2013) Variant ALDH2 is associated with accelerated progression of bone marrow failure in Japanese Fanconi anemia patients. Blood 122(18):3206–3209
35. Storms RW et al (1999) Isolation of primitive human hematopoietic progenitors on the basis of aldehyde dehydrogenase activity. Proc Natl Acad Sci U S A 96(16):9118–9123
36. Gasparetto M et al (2012) Aldehyde dehydrogenases are regulators of hematopoietic stem cell numbers and B-cell development. Exp Hematol 40(4):318–329.e2
37. Gasparetto M et al (2012) Varying levels of aldehyde dehydrogenase activity in adult murine marrow hematopoietic stem cells are associated with engraftment and cell cycle status. Exp Hematol 40(10):857–866.e5
38. Levi BP, Yilmaz OH, Duester G, Morrison SJ (2009) Aldehyde dehydrogenase 1a1 is dispensable for stem cell function in the mouse hematopoietic and nervous systems. Blood 113(8):1670–1680
39. Hernandez-Boluda JC et al (2012) A polymorphism in the XPD gene predisposes to leukemic transformation and new nonmyeloid malignancies in essential thrombocythemia and polycythemia vera. Blood 119(22):5221–5228
40. Minegishi Y et al (2007) Susceptibility to lung cancer and genetic polymorphisms in the alcohol metabolite-related enzymes alcohol dehydrogenase 3, aldehyde dehydrogenase 2, and cytochrome P450 2E1 in the Japanese population. Cancer 110(2):353–362

Chapter 21
The Effect of Alcohol on Sirt1 Expression and Function in Animal and Human Models of Hepatocellular Carcinoma (HCC)

Kyle J. Thompson, John R. Humphries, David J. Niemeyer, David Sindram, and Iain H. McKillop

Abstract *Introduction*: Chronic heavy alcohol use is an independent risk factor for developing hepatocellular carcinoma (HCC). Sirtuin-1 (Sirt1) is a NAD$^+$-dependent deacetylase implicated in alcohol-induced liver injury and overexpressed in human HCC. The aims of this study were to investigate Sirt1 expression in mouse models of HCC and chronic EtOH-feeding, and in human HCC cells expressing alcohol metabolizing enzymes.

Methods: C57BL/6 and B6C3 mice were injected with DEN and randomized to receive drinking water (DW) or EtOH-DW for 8 weeks at 36 weeks. Livers were analyzed for HCC incidence, size, and Sirt1 expression. In parallel, human HepG2 cells or HepG2 cells transfected to express ADH and CYP2E1 (VL-17a cells) were treated with alcohol (0–50 mM) and/or CAY10591 (Sirt1 activator) or EX-527 (Sirt1 inhibitor).

Results: B6C3 mice exhibited significantly elevated Sirt-1 expression vs. C57BL/6 mice and Sirt-1 expression was elevated in HCC vs. non-tumor liver. However, EtOH-feeding did not further affect Sirt1 expression in mice of either background despite EtOH increasing HCC size and incidence in B6C3 mice. In vitro, EtOH

This work was presented in part as a poster presentation at the 2nd NIH-NIAAA meeting on Alcohol and Cancer held in Breckenridge, CO (May 11th–15th, 2013).

K.J. Thompson, Ph.D. • D.J. Niemeyer, M.D. • D. Sindram, M.D. Ph.D.
Department of Surgery, Carolinas Medical Center, Charlotte, NC 28203, USA

J.R. Humphries, B.S.
Department of Biology, UNC Charlotte, Charlotte, NC 28223, USA

I.H. McKillop, Ph.D. (✉)
Department of Surgery, Carolinas Medical Center, Charlotte, NC 28203, USA

Department of Biology, UNC Charlotte, Charlotte, NC 28223, USA

Department of Surgery, Cannon Research Center,
Suite 402, 1000 Blythe Boulevard, Charlotte, NC 28203, USA
e-mail: iain.mckillop@carolinashealthcare.org

V. Vasiliou et al. (eds.), *Biological Basis of Alcohol-Induced Cancer*,
Advances in Experimental Medicine and Biology 815,
DOI 10.1007/978-3-319-09614-8_21

treatment significantly decreased Sirt1 expression in VL-17a-cells and stimulated cell growth, an effect not observed in HepG2 cells. The effects of ethanol on VL-17a cells were abrogated by pretreatment with CAY10591.

Conclusions: Sirt1 expression correlates with susceptibility to form HCC, but is not further affected by alcohol feeding. Conversely Sirt1 expression and function is impacted by alcohol metabolism capacity in human HCC cells in vitro. These discrepancies in Sirt1-expression-function may reflect differences in enzyme expression compared to activity, or more complex changes in genes targeted for deacetylation during tumor progression in the setting of chronic alcohol ingestion.

Keywords Hepatocellular carcinoma • Sirt1 • Ethanol • HepG2

21.1 Introduction

Primary tumors of the liver represent the third leading cause of cancer-related mortality worldwide and are among the most rapidly increasing types of cancer diagnosed [1]. Hepatocellular carcinoma (HCC) comprises approximately 80 % of all primary liver tumors and their incidence correlates strongly with exposure to known risk factors, including viral hepatitis B and C infection, aflatoxin exposure, obesity, and chronic alcohol consumption [2]. Despite advances in diagnosis and therapy for HCC late presentation and a lack of transplantable organs results in poor prognoses, 5-year survival for all HCC cases remaining at a dismal 15–20 % [1].

Alcohol consumption has long been noted to be associated with development of cancers, with 3.6 % of all cancers attributable to alcohol consumption, and men experiencing a 3:1 higher ratio of alcohol-related cancers than women [3, 4]. Moderate to high alcohol consumption (≤80 g/day) is a risk factor for a range of cancers, including the oral cavity, larynx, esophagus, and liver among other organs [5]. Although heavy consumption is most commonly associated with cancer risk, reports also suggest light alcohol consumption (<12.5 g/day; equivalent to approximately 1 drink/day) is also associated with elevated risk for oral-pharyngeal, breast, and esophageal cancers, but not liver cancer [6]. Furthermore, it is widely reported that alcohol acts synergistically with other established HCC risk factors [7, 8].

Following consumption, the majority of alcohol (~80 %) is absorbed in the small intestine and metabolized primarily by hepatocytes in the liver. Following short-term and acute alcohol consumption, alcohol dehydrogenase (ADH) metabolizes the majority of alcohol to acetaldehyde with concomitant reduction of nicotinamide adenine dinucleotide (NAD+) to NADH [9]. Acetaldehyde is then metabolized to acetate by acetaldehyde dehydrogenase (ALDHs), with acetate being further metabolized to acetyl-coA and entry into the citric acid cycle [10]. Following chronic, heavy alcohol consumption, induction of CYP2E1 occurs. In addition to acetaldehyde production, CYP2E1 metabolism of alcohol also yields reactive oxygen species (ROS), including hydroxyethyl radicals [9, 11]. Alcohol-dependent production of ROS and acetaldehyde can in turn lead to peroxidation and/or adduct formation occurring in multiple cellular components including DNA (e.g., N^2-ethyl-2′-deoxyguanosine, N^2-propano-2′-deoxyguanosine), protein, and lipids (e.g., malondialdehyde and 4-hydroxynonenal) [12, 13].

Alcohol metabolism can have additional consequences through the disruption of NAD+ reduction states. Sirtuin-1 (Sirtl) is an NAD+-dependent class III histone deacetylase that acts as a metabolic "sensor," and is implicated in regulating a wide range of intracellular processes including aging, DNA repair, apoptosis, inflammation, and energy production and storage [14]. Sirtl activity is regulated by the imbalance of the NAD+-NADH ratio, and accumulation of the NAD+ degradation product, nicotinamide (NAM) [15]. Impairment of Sirtl activity leads to sustained acetylation of Sirtl target transcription factors which acts to either increase (e.g., sterol regulatory element-binding protein-1c (SREBP-1c), nuclear factor kappa-light-chain-enhancer of activated B cells (NF-κB), and p53), or repress their activity (e.g., Peroxisome proliferator-activated receptor gamma co-activator 1-α (PGC-1α)) [16]. Within the liver, these actions can result in the increased lipogenesis and inflammation, concomitant with decreased gluconeogenesis, promoting steatosis and liver injury [16]. Furthermore, alcohol treatment has been shown to diminish Sirtl protein expression and activity, as well as the expression of known downstream Sirtl targets [17–19].

However, the role of Sirtl in HCC suggests increased Sirtl activity, may be, promote liver tumor progression and studies report a correlation between Sirtl overexpression in HCC patients and poor prognosis. Similarly, in an in vivo xenograft model of HCC tumor growth was impaired by inhibition of Sirtl [20–22]. These in vivo studies are further complimented by cell culture studies that report decreased cell proliferation following inhibition of Sirtl [20, 23]. The aims of the current studies were to examine the effect of alcohol and alcohol metabolism on expression and function of Sirtl during HCC progression using in vivo and in vitro models of hepatocarcinogenesis [24, 25].

21.2 Materials and Methods

Assurances: Male C57BL/6 (21 days) and B6C3 (21 days) were purchased from Jackson Laboratories (Bar Harbor, ME). All studies were approved by the Institutional Animal Care and Use Committee and conformed to the NIH Guidelines for Care and Use of Laboratory Animals.

Materials: Diethylnitrosamine (DEN) and ethanol were purchased from Sigma-Aldrich (St. Louis, MO). Antibodies against ADH, aldehyde dehydrogenase 1/2 (ALDH), cytochrome P450 2E1 (CYP 2E1), and β-actin were purchased from Santa Cruz Biotechnology (Santa Cruz, CA). An antibody against sirtuin-1 (Sirtl) was purchased from Abcam (Cambridge, MA). The Sirtl inhibitor EX-527 was purchased from Cayman Chemical (Ann Arbor, MI). Dulbecco's Modified Eagle Medium (DMEM), Penicillin/Streptomycin, Zeocin, G-418 and CyQuant Assay were purchased from Life Technologies (Grand Island, NY). Heat-inactivated Fetal

Bovine Serum (FBS) was purchased from Gemini Bio Products (West Sacramento, CA). EnzyChrom Ethanol Assay Kit was purchased from BioAssay Systems (Hayward, CA).

Animal models: The effects of chronic ethanol feeding on HCC progression were investigated in two strains of mice. Male C57BL/6 mice were injected intraperitoneally with either 5 mg/kg DEN (in 100 μL of sterile olive oil) or an equal volume of sterile olive oil alone at 21-day-old. In a parallel study series, male B6C3 mice were injected with either 3 mg/kg DEN in 100 μL of sterile olive oil, or an equal volume of sterile olive oil alone when 21-day-old. Mice were maintained with ad libitum access to standard rodent chow and on a 12 h light/dark cycle. At 35 weeks mice were randomized and weaned onto a 10 %/20 % (v/v) ethanol-drinking water (EtOH-DW) regiment (alternate days) or maintained on drinking water (DW) alone. At 42 weeks, mice were euthanized and whole blood was collected in Sodium-EDTA Vacutainers (BD, Franklin Lakes, NJ), and plasma separated by centrifugation. The liver was resected and examined grossly for visible lesions and representative sections taken from the left, right, median, and caudate lobes and either placed in neutral-buffered formalin for histology/immunohistochemistry or snap frozen in liquid nitrogen and stored (−80 °C) for subsequent analyses.

Immunohistochemistry: Tissue sections (4 μm) were de-paraffinized, rehydrated, subjected to antigen retrieval, and blocked with donkey serum in a humidified chamber [24]. Sections were incubated with a rabbit anti-Sirt1 antibody (1:1,000 dilution) and detection performed using an anti-goat IgG probe polymer kit with Betazoid DAB and counterstained with methyl green. Five random fields/section were examined microscopically (×200), photographed, and blind scored for Sirt1-positive staining.

Cell culture: The human HepG2 hepatoma cell line was purchased from ATCC (Manassas, VA). The HepG2 E47 cell line (E47) overexpresses human CYP 2E1 and was obtained as a generous gift from Dr. Arthur Cederbaum (Mt Sinai School of Medicine, NY, NY) [26]. The HegG2 VA-13 and VL-17a cell lines, that overexpress human ADH and human ADH/CYP 2E1 respectively, were obtained as a generous gift from Dr. Dahn Clemens (University of Nebraska Medical Center, Omaha, NE) [27, 28]. Cells were maintained in DMEM supplemented with 10 % (v/v) FBS and Penicillin/Streptomycin antibiotics at 37 °C in a 5 % CO_2 incubator. Media for E47 cells was supplemented with G-418 (400 μg/mL), VA-13 cells with Zeocin (400 μg/mL), and VL-17a with G-418 and Zeocin (400 μg/mL). For proliferation experiments, cells were seeded at 5,000 cells/well in 96-well plates and maintained overnight in growth media; however, selective antibiotics G-418 and Zeocin were withdrawn. Cells were then subjected to serum depletion for 24 h with 0.1 % (v/v) FBS.

To determine the effects of Sirt1 activity on EtOH-dependent HCC cell proliferation, cells were pretreated for 1 h with either a Sirt1 activator (CAY; CAY10591, 5 μM) or Sirt1 inhibitor (EX; EX-527, 5 μM) [29, 30]. Cells were then treated with or without EtOH (25 mM) for 24 h. One percent volume to volume FBS (FBS) was utilized as a positive control for proliferation. Proliferation was measured using the CyQuant assay was performed according to the manufacturer's specifications as previously [31].

Ethanol assay: Cells seeded in T25 screw-cap flasks were subjected to serum depletion in 0.1 % (v/v) FBS for 24 h. Cells were then treated with 50 mM EtOH for 24 h. A flask containing growth media plus 50 mM EtOH alone was used as a control for evaporation of EtOH. Following time point, 100 μL aliquots of growth media were taken and immediately flash-frozen and stored in screw-cap vials until EtOH assay was performed. EtOH assay was performed according to the manufacturer's specifications as previously [25].

Western blotting: Following treatment, cells were washed with phosphate-buffered saline (PBS, 4 °C) and whole-cell lysates were prepared using radioimmunoprecipitation assay buffer (RIPA; 1 % (v/v) NP-40, 0.5 % (v/v) deoxycholate, 0.1 % (w/v) SDS, 0.5 mM phenylmethylsulfonyl fluoride, 0.05 mM Na_3VO_4, 2 μg/mL aprotinin in PBS). For whole liver, lysates were homogenized in RIPA buffer and centrifuged to remove non-soluble debris (13,000 × g, 5 min, 4 °C). Protein concentrations were determined using Bradford Assay, pooled, and Western blot analysis and detection performed as previously reported [32]. Primary antibodies were diluted 1:1000 in 5 % (v/v) non-fat dry milk diluted in TBS Tween (NFDM-TBST) and incubated overnight (4 °C). Secondary antibodies were diluted 1:5,000 in 5 % NFDM-TTBS for 1 h at room temperature.

Statistical analysis: Data are expressed as mean ± SEM. Statistical analysis was performed using one-way ANOVA with Tukey's post-test or Fisher's Exact Test using GraphPad Prism V5.0b (La Jolla, CA). A p value < 0.05 was considered significant.

21.3 Results

Sirt1 expression in mouse models of EtOH-feeding and HCC: The expression and localization of Sirt1 was undertaken in two mouse models of HCC progression in the setting of chronic EtOH-feeding. DEN administration induced tumors in 85.7 % of B6C3 mice compared to 60 % of C57BL/6 mice. Induction and maintenance on an EtOH-DW regiment at 36 weeks of age further increased tumor incidence in B6C3 mice (92.9 %) (Fig. 21.1a). In contrast, we measured a statistically significant ($p < 0.05$) decrease in tumor incidence in C57BL/6 mice (44.4 %) maintained on EtOH-DW (Fig. 21.1a). No tumors were observed in mice injected with vehicle (sterile olive oil, i.p.), either with or without EtOH-DW, in either strain over the 44-week experimental period (Fig. 21.1a). Gross analysis of liver tissue revealed livers from DW-EtOH mice and from C57BL/6 mice had relatively small lesions as compared to similarly treated B6C3 mice. Induction and maintenance on the EtOH-DW regimen did not affect tumor size in C57BL/6 mice. In contrast 8 weeks

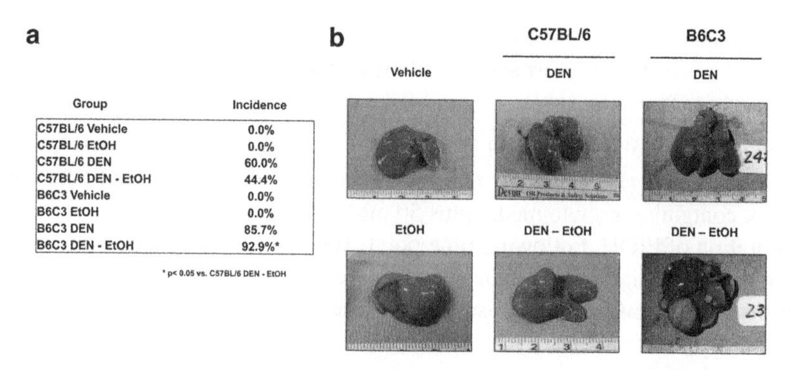

Group	Incidence
C57BL/6 Vehicle	0.0%
C57BL/6 EtOH	0.0%
C57BL/6 DEN	60.0%
C57BL/6 DEN - EtOH	44.4%
B6C3 Vehicle	0.0%
B6C3 EtOH	0.0%
B6C3 DEN	85.7%
B6C3 DEN - EtOH	92.9%*

* p< 0.05 vs. C57BL/6 DEN - EtOH

Fig. 21.1 Tumor incidence in two models of HCC and EtOH consumption. (**a**) Twenty-one-day-old male C57BL/6 or B6C3 mice were injected with vehicle (Veh; 100 μL sterile olive oil) or DEN (5 mg/kg, C57BL/6; 3 mg/kg B6C3). At 35 weeks mice were weaned onto either 10 %/20 % (v/v) alternate day ethanol in drinking water (EtOH) or drinking water alone (H_2O) until 42 weeks. Tumors formed in only 60 % of C57BL/6 mice, an effect that was mildly suppressed by EtOH (44.4 %). Tumor incidence was 85.7 % and 92.9 % respectively for B6C3 mice injected with DEN or DEN plus EtOH regimen, which was significantly higher than C57BL/6 mice. *$p<0.05$ vs. C57BL/6 EtOH DEN (**b**) Tumors were much larger in B6C3 DEN mice than C57BL/6 mice, an effect exacerbated by EtOH in B6C3 animals alone. No tumors were identified in Veh-injected mice

of ethanol consumption significantly promoted tumor progression in similarly DEN-initiated B6C3 mice (Fig. 21.1b). Immunohistological staining of liver sections demonstrated significantly increased Sirt1 nuclear staining in hepatocytes of B6C3 mice treated with DEN compared to vehicle-injected mice ($p<0.05$); however, EtOH-DW regimen had no additional effect (Fig. 21.2a, b). Western blot analysis of Sirt1 from pooled liver extracts showed poor correlation with IHC results, with the strongest Sirt1 expression being measured in Veh-DW B6C3 mice and DEN-DW C57BL/6 mice (Fig. 21.2c, d).

EtOH treatment on Sirt1 expression in HCC cell lines: To validate expression of EtOH-metabolizing enzymes, Western blot analysis was performed on lysates from HepG2, VA-13, E47 and VL-17a cell lines. As expected, ADH expression was strongest in VA-13 and VL-17a cells, and CYP 2E1 expression was strongest in E47 and VL-17a cells, although weak expression was detected in HepG2 and VA-13 cell lines (Fig. 21.3a).

Cells were treated with 50 mM EtOH for 24 h to evaluate ethanol-metabolizing capacity. HepG2 cells had minimal EtOH-metabolizing capacity while ADH-expressing VA-13 and VL-17a cells had highest EtOH-metabolizing capacity, and E47 cells displayed intermediate capacity (Fig. 21.3b).

Treatment of cells with increasing concentrations of EtOH (0–100 mM, 24 or 48 h) demonstrated no significant change in Sirt1 protein expression in HepG2 cells (Fig. 21.3b). However, treatment of VA-13 (24 and 48 h), E47 (48 h), and VL-17a cell lines (48 h) demonstrated decreased Sirt1 expression in a dose-dependent manner, increased effect (decreased Sirt1 protein) being detected at increasing doses of EtOH (Fig. 21.3c).

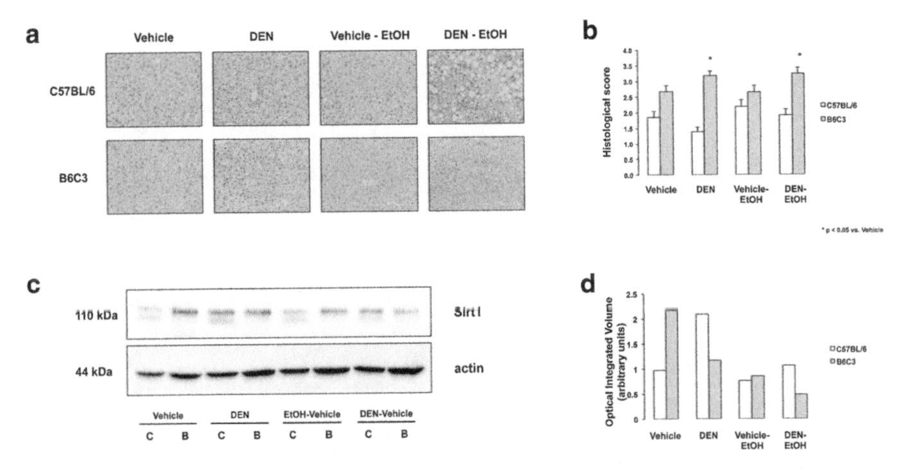

Fig. 21.2 Sirt1 expression in mouse models of HCC and EtOH consumption. (**a**) Representative Sirt1 IHC of hepatic sections (×200 magnification) from C57BL/6 and B6C3 mice injected with or without DEN and placed on EtOH or drinking water alone. Sirt1 expression was higher in B6C3 mice vs. C57BL/6 mice. DEN administration caused a significant increase in Sirt1 expression in B6C3 mice only. No effect due to EtOH was observed. (**b**) Data from Sirt1 IHC scoring in graph format. *$p<0.05$ vs. Veh-injection (B6C3 mice). (**c**) Representative Western blot for Sirt1 expression from pooled liver lysates. (**d**) Quantitative of bands following normalization with β-actin showed increased Sirt1 expression in B6C3 Veh-injected mice, in contrast to IHC findings. Likewise, increased Sirt1 expression was detected in DEN-injected C57BL/6 mice

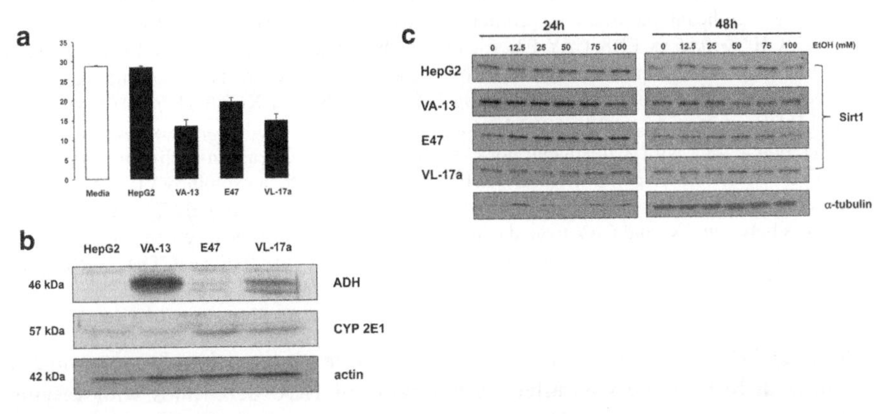

Fig. 21.3 EtOH metabolism and effect on Sirt1 expression in HCC cell lines varying in EtOH metabolism capacity. (**a**) Representative Western blots from HepG2, VA-13, E47 and VL-17a HCC cell lines. ADH expression was highest in VA-13 and VL-17a cells, with minimal expression in HepG2 and E47 cell lines. ALDH was expressed highest in E47 and VL-17a cell lines and CYP 2E1 had highest expression observed in E47 and VL-17a cells. (**b**) Sirt1 expression from HCC cell lines treated with 0–100 mM of EtOH for 24 and 48 h. Western blots for Sirt1 showed no effect on Sirt1 expression was seen in HepG2 cells; however, decreases were seen in VA-13 (24 and 48 h), E47 (48 h), and VL-17a cell lines (48 h)

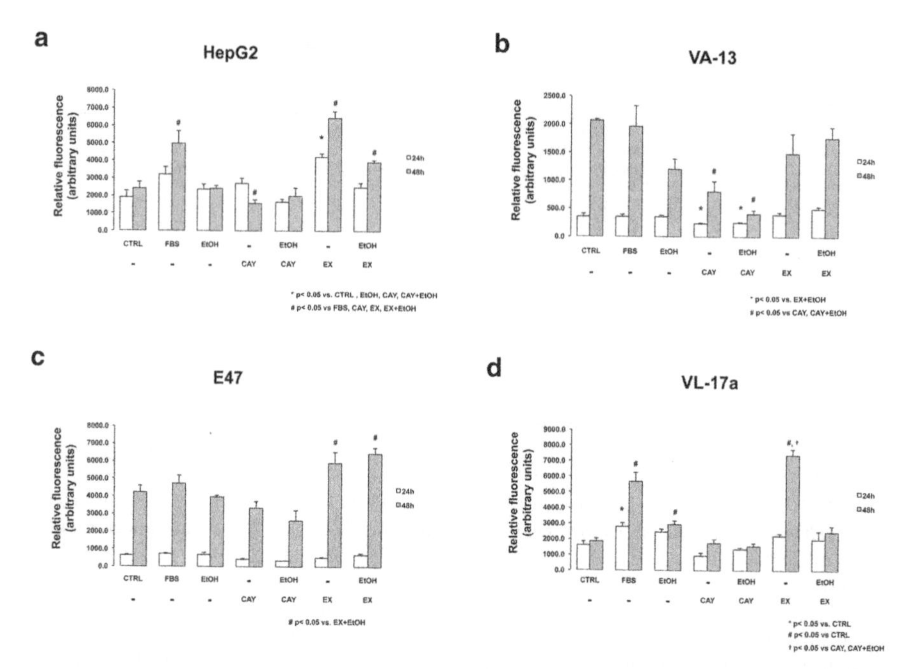

Fig. 21.4 Effect of Sirt1 modulation on proliferation in HCC cell lines with varying EtOH metabolism capacity. Cell lines were treated with or without 1 % FBS (FBS), 25 mM EtOH (EtOH), 5 μM CAY 10591 (CAY), 5 μM EX-527 (EX) or no treatment (CTRL) (**a**) HepG2 had no significant changes at 24 h except for treatment with the Sirt1 inhibitor EX-527, which significantly increased proliferation. At 48 h, there was a significant reduction in proliferation with CAY treatment, and significant increases in proliferation with EX. $^{*}p<0.05$ vs. CTRL, EtOH, CAY, CAY + EtOH $^{\#}p<0.05$ vs. FBS, CAY, EX, EX + EtOH (**b**) We next treated with VA-13 and observed reduction in proliferation with CAY at both 24 and 48 h; however, there were no significant changes in other treatment groups, including EX. $^{*}p<0.05$ vs. EX + EtOH $^{\#}p<0.05$ vs. CAY, CAY + EtOH (**c**) Treatment of the CYP 2E1-only overexpressing cell line, E47, showed increased proliferation at 48 h with X treatment; however, no significant decrease in proliferation was seen with CAY at either time point. $^{\#}p<0.05$ vs. EX + EtOH (**d**) We then examined VL-17a cells and detected significant differences in proliferation between EX-treated cells and CTRL cells at 48 h, as well as between EX- and CAY-treated cells. Additionally, VL-17a cells were the only cells that had increased proliferation in the presence of EtOH treatment $^{*}p<0.05$ vs. CTRL $^{\#}p<0.05$ vs. CTRL $^{\dagger}p<0.05$ vs. CAY, CAY + EtOH

Effects of EtOH and Sirt1 activity on HCC cell line proliferation: To determine if changes in Sirt1 expression affect proliferation of HCC cells lines with varying capacities to metabolize EtOH, HepG2, VA-13, E47, and VL-17a HCC cells were treated with or without EtOH (25 mM) in the absence or presence of a Sirt1 activator (CAY10591, 5 μM) or inhibitor (EX-527, 5 μM) (24 or 48 h). HepG2 cells, which have low EtOH-metabolizing capacity, had no significant change in proliferation at 24 h with the exception of cells treated EX-527, which significantly increased proliferation. At 48 h, there was a significant reduction in cell proliferation following CAY10591 treatment, while cells pretreated with EX-527 (Fig. 21.4a) exhibited significant increases in proliferation. We next treated VA-13 with the same regiment and found again, reduction in proliferation with Sirt1 agonist at both 24 and 48 h; however, there were no significant changes in other treatment groups, including the Sirt1 inhibitor group (Fig. 21.4b). Treatment of the CYP 2E1-only overexpressing cell line, E47, led to increased proliferation at 48 h with EX-527 treatment; however, no significant decrease in proliferation was measured with CAY10591 treatment at either time point (Fig. 21.4c). We then examined CYP 2E1 and ADH-expressing VL-17a cells and measured significant differences in proliferation between EX-527-treated cells and controls at 48 h, as well as between EX-527- and CAY10591-treated cells. Additionally, the CYP 2E1 and ADH-expression VL-17a cells were the only cells that had increased proliferation in the presence of EtOH treatment (Fig. 21.4d).

21.4 Discussion

We report increased Sirt1 expression in nuclei of hepatocytes in animals with significant liver tumor burden (DEN-injected) compared to Veh-injected animals on a B6C3 background (Fig. 21.1a, b); however, EtOH administration did not further enhance Sirt1 staining. These findings are somewhat paradoxical, as Sirt1 expression has been correlated to increased liver tumor progression/poor prognosis in humans; however, Sirt1 expression and activity are impaired by EtOH metabolism [18, 21]. Thus, future studies to assess acetylation of downstream Sirt1 targets such as NF-κB and SREBP-1c in HCC, including HCC in the setting of EtOH abuse would be of interest.

HCC is an increasingly diagnosed disease state in the liver for which alcohol is a leading risk factor in developed countries [33]. The deleterious effects of alcohol in the liver arise as a combination of the direct effects of alcohol metabolism within the liver, and the actions and interactions of the indirect/systemic effects of alcohol with other organs and physiological systems. Following hepatic metabolism much of the hepatic damage occurs as a result of the formation of reactive metabolic intermediates that promotes hepatocyte damage and, in doing so, the activation of hepatic stellate cells, inflammation, and fibrosis. If left unchecked, progression to hepatic cirrhosis occurs, a pathological state representing the setting for approximately 80 % of all HCC diagnosed [2, 34].

Sirt1 is a histone deacetylase that is intimately positioned as a cellular metabolic switch that participates in a diverse series of biological pathways. Within the liver, Sirt1 regulates acetylation status, and thus, influences the activity of numerous transcription factors associated with lipid and glucose metabolism, including SREBP-1c, cAMP-responsive element-binding protein (CREB) regulated transcription co-activator-2 (CRTC2) and peroxisome proliferator-activated receptor gamma co-activator 1-α (PGC-1α) [35, 36]. Sustained overnutrition and alcohol consumption deplete intracellular NAD+ stores, impairing Sirt1 activity, and thus promoting lipogenesis and steatosis. Sirt1 is also attractive in targeting alcoholic liver disease because of its role in hepatic inflammation. Indeed, overexpression of Sirt1 is reported to protect against liver inflammation whereas deletion of Sirt1 enhances inflammation [37, 38]. These findings at least in part are attributed to decreased deacetylation of nuclear factor-kappaB (NF-κB) at lysine 310 [39].

In the current study, we report increases in Sirt1 expression in DEN-treated animals on a B6C3 background, but not a C57BL/6 background (Fig. 21.1b). B6C3 animals are more susceptible to liver tumor formation than C57BL/6 mice and when fed EtOH-DW, have increased HCC progression (Fig. 21.1a) [24]. However, changes in Sirt1 expression levels do not necessarily correlate to increased activity, as Sirt1 activity is more closely tied to the availability of NAD+ as a cofactor in the deacetylation reaction [40]. Thus simply having high levels of Sirt1 may be a consequence, as opposed to a cause, of increased tumor promotion in the setting of EtOH-DW for B6C3 mice.

An increasing number of reports appear to suggest a deleterious role for Sirt1 in HCC and other cancers [41, 42]. These findings are attributed to in vitro findings linking Sirt1 activity and deacetylation (and thus inactivation) of the tumor suppressor p53 [23, 43]. These findings are also coupled with observations that Sirt1 can promote cell survival through telomere maintenance [21, 44]. Despite observation findings that Sirt1 expression is enhanced in liver tumors and correlated with poor prognosis, there is a lack of in vivo data that demonstrate a mechanistic role for Sirt1 in HCC [22].

Contrary to a role in tumor promoting tumor progression, reports indicate Sirt1 impairs Wnt/β-catenin pathway-associated proliferation. Constitutive expression of Wnt/β-catenin has been found in approximately 90 % of colorectal cancers and deacetylation of β-catenin by Sirt1 has been reported to prevent transcriptional activation by β-catenin in vitro [45, 46]. Additionally, Sirt1 plays an important role in DNA stability, as evidenced by knockout of Sirt1 leading to increased genomic instability and increased chromosomal fusions [47]. Indeed, overexpression of Sirt1 prevented evidence of metabolic syndrome and provided strong protection against liver cancer development in a DEN/high-fat diet mouse model of liver cancer [48].

We present additional findings to support an anticancer role for Sirt1 activity in an in vitro model of HCC progression (Fig. 21.4). Antagonism of Sirt1 expression, utilizing the Sirt1 inhibitor EX-527, promoted HCC cell line proliferation independent of EtOH metabolism, whereas activation of Sirt1 using the specific agonist CAY10591 inhibited proliferation in both the presence and absence of 25 mM EtOH. However, it is worth noting EtOH failed to promote HCC cell line proliferation except in VL-17a cell lines expressing both ADH and CYP 2E1 (Fig. 21.4d), despite previous findings that E47 cells proliferate in the presence of an equivalent dose of EtOH [25]. Lack of a proliferative response to EtOH by VA-13 cells is not surprising considering previous findings that ADH-expressing cells have diminished proliferative responses to EtOH treatment [28].

An important consideration to the contribution of Sirt1 to the promotion or prevention of HCC is the contribution of non-parenchymal liver cells (NPCs), including Kupffer cells, sinusoidal endothelial cells, and hepatic stellate cells. Currently, there is a dearth of studies examining Sirt1 expression and/or activity on the function of NPCs; however, one report suggested Sirt1-NF-κB signaling is involved in TNF-α production in the presence of alcohol [49]. Indeed additional studies to elucidate potential role(s) for Sirt1 in NPCs utilizing models of alcohol-induced liver injury and/or cancer are attractive and may help explain differences observed in in vitro and in vivo studies presented herein.

21.5 Conclusion

We report Sirt1 antagonism and activation promotes and represses HCC cell line proliferation respectively, regardless of the capacity to metabolize ethanol. These findings suggest that Sirt1 may impair HCC proliferation, despite the observation that Sirt1 expression is enhanced in a B6C3 model of EtOH-promoted HCC progression. To better elucidate the role of Sirt1 in ethanol-associated HCCs, in vivo experiments coupling liver-specific knockout of Sirt1, or use of Sirt1 activators in vivo, using a model of DEN/EtOH-DW are warranted.

Acknowledgements The HepG2 E47 cell line (E47) was obtained as a generous gift from Dr. Arthur Cederbaum (Icahn School of Medicine at Mount Sinai, New York, NY) [26]. The HegG2 VA-13 and VL-17a cell lines that overexpress human ADH and human ADH/CYP 2E1 respectively, were obtained as a generous gift from Dr. Dahn Clemens (Department of Internal Medicine, University of Nebraska Medical Center, Omaha, NE).

References

1. Siegel R et al (2013) Cancer statistics, 2013. CA Cancer J Clin 63(1):11–30
2. McKillop IH, Schrum LW (2009) Role of alcohol in liver carcinogenesis. Semin Liver Dis 29(2):222–232
3. Boffetta P et al (2006) The burden of cancer attributable to alcohol drinking. Int J Cancer 119(4):884–887
4. Rehm J et al (2013) Global burden of alcoholic liver diseases. J Hepatol 59(1):160–168
5. International Agency for Research on Cancer (2010) Alcohol consumption and ethyl carbamate. IARC monographs on the evaluation of carcinogenic risks to humans, vol 96. World Health Organization, Lyon, p 3–1383
6. Bagnardi V et al (2013) Light alcohol drinking and cancer: a meta-analysis. Ann Oncol 24(2):301–308
7. Berman K et al (2011) Hepatic and extrahepatic cancer in cirrhosis: a longitudinal cohort study. Am J Gastroenterol 106(5):899–906
8. Yuan JM et al (2004) Synergism of alcohol, diabetes, and viral hepatitis on the risk of hepatocellular carcinoma in blacks and whites in the U.S. Cancer 101(5):1009–1017
9. Zakhari S (2006) Overview: how is alcohol metabolized by the body? Alcohol Res Health 29(4):245–254
10. Deitrich RA et al (2007) Removal of acetaldehyde from the body. Novartis Found Symp 285:23–40; discussion 40–51, 198–199
11. Lieber CS (2004) The discovery of the microsomal ethanol oxidizing system and its physiologic and pathologic role. Drug Metab Rev 36(3–4):511–529
12. Brooks PJ, Theruvathu JA (2005) DNA adducts from acetaldehyde: implications for alcohol-related carcinogenesis. Alcohol 35(3):187–193

13. Seitz HK, Wang XD (2013) The Role of cytochrome P450 2E1 in ethanol-mediated carcinogenesis. Subcell Biochem 67:131–143
14. Sauve AA et al (2006) The biochemistry of sirtuins. Annu Rev Biochem 75:435–465
15. Canto C, Auwerx J (2012) Targeting sirtuin 1 to improve metabolism: all you need is NAD(+)? Pharmacol Rev 64(1):166–187
16. Stunkel W, Campbell RM (2011) Sirtuin 1 (SIRT1): the misunderstood HDAC. J Biomol Screen 16(10):1153–1169
17. You M et al (2008) Mammalian sirtuin 1 is involved in the protective action of dietary saturated fat against alcoholic fatty liver in mice. J Nutr 138(3):497–501
18. You M et al (2008) Involvement of mammalian sirtuin 1 in the action of ethanol in the liver. Am J Physiol Gastrointest Liver Physiol 294(4):G892–G898
19. Lieber CS et al (2008) Effect of chronic alcohol consumption on Hepatic SIRT1 and PGC-1alpha in rats. Biochem Biophys Res Commun 370(1):44–48
20. Portmann S et al (2013) Antitumor effect of SIRT1 inhibition in human HCC tumor models in vitro and in vivo. Mol Cancer Ther 12(4):499–508
21. Chen J et al (2011) Sirtuin 1 is upregulated in a subset of hepatocellular carcinomas where it is essential for telomere maintenance and tumor cell growth. Cancer Res 71(12):4138–4149
22. Jang KY et al (2012) SIRT1 and c-Myc promote liver tumor cell survival and predict poor survival of human hepatocellular carcinomas. PLoS One 7(9):e45119
23. Lee CW et al (2012) AMPK promotes p53 acetylation via phosphorylation and inactivation of SIRT1 in liver cancer cells. Cancer Res 72(17):4394–4404
24. Brandon-Warner E et al (2012) Chronic ethanol feeding accelerates hepatocellular carcinoma progression in a sex-dependent manner in a mouse model of hepatocarcinogenesis. Alcohol Clin Exp Res 36(4):641–653
25. Brandon-Warner E et al (2010) Silibinin inhibits ethanol metabolism and ethanol-dependent cell proliferation in an in vitro model of hepatocellular carcinoma. Cancer Lett 291(1): 120–129
26. Chen Q, Cederbaum AI (1998) Cytotoxicity and apoptosis produced by cytochrome P450 2E1 in Hep G2 cells. Mol Pharmacol 53(4):638–648
27. Osna NA et al (2003) Interferon gamma enhances proteasome activity in recombinant Hep G2 cells that express cytochrome P4502E1: modulation by ethanol. Biochem Pharmacol 66(5): 697–710
28. Clemens DL et al (2002) Relationship between acetaldehyde levels and cell survival in ethanol-metabolizing hepatoma cells. Hepatology 35(5):1196–1204
29. Gertz M et al (2013) Ex-527 inhibits sirtuins by exploiting their unique NAD+-dependent deacetylation mechanism. Proc Natl Acad Sci U S A 110(30):E2772–E2781
30. Nayagam VM et al (2006) SIRT1 modulating compounds from high-throughput screening as anti-inflammatory and insulin-sensitizing agents. J Biomol Screen 11(8):959–967
31. Sokolov E et al (2013) Lysophosphatidic acid receptor expression and function in human hepatocellular carcinoma. J Surg Res 180(1):104–113
32. Karaa A et al (2008) S-adenosyl-L-methionine attenuates oxidative stress and hepatic stellate cell activation in an ethanol-LPS-induced fibrotic rat model. Shock 30(2):197–205
33. Sanyal A et al (2010) Population-based risk factors and resource utilization for HCC: US perspective. Curr Med Res Opin 26(9):2183–2191
34. Kocabayoglu P, Friedman SL (2013) Cellular basis of hepatic fibrosis and its role in inflammation and cancer. Front Biosci (Schol Ed) 5:217–230
35. Schug TT, Li X (2011) Sirtuin 1 in lipid metabolism and obesity. Ann Med 43(3):198–211
36. Liu Y et al (2008) A fasting inducible switch modulates gluconeogenesis via activator/coactivator exchange. Nature 456(7219):269–273
37. Pfluger PT et al (2008) Sirt1 protects against high-fat diet-induced metabolic damage. Proc Natl Acad Sci U S A 105(28):9793–9798
38. Purushotham A et al (2009) Hepatocyte-specific deletion of SIRT1 alters fatty acid metabolism and results in hepatic steatosis and inflammation. Cell Metab 9(4):327–338

39. Yeung F et al (2004) Modulation of NF-kappaB-dependent transcription and cell survival by the SIRT1 deacetylase. EMBO J 23(12):2369–2380
40. Revollo JR, Li X (2013) The ways and means that fine tune Sirt1 activity. Trends Biochem Sci 38(3):160–167
41. Choi HN et al (2011) Expression and role of SIRT1 in hepatocellular carcinoma. Oncol Rep 26(2):503–510
42. Wang H et al (2012) SIRT1 promotes tumorigenesis of hepatocellular carcinoma through PI3K/PTEN/AKT signaling. Oncol Rep 28(1):311–318
43. Luo J et al (2001) Negative control of p53 by Sir2alpha promotes cell survival under stress. Cell 107(2):137–148
44. Palacios JA et al (2010) SIRT1 contributes to telomere maintenance and augments global homologous recombination. J Cell Biol 191(7):1299–1313
45. Morin PJ et al (1997) Activation of beta-catenin-Tcf signaling in colon cancer by mutations in beta-catenin or APC. Science 275(5307):1787–1790
46. Levy L et al (2004) Acetylation of beta-catenin by p300 regulates beta-catenin-Tcf4 interaction. Mol Cell Biol 24(8):3404–3414
47. Bakkenist CJ, Kastan MB (2003) DNA damage activates ATM through intermolecular autophosphorylation and dimer dissociation. Nature 421(6922):499–506
48. Herranz D et al (2010) Sirt1 improves healthy ageing and protects from metabolic syndrome-associated cancer. Nat Commun 1:3
49. Shen Z et al (2009) Role of SIRT1 in regulation of LPS- or two ethanol metabolites-induced TNF-alpha production in cultured macrophage cell lines. Am J Physiol Gastrointest Liver Physiol 296(5):G1047–G1053

Chapter 22
Transgenic Mouse Models for Alcohol Metabolism, Toxicity, and Cancer

Claire Heit, Hongbin Dong, Ying Chen, Yatrik M. Shah, David C. Thompson, and Vasilis Vasiliou

Abstract Alcohol abuse leads to tissue damage including a variety of cancers; however, the molecular mechanisms by which this damage occurs remain to be fully understood. The primary enzymes involved in ethanol metabolism include alcohol dehydrogenase (ADH), cytochrome P450 isoform 2E1, (CYP2E1), catalase (CAT), and aldehyde dehydrogenases (ALDH). Genetic polymorphisms in human genes encoding these enzymes are associated with increased risks of alcohol-related tissue damage, as well as differences in alcohol consumption and dependence. Oxidative stress resulting from ethanol oxidation is one established pathogenic event in alcohol-induced toxicity. Ethanol metabolism generates free radicals, such as reactive oxygen species (ROS) and reactive nitrogen species (RNS), and has been associated with diminished glutathione (GSH) levels as well as changes in other antioxidant mechanisms. In addition, the formation of protein and DNA adducts associated with the accumulation of ethanol-derived aldehydes can adversely affect critical biological functions and thereby promote cellular and tissue pathology. Animal models have proven to be valuable tools for investigating mechanisms underlying pathogenesis caused by alcohol. In this review, we provide a brief discussion on several animal models with genetic defects in alcohol-metabolizing enzymes and GSH-synthesizing enzymes and their relevance to alcohol research.

C. Heit • H. Dong • Y. Chen
Department of Pharmaceutical Sciences, School of Pharmacy,
University of Colorado Denver Anschutz Medical Campus,
12850 East Montview Boulevard, Aurora, CO 80045, USA

Y.M. Shah
Department of Molecular and Integrative Physiology, University of Michigan,
Ann Arbor, MI 48109, USA

D.C. Thompson
Department of Clinical Pharmacy, School of Pharmacy,
University of Colorado Anschutz Medical Campus, Aurora, CO 80045, USA

V. Vasiliou, Ph.D. (✉)
Department of Environmental Health Sciences, Yale School of Public Health,
New Haven, CT, USA
e-mail: Vasilis.Vasiliou@ucdenver.edu

V. Vasiliou et al. (eds.), *Biological Basis of Alcohol-Induced Cancer*,
Advances in Experimental Medicine and Biology 815,
DOI 10.1007/978-3-319-09614-8_22

22.1 Introduction

The Centers for Disease Control and Prevention reported that the annual number of alcohol-related deaths was 88,000 in the United States from 2006 to 2010. Alcohol is a causal factor in more than 60 human diseases and places a significant burden on the economy with healthcare costs estimated in 2006 as surpassing 223 billion dollars in the United States alone [1]. A comprehensive understanding of the mechanisms mediating alcohol toxicity is essential because it facilitates the development of therapies that prevent and/or treat the pathologies associated with alcohol consumption. The cellular and molecular mechanisms leading to alcohol-induced tissue damage are not fully understood. However, emerging evidence indicates that common mechanisms of cell injury, such as stress, inflammation, and alterations in signaling (including apoptosis) pathways, are all involved in the deleterious effects of alcohol.

Animals in which expression of specific proteins are repressed, the so-called knockout animals, represent an innovative and powerful research tool for scientific discovery. Genetic manipulation of proteins involved in the metabolism of ethanol or in the cellular defense mechanisms against alcohol-induced oxidative stress have allowed the exploration of their roles in alcohol-related pathologies, such as alcoholic liver disease, pancreatitis, cardiovascular disease, and diabetes mellitus, as well as in various cancers, including oral, colorectal, liver, pancreatic, aerodigestive, breast, and colon [2–7]. Currently available animal models will be outlined in the following review. In addition, double and triple knockout strains of these mice are currently being produced in our laboratory.

22.2 Clinical Significance of Human Polymorphisms of Genes Involved in Ethanol Metabolism

Ethanol is metabolized primarily *via* oxidation to acetaldehyde through the enzymatic activity of alcohol dehydrogenases (ADH), catalase (CAT), and cytochrome p450 2E1 (CYP2E1) (Fig. 22.1). Acetaldehyde is then oxidized to acetate by the aldehyde dehydrogenases (ALDHs). The role of ADH in ethanol metabolism is well established [8–10]. The human genome contains three Class I *ADH* genes (*ADH1A, ADH1B, ADH1C*); in contrast, rodents have only one *Adh1* gene [11]. Genetic polymorphisms in *ADH1* genes are associated with colon and breast cancers [4, 5, 12]. The role of CYP2E1 in ethanol metabolism, oxidative injury, and cancer is also well established [13–17] and genetic polymorphisms are associated with increased cancer risk [18–20]. Catalase appears to play an important role in ethanol metabolism in the brain [14, 15]. Nevertheless, polymorphisms in the catalase gene are associated with diabetes mellitus, hypertension, and vitiligo [21, 22]. ALDHs are a family of 19 human proteins that metabolize aldehydes, in which three isoforms are responsible for metabolizing acetaldehyde. Mitochondrial ALDH2 is the primary enzyme involved in the metabolism of acetaldehyde (Km $\leq$ 5 μM). The ALDH2*2 allele

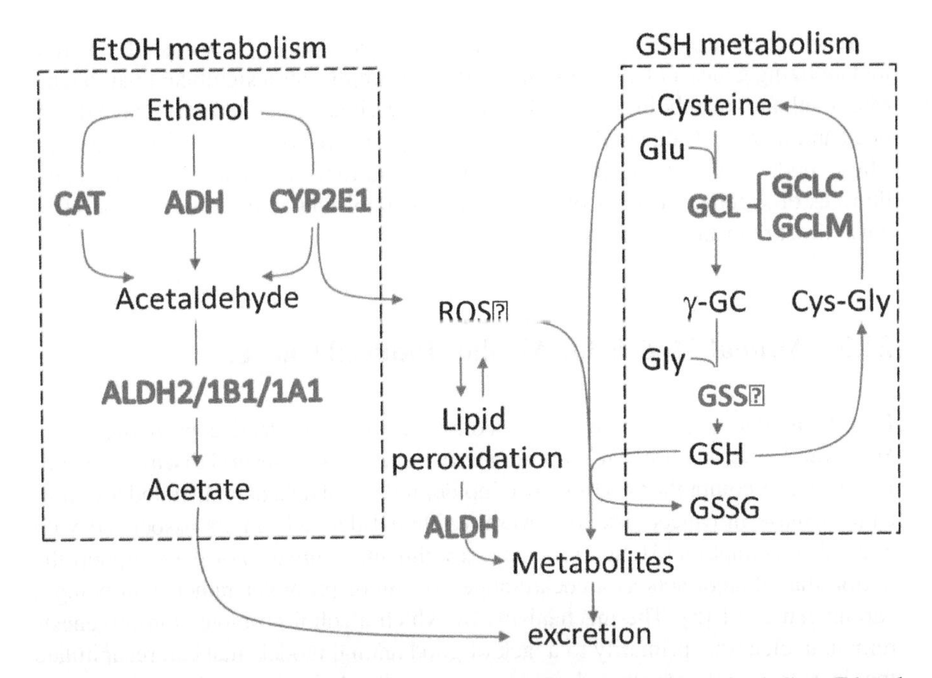

Fig. 22.1 Major enzymatic pathways involved in ethanol and glutathione metabolism. Ethanol (EtOH) is subject to metabolism by catalase (CAT), cytochrome P450 isoform 2E1 (CYP2E1), and alcohol dehydrogenase (ADH). Acetaldehyde is metabolized by aldehyde dehydrogenase (ALDH) isoforms 1A1, 1B1, and 2. In the glutathione (GSH) pathway, glutamate cysteine ligase (GCL), which includes two subunits the catalytic subunit (GCLC) and the modifier subunit (GCLM), catalyzes the synthesis of γ-glutamylcysteine (γ-GC). γ-GC is then coupled to glycine by glutathione synthetase (GSS) to form GSH. During oxidative processes, reactive oxygen species (ROS) form which can cause lipid peroxidation. ROS can be reduced by GSH, in the process forming GSSG, the oxidized form

(which appears restricted to an Asian genetic background) causes marked reductions in acetaldehyde metabolism that manifest clinically as flushing syndrome and ethanol avoidance in heterozygous and homozygous individuals. This polymorphism is also associated with alcohol-related cancers [3, 23]. ALDH1B1 has the next lowest Km for acetaldehyde (Km = 55 μM), implicating a role in acetaldehyde metabolism secondary to ALDH2. ALDH1B1 has been proposed as a biomarker for colon cancer [24]. Several ALDH1B1 polymorphisms have been found in humans and recent studies have linked these polymorphisms to drinking aversion, elevated systolic blood pressure, and frequent hypersensitivity reactions in Caucasians [25, 26]. ALDH1A1 has a role in acetaldehyde metabolism and drinking preference [27, 28] and a deficiency in this enzyme is associated with ethanol hypersensitivity in Caucasian subjects [29]. Polymorphisms in alcohol- and acetaldehyde-metabolizing enzymes have been closely linked with alcohol-related cancers. In a Japanese population, p53 accumulation, esophageal neoplasia, and esophageal squamous cell carcinomas were increased in subjects whose genes included the inactive heterozygous allele ALDH2 *1/*2 and the less active ADH1B *1/*1 [30, 31]. These polymorphisms also exhibited more frequent acetaldehyde-induced DNA damage [32].

Taken together, the association between human polymorphisms of ethanol-metabolizing genes and alcohol-related diseases implicates a significant pathogenic role of ethanol metabolism in alcohol toxicity. Various research groups have developed animal models that harbor genetic ablations of ethanol-metabolizing enzymes. These models can serve as important experimental tools to elucidate the mechanistic roles of specific enzymes or pathways in alcohol-related diseases and therefore have direct relevance to alcohol research.

22.3 Animal Models for Alcohol-Induced Cancer

Low to moderate consumption of alcohol has tissue-protective properties [33]. However, heavy alcohol consumption increases the risk of several diseases, including cancer. A comprehensive review of epidemiological data demonstrated a significant increase in cancer risk for several epithelial-derived tumors associated with ethanol consumption [3]. Studies using experimental animals, however, support the notion that ethanol acts as a cocarcinogen or tumor promoter rather than being a carcinogen itself [5]. The mechanisms by which alcohol promotes tumorigenesis remain unclear due primarily to a lack of good animal models that can recapitulate the increased risk of alcohol in carcinogenesis. Animal models analogous to inflammation-promoted cancers, such as the azoxymethane (AOM)/dextran sulfate sodium (DSS) model, are needed [34]. In this model, the carcinogen AOM by itself is administered at a dose that causes no dysplasia; however, when administered in combination with DSS (which induces inflammation), a synergistic increase in the number of tumors is observed. The creation of a similar model for alcohol will rely on a better understanding of the interactions between the genetic mutations (or carcinogen) with the duration, route, and concentration of ethanol for the epithelial tumor that is to be modeled. Moreover, diets that better mimic heavy alcohol consumption in humans are required. The interaction of tumor-promoting dietary components, such as alcohol, high-fat, and iron, may lead to more robust and precise models. Lastly, experimental evidence indicates that the metabolism of ethanol leading to the generation of acetaldehyde and free radicals is intimately involved in alcohol-associated carcinogenesis [3]. Therefore, a more comprehensive understanding of the enzymes required for alcohol metabolism in cancer are needed and the genetic animal models discussed in this review could represent unique opportunities to identify their roles in alcohol-induced cancers.

22.4 Glutathione in Alcoholic Tissue Injury

In the development of alcohol-induced tissue injury, it is apparent that numerous pathways in target organs are modulated by ethanol [35, 36]. Oxidative stress appears to play a central role in many of these pathways [37]. Ethanol metabolism, CYP2E1 induction, compromised antioxidant defense, mitochondrial injury, inflammation, hypoxia, and iron overload can all contribute to the alcohol-induced oxidative environment. Accumulation of the reactive molecules (including reactive oxygen species and electrophilic products, such as acetaldehyde and lipid peroxidation products) can be harmful to a biological system due to their propensity to inactivate enzymes and cause DNA damage, loss of protein functions and cell death [38].

Glutathione (GSH) plays an important role as an antioxidant by serving as a cofactor for antioxidant enzymes, such as glutathione peroxidase and glutathione *S*-transferases, or by directly scavenging free radicals [39]. It is the most abundant nonprotein thiol, attaining a concentration in the high millimolar range in the liver [40]. Because of its abundance, GSH plays a key role in maintaining cellular redox homeostasis and, therefore, enzymes that help generate GSH are critical in protecting cells against oxidative stress. GSH is a tripeptide composed of glutamate, cysteine, and glycine. It is synthesized in most types of cells by two successive enzymatic reactions. The first reaction couples glutamate and cysteine and is catalyzed by glutamate-cysteine ligase (GCL), resulting in the formation of γ-glutamylcysteine (γ-GC) [41] (Fig. 22.1). The second reaction, catalyzed by GSH synthetase, couples γ-GC with glycine. The formation of γ-GC by GCL is considered rate-limiting in GSH biosynthesis, and GCL has been the principal target of drugs designed to inhibit GSH biosynthesis [41] and to generate mice with GSH deficiency [42].

In higher eukaryotes, GCL in its most catalytically efficient form is a heterodimer composed of a catalytic (GCLC) and a modifier (GCLM) subunit, each of which is encoded by separate genes on different chromosomes. GCLC possesses all of the catalytic activity of γ-GC formation; GCLM optimizes the kinetic properties of the holoenzyme, thereby regulating tissue GSH levels [43]. GSH is exclusively synthesized in the cytoplasm [39] and further distributed into mitochondria, endoplasmic reticulum, and nuclei, where it plays a pivotal role in the normal functioning of these subcellular organelles [44]. During detoxication of free radicals, GSH is oxidized to glutathione disulfide (GSSG). Both GSH and GSSG can be transported outside the cell where it is broken down in sequence by γ-glutamyl transferases and dipeptidase, producing free cysteine and glycine for intracellular reutilization [45]. Depletion of hepatic GSH, particularly mitochondrial GSH, occurs as a result of excessive GSH consumption by free radicals and acetaldehyde generated during alcohol metabolism [46]. Given the above considerations, animal models exhibiting a GSH deficiency will serve as important tools to study GSH-regulated redox biology in ethanol metabolism and ethanol-induced tissue damage. As such, they are of direct relevance to alcohol research.

22.5 Mouse Models with Genetic Deficiencies in Ethanol-Metabolizing Enzymes

ADH1 global knockout: The *Adh1*$^{-/-}$ mouse line has been generated by Duester [47] (Table 22.1). It should be noted that human *ADH1* gene family consists of three genes, *viz*. *ADH1A*, *ADH1B*, and *ADH1C*, whereas the mouse genome has a single

Table 22.1 Transgenic mouse models

Strain	Genetic background	Phenotype	References
Adh1$^{-/-}$	C57BL6	• No gross abnormality • Reduction in blood ethanol clearance	Deltour et al. [9]
Cat$^{-/-}$	C57BL6	• No gross abnormality • Deficiency in brain mitochondrial respiration • Has not been used in ethanol toxicity studies	Ho et al. [49]
Cyp2e1$^{-/-}$	129/Sv	• No gross abnormality • Decreased sensitivity to acetaminophen hepatotoxicity • Resistance to ethanol-induced fatty liver and oxidant stress	Lee et al. [50], Lu et al. [54]
Cat$^{-/-}$ *Cyp2e1*$^{-/-}$ *double knockout*	C57BL6/129 mixed	• No gross abnormality • Has not been used in ethanol toxicity studies	Unpublished
Aldh2$^{-/-}$	C57BL6	• No gross abnormality • High susceptibility to ethanol toxicities by oral administration • High sensitivity to inhalation toxicities of acetaldehyde	Isse et al. [55], Oyama et al. [59, 61]
Aldh1a1$^{-/-}$	C57BL6	• Viable and fertile • Decreased susceptibility to diet-induced obesity and insulin resistance • Cataract development at age of 6-month • Has not been used in ethanol toxicity studies	Fan et al. [74], Ziouzenkova et al. (2007)
Aldh1b1$^{-/-}$	C57BL6	• No gross abnormality • Reduction in blood acetaldehyde clearance	Unpublished

A variety of transgenic strains are available for research. For each strain, the genetic background and phenotypes are provided

Adh1 gene [11]. *Adh1*$^{-/-}$ *mice* have limited capacity to oxidize ethanol and retinol. Pharmacokinetic studies show a reduction in blood ethanol clearance in these animals [9]. Following parenteral administration of ethanol, these mice displayed an increased sleep time and embryonic resorption was increased threefold [9]. While ADH1 is thought to be responsible for the majority of ethanol metabolism in the liver, new pharmacokinetic evidence suggests a role for other ADH isoforms as well [48]. Therefore, this model may be useful in determining the pathophysiological importance of compensatory ADH isoforms as well as elucidating the kinetics of these enzymes for ethanol.

Catalase global knockout: The *Cat*$^{-/-}$ mouse strain was developed and characterized by Ho and colleagues [49] (Table 22.1). These mice do not express catalase and develop normally, i.e., exhibit no gross abnormalities. However, brain mitochondria of these animals show deficiencies in mitochondrial respiration. To date, this knockout strain has not been subjected to ethanol toxicity studies. Given that earlier studies have shown a significant role of catalase in modulating ethanol sensitivity in the brain [14, 15], the *Cat*$^{-/-}$ mice would be anticipated to be a valuable animal model for examining ethanol drinking preference as well as alcohol toxicities.

CYP2E1 global knockout: CYP2E1 is an ethanol-inducible enzyme with a role in hepatic ethanol oxidation. By genetically ablating exon 2 of *Cyp2e1* gene, Gonzalez and colleagues developed *Cyp2e1*$^{-/-}$ mice [50] (Table 22.1). These mice do not express CYP2E1 enzyme and develop normally [50]. They also show lower sensitivity to the deleterious hepatic effects of acetaminophen [50]. As one of the primary xenobiotic/endobiotic-metabolizing p450s, CYP2E1 is a contributor to a variety of cellular toxicities induced by endogenous or exogenous pathogens. Using the *Cyp2e1*$^{-/-}$ mouse model, CYP2E1 has been shown to play a pivotal role in mediating hepatotoxicity making this an interesting model for alcohol-related liver toxicity [51, 52] *Cyp2e1*$^{-/-}$ and *Cyp2e1* knock-in mice have been used to examine the potentiation of ethanol-induced hypoxia. *Cyp2e1*$^{-/-}$ mice exhibited the lowest levels of hypoxia and HIF1-α [53]. Similarly, ethanol-induced fatty liver and oxidant stress are blunted in these mice [54]; this study confirmed the important role of CYP2E1 in ethanol-induced liver toxicities. *Cyp2e1*$^{-/-}$ mice also display longer ethanol-induced sleep time than do wild-type mice [15], confirming the relevance of the *Cyp2e1*$^{-/-}$ mouse line for the study of the CYP2E1 enzyme in ethanol toxicities and alcohol-induced drinking preference.

22.6 Mouse Models with Genetic Deficiencies in Acetaldehyde-Metabolizing Enzymes

ALDH2 global knockout: The *Aldh2*$^{-/-}$ strain was originally developed and characterized by Isse and colleagues [55, 56] (Table 22.1). *Aldh2*$^{-/-}$ mice do not express ALDH2 protein and have no detectable capacity to oxidize acetaldehyde, propionaldehyde, or methoxyacetaldehyde in liver mitochondrial fractions. Following oral administration of ethanol, *Aldh2*$^{-/-}$ mice exhibit higher ethanol and acetaldehyde levels and lower acetate levels in the blood, brain, and liver than *Aldh2*$^{+/+}$ mice [57, 58]. Further, they are more susceptible to ethanol-induced body weight loss [59], but show no change in mortality [60]. *Aldh2*$^{-/-}$ mice are more sensitive to the toxic effects of inhaled acetaldehyde [61] and exhibit more frequent mutations in the T cell receptor site than their corresponding wild-type [62]. A single oral dose of ethanol in *Aldh2*$^{-/-}$ downregulates the alcohol-metabolizing CYP2E1 mRNA [63], which suggests that there is compensation due to an abundance of acetaldehyde. This treatment has also been shown to decrease hepatic malondialdehyde and increase hepatic glutathione, both markers of oxidative stress, in *Aldh2*$^{-/-}$ mice [64]. Acetaldehyde adducts are also increased in *Aldh2*$^{-/-}$ mice. These mice have been used to determine ethanol- and acetaldehyde-induced cholinergic changes in the hippocampus. The null mice exhibit decreases in choline acetyltransferase mRNA and protein; however, neurotrophins (nerve growth factor or brain-derived neurotrophic factor) remain unaffected [65], indicating that aldehydes have a selective effect in the brain. *Aldh2*$^{-/-}$ mice also exhibit alcohol avoidance in a test of preference and difference in liver or brain acetaldehyde levels [55]. ALDH2 also appears to influence bone growth and cardiac function, as demonstrated by reductions in trabecular bone formation and cardiomyocyte function in *Aldh2*$^{-/-}$ mice treated with alcohol [66, 67]. Stomach DNA adducts are dramatically increased after chronic ethanol feeding of *Aldh2*$^{-/-}$ mice [68, 69] and acute ethanol treatment increases hepatic oxidative DNA adducts in null mice [70, 71]. The *Aldh2*$^{-/-}$ strain represents a valuable strain that can be used to identify functions of ALDH2 in ethanol metabolism and toxicity.

ALDH2 conditional knockout: A "knockout-first" conditional allele for *Aldh2* has been developed by Skarnes and colleagues [72]. These mice have been crossed with *FLP* mice to generate *Aldh2* floxed conditional knockout (*Aldh2*$^{f/f}$) mice, which can be further crossed with specific *CRE* mouse lines to generate cell-specific *Aldh2* knockout mice. As expected, *Aldh2*$^{f/f}$ mice develop normally and exhibit no observed phenotype (unpublished observation). To date, no ethanol studies have been conducted in these mice. This strain can be used to study tissue-specific contributions of ALDH2 in ethanol metabolism and toxicity.

ALDH1B1 global knockout: The *Aldh1b1*$^{-/-}$ strain was recently generated by Vasiliou and coworkers (*Singh S, Vasiliou V et al.*, manuscript in preparation) (Table 22.1). The *Aldh1b1*$^{-/-}$ mice develop normally and show no overt phenotype. In agreement with the catalytic properties of ALDH1B1 (i.e., the second lowest Km for acetaldehyde oxidation [73]), these mice exhibit higher blood concentrations of acetaldehyde following acute ethanol administration [manuscript in preparation]. The *Aldh1b1*$^{-/-}$ strain represents the first animal model for the study of ALDH1B1 enzyme in ethanol-induced tissue injury.

ALDH1A1 global knockout: The *Aldh1a1*$^{-/-}$ mouse line has been generated by Fan et al. [74] (Table 22.1). These mice are fertile and exhibit no overt phenotype, except that aged *Aldh1a1*$^{-/-}$ mice display ~2.4-fold higher cataract incidence than wild-type mice [75]. While ALDH1A1 primarily metabolizes retinaldehyde, it also plays a role in acetaldehyde metabolism. Genetic variants of *ALDH1A1* (that result in low enzyme activity) have been associated with increased alcohol sensitivity in Caucasians [29]. Therefore, the *Aldh1a1*$^{-/-}$ mouse line represents a useful animal model for investigation of the ALDH1A1 enzyme in ethanol toxicities.

22.7 Mouse Models with GSH Deficiency

GCLC conditional (Gclc $^{f/f}$*) knockout*: The global gene knockout of *Gclc* results in embryonic lethality, indicating an essential role of GSH in mouse development [76]. The *Gclc*$^{f/f}$ strain was developed and originally characterized by Chen and colleagues [77]. The in vivo role of hepatic GSH has been investigated using the hepatocyte-specific *Gclc* knockout (*Gclc*$^{h/h}$) mice created by intercrossing *Gclc*$^{f/f}$ and *Alb-Cre* mice [77]. *Gclc*$^{h/h}$ mice experience almost complete loss of hepatic GSH (~5 % of normal) and die from acute liver failure when mitochondrial failure occurs [77]. Chronic administration of *N*-acetylcysteine, a treatment that promotes only a mild increase in liver GSH levels (to 8 % of normal) but partially preserves mitochondrial function, allows *Gclc*$^{h/h}$ mice to survive to adulthood, albeit with the serious liver pathologies fibrosis and cirrhosis [78]. These studies demonstrate an essential role of GSH in normal functioning of the liver. The *Gclc*$^{f/f}$ mice represent a unique model that can be used to elucidate cell-specific functions of GSH in ethanol metabolism and toxicity.

GCLM global knockout: The *Gclm*$^{-/-}$ strain was developed and originally characterized by Yang and coworkers [79]. *Gclm*$^{-/-}$ mice are viable and fertile, despite having only 9–16 % of the normal GSH levels in liver, lung, pancreas, erythrocytes, and plasma [79]. Except when challenged with oxidant stress [80, 81], *Gclm*$^{-/-}$ mice exhibit no overt phenotype, making them a useful model for studying chronic GSH depletion. Interestingly, these mice show accelerated clearance of ethanol and acetaldehyde and are protected from alcohol-induced steatosis (*Chen Y, Vasiliou V et al, manuscript in preparation*). Thus, *Gclm*$^{-/-}$ mice represent a model wherein significant GSH depletion in the liver is associated with beneficial metabolic and stress responses to ethanol.

22.8 Concluding Remarks

Alcohol use is widespread and related to numerous diseases, including oral, colorectal, liver, pancreas, aerodigestive, breast, and colon cancers. Ethanol metabolism and resultant oxidative stress are primary pathogenic events mediating alcohol-induced organ damage and neurobehavioral changes, the molecular details of which are not yet fully understood. The knockout mouse models for enzymes metabolizing ethanol (ADH1, CAT, and CYP2E1), acetaldehyde (ALDH2, ALDH1A1, and ALDH1B1) and enzymes involved in GSH synthesis (GCLC and GCLM), which we have discussed briefly in this review, represent unique and highly relevant animal models for alcohol research. Utilization of these models will deliver valuable information about the fundamental mechanisms underlying ethanol toxicity. Such knowledge should accelerate the development of more effective, targeted therapies to both prevent and treat health issues associated with excessive alcohol consumption.

Acknowledgements This work was supported in part by the National Institutes of Health Grants No. R24AA022057, No. R01EY14390, No. T32AA007464, CA148828, and DK095201.

References

1. Bouchery EE et al (2011) Economic costs of excessive alcohol consumption in the U.S., 2006. Am J Prev Med 41(5):516–524
2. McKillop IH, Schrum LW (2009) Role of alcohol in liver carcinogenesis. Semin Liver Dis 29(2):222–232
3. Seitz HK, Stickel F (2007) Molecular mechanisms of alcohol-mediated carcinogenesis. Nat Rev Cancer 7(8):599–612
4. Visapaa JP et al (2004) Increased cancer risk in heavy drinkers with the alcohol dehydrogenase 1C*1 allele, possibly due to salivary acetaldehyde. Gut 53(6):871–876
5. Poschl G, Seitz HK (2004) Alcohol and cancer. Alcohol Alcohol 39(3):155–165
6. Boffetta P, Hashibe M (2006) Alcohol and cancer. Lancet Oncol 7(2):149–156
7. Reidy J, McHugh E, Stassen LF (2011) A review of the relationship between alcohol and oral cancer. Surgeon 9(5):278–283
8. Edenberg HJ (2007) The genetics of alcohol metabolism: role of alcohol dehydrogenase and aldehyde dehydrogenase variants. Alcohol Res Health 30(1):5–13
9. Deltour L, Foglio MH, Duester G (1999) Metabolic deficiencies in alcohol dehydrogenase Adh1, Adh3, and Adh4 null mutant mice. Overlapping roles of Adh1 and Adh4 in ethanol clearance and metabolism of retinol to retinoic acid. J Biol Chem 274(24):16796–16801
10. Scott DM, Taylor RE (2007) Health-related effects of genetic variations of alcohol-metabolizing enzymes in African Americans. Alcohol Res Health 30(1):18–21
11. Duester G et al (1999) Recommended nomenclature for the vertebrate alcohol dehydrogenase gene family. Biochem Pharmacol 58(3):389–395
12. Lilla C et al (2005) Alcohol dehydrogenase 1B (ADH1B) genotype, alcohol consumption and breast cancer risk by age 50 years in a German case-control study. Br J Cancer 92(11): 2039–2041
13. Lu Y, Cederbaum AI (2008) CYP2E1 and oxidative liver injury by alcohol. Free Radic Biol Med 44(5):723–738
14. Zimatkin SM et al (2006) Enzymatic mechanisms of ethanol oxidation in the brain. Alcohol Clin Exp Res 30(9):1500–1505
15. Vasiliou V et al (2006) CYP2E1 and catalase influence ethanol sensitivity in the central nervous system. Pharmacogenet Genomics 16(1):51–58
16. Millonig G et al (2011) Ethanol-mediated carcinogenesis in the human esophagus implicates CYP2E1 induction and the generation of carcinogenic DNA-lesions. Int J Cancer 128(3): 533–540
17. Wang Y et al (2009) Ethanol-induced cytochrome P4502E1 causes carcinogenic etheno-DNA lesions in alcoholic liver disease. Hepatology 50(2):453–461
18. Trafalis DT et al (2010) CYP2E1 and risk of chemically mediated cancers. Expert Opin Drug Metab Toxicol 6(3):307–319
19. Druesne-Pecollo N et al (2009) Alcohol and genetic polymorphisms: effect on risk of alcohol-related cancer. Lancet Oncol 10(2):173–180
20. Morita M et al (2009) Genetic polymorphisms of CYP2E1 and risk of colorectal cancer: the Fukuoka Colorectal Cancer Study. Cancer Epidemiol Biomarkers Prev 18(1):235–241
21. Goth LT, Nagy T (2012) Acatalasemia and diabetes mellitus. Arch Biochem Biophys 525(2):195–200
22. Goth L, Rass P, Pay A (2004) Catalase enzyme mutations and their association with diseases. Mol Diagn 8(3):141–149
23. Seitz HK et al (1990) Possible role of acetaldehyde in ethanol-related rectal cocarcinogenesis in the rat. Gastroenterology 98(2):406–413
24. Chen Y et al (2011) Aldehyde dehydrogenase 1B1 (ALDH1B1) is a potential biomarker for human colon cancer. Biochem Biophys Res Commun 405(2):173–179
25. Husemoen LL et al (2008) The association of ADH and ALDH gene variants with alcohol drinking habits and cardiovascular disease risk factors. Alcohol Clin Exp Res 32(11): 1984–1991

26. Linneberg A et al (2010) Genetic determinants of both ethanol and acetaldehyde metabolism influence alcohol hypersensitivity and drinking behaviour among Scandinavians. Clin Exp Allergy 40(1):123–130
27. Little RG 2nd, Petersen DR (1983) Subcellular distribution and kinetic parameters of HS mouse liver aldehyde dehydrogenase. Comp Biochem Physiol C 74(2):271–279
28. Bond SL, Wigle MR, Singh SM (1991) Acetaldehyde dehydrogenase (Ahd-2)-associated DNA polymorphisms in mouse strains with variable ethanol preferences. Alcohol Clin Exp Res 15(2):304–307
29. Eriksson CJ (2001) The role of acetaldehyde in the actions of alcohol (update 2000). Alcohol Clin Exp Res 25(5 Suppl ISBRA):15S–32S
30. Yokoyama A et al (2012) Development of squamous neoplasia in esophageal iodine-unstained lesions and the alcohol and aldehyde dehydrogenase genotypes of Japanese alcoholic men. Int J Cancer 130(12):2949–2960
31. Yokoyama A et al (2011) p53 protein accumulation, iodine-unstained lesions, and alcohol dehydrogenase-1B and aldehyde dehydrogenase-2 genotypes in Japanese alcoholic men with esophageal dysplasia. Cancer Lett 308(1):112–117
32. Yukawa Y et al (2012) Combination of ADH1B*2/ALDH2*2 polymorphisms alters acetaldehyde-derived DNA damage in the blood of Japanese alcoholics. Cancer Sci 103(9): 1651–1655
33. Arranz S et al (2012) Wine, beer, alcohol and polyphenols on cardiovascular disease and cancer. Nutrients 4(7):759–781
34. Tanaka T (2009) Colorectal carcinogenesis: review of human and experimental animal studies. J Carcinog 8:5
35. Altamirano J, Bataller R (2011) Alcoholic liver disease: pathogenesis and new targets for therapy. Nat Rev Gastroenterol Hepatol 8(9):491–501
36. Gao B, Bataller R (2011) Alcoholic liver disease: pathogenesis and new therapeutic targets. Gastroenterology 141(5):1572–1585
37. Albano E (2008) Oxidative mechanisms in the pathogenesis of alcoholic liver disease. Mol Aspects Med 29(1–2):9–16
38. Bergendi L et al (1999) Chemistry, physiology and pathology of free radicals. Life Sci 65(18–19):1865–1874
39. Meister A (1982) Metabolism and function of glutathione: an overview. Biochem Soc Trans 10(2):78–79
40. Kretzschmar M (1996) Regulation of hepatic glutathione metabolism and its role in hepatotoxicity. Exp Toxicol Pathol 48(5):439–446
41. Meister A (1988) Glutathione metabolism and its selective modification. J Biol Chem 263(33):17205–17208
42. Dalton TP et al (2004) Genetically altered mice to evaluate glutathione homeostasis in health and disease. Free Radic Biol Med 37(10):1511–1526
43. Chen Y et al (2005) Glutamate cysteine ligase catalysis: dependence on ATP and modifier subunit for regulation of tissue glutathione levels. J Biol Chem 280(40):33766–33774
44. Lu SC (1999) Regulation of hepatic glutathione synthesis: current concepts and controversies. FASEB J 13(10):1169–1183
45. Meister A (1995) Glutathione metabolism. Methods Enzymol 251:3–7
46. Wu D, Cederbaum AI (2009) Oxidative stress and alcoholic liver disease. Semin Liver Dis 29(2):141–154
47. Molotkov A et al (2002) Distinct retinoid metabolic functions for alcohol dehydrogenase genes Adh1 and Adh4 in protection against vitamin A toxicity or deficiency revealed in double null mutant mice. J Biol Chem 277(16):13804–13811
48. Haseba T, Ohno Y (2010) A new view of alcohol metabolism and alcoholism—role of the high-Km Class III alcohol dehydrogenase (ADH3). Int J Environ Res Public Health 7(3): 1076–1092
49. Ho YS et al (2004) Mice lacking catalase develop normally but show differential sensitivity to oxidant tissue injury. J Biol Chem 279(31):32804–32812

50. Lee SS et al (1996) Role of CYP2E1 in the hepatotoxicity of acetaminophen. J Biol Chem 271(20):12063–12067
51. Valentine JL et al (1996) Reduction of benzene metabolism and toxicity in mice that lack CYP2E1 expression. Toxicol Appl Pharmacol 141(1):205–213
52. Wong FW, Chan WY, Lee SS (1998) Resistance to carbon tetrachloride-induced hepatotoxicity in mice which lack CYP2E1 expression. Toxicol Appl Pharmacol 153(1):109–118
53. Wang X et al (2013) Cytochrome P450 2E1 potentiates ethanol induction of hypoxia and HIF-1alpha in vivo. Free Radic Biol Med 63:175–186
54. Lu Y et al (2008) Cytochrome P450 2E1 contributes to ethanol-induced fatty liver in mice. Hepatology 47(5):1483–1494
55. Isse T et al (2002) Diminished alcohol preference in transgenic mice lacking aldehyde dehydrogenase activity. Pharmacogenetics 12(8):621–626
56. Yu HS et al (2009) Characteristics of aldehyde dehydrogenase 2 (Aldh2) knockout mice. Toxicol Mech Methods 19(9):535–540
57. Kiyoshi A et al (2009) Ethanol metabolism in ALDH2 knockout mice–blood acetate levels. Leg Med (Tokyo) 11(Suppl 1):S413–S415
58. Isse T et al (2005) Aldehyde dehydrogenase 2 gene targeting mouse lacking enzyme activity shows high acetaldehyde level in blood, brain, and liver after ethanol gavages. Alcohol Clin Exp Res 29(11):1959–1964
59. Oyama T et al (2007) A pilot study on subacute ethanol treatment of ALDH2 KO mice. J Toxicol Sci 32(4):421–428
60. Isse T et al (2005) Aldehyde dehydrogenase 2 activity affects symptoms produced by an intraperitoneal acetaldehyde injection, but not acetaldehyde lethality. J Toxicol Sci 30(4):315–328
61. Oyama T et al (2007) Susceptibility to inhalation toxicity of acetaldehyde in Aldh2 knockout mice. Front Biosci 12:1927–1934
62. Kunugita N et al (2008) Increased frequencies of micronucleated reticulocytes and T-cell receptor mutation in Aldh2 knockout mice exposed to acetaldehyde. J Toxicol Sci 33(1): 31–36
63. Matsumoto A et al (2007) Single-dose ethanol administration downregulates expression of cytochrome p450 2E1 mRNA in aldehyde dehydrogenase 2 knockout mice. Alcohol 41(8): 587–589
64. Matsumoto A et al (2007) Lack of aldehyde dehydrogenase ameliorates oxidative stress induced by single-dose ethanol administration in mouse liver. Alcohol 41(1):57–59
65. Jamal M et al (2013) Ethanol- and acetaldehyde-induced cholinergic imbalance in the hippocampus of Aldh2-knockout mice does not affect nerve growth factor or brain-derived neurotrophic factor. Brain Res 1539:41–47
66. Ma H et al (2010) Aldehyde dehydrogenase 2 knockout accentuates ethanol-induced cardiac depression: role of protein phosphatases. J Mol Cell Cardiol 49(2):322–329
67. Shimizu Y et al (2011) Reduced bone formation in alcohol-induced osteopenia is associated with elevated p21 expression in bone marrow cells in aldehyde dehydrogenase 2-disrupted mice. Bone 48(5):1075–1086
68. Nagayoshi H et al (2009) Increased formation of gastric N(2)-ethylidene-2′-deoxyguanosine DNA adducts in aldehyde dehydrogenase-2 knockout mice treated with ethanol. Mutat Res 673(1):74–77
69. Matsumoto A et al (2008) Effects of 5-week ethanol feeding on the liver of aldehyde dehydrogenase 2 knockout mice. Pharmacogenet Genomics 18(10):847–852
70. Kim YD et al (2007) Ethanol-induced oxidative DNA damage and CYP2E1 expression in liver tissue of Aldh2 knockout mice. J Occup Health 49(5):363–369
71. Matsuda T et al (2007) Increased formation of hepatic N2-ethylidene-2′-deoxyguanosine DNA adducts in aldehyde dehydrogenase 2-knockout mice treated with ethanol. Carcinogenesis 28(11):2363–2366
72. Skarnes WC et al (2011) A conditional knockout resource for the genome-wide study of mouse gene function. Nature 474(7351):337–342

73. Stagos D et al (2010) Aldehyde dehydrogenase 1B1: molecular cloning and characterization of a novel mitochondrial acetaldehyde-metabolizing enzyme. Drug Metab Dispos 38(10): 1679–1687
74. Fan X et al (2003) Targeted disruption of Aldh1a1 (Raldh1) provides evidence for a complex mechanism of retinoic acid synthesis in the developing retina. Mol Cell Biol 23(13): 4637–4648
75. Lassen N et al (2007) Multiple and additive functions of ALDH3A1 and ALDH1A1: cataract phenotype and ocular oxidative damage in Aldh3a1(-/-)/Aldh1a1(-/-) knock-out mice. J Biol Chem 282(35):25668–25676
76. Shi ZZ et al (2000) Glutathione synthesis is essential for mouse development but not for cell growth in culture. Proc Natl Acad Sci U S A 97(10):5101–5106
77. Chen Y et al (2007) Hepatocyte-specific Gclc deletion leads to rapid onset of steatosis with mitochondrial injury and liver failure. Hepatology 45(5):1118–1128
78. Chen Y et al (2010) Oral N-acetylcysteine rescues lethality of hepatocyte-specific Gclc-knockout mice, providing a model for hepatic cirrhosis. J Hepatol 53(6):1085–1094
79. Yang Y et al (2002) Initial characterization of the glutamate-cysteine ligase modifier subunit Gclm(-/-) knockout mouse. Novel model system for a severely compromised oxidative stress response. J Biol Chem 277(51):49446–49452
80. McConnachie LA et al (2007) Glutamate cysteine ligase modifier subunit deficiency and gender as determinants of acetaminophen-induced hepatotoxicity in mice. Toxicol Sci 99(2): 628–636
81. Chen Y et al (2012) Glutathione-deficient mice are susceptible to TCDD-Induced hepatocellular toxicity but resistant to steatosis. Chem Res Toxicol 25(1):94–100

Chapter 23
Fetal Alcohol Exposure Increases Susceptibility to Carcinogenesis and Promotes Tumor Progression in Prostate Gland

Dipak K. Sarkar

Abstract The idea that exposure to adverse environmental conditions and lifestyle choices during pregnancy can result in fetal programming that underlies disease susceptibility in adulthood is now widely accepted. Fetal alcohol exposed offspring displays many behavioral and physiological abnormalities including neuroendocrine–immune functions, which often carry over into their adult life. Since the neuroendocrine–immune system plays an important role in controlling tumor surveillance, fetal alcohol exposed offspring can be vulnerable to develop cancer. Animal studies have recently showed increased cancer growth and progression in various tissues of fetal alcohol exposed offspring. I will detail in this chapter the recent evidence for increased prostate carcinogenesis in fetal alcohol exposed rats. I will also provide evidence for a role of excessive estrogenization during prostatic development in the increased incidence of prostatic carcinoma in these animals. Furthermore, I will discuss the additional possibility of the involvement of impaired stress regulation and resulting immune incompetence in the increased prostatic neoplasia in the fetal alcohol exposed offspring.

Keywords Fetal alcohol • Prostate cancer • Estrogen • Beta-endorphin • Stress axis • Innate immune system

---

D.K. Sarkar (✉)
Endocrinology Program and Department of Animal Sciences, Rutgers, The State University of New Jersey, 67 Poultry Farm Road, New Brunswick, NJ 08901, USA
e-mail: sarkar@aesop.rutgers.edu

V. Vasiliou et al. (eds.), *Biological Basis of Alcohol-Induced Cancer*,
Advances in Experimental Medicine and Biology 815,
DOI 10.1007/978-3-319-09614-8_23

23.1 Introduction

It is well established that there is a positive association between excessive alcohol consumption and increased risk of cancers of various organs [1, 6, 66]. A large number of studies in alcoholic patients showed a strong positive association between heavy alcohol consumption and cancer of the upper digestive tract, colon, lung, and liver [34, 64, 77, 78] and breast tissues [15, 18, 80]. There are also reports showing deleterious effects of repetitive alcohol drinking on prostate epithelial cell in humans and animal models [21, 79]. Clinically, increased risk of prostate cancer has been seen in a large cohort study of heavy alcohol abusers in Denmark [78] and among alcoholics in Sweden who are less than 65 years of age [1]. A similar association of heavy alcohol drinking and increased prostate cancer risk was observed in a recent study conducted in San Francisco with 2,129 participants in the Prostate Cancer Prevention Trial (PCPT) [27]. A prospective associations study between quantity and frequency of alcohol consumption and cancer risk using 323,354 male participants identified a significant positive association between 3 or more drinks on drinking days and the incidence of prostate cancer [15].

Fetal alcohol exposure has also been shown to increase cancer susceptibility of offspring. Clinicians reported that there were many cases of children that came to the hospital with fetal alcohol syndrome in conjunction with benign or malignant tumors, while the clinical cases found didn't show uniformity as to the tumor type [20]. Fetal alcohol exposed children often show mild hyperplasia such as tibial exostoses [5] as well as malignant cancer like embryonal rhabdomyosarcoma of urinary bladder and prostate cancer [8]. A case-control study showed some evidence of an increased risk of childhood acute myeloid leukemia in fetal alcohol exposed children [35]. Maternal consumption of alcohol during pregnancy was also found to be associated with development of testicular cancer in the sons [44]. Several case reports also identified incidence of childhood acute lymphoblastic leukemia, acute myeloid leukemia, brain tumors, neuroblastoma, prostate, and testicular cancer in children born from mothers who abused alcohol during pregnancy (Table 23.1). Animal research also showed that fetal alcohol exposure increases the incidence of tobacco-related pancreatic cancer [73], β-estradiol induced prolactinoma [24, 28], and carcinogen-induced mammary tumor growth [28, 32, 52, 76]. Studies using animal models showed that maternal alcohol consumption during pregnancy advanced the occurrence of malignant mammary tumor phenotype to the offspring [32, 52]. Interestingly, breast and prostate cancer share many similarities, in terms of geographical distribution, risk factors, biomolecular determinants, and natural history of disease [19].

23.2 Fetal Alcohol Promotion of Prostate Cancer

We recently tested whether an analogous circumstance occurs in the prostate because diet and hormones, especially estrogen and androgen, are crucial and interactive players in many biological and pathological processes of tumorigenesis in

Table 23.1 Incidence of cancer in FASD patients

Reference	Gender	Age	FASD	Cancer type
[7]	M	28 months	Yes	Neuroblastoma
[11]	F	3 years 9 months	Yes, with hydantoin	Hodgkin disease
[33]	F	12years 11 months	Yes	Adrenal carcinoma
[37]	M	27 months	Yes	Hepatoblastoma
[38]	M	21 months	Yes	Malignant disease
[39]		25 months	Yes	Ganglioneuroblastoma
[40]	M	110 days	Yes	Neuroblastoma Paravertebral found near kidney
[63]	M	35 months	Yes, and hydantoin	Neuroblastoma
[72]	M	35 months	Yes, and hydantoin	Ganglioneuroblastoma
[83]	M	21 months	Yes	Rhabdomyosarcoma
	F	6 years	Yes	Nephroblastoma
	F	16 months	Yes	Leukemia

breast and prostate. We found that prostates of noncarcinogen-treated animals which were alcohol (6.7 % v/v) exposed during the prenatal period (day 11–21) demonstrated inflammatory cell infiltration and epithelial atypia and increased number of proliferative cells in the ventral lobe of this gland, but the prostate of control animal showed normal cytoarchitecture [46]. Prenatally ethanol exposed rats, when treated with a carcinogen N-nitrosomethylurea (NMU) and testosterone, showed histological evidence for high-grade prostatic intraepithelial neoplasia (PIN) primarily in the ventral prostate, whereas control animals showed only low-grade PIN. Prenatally ethanol exposed rats treated with NMU and testosterone also showed increased number of proliferative cells in the ventral prostate (Fig. 23.1; [46]). These results suggest that prenatal ethanol exposure induces histophysiological changes in the prostate as well as it increases the susceptibility of the prostate to develop neoplasia during adulthood.

23.3 Fetal Alcohol, Prostate Estrogenization, and Cancer

23.3.1 Evidence for an Increased Estrogen Production in the Prostate

Prostate gland development from urogenital sinus (UGS) during fetal life occurs at around 10–12 weeks of gestation in humans and 18.5 embryonic days in rats [57]. The proper development of prostate is largely dependent on the constant supply and binding of circulating testosterone or its more potent metabolite, 5α-dihydrotestosterone (DHT), to androgen receptors in the UGS mesenchyme and further activation of UGS epithelium by mesenchyme. A variation in androgen

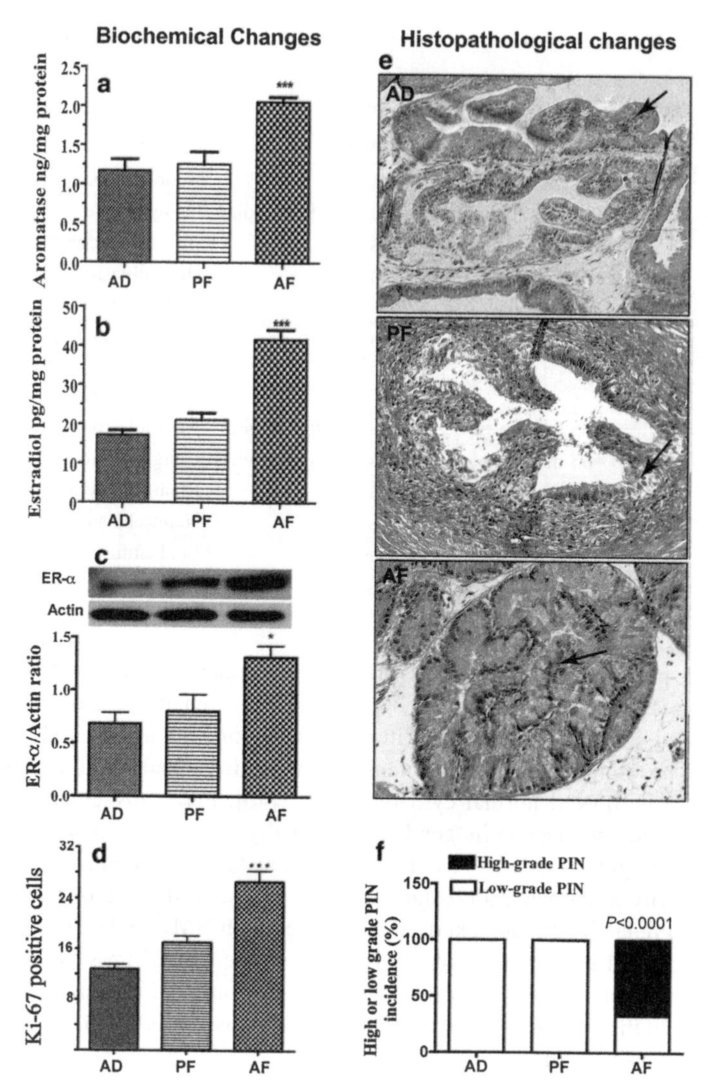

Fig. 23.1 Fetal alcohol exposures increase estrogen production in the prostate gland and promote prostate's susceptibility to carcinogen. (**a–c**) Showing changes in aromatase, estradiol, and estrogen receptor (ER)-α in the ventral prostatic lobe from prenatally fed with control diet (AD, PF) or alcohol diet (AF) and sacrificed at adult age. The mean + SEM tissue concentration of aromatase and estradiol in the ventral prostate measured using ELISA are shown as histograms (**a**, **b**, respectively). Immunoblot of ER-α and actin were presented in represented gel blots and mean + SEM values as histograms (**c**). $N=6$–7 rats. ***, $P<0.001$ vs. AD and AF. (**d–f**) Changes in Ki-67 immunostaining and histopathology in ventral prostatic lobes of adult (7–8 months of age) fetal alcohol exposed offspring treated with NMU and testosterone (treatment details are described in [46]). The average number of Ki-67 from 5 representative slides containing 3 serial sections from each part of the ventral lobe in each group was presented in histograms (**d**). Representative H&E-stained photomicrographs of ventral prostatic lobes of AD, PF, and AF rat offspring (**e**). The ventral lobe of AD and PF rats showed low-grade prostatic intraepithelial neoplasia (PIN; *arrows*), while similar prostatic lobe of AF rats showed high-grade PIN. All the images were captured at 40× magnification. Mean + SEM values of percentage ratio of low-grade and high-grade PIN in animals after carcinogen treatment in AD, PF, and AF groups are shown as histograms (**f**). $N=9$–14 rats. ***, $P<0.001$ vs. AD and AF. Adopted from Murugan S, Zhang C, Mojtahedzadeh S, Sarkar DK (2013) Alcohol Clin Exp Res doi: 10.1111/acer.12171

synthesis and secretion and/or delivery to the UGS has the potential to perturbate normal development [57]. Studies have shown that estrogens are equally involved in the normal and abnormal growth of prostate in human and many rodents [10, 49]. It has been shown that exposure to estrogenic compounds during the neonatal period leads to permanent alterations in prostatic growth and also produced lobe-specific histological changes later in life [22].

Neonatal exposure of rodents to high doses of estrogen permanently imprints the growth and function of the prostate and predisposes the gland to hyperplasia and severe dysplasia analogous to PIN during aging [22]. Because prostatic levels of steroid hormones and their receptors are known to be critically involved in prostate maintenance and pathogenesis [65], we measured the levels of aromatase, an enzyme essential for production of estradiol from testosterone, estrogen, and estrogen receptor-α (ER-α) in the ventral lobe of the prostate of alcohol-fed and control fed rats. We found that the aromatase activity, ER-α immunoreactivity, and estrogen level were higher in the ventral prostate of fetal alcohol exposed offspring (Fig. 23.1). These findings indicate that the prostate gland of fetal alcohol exposed rats produce more estrogen than the control rats.

The initiation of prostatic development is known to be dependent on androgens produced by the fetal testes [54]. In addition, the developing prostate is particularly sensitive to estrogens [55, 67]. Estrogen receptor α (ERα) localizes to proximal mesenchymal cells during early development in rodents and declines as morphogenesis proceeds implicating a specific developmental role for estrogens [55]. ERβ is induced in luminal epithelial cells upon cytodifferentiation suggesting a role for ERβ in differentiated function [41, 58]. In humans, under the influence of maternal estrogens, all fetal males contain marked prostatic squamous metaplasia which sloughs at birth [87]. Additionally, maternal exposure to pharmacological levels of diethylstilbestrol (DES) has been shown to induce prostatic abnormalities in human offspring [23].

Consequently, it is proposed that excessive estrogenization during prostatic development may contribute to the high incidence of prostatic carcinoma observed in the aging male population [67]. Previous works from various laboratories addressing this hypothesis have documented that brief exposure of rodents to high doses of natural or synthetic estrogen early in life results in permanent alterations of the prostate gland, a phenomenon referred to as estrogen imprinting or developmental estrogenization [55, 59, 62]. Organ culture studies confirm a direct effect since the developmental aberrations are recapitulated upon estrogen addition [61]. Using the Sprague–Dawley rat model, it has been shown that brief exposure to high levels of estrogens during the neonatal period (days 1–5) causes marked developmental and cellular differentiation defects [56], dose-dependent reductions in adult prostate size [29, 48, 62], and compromised secretory capacity in adulthood [31, 62]. Upon aging, prostatic inflammation, epithelial hyperplasia, PIN, and eventually adenocarcinoma are prominent [55]. Thus, it appears that developmental estrogen exposures may predispose to neoplasia in adulthood.

23.3.2 Connection between the Developmental Estrogenization and Prostatic Neoplasia

Alcohol consumption has been shown to be associated with a 20 % increase in serum estrogen levels in women [43], and this finding has been supported by rodent studies [32]. Exposure to estrogens in utero [14], or synthetic estrogen diethylstilbestrol (DES) treatment has also been associated with a higher risk for breast cancer in female offspring and prostate cancer in male offspring later in life [23]. It is interesting to note that although the initiation of prostatic development is dependent on androgens produced by the fetal testes [54], the developing prostate is particularly sensitive to estrogens [55, 67]. In the estrogen receptor (ER) positive cell line MCF-7, alcohol treatment resulted in a dose-dependent increase in cell proliferation. In contrast, no increase in proliferation was observed in ER negative cell lines. In the same study, it was shown that alcohol also induced production of ERα protein in a dose-dependent manner [74]. These data suggest that ER-α may be one of the factors involved in alcohol-induced tumorigenesis. It is also interesting to note that combined treatment of estrogen and testosterone induces biochemical and histological carcinogenesis in ER-β-knockout (β-ERKO) mice but not in ER-α-knockout (α-ERKO) mice, suggesting a role for ER-α in prostate carcinogenesis [65]. In humans, under the influence of maternal estrogens, all fetal males contain marked prostatic squamous metaplasia which sloughs at birth [87]. Additionally, maternal exposure to pharmacological levels of diethylstilbestrol (DES) has been shown to induce prostatic abnormalities in human offspring [23].

Consequently, it is proposed that excessive estrogenization during prostatic development may contribute to the high incidence of prostatic carcinoma observed in the aging male population [67]. In this regard, fetal exposure to alcohol has been shown to alter pubertal mammary gland development [53] due to excessive estrogenization [32, 52] and increases aromatase and ER-a levels in prostate [46]. Furthermore, prolonged adult exposure to estradiol at levels within a physiologic range is capable of driving prostatic carcinogenesis in the Noble rat model [82]. Thus, the possibility exists that developmental estrogenic exposures within the prostate may lead to cellular abnormalities and carcinogenesis in the adult life.

23.4 Fetal Alcohol, Neuroimmune Axis Abnormalities, and Cancer

23.4.1 Neuroendocrine–Immune System and Tumor Surveillance

As discussed earlier fetal alcohol exposure increases the incidence of various hormone-dependent and independent cancers. The question arises why the fetal alcohol exposed offspring shows higher tumor incidence in many tissues? It could be hypothesized that many types of cancers in fetal alcohol exposed patients might have been promoted due to an abnormality in the physiological process(s) that prevents cancer development as well as molecular processes (e.g., prostate estrogenization) that control growth and differentiation of a specific cell population. There are two biological systems that are significantly altered in fetal alcohol exposed patients. One relates to stress regulation and the other is the immune system function. Children who are exposed to alcohol during fetal life often show behavioral and physiological changes such as depression, anxiety, hyperactivity, and an inability to deal with stressful situations [30, 75]. Fetal alcohol exposed patients and prenatal ethanol exposed rats often show elevated basal and stimulated levels of the adrenal hormone glucocorticoid, pituitary hormone adrenocorticotropic hormone (ACTH), and hypothalamic hormone corticotropin-releasing hormone (CRH) in response to stressors such as repeated restraint, foot shock, and immune challenges [13, 42, 85]. Children prenatally exposed to alcohol have lower cell counts of eosinophils and neutrophils, decreased circulating E-rosette-forming lymphocytes, reduced mitogen-stimulated proliferative responses by peripheral blood leukocytes, and hypo-c-globulinemia [35]. In animal models, fetal alcohol exposure is shown to negatively affect lymphoid tissue development, immune cell function, humoral immunity, and cytokine productions [4]. Activity of NK cells is also suppressed in these animals [13]. Children prenatally exposed to alcohol often have an increased incidence of bacterial infections such as urinary tract and upper respiratory tract infections [35].

Studies from human and animal models have shown that acute and chronic stressful events have adverse effects on a variety of immunological mechanisms, such as trafficking of neutrophils, macrophages, antigen-presenting cells, natural killer (NK) cells, and T and B lymphocytes [9, 16]. Exposure to stress modulates cell-mediated immunity by suppressing lymphocyte proliferation and NK activation, lowering the number of CD4+ cells in the peripheral blood and altering CD4/CD8 T cell ratios [9, 81]. Studies have also shown that depression and stress might have effects on carcinogenesis indirectly, through the poorer destruction or elimination of abnormal cells by reduced NK cell activity. Decreased NK cell activity is also associated with growth and progression of a variety of cancers in animals and humans, because NK cells appear to represent a first line of defense against the metastatic spread of tumor cells [50]. Stress is also associated with altered inflammatory and anti-inflammatory cytokine ratios in systemic circulation, increases expression of interleukin-1 beta (IL-1ß) and tumor necrosis factor-alpha (TNF-α), and reduced expression of IL-2 and interferon-gamma (IFN-γ) [2]. Sustained elevation of TNF-α is known to inhibit the activity of protein tyrosine phosphatase (PTPase), causing reduced production of the MHC class I antigen of the cell surface, and leading to malignant cells escaping immune surveillance [45]. Although there are many specific details yet to be delineated, it is becoming increasingly clear that stressful life events can impact cancer growth, progression, and metastasis by modulating nervous, endocrine, and immune systems of the body. Thus, reduced immune function and increased susceptibility to cancer may be a consequence of the stress axis abnormalities in fetal alcohol exposed offspring.

23.4.2 *Neuroendocrine–Immune System Abnormalities and Prostatic Neoplasia*

As discussed earlier, studies in animals indicated that fetal alcohol exposed offspring show elevated basal and hyper stress response to stressors and immune challenges [13, 35, 42, 85]. Can these changes of the stress axis relate to an abnormality in the feedback regulation of the axis? It is known that beta-endorphin (BEP), a hormone release from a set of neurons in the hypothalamus inhibits CRH secretion and downregulates the HPA axis functions. Fetal alcohol decreases the levels of BEP in the hypothalamus during adult life [17, 68]. Furthermore, transplantation of BEP neurons in the hypothalamus normalizes the HPA dysfunction in fetal alcohol exposed offspring [13]. Therefore, the stress hyperactivity in fetal alcohol exposed offspring may have resulted from a decreased BEP neuronal function Fig. 23.2.

BEP is an endogenous opioid polypeptide, a cleavage product of proopiomelanocortin (POMC), which is also the precursor hormone for adrenocorticotrophic hormone (ACTH) and α-melanocyte-stimulating hormone (α-MSH). POMC products are produced and secreted by the hypothalamus (BEP and α-MSH) and pituitary gland (ACTH, α-MSH and BEP) in vertebrates during exercise, excitement, pain, and orgasm, and they resemble the opiates in their abilities to produce analgesia and a feeling of well-being [3, 26].

BEP neuronal cell bodies are primarily localized in the arcuate nuclei of the hypothalamus, and its terminals are distributed throughout the CNS, including the paraventricular nucleus of the hypothalamus [69]. In the paraventricular nucleus these neurons innervate CRH neurons and inhibit CRH release [36]. During stress, secretion of CRH and catecholamine stimulate secretion of hypothalamic BEP and other POMC-derived peptides, which in turn inhibit the activity of the stress system [51]. BEP is known to bind to δ- and μ-opioid receptors and modulate the neurotransmission in sympathetic neurons via neuronal circuitry within the paraventricular nucleus to alter NK cell cytolytic functions in the spleen [12, 13]. Low levels of hypothalamic BEP are correlated with a higher incidence of cancers and infections in patients with schizophrenia, depression, and fetal alcohol syndrome and in obese patients (reviewed in [84]). Hence, hypothalamic BEP inhibits CRH secretion and sympathetic outflow to the lymphoids and stimulates immunity.

We have recently shown that in vitro produced BEP neurons, when transplanted in the paraventricular nucleus of the hypothalamus, remained at the site of transplantation, and increase NK cell cytolytic function and production of anti-inflammatory cytokines in response to an immune challenge [69, 70]. BEP transplantation in the paraventricular nucleus of the hypothalamus also suppresses carcinogen-induced prostate and mammary tumorigenesis. Importantly, when the BEP transplants were given at an early stage of tumor development, many tumors were destroyed possibly due to increased innate immune activity, and the surviving tumors lost their ability to progress to high-grade cancer due to BEP cells' suppressive effects on epithelial–mesenchymal transition (EMT) regulators. Another remarkable effect of the BEP transplantation was that it promoted the activation of the innate immune activity following tumor cell invasion to such an extent that

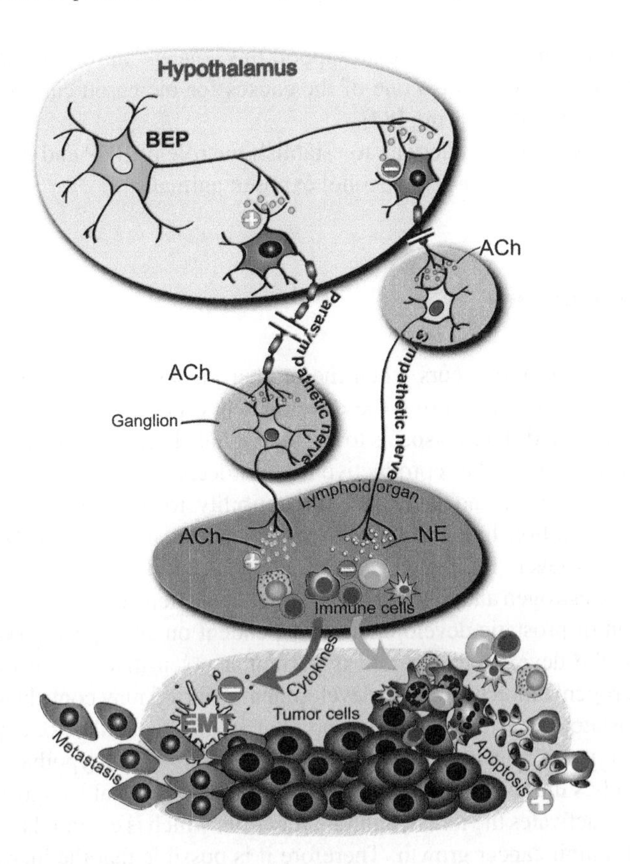

Fig. 23.2 Beta-endorphin (BEP) neuron in the hypothalamus controls the growth and progression of tumor cells by modulating the neurotransmission in the autonomic nervous system and activating innate immune system. Effects include the stimulation of parasympathetic nervous system and release of acetyl choline (Ach) and suppression of the sympathetic nervous system and release of norepinephrine (NE) leading to activation of innate immune cells (including macrophages and natural killer cells) of the lymphoid organ and an increase in cytotoxic immune cells and anti-inflammatory cytokine levels in the circulation. In a tumor microenvironment these immune cell and cytokine changes increase apoptotic death of tumor cells and reduce inflammation-mediated epithelial-mesenchymal transition (EMT), and thereby suppress cancer growth and metastasis. Collectively, these effects create an unfavorable environment for tumor initiation, growth, and progression. From: Zhang C, Sarkar DK (2012) Oncoimmunology 1(4):552-554

tumor cell migration to another site was completely halted. The cellular mechanism involved in the cancer preventive effects of BEP appears to involve alteration of the autonomic nervous system functioning, leading to activation of innate immunity and reduction in systemic levels of inflammatory and anti-inflammatory cytokine ratios. Since fetal alcohol exposed animals have BEP neuronal deficiencies and higher incidence of aggressive prostate tumors following a carcinogen treatment, the possibility is raised that the stress abnormalities and resulting immune incompetence might have contributed to the increased prostatic neoplasia in these animals. Our recent preliminary data support this concept and show that BEP neural functional abnormality is also one of the causes for increased cancer incidence in fetal alcohol exposed offspring [84].

Studies are currently underway to establish the role of BEP and its potential use in prevention of cancers in fetal alcohol exposed animals.

23.5 Conclusions

Fetal alcohol exposure occurs when the mother drinks alcohol during pregnancy. This exposure to alcohol during the developmental period as a fetus will cause a series of defects in different aspects to the individual. These defects include elevated estradiol production in the reproductive tissues, decreased BEP neuronal numbers, increased stress level, and increased susceptibility to different types of cancers. Alcohol consumption has been shown to be associated with a 20 % increase in serum estrogen levels in women. Fetal alcohol exposed offspring also appears to produce more estrogen and ER-α in the prostate. It is interesting to note that although the initiation of prostatic development is dependent on androgens produced by the fetal testes, the developing prostate is particularly sensitive to estrogens. Hence, prostate estrogenization during the developmental period may contribute to a higher risk for prostate cancer in male offspring later in life. Fetal alcohol exposures also alter stress axis function partly by killing BEP neurons in the hypothalamus, and by reducing body's immunity. Previous studies in rats showed that increasing BEP neuronal activity activates the innate immune function, which is essential to fight against cancer and inhibit cancer growth. Therefore it is possible that the increased cancer susceptibility in fetal alcohol rats is also partially due to the decrease. BEP neuronal number leads to a hyperactive stress axis and a suppressed immune system. Further studies are needed to determine the exact cause(s) for the higher incidence of neoplasm in the prostate and other tissues in fetal alcohol exposed animals and patients.

Acknowledgment This work was partly supported by National Institute of Health grants R37AA08757 and R01AA11591. Prostate research works were conducted in our laboratory by Drs. Sengottuvelan Murugan, Changqing Zhang, Sepideh Mojtahedzadeh, and Nadka Boyadjieva.

References

1. Adami HO, McLaughlin JK, Hsing AW, Wolk A, Ekbom A, Holmberg L, Persson I (1992) Alcoholism and cancer risk: a population-based cohort study. Cancer Causes Control 3(5):419–425
2. Adler UC, Marques AH, Calil HM (2008) Inflammatory aspects of depression. Inflamm Allergy Drug Targets 7(1):19–23
3. Akil H, Watson SJ, Young E et al (1984) Endogenous opioids: biology and function. Annu Rev Neurosci 7:223–255

4. Arjona A, Boyadjieva N, Kuhn P, Sarkar DK (2006) Fetal ethanol exposure disrupts the daily rhythms of splenic granzyme B, IFN-gamma, and NK cell cytotoxicity in adulthood. Alcohol Clin Exp Res 30(6):1039–1104
5. Azouz EM, Kavianian G, Der Kaloustian VM (1993) Fetal alcohol syndrome and bilateral tibial exostoses. A case report. Pediatr Radiol 23(8):615–616
6. Bagnardi V, Blangiardo M, La Vecchia C, Corrao G (2001) A meta-analysis of alcohol drinking and cancer risk. Br J Cancer 85(11):1700–1705
7. Battisti L, Degani D, Rugolotto S, Borgna-Pignatti C (1993) Fetal alcohol syndrome and malignant disease: a case report. Am J Pediatr Hematol Oncol 5(1):136–137
8. Becker H, Zaunschirm A, Muntean W, Domej W (1982) Fetal alcohol syndrome and malignant tumors. Wien Klin Wochenschr 94(14):364–365
9. Ben-Eliyahu S (2013) The promotion of tumor metastasis by surgery and stress: immunological basis and implications for psychoneuroimmunology. Brain Behav Immun 17(Suppl 1):S27–S36
10. Bonkhoff H, Berges R (2009) The evolving role of oestrogens and their receptors in the development and progression of prostate cancer. Eur Urol 55(3):533–542
11. Bostrom B, Nesbit ME Jr (1983) Hodgkin disease in a child with fetal alcohol-hydantoin syndrome. J Pediatr 103(5):760–762
12. Boyadjieva N, Advis JP, Sarkar DK (2006) Role of beta-endorphin, corticotropin-releasing hormone and autonomic nervous system in mediation of the effect of chronic ethanol on natural killer cell cytolytic activity. Alcohol Clin Exp Res 30(10):1761–1767
13. Boyadjieva NI, Ortigüela M, Arjona A et al (2009) Beta-endorphin neuronal cell transplant reduces corticotropin releasing hormone hyperresponse to lipopolysaccharide and eliminates natural killer cell functional deficiencies in fetal alcohol exposed rats. Alcohol Clin Exp Res 33(5):931–937
14. Braun MM, Ahlbom A, Floderus B et al (1995) Effect of twinship on incidence of cancer of the testis, breast, and other sites (Sweden). Cancer Causes Control 6(6):519–524
15. Breslow RA, Chen CM, Graubard BI, Mukamal KJ (2011) Prospective study of alcohol consumption quantity and frequency and cancer-specific mortality in the US population. Am J Epidemiol 174(9):1044–1053
16. Capuron L, Miller AH (2011) Immune system to brain signaling: neuropsychopharmacological implications. Pharmacol Ther 130(2):226–238
17. Chen CP, Kuhn P, Advis JP, Sarkar DK (2006) Prenatal ethanol exposure alters the expression of period genes governing the circadian function of beta-endorphin neurons in the hypothalamus. J Neurochem 97(4):1026–1033
18. Chen WY, Rosner B, Hankinson SE et al (2011) Moderate alcohol consumption during adult life, drinking patterns, and breast cancer risk. JAMA 306(17):1884–1890
19. Coffey DS (2001) Similarities of prostate and breast cancer: evolution, diet, and estrogens. Urology 57(4 Suppl 1):31–38
20. Cohen MM Jr (1981) Neoplasia and the fetal alcohol and hydantoin syndromes. Neurobehav Toxicol Teratol 3(2):161–162
21. Crispo A, Talamini R, Gallus S et al (2004) Alcohol and the risk of prostate cancer and benign prostatic hyperplasia. Urology 64(4):717–722
22. Cunha GR, Wang YZ, Hayward SW, Risbridger GP (2001) Estrogenic effects on prostatic differentiation and carcinogenesis. Reprod Fertil Dev 13(4):285–296
23. Driscoll SG, Taylor SH (1980) Effects of prenatal maternal estrogen on the male urogenital system. Obstet Gynecol 56(5):537–542
24. Gangisetty O, Wynne O, Radl DB, Maglakelidze G, Sarkar DK (2013) Fetal alcohol exposures epigenetically program the growth regulatory mechanism and increase the susceptibility to prolactin-secreting pituitary tumors. Alcohol Clin Exp Res 37:64A
25. Giovannucci E, Michaud D (2007) The role of obesity and related metabolic disturbances in cancers of the colon, prostate, and pancreas. Gastroenterology 132(6):2208–2225
26. Goldfarb AH, Jamurtas AZ (1997) Beta-endorphin response to exercise. An update. Sports Med 24(1):8–16

27. Gong Z, Kristal AR, Schenk JM et al (2009) Alcohol consumption, finasteride, and prostate cancer risk: results from the Prostate Cancer Prevention Trial. Cancer 115(16):3661–3669
28. Gottesfeld Z, Trippe K, Wargovich MJ, Berkowitz AS (1992) Fetal alcohol exposure and adult tumorigenesis. Alcohol 9(6):465–471
29. Greene DR, Wheeler TM, Egawa S et al (1991) Relationship between clinical stage and histological zone of origin in early prostate cancer: morphometric analysis. Br J Urol 68(5): 499–509
30. Hellemans KG, Sliwowska JH, Verma P, Weinberg J (2010) Prenatal alcohol exposure: fetal programming and later life vulnerability to stress, depression and anxiety disorders. Neurosci Biobehav Rev 34(6):791–807
31. Higgins SJ, Brooks DE, Fuller FM et al (2007) Steroid hormones and carcinogenesis of the prostate: the role of estrogens. Differentiation 75(9):871–882
32. Hilakivi-Clarke L, Cabanes A, de Assis S et al (2004) In utero alcohol exposure increases mammary tumorigenesis in rats. Br J Cancer 90(11):2225–2231
33. Hornstein L, Crowe C, Gruppo R (1977) Adrenal carcinoma in child with history of fetal alcohol syndrome. Lancet 2(8051):1292–1293
34. Infante-Rivard C, El-Zein M (2007) Parental alcohol consumption and childhood cancers: a review. J Toxicol Environ Health B Crit Rev 10(1–2):101–129
35. Johnson S, Knight R, Marmer DJ, Steele RW (1981) Immune deficiency in fetal alcohol syndrome. Pediatr Res 15(6):908–911
36. Kawano H, Masuko S (2007) Beta-endorphin-, adrenocorticotrophic hormone- and neuropeptide y-containing projection fibers from the arcuate hypothalamic nucleus make synaptic contacts on to nucleus preopticus medianus neurons projecting to the paraventricular hypothalamic nucleus in the rat. Neuroscience 98(3):555–565
37. Khan A, Bader JL, Hoy GR, Sinks LF (1979) Hepatoblastoma in child with fetal alcohol syndrome. Lancet 1(8131):1403–1404
38. Kiess W, Linderkamp O, Hadorn HB, Haas R (1984) Fetal alcohol syndrome and malignant disease. Eur J Pediatr 143(2):160–161
39. Kiley VA, Lazerson J (1986) Agenesis of the corpus callosum and neural crest tumors. Pediatr Hematol Oncol 3(2):179–182
40. Kinney H, Faix R, Brazy J (1980) The fetal alcohol syndrome and neuroblastoma. Pediatrics 66(1):130–132
41. Lau KM, Leav I, Ho SM (1998) Rat estrogen receptor-α and -β, and progesterone receptor mRNA expression in various prostatic lobes and microdissected normal and dysplastic epithelial tissues of the Noble rats. Endocrinology 139(1):424–427
42. Lee S, Schmidt D, Tilders F, Rivier C (2000) Increased activity of the hypothalamic-pituitary adrenal axis of rats exposed to alcohol in utero: role of altered pituitary and hypothalamic function. Mol Cell Neurosci 16(4):515–528
43. Maskarinec G, Morimoto Y, Takata Y et al (2006) Alcohol and dietary fibre intakes affect circulating sex hormones among premenopausal women. Public Health Nutr 9(7):875–881
44. Mongraw-Chaffin ML, Cohn BA, Anglemyer AT et al (2009) Maternal smoking, alcohol, and coffee use during pregnancy and son's risk of testicular cancer. Alcohol 43(3):241–245
45. Moreno-Smith M, Lutgendorf SK, Sood AK (2010) Impact of stress on cancer metastasis. Future Oncol 6(12):1863–1881
46. Murugan S, Zhang C, Mojtahedzadeh S, Sarkar DK (2013) Alcohol exposure in utero increases susceptibility to prostate tumorigenesis in rat offspring. Alcohol Clin Exp Res 37:1901–1909. doi:10.1111/acer.12171
47. Narod SA (2011) Alcohol and risk of breast cancer. JAMA 306(17):1920–1921
48. Naslund MJ, Coffey DS (1986) The differential effects of neonatal androgen, estrogen and progesterone on adult rat prostate growth. J Urol 136(5):1136–1140
49. Nicholson TM, Ricke WA (2011) Androgens and estrogens in benign prostatic hyperplasia: past, present and future. Differentiation 82(4–5):184–199

50. Padgett DA, Glaser R (2003) How stress influences the immune response. Trends Immunol 24:444–448
51. Plotsky PM, Thrivikraman KV, Meaney MJ (1993) Central and feedback regulation of hypothalamic corticotropin-releasing factor secretion. Ciba Found Symp 172:59–75
52. Polanco TA, Crismale-Gann C, Reuhl KR et al (2010) Fetal alcohol exposure increases mammary tumor susceptibility and alters tumor phenotype in rats. Alcohol Clin Exp Res 34(11):1879–1887
53. Polanco TA, Crismale-Gann C, Cohick WS (2011) Alcohol exposure in utero leads to enhanced prepubertal mammary development and alterations in mammary IGF and estradiol systems. Horm Cancer 2(4):239–248
54. Price D (1963) Comparative aspects of development and structure in the prostate. In: Vollmer EP (ed) Biology of the prostate and related tissues. National Cancer Institute, Washington DC, pp 1–27
55. Prins GS (1997) Developmental estrogenization of the prostate gland. In: Naz RK (ed) Prostate: basic and clinical aspects. CRC Press, Boca Raton, pp 247–265
56. Prins GS, Birch L (1995) The developmental pattern of androgen receptor expression in rat prostate lobes is altered after neonatal exposure to estrogen. Endocrinology 136(3):1303–1314
57. Prins GS, Putz O (2008) Molecular signaling pathways that regulate prostate gland development. Differentiation 76(6):641–659
58. Prins GS, Marmer M, Woodham C et al (1998) Estrogen receptor-β messenger ribonucleic acid ontogeny in the prostate of normal and neonatally estrogenized rats. Endocrinology 139(3):874–883
59. Prins GS, Birch L, Habermann H et al (2001) Influence of neonatal estrogens on rat prostate development. Reprod Fertil Dev 13(4):241–252
60. Putz O, Prins GS (2002) Prostate gland development and estrogenic imprinting. In: Burnstein KL (ed) Steroid hormones and cell cycle regulation. Kluwer Academic Publishers, Boston, MA, pp 73–89
61. Pylkkanen L, Makela S, Valve E et al (1993) Prostatic dysplasia associated with increased expression of C-MYC in neonatally estrogenized mice. J Urol 149(6):1593–1601
62. Rajfer J, Coffey DS (1978) Sex steroid imprinting of the immature prostate. Invest Urol 16(3):186–190
63. Ramilo J, Harris VJ (1979) Neuroblastoma in a child with the hydantoin and fetal alcohol syndrome. The radiographic features. Br J Radiol 52(624):993–935
64. Rehm J, Baliunas D, Borges GL et al (2010) The relation between different dimensions of alcohol consumption and burden of disease: an overview. Addiction 105(5):817–843
65. Ricke WA, McPherson SJ, Bianco JJ et al (2008) Prostatic hormonal carcinogenesis is mediated by in situ estrogen production and estrogen receptor alpha signaling. FASEB J 22(5):1512–1520
66. Rothman KJ (1980) The proportion of cancer attributable to alcohol consumption. Prev Med 9(2):174–179
67. Santti R, Newbold RR, Makela S et al (1994) Developmental estrogenization and prostatic neoplasia. Prostate 24(2):67–78
68. Sarkar DK, Kuhn P, Marano J et al (2007) Alcohol exposure during the developmental period induces beta-endorphin neuronal death and causes alteration in the opioid control of stress axis function. Endocrinology 148(6):2828–2834
69. Sarkar DK, Boyadjieva NI, Chen CP et al (2008) Cyclic adenosine monophosphate differentiated beta-endorphin neurons promote immune function and prevent prostate cancer growth. Proc Natl Acad Sci U S A 105(26):9105–9110
70. Sarkar DK, Zhang C, Murugan S (2011) Transplantation of β-endorphin neurons into the hypothalamus promotes immune function and restricts the growth and metastasis of mammary carcinoma. Cancer Res 71(19):6282–6291
71. Sarkar DK, Murugan S, Zhang C, Boyadjieva N (2012) Regulation of cancer progression by β- endorphin neuron. Cancer Res 72(4):836–480

72. Seeler RA, Israel JN, Royal JE, Kaye CI, Rao S, Abulaban M (1979) Ganglioneuroblastoma and fetal hydantoin-alcohol syndromes. Pediatrics 63(4):524–527
73. Severson RK, Buckley JD, Woods WG et al (1993) Cigarette smoking and alcohol consumption by parents of children with acute myeloid leukemia: an analysis within morphological subgroups—a report from the Children's Cancer Group. Cancer Epidemiol Biomarkers Prev 2(5):433–439
74. Singletary KW, Frey RS, Yan W (2001) Effect of ethanol on proliferation and estrogen receptor-alpha expression in human breast cancer cells. Cancer Lett 165(2):131–137
75. Steinhausen HC (1995) Children of alcoholic parents. A review. Eur Child Adolesc Psychiatry 4(3):143–152
76. Taylor AN, Ben-Eliyahu S, Yirmiya R et al (1993) Action of alcohol on immunity and neoplasia in fetal alcohol exposed and adult rats. Alcohol Alcohol Suppl 2:69–74
77. Thygesen LC, Albertsen K, Johansen C, Grønbaek M (2005) Cancer incidence among Danish brewery workers. Int J Cancer 116(5):774–778
78. Tønnesen H, Moller H, Andersen JR et al (1994) Cancer morbidity in alcohol abusers. Br J Cancer 69(2):327–332
79. Velicer CM, Kristal A, White E (2006) Alcohol use and the risk of prostate cancer: results from the VITAL cohort study. Nutr Cancer 56(1):50–56
80. Winstanley MH, Pratt IS, Chapman K (2011) Alcohol and cancer: a position statement from Cancer Council Australia. Med J Aust 194(9):479–482
81. Witek-Janusek L, Gabram S, Mathews HL (2007) Psychologic stress, reduced NK cell activity, and cytokine dysregulation in women experiencing diagnostic breast biopsy. Psychoneuroendocrinology 32(1):22–35
82. Yuen MT, Leung LK, Wang J et al (2005) Enhanced induction of prostatic dysplasia and carcinoma in Noble rat model by combination of neonatal estrogen exposure and hormonal treatments at adulthood. Int J Oncol 27(6):1685–1695
83. Zaunschirm A, Muntean W (1984) Fetal alcohol syndrome and malignant disease. Eur J Pediatr 141(4):256
84. Zhang C, Sarkar DK (2012) β-endorphin neuron transplantation: a possible novel therapy for cancer prevention. Oncoimmunology 1(4):552–554
85. Zhang X, Sliwowska JH, Weinberg J (2005) Prenatal alcohol exposure and fetal programming effects on neuroendocrine and immune function. Exp Biol Med (Maywood) 230(6):376–388
86. Zhang C, Boyadjieva N, Ortiguela M et al (2012) Fetal alcohol exposure increases tumor colonization in the lung: role of hypothalamic beta-endorphin. Alcohol Clin Exp Res 36:198A
87. Zondek T, Mansfield MD, Attree SL, Zondek LH (1986) Hormone levels in the fetal and neonatal prostate. Acta Endocr 112(3):447–456

Chapter 24
Fetal Alcohol Exposure and Mammary Tumorigenesis in Offspring: Role of the Estrogen and Insulin-Like Growth Factor Systems

Wendie S. Cohick, Catina Crismale-Gann, Hillary Stires, and Tiffany A. Katz

Abstract Fetal alcohol spectrum disorders affect a significant number of live births each year, indicating that alcohol consumption during pregnancy is an important public health issue. Environmental exposures and lifestyle choices during pregnancy may affect the offspring's risk of disease in adulthood, leading to the idea that a woman's risk of breast cancer may be pre-programmed prior to birth. Exposure of pregnant rats to alcohol increases tumorigenesis in the adult offspring in response to mammary carcinogens. The estrogen and insulin-like growth factor (IGF-I) axes occupy central roles in normal mammary gland development and breast cancer. 17-β estradiol (E2) and IGF-I synergize to regulate formation of terminal end buds and ductal elongation during pubertal development. The intracellular signaling pathways mediated by the estrogen and IGF-I receptors cross-talk at multiple levels through both genomic and non-genomic mechanisms. Several components of the E2 and IGF-I systems are altered in early development in rat offspring exposed to alcohol in utero, therefore, these changes may play a role in the enhanced susceptibility to mammary carcinogens observed in adulthood. Alcohol exposure in utero induces a number of epigenetic alterations in non-mammary tissues in the offspring and other adverse in utero exposures induce epigenetic modifications in the mammary gland. Future studies will determine if fetal alcohol exposure can induce epigenetic modifications in genes that regulate E2/IGF action at key phases of mammary development, ultimately leading to changes in susceptibility to carcinogens.

Keywords Fetal alcohol • Mammary tumorigenesis • Fetal origins • IGF-I • Estrogen

W.S. Cohick (✉) • C. Crismale-Gann • H. Stires
Department of Animal Sciences, Rutgers, The State University of New Jersey, New Brunswick, NJ 08901-8520, USA
e-mail: cohick@aesop.rutgers.edu; ccrismal@scarletmail.rutgers.edu; h.stires@gmail.com

T.A. Katz
Hillman Cancer Center, University of Pittsburg Cancer Institute, Pittsburgh, PA 15232, USA
e-mail: polancot@upmc.edu

V. Vasiliou et al. (eds.), *Biological Basis of Alcohol-Induced Cancer*, Advances in Experimental Medicine and Biology 815,
DOI 10.1007/978-3-319-09614-8_24

24.1 Introduction

It has long been recognized that prenatal alcohol exposure can lead to fetal alcohol spectrum disorder (FASD), the most severe case being fetal alcohol syndrome (FAS) [1]. FASDs are characterized by facial abnormalities, growth deficiencies, central nervous system dysfunction, intellectual impairment, and behavioral problems, leading to long-term adverse outcomes [2, 3]. It is estimated that FASDs affect at least 1 % of all live births in the United States each year [4]. In response to these data, the US Surgeon General issued an advisory in 2005 stating that women who are pregnant or who are planning on becoming pregnant should abstain from drinking alcohol [5]. However, despite this information, many women continue to consume alcohol during pregnancy. The Centers for Disease Control report that 51.5 % of women who are of child-bearing age and 7.6 % of pregnant women consume alcohol. Of pregnant women surveyed, 1.4 % self-reported binge drinking in the United States [6]. In a study conducted in Australia, New Zealand, Ireland, and the UK 60 % of women reported drinking during pregnancy, with 16 % reporting moderate (8–14 units per week) or high alcohol (14 or more units per week) intake [7]. Collectively these statistics indicate that alcohol consumption during pregnancy is an important public health issue worldwide.

Breast cancer is the most common form of cancer among women and the National Cancer Institute estimates that there will be 232,340 new cases in the United States in 2013 [8]. Many factors contribute to a woman's risk of breast cancer including reproductive parameters (e.g., age at menarche and menopause, age at first full-term pregnancy, parity), genetics, and obesity [9–11]. In addition, epidemiological analyses and animal studies have shown that alcohol consumption is associated with a higher risk of breast cancer [12–16]. A newly emerging concept is that environmental exposures and lifestyle choices during pregnancy may affect the offspring's risk of disease in adulthood [17–20]. This has led to the idea that a woman's breast cancer risk may be pre-programmed prior to birth [21–23]. In support of this hypothesis, Hilakivi-Clarke and colleagues found that alcohol exposure in utero increases mammary tumor susceptibility to the carcinogen 7,12-dimethylbenz(a) anthracene (DMBA) in adult offspring [24]. In this study pregnant Sprague Dawley rats were fed a liquid diet containing either low or moderate levels of alcohol from days 7–19 of pregnancy or pair-fed an isocaloric liquid diet. Offspring exposed to the moderate, but not low, levels of alcohol developed more tumors as a group relative to the isocaloric pair-fed controls (Fig. 24.1a). Given the number of children born with FASD each year, which is associated with consuming more alcohol than used in the Hilakivi-Clarke study, we fed pregnant Sprague Dawley dams a high level of alcohol. Dams were acclimated to the alcohol diet by feeding a liquid diet containing 2.2 % ethanol on days 7 and 8 and 4.4 % ethanol on days 9 and 10 of gestation. Once acclimated, dams were fed the liquid diet containing 6.7 % ethanol, representing 35 % of total calories, from days 11–21 of pregnancy [25]. Control dams were fed an isocaloric liquid diet with the alcohol calories replaced by maltose-dextrin (pair-fed) or rat chow ad libitum (ad lib-fed). At birth, female offspring were cross-fostered to ad lib-fed mothers so that female offspring were

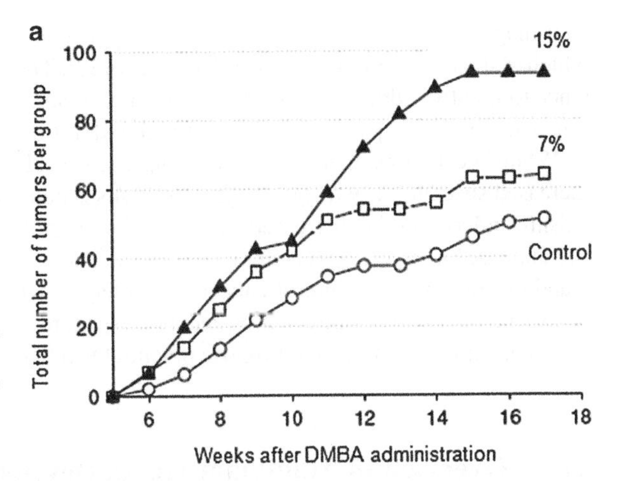

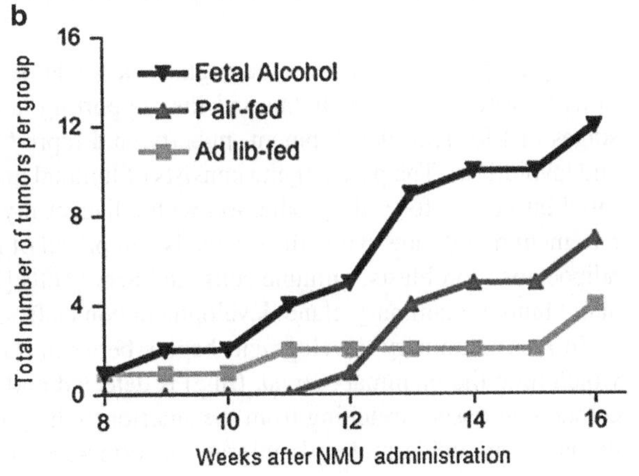

Fig. 24.1 Alcohol exposure in utero increases total number of mammary tumors in offspring. (**a**) Rats exposed in utero to low (7 % of total calories) or moderate (15 % of total calories) alcohol or an isocaloric liquid diet (control) were injected with DMBA at 47 days of age. Adapted by permission from Macmillan Publishers Ltd: on behalf of Cancer Research UK [24]. (**b**). Rats exposed to high alcohol (35 % of total calories), an isocaloric liquid diet (pair-fed), or ad libitum rat chow (ad lib-fed) were treated with NMU at approximately 50 days of age. Reproduced with kind permission from Springer Science + Business Media [131]

only exposed to alcohol in the fetal environment. Offspring were exposed to 50 mg *N*-nitroso-*N*-methylurea (NMU)/kg body weight at day 50 of life by intraperitoneal injection [26]. Similar to Hilakivi-Clarke et al. [24] we found an increased number of tumors in the alcohol-exposed offspring at 16 weeks post-NMU injection compared to either control group (Fig. 24.1b). The increased number of tumors per group was a function of both increased incidence (i.e., number of animals with tumors) as well as increased multiplicity (more tumors per animal). Furthermore, in this study the tumors from the alcohol-exposed offspring developed earlier and had a more malignant phenotype (i.e., more adenocarcinomas and more estrogen receptor (ER)-α-negative tumors) [25].

These studies suggest that an additional adverse outcome of alcohol consumption during pregnancy could be that women born to these mothers have increased risk of developing breast cancer as adults. Fetal exposure to alcohol has been shown to increase risk for childhood leukemia, suggesting that findings in rodent models may be relevant to human disease [27]. There is considerable discussion surrounding the over-diagnosis and overtreatment of breast atypia due to the difficulty in discerning which of these will progress to invasive cancer [28]. Therefore, it is of utmost importance to identify additional risk factors for breast cancer so that we can clearly distinguish which groups of women are at high risk for developing the disease.

While the finding that rodent offspring born to alcohol-exposed dams exhibit increased susceptibility to carcinogens as adults is consistently observed, the mechanisms underlying this increased susceptibility are not well understood. The estrogen and insulin-like growth factor (IGF) axes occupy central roles in both mammary gland development and breast cancer. This chapter will discuss evidence that this endocrine axis may be affected by alcohol exposure in utero leading to increased mammary tumorigenesis and propose potential mechanisms for these alterations.

24.2 Overview of Mammary Gland Development

The development of the mammary gland depends on complex molecular interactions between the parenchyma and the supporting stroma throughout three main stages of life: fetal development, puberty, and reproduction (pregnancy, lactation, and involution). The parenchyma consists of luminal and basal epithelial cells structured into ducts for milk production with adjacent myoepithelial cells that secrete basement membrane, while the stroma is comprised of multiple cell types including adipocytes, fibroblasts, immune cells, and nerve cells [29]. Much of what is known about human mammary gland development comes from rodent models [30].

In rats, mammary development begins between day 10 and 11 of gestation at which time the mammary streak (line) is detected and consists of a single layer of ectodermal tissue stretching from the anterior to the posterior limb buds [31]. Over the next day, epidermal cells migrate to form several layers of cells, creating the mammary placodes [32, 33]. In rats there are six pairs of mammary placodes which correspond to the number of mammary glands in the adult. Each mammary placode invaginates into the underlying mesenchyme to become a bulb-shaped mammary bud. The surrounding mesenchymal cells reorient themselves into a radial pattern surrounding the parenchyma to form a pad of fatty connective tissue (the fat pad precursor). After a 3–4 day morphological quiescence there is a period of rapid cellular proliferation, which results in the mammary sprout. This elongates, forming ducts that penetrate the fat pad precursor such that at birth, the gland is comprised of a limited number of branching ducts [31, 34, 35].

After birth the mammary gland continues to grow isometrically with the rest of the body until the onset of ovarian hormones during puberty, when growth becomes allometric. The leading edge of each duct enlarges to form a club-shaped structure called a terminal end bud (TEB), which contains highly proliferative cells. Once the ducts have penetrated the fat pad and undergone initial bifurcations, the complexity of the milk duct system increases through side branching during each estrous cycle. During pregnancy, ductal side branching becomes even more extensive. Subsequently during lactation the gland becomes mature and fully functional. The epithelial component of the gland proliferates and fills the fat pad with lobular alveolar structures. The alveoli differentiate to produce milk proteins which are secreted into the ductal lumen and transported to the nipple. Once suckling ceases the gland undergoes involution, during which epithelial cells undergo apoptosis and the alveolar structures dedifferentiate [35].

24.3 Role of Estrogen in Mammary Gland Development and Breast Cancer

Estrogens are steroid hormones that are produced mainly in the ovaries in pre-menopausal women, with minor contributions from adipose tissue and the adrenal glands. 17β-estradiol (E2) is the major circulating estrogen, but minor forms include estrone (E1) and estriol (E3), with E3 produced by the placenta during pregnancy. The hypothalamic-pituitary-gonadal (HPG) axis regulates ovarian production of estrogen. The hypothalamic peptide gonadotropin-releasing hormone (GnRH) induces release of luteinizing hormone (LH) and follicle-stimulating hormone (FSH) which simulate androgen production and aromatase activity, respectively. Aromatase catalyzes the aromatization of testosterone to E2, the final rate-limiting step of E2 biosynthesis. The high levels of E2 released during ovulation feedback to inhibit the release of GnRH, LH, and FSH.

The classical genomic mechanism of E2 action is initiated when E2 diffuses across the plasma membrane and binds to ER in the cytoplasm, where it resides bound to heat-shock proteins (Fig. 24.2). Binding of E2 to ER results in its dissociation from these proteins, leading to dimerization of the receptor and translocation to the nucleus where the complex functions as a ligand-activated transcription factor by binding to estrogen response elements (EREs) to initiate expression of multiple target genes [36–38]. Additionally, non-genomic effects of E2 have been described and will be discussed more below [39, 40].

Estrogens act primarily through two nuclear ERs, ER-α and ER-β. E2 is required for the initial formation of the TEBs during pubertal mammary gland development. Ovariectomy of mice around the onset of puberty causes regression of TEBs which reappear when E2 pellets are implanted [41]. Additionally, ER-α knockout mice (ERKO) have mammary glands that appear rudimentary and lack any branching or TEBs [42]. Transplantation studies indicate that while ER-α is expressed in both the stroma and the epithelium of the developing gland [41], ER-α in the epithelial cells is most important for growth [42]. Deletion of ER-β does not affect ductal growth of the mammary gland [43] though it does interfere with terminal differentiation during lactation [44].

A role for E2 in the etiology and progression of breast cancer was first recognized when bilateral ovariectomy in pre-menopausal breast cancer patients was found to result in cancer remission [45]. This finding led to investigation into the role of ovarian hormones and their receptors in breast cancer. Since approximately 70 % of all breast cancers are ER-α positive and require E2 for growth, an effective

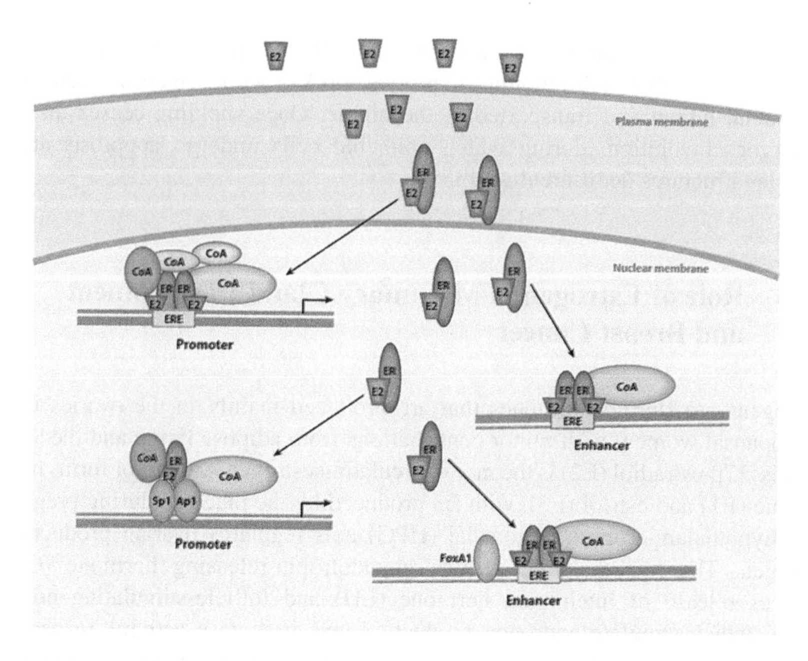

Fig. 24.2 Classical estrogen signaling pathway in the cell. 17-β estradiol (E2) diffuses across the plasma membrane and binds to ER in the cytoplasm. The E2/ER complex translocates to the nucleus and binds to estrogen response elements (EREs) to initiate gene expression. Reproduced with permission of Annual Reviews in the format Republish in a book via Copyright Clearance Center [38]

treatment strategy in women with ER-α positive tumors has been to block the E2 system with either selective ER modulators (SERMs) that bind to and block ER-α, or aromatase inhibitors, which act by inhibiting E2 production [46]. For this reason, ER-α has become a prognostic and predictive marker for breast cancer.

The fact that a number of breast cancer risk factors correspond with an increase in exposure of the mammary gland to E2 (e.g., women who experience menarche at an early age and/or menopause at a later age) has led to the idea that lifetime exposure to this hormone is a primary risk factor for the disease [47]. Increased exposure to E2 across a women's lifetime is thought to indirectly affect carcinogenesis by increasing the number of cell cycles and thus the chances of genetic mutations. In addition, many studies have reported that increased serum E2 is positively associated with breast cancer risk in postmenopausal women [48–50]. However, recent meta-analyses of multiple epidemiological studies suggest that postmenopausal hormone concentrations are not strongly related to age at menarche or first full-term pregnancy, parity or family history [51, 52], suggesting that these risk factors are not likely mediated through postmenopausal hormone levels. These risk factors may instead operate through long-term effects of sex hormone levels in pre-menopausal women such as changes in the duration rather than the level of long-term exposure to sex hormones, by permanent changes in breast structure induced by pregnancy and lactation, or by other mechanisms [51].

A second mechanism by which E2 may impact carcinogenesis is via direct carcinogenic properties of E2 metabolites. The conversion of E2 to 2- or 4-OH-E2 results in reactive oxygen species byproducts that are carcinogenic [53]. Also, the metabolites themselves can directly bind DNA and form adducts leading to mutagenesis [54, 55]. The ERKO/Wnt-1 transgenic mouse demonstrates the importance of these metabolites in tumor initiation. In this study treating ERKO/Wnt-1 transgenic mice with E2 increased mammary carcinogenesis in the absence of ER-α, thus implicating the metabolites as the culprits [56].

24.4 Role of the IGF System in Normal Mammary Gland Biology and Breast Cancer

The IGF family is comprised of two ligands (IGF-I and IGF-II) and six IGF-binding proteins (IGFBP-1 through IGFBP-6) that modulate the actions of IGFs. They exert their biological effects primarily through the IGF type I receptor (IGF-IR), a heterodimeric intrinsic tyrosine kinase receptor [57, 58]. The *IGF-IR* gene encodes a protein consisting of an extracellular α ligand binding domain and a β subunit containing the intracellular tyrosine kinase domain which dimerizes with a second unit to form the functional receptor. The insulin receptors A and B are closely related to IGF-IR and it has recently been recognized that IGF-insulin hybrid receptors may also influence IGF action in the mammary gland [59, 60]. Binding of IGF ligand induces autophosphorylation of the receptor, followed by activation of adapter molecules such as insulin-receptor substrate 1 (IRS-1) and Shc, setting in motion activation of the phosphatidyl inositol-3-kinase (PI3K) and mitogen activated protein kinase (MAPK) signal transduction cascades (Fig. 24.3). These signaling molecules activate downstream factors that stimulate multiple processes that regulate normal growth and development of the mammary gland, including cellular proliferation, survival, angiogenesis, and migration. These processes also underlie abnormal growth processes that lead to breast cancer, including tumorigenesis and metastasis.

Development of the mammary gland is regulated by complex hormonal interactions involving the pituitary hormones, growth hormone (GH) and prolactin, and the ovarian hormones, estrogen and progesterone. The postnatal effects of GH in mammary gland development are predominately mediated through IGF-I [61]. The importance of IGF-I in mammary gland development is demonstrated by the finding that IGF-I knockout mice have dramatically impaired TEB formation and ductal development [62]. Understanding the role of the IGF system in normal mammary gland development is complicated by the fact that IGF-I exerts it effects on the mammary gland through both endocrine and autocrine/paracrine mechanisms [58]. The majority of circulating IGF-I is produced in the liver, while most tissues, including the mammary gland, also produce IGF-I at the local level. GH can increase circulating IGF-I by binding to hepatic GH receptors and regulating hepatic production and/or it can bind to GH receptors in the mammary gland to increase local production [63]. Mice that exhibit reduced local and circulating levels of IGF-I

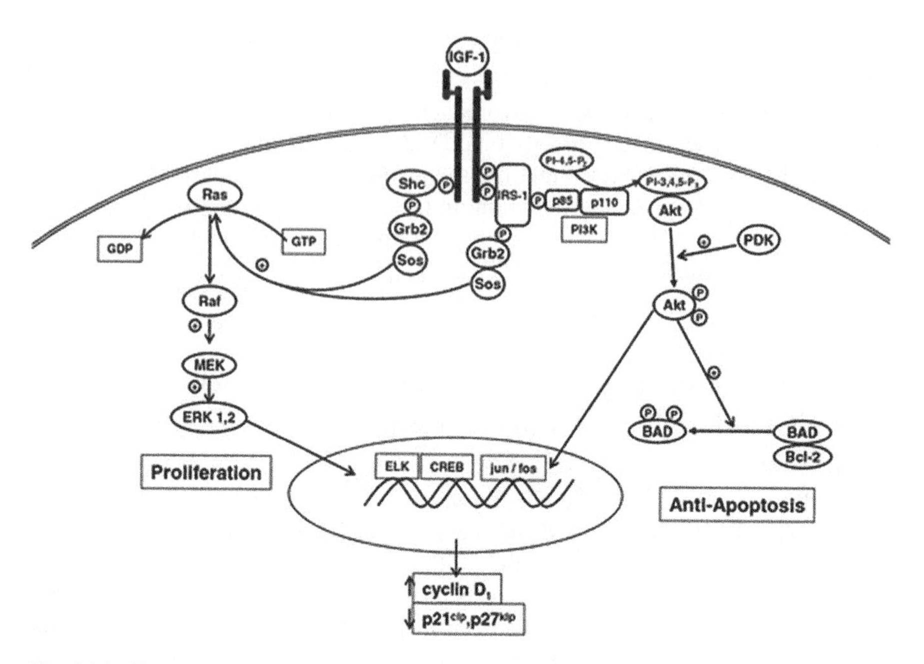

Fig. 24.3 Signal transduction pathways activated by IGF-I via the IGF-I receptor in the cell Reproduced with permission [131]

(IGF-I midi mice) have normal ductal elongation into the fat pad but fewer branching structures than normal mice. However, mice with a liver-specific deletion of IGF-I (LID mice) exhibit a 75 % reduction in circulating IGF-I, yet maintain normal mammary ductal branching [64], suggesting that locally produced IGF-I acts in a paracrine fashion to regulate branching morphogenesis. Studies with transgenic mice that overexpress hepatic IGF-I (HIT) but have no local IGF-I production (IGF-I knockout/HIT) show that circulating IGF-I can maintain normal mammary gland development in the absence of local IGF-I. In addition, HIT mice exhibit enhanced mammary gland proliferation compared to control mice, indicating that elevated circulating IGF-I in the presence of local IGF-I production can accelerate mammary gland growth [65]. Collectively these studies show that both local and endocrine actions of IGF-I are important in mammary gland development.

Both circulating and local production of IGF-I are also important in mammary tumorigenesis and progression. In LID mice, administration of the carcinogen DMBA results in a lower incidence of mammary tumors and delays tumor onset [66], supporting a role for circulating IGF-I in breast tumor progression. Since IGF-I is predominately made in the stromal tissue in response to GH, a transgenic model was generated that mimics paracrine exposure of breast epithelial to stromal IGF-I by placing IGF-I under control of the keratinocyte 5 promoter, leading to overexpression of IGF-I in myoepithelial cells. These mice show increased ductal proliferation prepubertally compared with wild-type mice and exhibit increased susceptibility to carcinogen-induced tumorigenesis as adults [67].

As mentioned above, the biological activity of IGF-I is modulated by a family of six high affinity IGFBPs. Of these IGFBPs, a large body of evidence supports a role for IGFBP-5 as a growth-inhibitory and/or pro-apoptotic factor in the mammary gland [68]. There is a positive relationship between IGFBP-5 expression and cell death during mammary involution. Transgenic overexpression of IGFBP-5 in mammary tissue causes increases in apoptotic death of epithelial cells and reduces invasion of the mammary fat pad, while addition of exogenous IGFBP-5 to murine mammary epithelial cells suppresses IGF-I mediated survival [69, 70]. Implantation of IGFBP-5 into one mammary gland of hypophysectomized, oophorectomized Sprague Dawley rats decreases the ability of GH and E2 to stimulate TEB formation after 7 days [71]. Likewise, when IGFBP-5 is injected into wild-type intact mice for 7 days beginning on day 16 of pregnancy, the proportion of secretory tissue in the gland is decreased and lactation is impaired [72]. Many of these pro-apoptotic, growth-inhibitory effects are thought to be related to the ability of IGFBP-5 to sequester IGF-I and prevent its pro-survival and growth-stimulatory effects.

24.5 Cross-Talk Between the E2 and IGF Systems

The E2 system interacts extensively with the IGF system in modulating mammary gland development and tumorigenesis [73–75]. During mammary gland development, administration of E2 or IGF-I alone to ovariectomized, hypophysectomized rats does not restore full ductal development; however, the gland develops normally when these hormones are administered together [76]. Likewise, both E2 and IGF-I are required to restore normal mammary gland development in the IGF-I knockout mouse [62]. At the cellular level, normal breast epithelial cells that are positive for ER-α do not stain positive for Ki67 or bromodeoxyuridine (BrdU) incorporation, indicating that E2 is not directly driving proliferation [77–79]. E2 is believed to stimulate the release of IGF-I, which acts in a paracrine manner to drive growth of adjacent cells [80].

In breast cancer cells, the proliferative effect of E2 is accounted for, in part, by its ability to upregulate components of the IGF signaling pathway, including IGF ligands, IGF-IR, and IRS-1 [81–84]. Reciprocally, IGF ligands can act via the IGF-IR to enhance transcriptional activation of ER-α by promoting its binding to EREs or by phosphorylating ER-α via activation of the PI3K and MAPK pathways [85–88]. The interactions between ER and IGF-IR may be mediated through a small pool of ERs localized to the plasma membrane, which can also occur in the absence of sex steroid [37, 89, 90]. These include not only ER-α and ER-β, but also ER-α transcript variants (including ER-α36) and the G-protein-coupled ER 1 (GPER, formerly known as GPR30), a G-protein coupled receptor that is structurally different from the classical ERs [91, 92]. A complex relationship between the membrane and nuclear effects of E2 also involves membrane-initiated phosphorylation of coactivators, which are then recruited to the nucleus. In addition to localizing to the nucleus and the plasma membrane, some ER may also localize to mitochondria

[37]. Therefore, the integration of effects that are mediated by ERs at distinct cellular locations plays a major role in regulating cellular outcomes in the cell. The finding that the E2/ER/IGF systems interact so extensively has led to the development of an array of therapeutic agents to target the IGF-IR signaling system in cancer, including breast cancer. However, despite promising results in early phase trials, randomized phase III trials have not yet been successful, demonstrating the complexity of the IGF signaling system [93, 94].

24.6 Evidence for Alterations in E2 and IGF Systems in Alcohol-Exposed Offspring

Several lines of evidence from rodent studies conducted in our laboratory show that alcohol-exposed offspring exhibit alterations in various components of the E2/IGF axis. Few mutations in components of the E2/IGF axis have been reported in humans which demonstrate the primary importance of these system components to normal growth and development. Therefore it is possible that alterations in the E2/IGF system are caused by epigenetic modifications, defined as heritable but reversible changes in gene function that occur without changes in nucleotide sequence. In the sections below, we outline evidence indicating that exposure to alcohol in utero induces epigenetic events. It is plausible that there could be epigenetic changes to E2/IGF hormone action in response to these alcohol-induced modifications. This regulation could occur at one or more levels to affect hormone synthesis, circulating and/or target tissue hormone levels, and/or target-organ responsiveness [95].

To determine if prenatal alcohol exposure induces early changes in mammary gland morphology that might enhance the susceptibility to a carcinogen in adulthood, we used the same in utero model of alcohol exposure described in the introduction. Instead of injecting NMU at postnatal day (PND) 50, we euthanized animals at three developmental time points: PND 20, a prepubertal time point when the ductal structures in the mammary gland are highly proliferative, PND 40, near puberty when the HPG hormonal axis has been activated, and PND 80, when the mammary gland is mature. Offspring were injected with BrdU prior to euthanization to quantitate cell proliferation. At 20 and 40 days of age, animals exposed to alcohol in utero had an increased proliferative index compared to PF controls [26]. The increases between days 20 and 40 indicate that in utero alcohol exposure causes early changes in programming of the mammary gland that may contribute to the enhanced tumorigenesis observed in response to NMU.

Given the increase in the proliferative state of the mammary gland at this early age, we hypothesized that alterations in the IGF-I axis might contribute to this effect. Hepatic IGF-I mRNA levels were increased at all three time points in alcohol-exposed offspring, while IGF-I mRNA levels were also increased in the mammary gland at PND 20. Interestingly, IGFBP-5 mRNA levels were significantly lower at PND 40 in animals exposed to prenatal alcohol relative to pair-fed controls (Fig. 24.4). Therefore, a decrease in IGFBP-5 expression in tumors of rats exposed

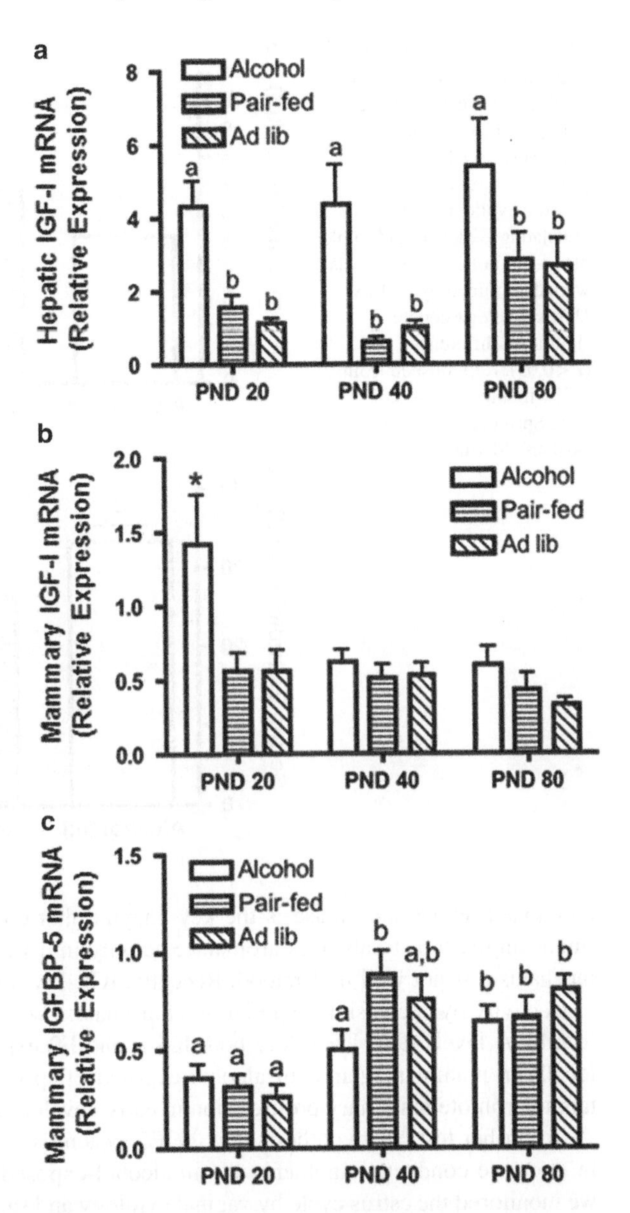

Fig. 24.4 Alcohol exposure in utero increases hepatic and mammary IGF-I expression and delays the developmental increase in mammary IGFBP-5 expression. Following alcohol exposure in utero, liver and mammary glands were collected at sacrifice on postnatal day (PND) 20, 40, or 80. RNA was isolated and analyzed by qRT-PCR. (**a**, **c**) *Different letters* denote significant differences ($P<0.05$); (**b**) *Asterisk* denotes significance at $P<0.001$. Adapted from [26] with kind permission from Springer Science + Business Media

to alcohol in utero may allow more free IGF-I to access the IGF-IR and promote tumorigenesis. IGFBP-5 is reduced in epithelial tumors such as head and neck squamous cell carcinoma and cervical carcinoma and has been proposed as a tumor suppressor gene [96, 97]. Therefore, the increased proliferation we observed in alcohol-exposed offspring could also be related to a decrease in a tumor suppressor role of IGFBP-5, which could be independent of IGF-I.

In addition to changes in the IGF axis, aromatase expression was increased in the mammary glands of alcohol-exposed offspring at both PND 20 and 40 [26]. As

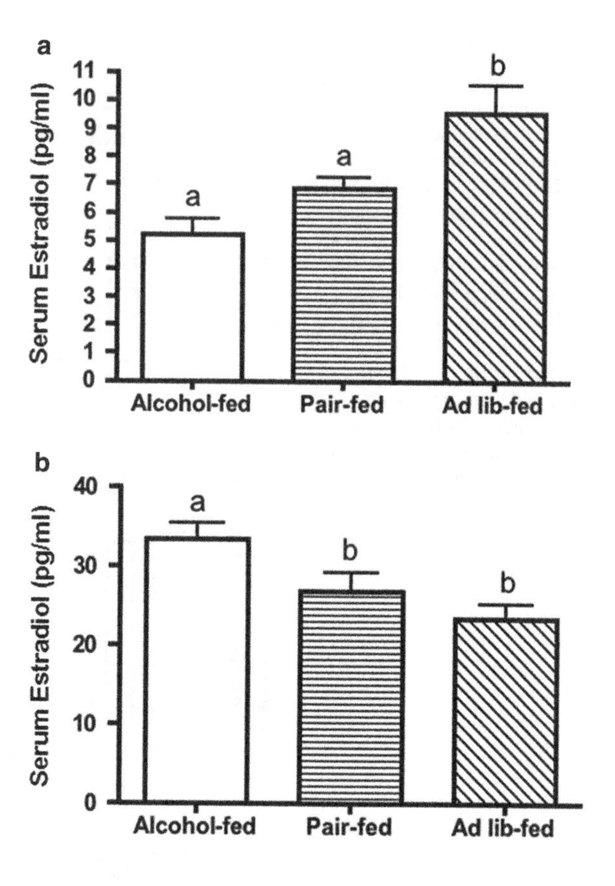

Fig. 24.5 Alcohol exposure in utero increases circulating levels of 17-β estradiol (E2) during proestrus in adulthood. Pups were sacrificed during proestrus (ages 62–76 days) to avoid cyclic differences in circulating E2 levels and trunk blood was collected. E2 levels were determined by ELISA. *Different letters* denote significant differences ($P < 0.05$). Republished from [25] with kind permission from Springer Science + Business Media

mentioned above, aromatase is the key enzyme that converts testosterone to E2. Interestingly, IGF-I enhances aromatase activity in a variety of cells although the mechanism is not well understood. Recently, IGF-I was shown to increase aromatase activity by decreasing degradation of aromatase with no corresponding change in mRNA levels [98]. Therefore both IGF-I and E2 expression may be increased locally in mammary glands of alcohol-exposed offspring, where they may cross-talk to promote mammary proliferation in early development.

To further test our hypothesis that the E2 system is altered by alcohol exposure in utero we conducted another study in alcohol-exposed adult offspring in which we monitored the estrus cycle by vaginal cytology and sacrificed all animals during proestrus. Animals exposed to alcohol in utero exhibited significantly higher circulating levels of E2 relative to either pair-fed or ad libitum control groups (Fig. 24.5). These data may be related to findings that in utero alcohol exposure alters the development and maturation of the HPG axis in female rats [99].

In addition to alcohol, exposure to a wide variety of toxic compounds during fetal or neonatal development affects postnatal mammary gland development and/or leads to increased susceptibility to mammary carcinogens in adulthood, supporting the fetal origins of disease hypothesis [100]. Many of these toxins are endocrine disruptors, which act by interfering with E2 action. Furthermore, rodent studies have shown that maternal exposures to dietary factors such as high fat leads to alterations in mammary morphology during development and/or an increased risk of mammary cancer in the offspring [101, 102]. Feeding a high fat diet also increases circulating concentrations of E2 in the dams [101]. Exposure to estrogens in utero due to female twin-ship [103] or DES treatment [104] is associated with a higher risk for breast cancer later in life. Based on the rationale that alterations in the fetal estrogenic environment may change the susceptibility of the mammary gland to later exposure to carcinogens, Hilakivi-Clarke et al. [24] measured E2 in the circulation of pregnant dams exposed to alcohol. They found an increase in E2 concentrations in the dams fed the lower level of alcohol (7 % of total calories) but not in the dams fed the intermediate level of alcohol (15 % of total calories). Interestingly, tumor number was only increased in the dams fed the intermediate level of alcohol. Whether circulating E2 levels are increased in dams fed high concentrations of alcohol has not yet been determined.

24.7 Fetal Alcohol and Epigenetics

As mentioned above, epigenetics involves heritable changes in gene expression that occur without an alteration in the primary nucleotide sequence of a gene. Epigenetic modifications to nuclear chromatin structure alter DNA, histones, and non-histone proteins. These modifications limit or enhance the accessibility and binding of the transcriptional machinery or recruit repressor complexes, resulting in changes in gene expression. Epigenetic mechanisms include DNA methylation of promoter and/or non-promoter CpG islands, covalent histone modifications (methylation, acetylation, phosphorylation, ubiquitination or sumoylation), microRNAs, and the more recently described long noncoding RNAs [95, 105–107].

DNA methylation represents a major epigenetic regulatory pathway that is catalyzed by DNA methyltransferases (DNMTs). These enzymes add methyl groups from *S*-adenosylmethionines to carbon 5 positions of cytosines [108]. Three important family members of DNMTs have been reported: DNMT1, DNMT3A, and DNMT3B. DNMT1 is considered the maintenance DNA methyltransferase, whereas DNMT3A and DNMT3B are primarily involved in de novo DNA methylation [95, 109]. Covalent histone modifications, including acetylation and methylation, are a second major pathway. Histone acetylation and deacetylation are regulated by histone acetyltransferases (HATs) and histone deacetylases (HDACs), respectively. Histone methylation and demethylation are regulated by histone methyltransferases (HMTs) and histone demethylases (HDMs), respectively [110]. Lysine residues on histone tails can be regulated by both methylation and acetylation. Modifications of lysines on histone H3 can be repressive (e.g., H3K9 di- or trimethylation, H3K27 trimethylation) or activating (e.g., H3K9 acetylation, H3K4 trimethylation) [95, 110–112]. HATs do not bind to DNA promoters directly, but are recruited by DNA-bound transcription factors. Overall, hypomethylation of DNA and histone acetylation cause a more relaxed chromatin state, allowing easier interaction between the DNA and the transcriptional machinery, thus resulting in increased gene transcription. DNA hypermethylation and histone deacetylation result in condensed chromatin structure, causing a decrease in gene transcription [95].

The epigenome is particularly susceptible to environmental factors during fetal development as the DNA synthetic rate is high and the complex DNA methylation patterning and chromatin structure required for normal tissue development is established at this time [113]. DNA methylation and histone modifying pathways cross-talk and both are necessary for normal genomic imprinting during development [114]. Epigenetic reprogramming involves genome-wide erasure and re-establishment of DNA methylation and histone modifications during normal mammalian development (immediately after fertilization until pre-implantation of the embryo and also during primordial germ cell development) [109, 111, 115–118]. Aberrant epigenetic regulation is also implicated in cancer, whereby oncogenes are expressed, and tumor suppressor genes are silenced [112, 119].

In 1991, Garro et al. showed that administration of alcohol to pregnant mice on days 9–11 of pregnancy resulted in genome-wide hypomethylation in 11-day old fetuses. The effect was proposed to be mediated by acetaldehyde, a product of alcohol metabolism, which inhibits DNA methyltransferase in vitro [120]. In the last few years numerous studies have confirmed that alcohol exposure in utero induces epigenetic alterations in the offspring. Studies with the Agouti mouse model demonstrate that alcohol exposure during gestation can affect adult phenotype through epigenetic mechanisms [121]. Microarray analysis of murine embryos exposed to alcohol on gestational day 9 shows altered expression of a subset of genes involved in methylation and chromatin remodeling [122]. In support of this finding, analysis of whole-embryo murine cultures treated with alcohol show increased and decreased DNA methylation of 1,028 and 1,136 genes, respectively. Greater than 200 of these methylation alterations are found on chromosomes 7, 10, and X, which are chromosomes that contain many imprinted genes as well as genes prone to aberrant epigenetic silencing. Additionally, changes in DNA methylation correspond with actual changes in gene expression in 84 genes, indicating that fetal alcohol can impact gene expression in the developing embryo through an epigenetic mechanism [123].

In addition to global methylation analyses, several studies have focused on methylation changes in specific genes. Exposure of mice to alcohol between gestational days 10 and 18 affects methylation of the paternally imprinted H19 gene in the offspring's sperm, resulting in a 3 % decrease in the number of methylated CpGs in this gene at 8 weeks of age. The CCCTC-factor DNA-binding sites of H19 play a role in regulating IGF-II expression [124]. Due to the well-defined neurological consequences of developmental alcohol exposure, recent studies have focused on epigenetic regulators expressed specifically in the brain. Perinatal alcohol exposure (gestational day 1 through PND 10) induces gene expression changes in epigenetic regulators (DNMT1, DNMT3a, and methyl-CpG binding protein 2) in the hippocampus as well as increases in DNMT activity [125]. Alcohol exposure of fetal cerebral cortical neuroepithelial stem cells causes a decrease in trimethylation of H3K4 (H3K4me3) and H3K27 (H3K27me3) in developmental genes that play an integral role in regulating neural stem cells as well as neural differentiation [126]. H3K4me3 is usually associated with gene activation, whereas H3K27me3 is considered a repressive histone mark [95, 126]. Expression of epigenetic-modifying enzymes such as Dnmt1, Uhrf1, Ash2L1, Wdr5, Ehmt1, and Kdm1b are also observed [126]. DNMT expression and histone modifications are also altered in the hypothalamus of offspring exposed to fetal alcohol. In utero alcohol exposure from gestational day 7–21 results in a decrease in H3K4 di- and trimethylation as well as H3K9 acetylation, and an increase in H3K9 dimethylation as well as expression of DNMT1, DNMT3a, and methyl-CpG-binding protein [127]. These studies raise the possibility that alcohol exposure in utero will affect epigenetic regulators in other tissues as well, including the ovaries and mammary gland.

While epigenetic alterations induced by fetal alcohol exposure in the mammary gland have not been investigated to date, other in utero exposures cause epigenetic modifications in the mammary gland. High fat or ethinyl-oestradiol (EE2) exposure in utero increases DNMT1 expression in PND 50 rat mammary glands [102]. Global methylation analysis of PND 50 mammary glands from EE2 exposed rats reveals 375 differentially methylated promoter regions, of which 21 are hypermethylated and 161 are hypomethylated [102]. In utero exposure to DES or bisphenol-A (BPA) also induces epigenetic alterations in the adult mammary gland [128]. DES exposure between gestational days 9 and 26 increases expression of the histone methyltransferase EZH2 in mammary glands of 6-week-old female mice. BPA or DES exposure also lead to increased histone H3K27 trimethylation, an EZH2 target, in the mammary gland indicating that EZH2 methyltransferase activity increases. Overexpression of EZH2 has been documented in breast cancer and is associated with aggressive forms of the disease [128]. In summary, in utero exposures can cause epigenetic modifications specifically in the mammary gland and these epigenetic alterations can potentially affect the offspring's risk of mammary cancer.

The E2 signaling axis can also be epigenetically affected by in utero exposure. Treatment of mice with BPA during gestational days 9–16 leads to decreased DNA methylation in the homeobox gene *Hoxa10* in the uteri of 2 week old mice, subsequently resulting in increased *Hoxa10* mRNA and protein expression [129]. The decrease in *Hoxa10* DNA methylation leads to increased binding of ER-α to the EREs located on the *Hoxa10* promoter. Furthermore, transfection of MCF-7 breast cancer cells with the unmethylated *Hoxa10* promoter increases luciferase activity in response to E2 compared to MCF-7 cells transfected with the methylated promoter, indicating that the unmethylated *Hoxa10* promoter is more estrogen responsive, leading to an increase in ERE-driven gene expression [129]. It is possible that this epigenetic effect on the E2 signaling axis is not specific to BPA or endocrine disruptors in general, but may be an effect that can be initiated by a variety of in utero exposures such as alcohol.

The studies outlined above indicate that prenatal alcohol exposure induces epigenetic modifications. Therefore, it is plausible that alcohol exposure in utero could affect epigenetic regulation of the E2/IGF system. Recent evidence has indicated that steroidogenic enzymes, nuclear receptors, and transcription factors involved in steroid hormone synthesis and action can be epigenetically regulated [130]. It is therefore possible that alcohol exposure in utero can disrupt the normal epigenetic regulation of these key molecules involved in hormone signaling. Changes in the steroid signaling pathway induced by epigenetic modification could then interface with the IGF system as described above. Further studies will determine if fetal alcohol exposure induces epigenetic modifications in genes that can regulate hormone action at key phases of mammary development, ultimately leading to changes in susceptibility to carcinogens.

References

1. Nayak RB, Murthy P (2008) Fetal alcohol spectrum disorder. Indian Pediatr 45:977–983
2. Chaudhuri JD (2000) Alcohol and the developing fetus–a review. Med Sci Monit 6:1031–1041
3. Streissguth AP, Bookstein FL, Barr HM, Sampson PD, O'Malley K, Young JK (2004) Risk factors for adverse life outcomes in fetal alcohol syndrome and fetal alcohol effects. J Dev Behav Pediatr 25:228–238
4. May PA, Gossage JP (2001) Estimating the prevalence of fetal alcohol syndrome. A summary. Alcohol Res Health 25:159–167
5. U.S. Department of Health and Human Services (2005) US Surgeon General releases advisory on alcohol use in pregnancy. Available at http://www.cdc.gov/ncbddd/fasd/dcouments/sg-advisory.pdf. Accessed Oct. 9, 2013
6. Center for Disease Control and Prevention (2012) Morbidity and Mortality Weekly Report. Alcohol Use and Binge Drinking Among Women of Childbearing Age – US, 2006–2010, 61(28):534–538
7. McCarthy FP, O'Keeffe LM, Khashan AS, North RA, Poston L, McCowan LM et al (2013) Association between maternal alcohol consumption in early pregnancy and pregnancy outcomes. Obstet Gynecol 122:830–837
8. Howlader N, Noone AM, Krapcho M, Garshell J, Neyman N, Altekruse SF et al (eds) (2013) SEER Cancer Statistics Review, 1975–2010, National Cancer Institute. Bethesda, MD http://seer.cancer.gov/csr/1975_2010/, based on November 2012 SEER data submission, posted to the SEER web site, April 2013
9. Simpson ER, Brown KA (2013) Minireview: obesity and breast cancer: a tale of inflammation and dysregulated metabolism. Mol Endocrinol 27:715–725
10. Faupel-Badger JM, Arcaro KF, Balkam JJ, Eliassen AH, Hassiotou F, Lebrilla CB et al (2012) Postpartum remodeling, lactation, and breast cancer risk: summary of a National Cancer Institute-sponsored workshop. J Natl Cancer Inst 105:166–174
11. Collaborative Group on Hormonal Factors in Breast Cancer (2012) Menarche, menopause, and breast cancer risk: individual participant meta-analysis, including 118,964 women with breast cancer from 117 epidemiological studies. Lancet Oncol 13:1141–1151
12. Smith-Warner SA, Spiegelman D, Yaun SS, van den Brandt PA, Folsom AR, Goldbohm RA et al (1998) Alcohol and breast cancer in women: a pooled analysis of cohort studies. JAMA 279:535–540
13. Hamajima N, Hirose K, Tajima K, Rohan T, Calle EE, Heath CW Jr et al (2002) Alcohol, tobacco and breast cancer–collaborative reanalysis of individual data from 53 epidemiological studies, including 58,515 women with breast cancer and 95,067 women without the disease. Br J Cancer 87:1234–1245
14. Chen WY, Rosner B, Hankinson SE, Colditz GA, Willett WC (2011) Moderate alcohol consumption during adult life, drinking patterns, and breast cancer risk. JAMA 306:1884–1890
15. Park SY, Kolonel LN, Lim U, White KK, Henderson BE, Wilkens LR (2014) Alcohol consumption and breast cancer risk among women from five ethnic groups with light to moderate intakes: The multiethnic cohort study. Int J Cancer 134(6):1504–1510
16. Warren Andersen S, Trentham-Dietz A, Gangnon R, Hampton J, Figueroa J, Skinner H et al (2013) The associations between a polygenic score, reproductive and menstrual risk factors and breast cancer risk. Breast Cancer Res Treat 140:427–434

17. Barker DJ (2007) The origins of the developmental origins theory. J Intern Med 261:412–417
18. Jones RH, Ozanne SE (2007) Intra-uterine origins of type 2 diabetes. Arch Physiol Biochem 113:25–29
19. Tang WY, Ho SM (2007) Epigenetic reprogramming and imprinting in origins of disease. Rev Endocr Metab Disord 8:173–182
20. Simmen FA, Simmen RC (2011) The maternal womb: a novel target for cancer prevention in the era of the obesity pandemic? Eur J Cancer Prev 20:539–548
21. Trichopoulos D (1990) Hypothesis: does breast cancer originate in utero? Lancet 335:939–940
22. Hilakivi-Clarke L, de Assis S (2006) Fetal origins of breast cancer. Trends Endocrinol Metab 17:340–348
23. Soto AM, Vandenberg LN, Maffini MV, Sonnenschein C (2008) Does breast cancer start in the womb? Basic Clin Pharmacol Toxicol 102:125–133
24. Hilakivi-Clarke L, Cabanes A, de Assis S, Wang M, Khan G, Shoemaker WJ et al (2004) In utero alcohol exposure increases mammary tumorigenesis in rats. Br J Cancer 90:2225–2231
25. Polanco TA, Crismale-Gann C, Reuhl KR, Sarkar DK, Cohick WS (2010) Fetal alcohol exposure increases mammary tumor susceptibility and alters tumor phenotype in rats. Alcohol Clin Exp Res 34:1–9
26. Polanco TA, Crismale-Gann C, Cohick WS (2011) Alcohol exposure in utero leads to enhanced prepubertal mammary development and alterations in mammary IGF and estradiol systems. Horm Cancer 2:239–248
27. Latino-Martel P, Chan DS, Druesne-Pecollo N, Barrandon E, Hercberg S, Norat T (2010) Maternal alcohol consumption during pregnancy and risk of childhood leukemia: systematic review and meta-analysis. Cancer Epidemiol Biomarkers Prev 19:1238–1260
28. Esserman LJ, Thompson IM Jr, Reid B (2013) Overdiagnosis and overtreatment in cancer: an opportunity for improvement. JAMA 310:797–798
29. Masso-Welch PA, Darcy KM, Stangle-Castor NC, Ip MM (2000) A developmental atlas of rat mammary gland histology. J Mammary Gland Biol Neoplasia 5:165–185
30. Gusterson BA, Stein T (2012) Human breast development. Semin Cell Dev Biol 23:567–573
31. Sakakura T, Kusano I, Kusakabe M, Inaguma Y, Nishizuka Y (1987) Biology of mammary fat pad in fetal mouse: capacity to support development of various fetal epithelia in vivo. Development 100:421–430
32. Balinsky BI (1950) On the prenatal growth of the mammary gland rudiment in the mouse. J Anat 84:227–235
33. Propper AY (1978) Wandering epithelial cells in the rabbit embryo milk line. A preliminary scanning electron microscope study. Dev Biol 67:225–231
34. Hens JR, Wysolmerski JJ (2005) Key stages of mammary gland development: molecular mechanisms involved in the formation of the embryonic mammary gland. Breast Cancer Res 7:220–224
35. Macias H, Hinck L (2012) Mammary gland development. Wiley Interdiscip Rev Dev Biol 1:533–557
36. Schiff R, Osborne CK (2005) Endocrinology and hormone therapy in breast cancer: new insight into estrogen receptor-alpha function and its implication for endocrine therapy resistance in breast cancer. Breast Cancer Res 7:205–211
37. Levin ER (2005) Integration of the extranuclear and nuclear actions of estrogen. Mol Endocrinol 19:1951–1959
38. Liang J, Shang Y (2013) Estrogen and cancer. Annu Rev Physiol 75:225–240
39. Hammes SR, Levin ER (2011) Minireview: Recent advances in extranuclear steroid receptor actions. Endocrinology 152:4489–4495
40. Manavathi B, Dey O, Gajulapalli VN, Bhatia RS, Bugide S, Kumar R (2013) Derailed estrogen signaling and breast cancer: an authentic couple. Endocr Rev 34:1–32
41. Daniel CW, Silberstein GB, Strickland P (1987) Direct action of 17 beta-estradiol on mouse mammary ducts analyzed by sustained release implants and steroid autoradiography. Cancer Res 47:6052–6057

42. Mallepell S, Krust A, Chambon P, Brisken C (2006) Paracrine signaling through the epithelial estrogen receptor alpha is required for proliferation and morphogenesis in the mammary gland. Proc Natl Acad Sci U S A 103:2196–2201
43. Krege JH, Hodgin JB, Couse JF, Enmark E, Warner M, Mahler JF et al (1998) Generation and reproductive phenotypes of mice lacking estrogen receptor beta. Proc Natl Acad Sci U S A 95:15677–15682
44. Forster C, Makela S, Warri A, Kietz S, Becker D, Hultenby K et al (2002) Involvement of estrogen receptor beta in terminal differentiation of mammary gland epithelium. Proc Natl Acad Sci U S A 99:15578–15583
45. Beatson G (1896) On the treatment of inoperable cases of carcinoma of the mamma: suggestions for anew method of treatment, with illustrative cases. Lancet 148:104–107
46. Schiff R, Chamness GC, Brown PH (2003) Advances in breast cancer treatment and prevention: preclinical studies on aromatase inhibitors and new selective estrogen receptor modulators (SERMs). Breast Cancer Res 5:228–231
47. Russo J, Russo IH (2006) The role of estrogen in the initiation of breast cancer. J Steroid Biochem Mol Biol 102:89–96
48. Key T, Appleby P, Barnes I, Reeves G (2002) Endogenous sex hormones and breast cancer in postmenopausal women: reanalysis of nine prospective studies. J Natl Cancer Inst 94: 606–616
49. Kaaks R, Rinaldi S, Key TJ, Berrino F, Peeters PH, Biessy C et al (2005) Postmenopausal serum androgens, oestrogens and breast cancer risk: the European prospective investigation into cancer and nutrition. Endocr Relat Cancer 12:1071–1082
50. Missmer SA, Eliassen AH, Barbieri RL, Hankinson SE (2004) Endogenous estrogen, androgen, and progesterone concentrations and breast cancer risk among postmenopausal women. J Natl Cancer Inst 96:1856–1865
51. Key TJ, Appleby PN, Reeves GK, Roddam AW, Helzlsouer KJ, Alberg AJ et al (2011) Circulating sex hormones and breast cancer risk factors in postmenopausal women: reanalysis of 13 studies. Br J Cancer 105:709–722
52. Britt K (2012) Menarche, menopause, and breast cancer risk. Lancet Oncol 13:1071–1072
53. Yager JD, Davidson NE (2006) Estrogen carcinogenesis in breast cancer. N Engl J Med 354:270–282
54. Cavalieri E, Rogan E (2006) Catechol quinones of estrogens in the initiation of breast, prostate, and other human cancers: keynote lecture. Ann N Y Acad Sci 1089:286–301
55. Santen R, Cavalieri E, Rogan E, Russo J, Guttenplan J, Ingle J et al (2009) Estrogen mediation of breast tumor formation involves estrogen receptor-dependent, as well as independent, genotoxic effects. Ann N Y Acad Sci 1155:132–140
56. Yue W, Wang JP, Li Y, Fan P, Liu G, Zhang N et al (2010) Effects of estrogen on breast cancer development: Role of estrogen receptor independent mechanisms. Int J Cancer 127:1748–1757
57. LeRoith D, Roberts CT Jr (2003) The insulin-like growth factor system and cancer. Cancer Lett 195:127–137
58. Kleinberg DL, Wood TL, Furth PA, Lee AV (2009) Growth hormone and insulin-like growth factor-I in the transition from normal mammary development to preneoplastic mammary lesions. Endocr Rev 30:51–74
59. Belfiore A, Frasca F, Pandini G, Sciacca L, Vigneri R (2009) Insulin receptor isoforms and insulin receptor/insulin-like growth factor receptor hybrids in physiology and disease. Endocr Rev 30:586–623
60. Rowzee AM, Ludwig DL, Wood TL (2009) Insulin-like growth factor type 1 receptor and insulin receptor isoform expression and signaling in mammary epithelial cells. Endocrinology 150:3611–3619
61. Kleinberg DL, Ruan W (2008) IGF-I, GH, and sex steroid effects in normal mammary gland development. J Mammary Gland Biol Neoplasia 13:353–360
62. Ruan W, Kleinberg DL (1999) Insulin-like growth factor I is essential for terminal end bud formation and ductal morphogenesis during mammary development. Endocrinology 140:5075–5081

63. Kaplan SA, Cohen P (2007) The somatomedin hypothesis 2007: 50 years later. J Clin Endocrinol Metab 92:4529–4535
64. Richards RG, Klotz DM, Walker MP, Diaugustine RP (2004) Mammary gland branching morphogenesis is diminished in mice with a deficiency of insulin-like growth factor-I (IGF-I), but not in mice with a liver-specific deletion of IGF-I. Endocrinology 145:3106–3110
65. Cannata D, Lann D, Wu Y, Elis S, Sun H, Yakar S et al (2010) Elevated circulating IGF-I promotes mammary gland development and proliferation. Endocrinology 151:5751–5761
66. Wu Y, Cui K, Miyoshi K, Hennighausen L, Green JE, Setser J et al (2003) Reduced circulating insulin-like growth factor I levels delay the onset of chemically and genetically induced mammary tumors. Cancer Res 63:4384–4388
67. de Ostrovich KK, Lambertz I, Colby JK, Tian J, Rundhaug JE, Johnston D et al (2008) Paracrine overexpression of insulin-like growth factor-1 enhances mammary tumorigenesis in vivo. Am J Pathol 173:824–834
68. Beattie J, Allan GJ, Lochrie JD, Flint DJ (2006) Insulin-like growth factor-binding protein-5 (IGFBP-5): a critical member of the IGF axis. Biochem J 395:1–19
69. Marshman E, Green KA, Flint DJ, White A, Streuli CH, Westwood M (2003) Insulin-like growth factor binding protein 5 and apoptosis in mammary epithelial cells. J Cell Sci 116:675–682
70. Tonner E, Barber MC, Allan GJ, Beattie J, Webster J, Whitelaw CB et al (2002) Insulin-like growth factor binding protein-5 (IGFBP-5) induces premature cell death in the mammary glands of transgenic mice. Development 129:4547–4557
71. Ruan W, Fahlbusch F, Clemmons DR, Monaco ME, Walden PD, Silva AP et al (2006) SOM230 inhibits insulin-like growth factor-I action in mammary gland development by pituitary independent mechanism: mediated through somatostatin subtype receptor 3? Mol Endocrinol 20:426–436
72. Allan GJ, Beattie J, Flint DJ (2004) The role of IGFBP-5 in mammary gland development and involution. Domest Anim Endocrinol 27:257–266
73. Hamelers IH, Steenbergh PH (2003) Interactions between estrogen and insulin-like growth factor signaling pathways in human breast tumor cells. Endocr Relat Cancer 10:331–345
74. Thorne C, Lee AV (2003) Cross talk between estrogen receptor and IGF signaling in normal mammary gland development and breast cancer. Breast Dis 17:105–114
75. Bartella V, De Marco P, Malaguarnera R, Belfiore A, Maggiolini M (2012) New advances on the functional cross-talk between insulin-like growth factor-I and estrogen signaling in cancer. Cell Signal 24:1515–1521
76. Ruan W, Catanese V, Wieczorek R, Feldman M, Kleinberg DL (1995) Estradiol enhances the stimulatory effect of insulin-like growth factor-I (IGF-I) on mammary development and growth hormone-induced IGF-I messenger ribonucleic acid. Endocrinology 136:1296–1302
77. Clarke RB, Howell A, Potten CS, Anderson E (1997) Dissociation between steroid receptor expression and cell proliferation in the human breast. Cancer Res 57:4987–4991
78. Zeps N, Bentel JM, Papadimitriou JM, D'Antuono MF, Dawkins HJ (1998) Estrogen receptor-negative epithelial cells in mouse mammary gland development and growth. Differentiation 62:221–226
79. Russo J, Ao X, Grill C, Russo IH (1999) Pattern of distribution of cells positive for estrogen receptor alpha and progesterone receptor in relation to proliferating cells in the mammary gland. Breast Cancer Res Treat 53:217–227
80. Clarke RB, Howell A, Anderson E (1997) Type I insulin-like growth factor receptor gene expression in normal human breast tissue treated with oestrogen and progesterone. Br J Cancer 75:251–257
81. Stewart AJ, Johnson AD, May FEB, Westley BR (1990) Role of the insulin-like growth factors and the type-I insulin-like growth factor receptor in the estrogen-stimulated proliferation of human breast cancer cells. J Biol Chem 265:21172–21178
82. Lee AV, Darbre P, King RJ (1994) Processing of insulin-like growth factor-II (IGF-II) by human breast cancer cells. Mol Cell Endocrinol 99:211–220

83. Umayahara Y, Kawamori R, Watada H, Imano E, Iwama N, Morishima T et al (1994) Estrogen regulation of the insulin-like growth factor I gene transcription involves an AP-1 enhancer. J Biol Chem 269:16433–16442
84. Lee AV, Jackson JG, Gooch JL, Hilsenbeck SG, Coronado-Heinsohn E, Osborne CK et al (1999) Enhancement of insulin-like growth factor signaling in human breast cancer: estrogen regulation of insulin receptor substrate-1 expression in vitro and in vivo. Mol Endocrinol 13:787–796
85. Kato S, Endoh H, Masuhiro Y, Kitamoto T, Uchiyama S, Sasaki H et al (1995) Activation of the estrogen receptor through phosphorylation by mitogen-activated protein kinase. Science 270:1491–1494
86. Lee AV, Weng CN, Jackson JG, Yee D (1997) Activation of estrogen receptor-mediated gene transcription by IGF-I in human breast cancer cells. J Endocrinol 152:39–47
87. Cascio S, Bartella V, Garofalo C, Russo A, Giordano A, Surmacz E (2007) Insulin-like growth factor 1 differentially regulates estrogen receptor-dependent transcription at estrogen response element and AP-1 sites in breast cancer cells. J Biol Chem 282:3498–3506
88. Becker MA, Ibrahim YH, Cui X, Lee AV, Yee D (2011) The IGF pathway regulates ERalpha through a S6K1-dependent mechanism in breast cancer cells. Mol Endocrinol 25:516–528
89. Kahlert S, Nuedling S, van Eickels M, Vetter H, Meyer R, Grohe C (2000) Estrogen receptor alpha rapidly activates the IGF-1 receptor pathway. J Biol Chem 275:18447–18453
90. Song RX, Chen Y, Zhang Z, Bao Y, Yue W, Wang JP et al (2009) Estrogen utilization of IGF-1-R and EGF-R to signal in breast cancer cells. J Steroid Biochem Mol Biol 118:219–230
91. Prossnitz ER, Barton M (2011) The G-protein-coupled estrogen receptor GPER in health and disease. Nat Rev Endocrinol 7:715–726
92. Kang L, Zhang X, Xie Y, Tu Y, Wang D, Liu Z et al (2010) Involvement of estrogen receptor variant ER-alpha36, not GPR30, in nongenomic estrogen signaling. Mol Endocrinol 24:709–721
93. Yang Y, Yee D (2012) Targeting insulin and insulin-like growth factor signaling in breast cancer. J Mammary Gland Biol Neoplasia 17:251–261
94. Pollak M (2012) The insulin and insulin-like growth factor receptor family in neoplasia: an update. Nat Rev Cancer 12:159–169
95. Zhang X, Ho SM (2011) Epigenetics meets endocrinology. J Mol Endocrinol 46:R11–R32
96. Miyatake T, Ueda Y, Nakashima R, Yoshino K, Kimura T, Murata T et al (2007) Down-regulation of insulin-like growth factor binding protein-5 (IGFBP-5): novel marker for cervical carcinogenesis. Int J Cancer 120:2068–2077
97. Hung PS, Kao SY, Shih YH, Chiou SH, Liu CJ, Chang KW et al (2008) Insulin-like growth factor binding protein-5 (IGFBP-5) suppresses the tumourigenesis of head and neck squamous cell carcinoma. J Pathol 214:368–376
98. Zhang B, Shozu M, Okada M, Ishikawa H, Kasai T, Murakami K et al (2010) Insulin-like growth factor I enhances the expression of aromatase P450 by inhibiting autophagy. Endocrinology 151:4949–4958
99. Lan N, Yamashita F, Halpert AG, Sliwowska JH, Viau V, Weinberg J (2009) Effects of prenatal ethanol exposure on hypothalamic-pituitary-adrenal function across the estrous cycle. Alcohol Clin Exp Res 33:1075–1088
100. Macon MB, Fenton SE (2013) Endocrine disruptors and the breast: early life effects and later life disease. J Mammary Gland Biol Neoplasia 18:43–61
101. Hilakivi-Clarke L, Clarke R, Onojafe I, Raygada M, Cho E, Lippman M (1997) A maternal diet high in n-6 polyunsaturated fats alters mammary gland development, puberty onset, and breast cancer risk among female rat offspring. Proc Natl Acad Sci U S A 94:9372–9377
102. de Assis S, Warri A, Cruz MI, Laja O, Tian Y, Zhang B et al (2012) High-fat or ethinyl-oestradiol intake during pregnancy increases mammary cancer risk in several generations of offspring. Nat Commun 3:1053
103. Braun MM, Ahlbom A, Floderus B, Brinton LA, Hoover RN (1995) Effect of twinship on incidence of cancer of the testis, breast, and other sites (Sweden). Cancer Causes Control 6:519–524

104. Palmer JR, Hatch EE, Rosenberg CL, Hartge P, Kaufman RH, Titus-Ernstoff L et al (2002) Risk of breast cancer in women exposed to diethylstilbestrol in utero: preliminary results (United States). Cancer Causes Control 13:753–758
105. Dalvai M, Bystricky K (2010) The role of histone modifications and variants in regulating gene expression in breast cancer. J Mammary Gland Biol Neoplasia 15:19–33
106. Tsai MC, Manor O, Wan Y, Mosammaparast N, Wang JK, Lan F et al (2010) Long noncoding RNA as modular scaffold of histone modification complexes. Science 329:689–693
107. Alves CP, Fonseca AS, Muys BR, de Barros ELBR, Burger MC, de Souza JE et al (2013) The lincRNA Hotair is required for epithelial-to-mesenchymal transition and stemness maintenance of cancer cells lines. Stem Cells 31:2827–2832
108. Li S, Hursting SD, Davis BJ, McLachlan JA, Barrett JC (2003) Environmental exposure, DNA methylation, and gene regulation: lessons from diethylstilbesterol-induced cancers. Ann N Y Acad Sci 983:161–169
109. Feng S, Jacobsen SE, Reik W (2010) Epigenetic reprogramming in plant and animal development. Science 330:622–627
110. Cloos PA, Christensen J, Agger K, Helin K (2008) Erasing the methyl mark: histone demethylases at the center of cellular differentiation and disease. Genes Dev 22:1115–1140
111. Turndrup Pedersen M, Helin K (2010) Histone demethylases in development and disease. Trends Cell Biol 20:662–671
112. Rodriguez-Paredes M, Esteller M (2011) Cancer epigenetics reaches mainstream oncology. Nat Med 17:330–339
113. Dolinoy DC, Weidman JR, Jirtle RL (2007) Epigenetic gene regulation: linking early developmental environment to adult disease. Reprod Toxicol 23:297–307
114. Ciccone DN, Su H, Hevi S, Gay F, Lei H, Bajko J et al (2009) KDM1B is a histone H3K4 demethylase required to establish maternal genomic imprints. Nature 461:415–418
115. Ungerer M, Knezovich J, Ramsay M (2013) In utero alcohol exposure, epigenetic changes, and their consequences. Alcohol Res 35:37–46
116. Reik W, Dean W, Walter J (2001) Epigenetic reprogramming in mammalian development. Science 293:1089–1093
117. Kafri T, Ariel M, Brandeis M, Shemer R, Urven L, McCarrey J et al (1992) Developmental pattern of gene-specific DNA methylation in the mouse embryo and germ line. Genes Dev 6:705–714
118. Santos F, Dean W (2004) Epigenetic reprogramming during early development in mammals. Reproduction 127:643–651
119. Verma M, Srivastava S (2002) Epigenetics in cancer: implications for early detection and prevention. Lancet Oncol 3:755–763
120. Garro AJ, McBeth DL, Lima V, Lieber CS (1991) Ethanol consumption inhibits fetal DNA methylation in mice: implications for the fetal alcohol syndrome. Alcohol Clin Exp Res 15:395–398
121. Kaminen-Ahola N, Ahola A, Maga M, Mallitt KA, Fahey P, Cox TC et al (2010) Maternal ethanol consumption alters the epigenotype and the phenotype of offspring in a mouse model. PLoS Genet 6:e1000811
122. Downing C, Flink S, Florez-McClure ML, Johnson TE, Tabakoff B, Kechris KJ (2012) Gene expression changes in C57BL/6J and DBA/2J mice following prenatal alcohol exposure. Alcohol Clin Exp Res 36:1519–1529
123. Liu Y, Balaraman Y, Wang G, Nephew KP, Zhou FC (2009) Alcohol exposure alters DNA methylation profiles in mouse embryos at early neurulation. Epigenetics 4:500–511
124. Stouder C, Somm E, Paoloni-Giacobino A (2011) Prenatal exposure to ethanol: a specific effect on the H19 gene in sperm. Reprod Toxicol 31:507–512
125. Perkins A, Lehmann C, Lawrence RC, Kelly SJ (2013) Alcohol exposure during development: Impact on the epigenome. Int J Dev Neurosci 31:391–397
126. Veazey KJ, Carnahan MN, Muller D, Miranda RC, Golding MC (2013) Alcohol-induced epigenetic alterations to developmentally crucial genes regulating neural stemness and differentiation. Alcohol Clin Exp Res 37:1111–1122

127. Bekdash RA, Zhang C, Sarkar DK (2013) Gestational choline supplementation normalized fetal alcohol-induced alterations in histone modifications, DNA methylation, and proopiomelanocortin (POMC) gene expression in beta-endorphin-producing POMC neurons of the hypothalamus. Alcohol Clin Exp Res 37:1133–1142
128. Doherty LF, Bromer JG, Zhou Y, Aldad TS, Taylor HS (2010) In utero exposure to diethylstilbestrol (DES) or bisphenol-A (BPA) increases EZH2 expression in the mammary gland: an epigenetic mechanism linking endocrine disruptors to breast cancer. Horm Cancer 1:146–155
129. Bromer JG, Zhou Y, Taylor MB, Doherty L, Taylor HS (2010) Bisphenol-A exposure in utero leads to epigenetic alterations in the developmental programming of uterine estrogen response. FASEB J 24:2273–2280
130. Martinez-Arguelles DB, Papadopoulos V (2010) Epigenetic regulation of the expression of genes involved in steroid hormone biosynthesis and action. Steroids 75:467–476
131. Bikle DD (2008) Integrins, insulin like growth factors, and the skeletal response to load. Osteoporos Int 19:1237–1246

Index

V. Vasiliou et al. (eds.), *Biological Basis of Alcohol-Induced Cancer*,
Advances in Experimental Medicine and Biology 815,
DOI 10.1007/978-3-319-09614-8

H

M

N

CPSIA information can be obtained at www.ICGtesting.com
Printed in the USA
BVOW10*2150301214

381424BV00001B/1/P

9 783319 096131